**SCHAUM'S
OUTLINE OF**

# Mathematical Methods for Business, Economics, and Finance

SCHAUM'S
OUTLINE OF

# Mathematical Methods for Business, Economics, and Finance

**Second Edition**

Luis Moisés Peña-Lévano, PhD

**Schaum's Outline Series**

New York   Chicago   San Francisco   Athens
London   Madrid   Mexico City   Milan   New Delhi
Singapore   Sydney   Toronto

**Luis Peña-Lévano, PhD,** (St. Paul, MN) is Assistant Professor of Agricultural Economics at the University of Wisconsin River-Falls. He received his MS in applied economics from the University of Georgia and his PhD is from Purdue University. As a teacher, he has taught mathematics for economics, microeconomics, international trade and policy, environmental and natural resource economics, finance, and contemporary issues in business. Dr. Peña-Lévano is an award-winning professional and has been recognized globally. In 2020, he received the EAERE Award for the Best Paper Published in Environmental and Resource Economics. He is also member of the Global Trade Analysis Project, and currently serves as treasurer and secretary of the Teaching, Learning, and Communication section of the Agricultural and Applied Economic Association, and has served as reviewer and guest editor for multiple peer-reviewed journals.

# Preface

Students of undergraduate business, finance, and economics, as well as candidates for the MBA and MA degrees in economics today need a variety of mathematical skills to successfully complete their degree requirements and compete effectively in their chosen careers. Unfortunately, the requisite mathematical competence is not the subject of a single course in mathematics such as Calculus I or Linear Algebra I, and many students, pressed with the demands from business and economics courses, do not have space in their schedules for a series of math courses. *Mathematical Methods for Business, Economics, and Finance,* Second Edition is designed to cull the mathematical tools, topics, and techniques essential for success in business and economics today. It is suitable for a one- or two-semester course in business mathematics, depending on the previous background of the students. It can also be used profitably in an introductory calculus or linear algebra course by professors and students interested in business connections and applications.

The theory-and-solved-problem format of each chapter provides concise explanations illustrated by examples, plus numerous problems with fully worked-out solutions. No mathematical proficiency beyond the high-school level is assumed. The learning-by-doing pedagogy will enable students to progress at their own rate and adapt the book to their own needs.

*Mathematical Methods for Business, Economics, and Finance,* Second Edition can be used by itself or as a supplement to other texts for undergraduate and graduate students in business and economics. It is largely self-contained. Starting with a basic review of high-school algebra in Chapter 1, the book consistently explains all the concepts and techniques needed for the material in subsequent chapters.

This edition introduces one new chapter on sequences and series, and two sections on computing linear programming and series using Microsoft Excel. The end of book also presents a new section of problems called Additional Practice Problems, which provides a template that can be used to test students' knowledge on each chapter, with solutions in the subsequent pages (called 'Solution to Practice Problems').

This book contains over 1,200 problems, all of them solved in considerable detail. To derive the most from the book, students should strive as soon as possible to work independently of the solutions. This can be done by solving problems on individual sheets of paper with the book closed. If difficulties arise, the solution can then be checked in the book.

For best results, students should never be satisfied with passive knowledge—the capacity merely to follow or comprehend the various steps presented in the book. Mastery of the subject and doing well on exams require active knowledge—the ability to solve any problem, in any order, without the aid of the book.

Experience has proved that students of very different backgrounds and abilities can be successful in handling the subject matter of this text when the material is presented in the current format.

I wish to express my gratitude to father Alberto, my mother Betty, my sister Mirella, and my brother Willy, for raising me to be the man I am today. I also want to thank Ernesto, Mario, Daniele, Shaheer, Jared, Jose, Víctor, Ahmed, Edward, Irvin, and Andres, my 11 best friends who are part of my family and have encouraged me and helped me improve my skills and abilities over these last 15 years. I would also like to dedicate this book to my goddaughters Briggitte, Ashley, and Valentina, my nephew Fitzgerald, and my godson Adrian, who I envision to be the future of the upcoming generations.

Luis Moisés Peña Lévano

# Contents

# Chapter 1

# Review

## 1.1 EXPONENTS

Given a positive integer $n$, $x^n$ signifies that $x$ is multiplied by itself $n$ number of times. Here $x$ is referred to as the *base*; $n$ is called an *exponent*. By convention an exponent of one is not expressed: $x^{(1)} = x$, $8^{(1)} = 8$. By definition any nonzero number or variable raised to the zero power is equal to 1: $x^0 = 1$, $3^0 = 1$, and $0^0$ is undefined. Assuming $a$ and $b$ are positive integers and $x$ and $y$ are real numbers for which the following exist, the *rules of exponents* are presented below, illustrated in Examples 1 and 2, and treated in Problems 1.1, 1.24, 1.26, and 1.27.

1. $x^a \cdot x^b = x^{a+b}$

2. $\dfrac{x^a}{x^b} = x^{a-b}$

3. $(x^a)^b = x^{ab}$

4. $(xy)^a = x^a y^a$

5. $\left(\dfrac{x}{y}\right)^a = \dfrac{x^a}{y^a}$

6. $\dfrac{1}{x^a} = x^{-a}$

7. $\sqrt{x} = x^{1/2}$

8. $\sqrt[a]{x} = x^{1/a}$

9. $x^0 = 1 \qquad (x \neq 0)$

**EXAMPLE 1.** In multiplication, exponents of the same variable are added; in division, exponents of the same variable are subtracted; when raised to a power, the exponents of a variable are multiplied, as indicated by the rules above and shown in the examples below followed by illustrations.

(a) $x^2 \cdot x^5 = x^{2+5} = x^7 \neq x^{10}$       (Rule 1)

$$x^2 \cdot x^5 = (x \cdot x)(x \cdot x \cdot x \cdot x \cdot x) = x^7$$

(b) $\dfrac{x^8}{x^2} = x^{8-2} = x^6 \neq x^4$       (Rule 2)

$$\frac{x^8}{x^2} = \frac{x \cdot x \cdot x \cdot x \cdot x \cdot x \cdot x \cdot x}{x \cdot x} = x \cdot x \cdot x \cdot x \cdot x \cdot x = x^6$$

(c) $(x^3)^2 = x^{3 \cdot 2} = x^6 \neq x^9$ or $x^5$       (Rule 3)

$$(x^3)^2 = (x \cdot x \cdot x)(x \cdot x \cdot x) = x^6$$

(d) $(xy)^3 = x^3 y^3 \neq xy^3$ or $x^3 y$       (Rule 4)

$$(xy)^3 = (xy)(xy)(xy) = (x \cdot x \cdot x)(y \cdot y \cdot y) = x^3 y^3$$

(e) $\left(\dfrac{x}{y}\right)^5 = \dfrac{x^5}{y^5} \neq \dfrac{x^5}{y}$ or $\dfrac{x}{y^5}$       (Rule 5)

$$\left(\frac{x}{y}\right)^5 = \frac{x \cdot x \cdot x \cdot x \cdot x}{y \cdot y \cdot y \cdot y \cdot y} = \frac{x^5}{y^5}$$

(f) $\dfrac{x^2}{x^3} = x^{2-3} = x^{-1} = \dfrac{1}{x} \neq x^{2/3}$       (Rules 2 and 6)

$$\frac{x^2}{x^3} = \frac{x \cdot x}{x \cdot x \cdot x} = \frac{1}{x}$$

(g) $\sqrt{x} = x^{1/2}$       (Rule 7)

Since $\sqrt{x} \cdot \sqrt{x} = x$ and from Rule 1 exponents of a common base are added in multiplication, the exponent of $\sqrt{x}$, when added to itself, must equal 1. With $\frac{1}{2} + \frac{1}{2} = 1$, the exponent of $\sqrt{x}$ must equal $\frac{1}{2}$. Thus, $\sqrt{x} \cdot \sqrt{x} = x^{1/2} \cdot x^{1/2} = x^{1/2+1/2} = x^1 = x$.

(h)  $\sqrt[4]{x} = x^{1/4}$                                                                    (Rule 8)

   Just as $\sqrt[4]{x} \cdot \sqrt[4]{x} \cdot \sqrt[4]{x} \cdot \sqrt[4]{x} = x$, so $x^{1/4} \cdot x^{1/4} \cdot x^{1/4} \cdot x^{1/4} = x^{1/4+1/4+1/4+1/4} = x^1 = x$.

See Problems 1.1, 1.24, 1.26, and 1.27.

**EXAMPLE 2.**  From Rule 2, it can easily be seen why any variable or nonzero number raised to the zero power equals one. For example, $x^3/x^3 = x^{3-3} = x^0 = 1$; $8^5/8^5 = 8^{5-5} = 8^0 = 1$.

## 1.2  POLYNOMIALS

Given an expression such as $9x^5$, $x$ is called a *variable* because it can assume any number of different values, and 9 is referred to as the *coefficient* of $x$. Expressions consisting simply of a real number or of a coefficient times one or more variables raised to the power of a positive integer are called *monomials*. Monomials can be added or subtracted to form *polynomials*. Each monomial constituting a polynomial is called a *term*. Terms that have the same variables and respective exponents are called *like terms*. The *degree* of a monomial is the sum of the exponents of its variables. The degree of a polynomial is the degree of its highest term. Rules for adding, subtracting, multiplying, and dividing polynomials are explained below, illustrated in Examples 3 to 5, and treated in Problems 1.3 and 1.4.

### 1.2.1  Addition and Subtraction of Polynomials

Like terms in polynomials can be added or subtracted by adding or subtracting their coefficients. Unlike terms cannot be so added or subtracted.

**EXAMPLE 3.**

(a)  $6x^3 + 15x^3 = 21x^3$                                    (b)  $18xy - 7xy = 11xy$

(c)  $(4x^3 + 13x^2 - 7x) + (11x^3 - 8x^2 - 9x) = 15x^3 + 5x^2 - 16x$

(d)  $(22x - 19y) + (7x + 6z) = 29x - 19y + 6z$

See also Problem 1.3.

### 1.2.2  Multiplication and Division of Terms

Like and unlike terms can be multiplied or divided by multiplying or dividing both the coefficients and variables.

**EXAMPLE 4.**

(a)  $20x^4 \cdot 7y^6 = 140x^4y^6$                                    (b)  $6x^2y^3 \cdot 8x^4y^6 = 48x^6y^9$

(c)  $12x^3y^2 \cdot 5y^4z^5 = 60x^3y^6z^5$                            (d)  $3x^3y^2z^5 \cdot 15x^4y^3z^4 = 45x^7y^5z^9$

(e)  $\dfrac{24x^5y^3z^7}{6x^3y^2z^4} = 4x^2yz^3$                      (f)  $\dfrac{35x^2y^7z^5}{5x^6y^4z^8} = 7x^{-4}y^3z^{-3} = \dfrac{7y^3}{x^4z^3}$

### 1.2.3  Multiplication of Polynomials

To multiply two polynomials, multiply each term in the first polynomial by each term in the second polynomial and then add their products together.

**EXAMPLE 5.**

$(a)$   $(5x + 8y)(3x + 7y) = 15x^2 + 35xy + 24xy + 56y^2$

$$= 15x^2 + 59xy + 56y^2$$

$(b)$   $(4x + 5y)(2x - 7y - 3z) = 8x^2 - 28xy - 12xz + 10xy - 35y^2 - 15yz$

$$= 8x^2 - 18xy - 12xz - 15yz - 35y^2$$

See also Problem 1.4.

## 1.3   FACTORING

*Factoring* reverses the process of polynomial multiplication in order to express a given polynomial as a product of simpler polynomials called *factors*. A *monomial* such as the number 14 is easily factored by expressing it as a product of its integer factors $1 \cdot 14$, $2 \cdot 7$, $(-1) \cdot (-14)$, or $(-2) \cdot (-7)$. A *binomial* such as $5x^4 - 45x^3$ is easily factored by dividing or *factoring out* the *greatest common factor*, here $5x^3$, to obtain $5x^3(x - 9)$. Factoring a *trinomial* such as $mx^2 + nx + p$, however, generally requires the following rules:

1.   Given $(mx^2 + nx + p)$, the factors are $(ax + c)(bx + d)$, where (1) $ab = m$; (2) $cd = p$; and (3) $ad + bc = n$.
2.   Given $(mx^2 + nxy + py^2)$, the factors are $(ax + cy)(bx + dy)$, where (1) $ab = m$; (2) $cd = p$; and (3) $ad + bc = n$, exactly as above. For proof of these rules, see Problems 1.28 and 1.29.

**EXAMPLE 6.**   To factor $(x^2 + 11x + 24)$, where in terms of Rule 1 (above) $m = 1$, $n = 11$, and $p = 24$, we seek integer factors such that:

1)   $a \cdot b = 1$. Integer factors: $1 \cdot 1$, $(-1) \cdot (-1)$. For simplicity we shall consider only positive sets of integer factors here and in step 2.

2)   $c \cdot d = 24$. Integer factors: $1 \cdot 24$, $2 \cdot 12$, $3 \cdot 8$, $4 \cdot 6$, $6 \cdot 4$, $8 \cdot 3$, $12 \cdot 2$, $24 \cdot 1$.

3)   $ad + bc = 11$. With $a = b = 1$, $c + d$ must equal 11.

Adding the different combinations of factors from step 2, we have $1 + 24 = 25$, $2 + 12 = 14$, $3 + 8 = 11$, $4 + 6 = 10$, $6 + 4 = 10$, $8 + 3 = 11$, $12 + 2 = 14$, and $21 + 1 = 25$. Since only $3 + 8$ and $8 + 3 = 11$ in step 3, 3 and 8 are the only candidates for $c$ and $d$ from step 2 which, when used with $a = b = 1$ from step 1, fulfill all the above requirements, and the order does not matter. Hence

$$(x^2 + 11x + 24) = (x + 3)(x + 8)  \qquad \text{or} \qquad (x + 8)(x + 3)$$

See Problems 1.5 to 1.13. For derivation of the rules, see Problems 1.28 and 1.29.

## 1.4   FRACTIONS

*Fractions*, or *rational numbers*, consist of polynomials in both numerator and denominator, assuming always that the denominator does not equal zero. *Reducing* a fraction *to lowest terms* involves the cancellation of all common factors from both the numerator and the denominator. *Raising* a fraction *to higher terms* means multiplying the numerator and denominator by the same nonzero polynomial. Assuming that $A$, $B$, $C$, and $D$ are polynomials and $C$ and $D \neq 0$, fractions are governed by the following rules:

1.   $\dfrac{A}{C} \cdot \dfrac{D}{D} = \dfrac{A}{C}$          2.   $\dfrac{A}{C} \cdot \dfrac{B}{D} = \dfrac{AB}{CD}$

3.   $\dfrac{A}{C} \div \dfrac{B}{D} = \dfrac{A}{C} \cdot \dfrac{D}{B}$      $B \neq 0$          4.   $\dfrac{A}{C} \pm \dfrac{B}{C} = \dfrac{A \pm B}{C}$

5.   $\dfrac{A}{C} \pm \dfrac{B}{D} = \left( \dfrac{A}{C} \cdot \dfrac{D}{D} \right) \pm \left( \dfrac{B}{D} \cdot \dfrac{C}{C} \right) = \dfrac{AD \pm BC}{CD}$

The properties of fractions are illustrated in Example 7 and treated in Problems 1.14 to 1.21.

**EXAMPLE 7.**

(a) Multiplying or dividing both the numerator and the denominator of a fraction by the same nonzero number or polynomial leaves the value of the fraction unchanged.

$$\frac{2}{3} \cdot \frac{x}{x} = \frac{2x}{3x} = \frac{2\cancel{x}}{3\cancel{x}} = \frac{2}{3} \qquad \text{(Rule 1)}$$

Rule 1 provides the basis for reducing a fraction to its lowest terms as well as for raising a fraction to higher terms.

(b) To multiply fractions, simply multiply the numerators and the denominators separately. The product of the numerators then forms the numerator of the product and the product of the denominators forms the denominator of the product.

$$\frac{5}{x+6} \cdot \frac{x-9}{x-4} = \frac{5(x-9)}{(x+6)(x-4)} = \frac{5x-45}{x^2+2x-24} \qquad \text{(Rule 2)}$$

(c) To divide fractions, simply invert the divisor and multiply.

$$\frac{16}{y} \div \frac{7}{y^2-3} = \frac{16}{y} \cdot \frac{y^2-3}{7} = \frac{16y^2-48}{7y} \qquad \text{(Rule 3)}$$

(d) Fractions can be added or subtracted only if they have exactly the same denominator, called a *common denominator*. If a common denominator is present, simply add or subtract the numerators and set the result over the common denominator. Remember always to subtract all the terms within a given set of parentheses.

$$\frac{6z}{z+5} - \frac{4z+9}{z+5} = \frac{6z-(4z+9)}{z+5} = \frac{2z-9}{z+5} \qquad \text{(Rule 4)}$$

(e) To add or subtract fractions with different denominators, a common denominator must first be found. Multiplication of one denominator by the other will always produce a common denominator. Each fraction can then be restated in terms of the common denominator using Rule 1 and the numerators added as in (d).

$$\frac{1}{3} + \frac{3}{4} = \left(\frac{1}{3} \cdot \frac{4}{4}\right) + \left(\frac{3}{4} \cdot \frac{3}{3}\right) = \frac{4}{12} + \frac{9}{12} = \frac{4+9}{12} = \frac{13}{12} \qquad \text{(Rule 5)}$$

(f) Similarly,

$$\frac{x}{5} - \frac{3}{7x} = \frac{x(7x)-3(5)}{5(7x)} = \frac{7x^2-15}{35x} \qquad \text{(Rule 5)}$$

The *least common denominator* (LCD) of two or more fractions is the polynomial of lowest degree and smallest coefficient that is exactly divisible by the denominators of the original fractions. Use of the LCD helps simplify the final sum or difference. See Problems 1.19 to 1.21. Fractions are reviewed in Problems 1.14 to 1.21.

## 1.5 RADICALS

If $b^n = a$, where $b > 0$, then by taking the $n$th root of both sides of the equation, $b = \sqrt[n]{a}$, where $\sqrt{\ }$ is a *radical (sign)*, $a$ is the *radicand*, and $n$ is the *index*. For square roots, the index 2 is not expressed. Thus, $\sqrt[2]{\ } = \sqrt{\ }$. From Rules 7 and 8 in Section 1.1, we should also be aware that $\sqrt{a} = a^{1/2}$ and $\sqrt[n]{a} = a^{1/n}$.

Assuming $x$ and $y$ are real nonnegative numbers and $m$ and $n$ are positive integers such that $\sqrt[m]{x}$ and $\sqrt[n]{y}$ exist, the *rules of radicals* are given below. For proof of Rule 1, see Problem 1.30.

1. $\left(\sqrt[n]{x}\right)^n = x$      2. $\sqrt[m]{\sqrt[n]{x}} = \sqrt[mn]{x}$

3. $\sqrt[n]{x} \cdot \sqrt[n]{y} = \sqrt[n]{xy}$      4. $\dfrac{\sqrt[n]{x}}{\sqrt[n]{y}} = \sqrt[n]{\dfrac{x}{y}}, \qquad y \neq 0$

**EXAMPLE 8.**   The laws of radicals are used to simplify the following expressions. Note that for even-numbered roots, positive and negative answers are equally valid.

(a)   $(\sqrt[3]{27})^3 = \sqrt[3]{27} \cdot \sqrt[3]{27} \cdot \sqrt[3]{27} = 3 \cdot 3 \cdot 3 = 27$                                                    (Rule 1)

(b)   $\sqrt{\sqrt[3]{64}} = \sqrt[6]{64} = \pm 2$,  since  $\sqrt{\phantom{x}} = \sqrt[2]{\phantom{x}}$                                      (Rule 2)
       With $\sqrt[3]{64} = 4$, $\sqrt{\sqrt[3]{64}} = \sqrt{4} = \pm 2$

(c)   $\sqrt{8} \cdot \sqrt{18} = \sqrt{144} = \pm 12$                                                                          (Rule 3)

(d)   $\dfrac{\sqrt[4]{1782}}{\sqrt[4]{22}} = \sqrt[4]{\dfrac{1782}{22}} = \sqrt[4]{81} = \pm 3$                                                   (Rule 4)

See also Problems 1.22, 1.23, and Problems 1.25 to 1.27.

## 1.6   ORDER OF MATHEMATICAL OPERATIONS

Given an expression involving multiple mathematical operations, computations within parentheses are performed first. If there are parentheses within parentheses, computations on the innermost set take precedence. Within parentheses, all constants and variables are first raised to the powers of their respective exponents. Multiplication and division are then performed before addition and subtraction. In carrying out operations of the same priority, the procedure is from left to right. In sum,

1.   Start within parentheses, beginning with the innermost.
2.   Raise all terms to their respective exponents.
3.   Multiply and divide before adding and subtracting.
4.   For similar priorities, move from left to right.

**EXAMPLE 9.**   The following steps are performed to solve

$$\frac{(5^2 \cdot 6)}{10} - 8$$

1.   $5^2 = 25$
2.   $25 \cdot 6 = 150$
3.   $\dfrac{150}{10} = 15$
4.   $15 - 8 = 7$

Thus                                                      $\dfrac{(5^2 \cdot 6)}{10} - 8 = 7$

## 1.7   USE OF A POCKET CALCULATOR

Pocket calculators are helpful for checking one's ordinary calculations and performing arduous or otherwise time-consuming computations. Rules for the different mathematical operations are set forth and illustrated below, including some rules which will not be used or needed until later in the text.

### 1.7.1   Addition of Two Numbers

To add two numbers, enter the first number, press the $\boxed{+}$ key, and enter the second number. Then press the $\boxed{=}$ key to find the total.

**EXAMPLE 10.**

(a)   To find $139 + 216$, enter 139, press the $\boxed{+}$ key, enter 216, and press the $\boxed{=}$ key to find $139 + 216 = 355$.

(b)   To find $1025 + 38.75$, enter 1025, press the $\boxed{+}$ key, then enter 38.75, and hit the $\boxed{=}$ key to find $1025 + 38.75 = 1063.75$. Practice this and subsequent examples using simple numbers to which you already know the answers to see if you are doing the procedure correctly.

### 1.7.2   Addition of More Than Two Numbers

To add more than two numbers, simply follow each entry of a number by pressing the $\boxed{+}$ key until all the numbers have been entered. Then press the $\boxed{=}$ key to find the total. Pressing the $\boxed{=}$ key at any time after a number will give the subtotal at that point.

**EXAMPLE 11.**   To find $139 + 216 + 187$, enter 139, press the $\boxed{+}$ key, enter 216, press the $\boxed{+}$ key again, enter 187, and hit the $\boxed{=}$ key to find $139 + 216 + 187 = 542$. Hitting the $\boxed{=}$ key after 216 would reveal the subtotal of $139 + 216$ is 355, as in the example above.

### 1.7.3   Subtraction

To find the difference $A - B$, enter $A$, press the $\boxed{-}$ key, and enter $B$. Then press the $\boxed{=}$ key to find the remainder. Multiple subtractions can be done as multiple additions in 1.7.2 above, with the $\boxed{-}$ key substituted for the $\boxed{+}$ key.

**EXAMPLE 12.**

(a)   To find $315 - 708$, enter 315, press the $\boxed{-}$ key, then enter 708 followed by the $\boxed{=}$ key to find $315 - 708 = -393$.

(b)   To find $528 - 79.62$, enter 528, hit the $\boxed{-}$ key, then enter 79.62 followed by the $\boxed{=}$ key to find $528 - 79.62 = 448.38$.

### 1.7.4   Multiplication

To multiply two numbers, enter the first number, press the $\boxed{x}$ key, enter the second number, and press the $\boxed{=}$ key to find the product. Serial multiplications can be done in the same way as multiple additions in 1.7.2. with the $\boxed{x}$ key substituted for the $\boxed{+}$ key.

**EXAMPLE 13.**

(a)   To find $486 \cdot 27$, enter 486, press the $\boxed{x}$ key, then enter 27, and hit the $\boxed{=}$ key to learn $486 \cdot 27 = 13{,}122$.

(b)   To find $149 \cdot -35$, enter 149, press the $\boxed{x}$ key, then enter 35 followed by the $\boxed{\pm}$ key to make it negative, and hit the $\boxed{=}$ key to learn that $149 \cdot -35 = -5215$.
      *Note*: Be aware of the distinction between the $\boxed{-}$ key and the $\boxed{\pm}$ key. The $\boxed{-}$ key initiates the process of subtraction; the $\boxed{\pm}$ key simply changes the value of the previous entry from positive to negative or negative to positive.

### 1.7.5   Division

Dividing $A$ by $B$ is accomplished by entering $A$, pressing the $\boxed{\div}$ key, then entering $B$ and pressing the $\boxed{=}$ key.

**EXAMPLE 14.**

(a)   To find $6715 \div 79$, enter 6715, hit the $\boxed{\div}$ key, then enter 79 followed by the $\boxed{=}$ key. The display will show 85, indicating that $6715 \div 79 = 85$.

(b)   To find $-297.36 \div 72.128$, enter 297.36 followed by the $\boxed{\pm}$ key to make it negative, then press the $\boxed{\div}$ key, enter 72.128, and hit the $\boxed{=}$ key to find $-297.36 \div 72.128 = -4.1226708$.

### 1.7.6   Raising to a Power

To raise a number to a power, enter the number, hit the $\boxed{y^x}$ key, then enter the exponent and press the $\boxed{=}$ key.

**EXAMPLE 15.**

(a)   To find $8^5$, enter 8, press the $\boxed{y^x}$ key, then enter 5 followed by the $\boxed{=}$ key to learn that $8^5 = 32,768$. Continue to practice these and subsequent exercises by using simple numbers for which you already know the answers.

(b)   To find $36^{0.25}$, enter 36, hit the $\boxed{y^x}$ key, then enter 0.25, and press the $\boxed{=}$ key to see $36^{0.25} = 2.4494897$.

(c)   To find $2^{-3}$, enter 2, hit the $\boxed{y^x}$ key, then press 3 followed by the $\boxed{\pm}$ key to make it negative, and hit the $\boxed{=}$ key to discover $2^{-3} = 0.125$.

See also Problem 1.24.

### 1.7.7   Finding a Square Root

To find the square root of a number, enter the number, then press the $\boxed{\sqrt{x}}$ key to find the square root immediately without having to press the $\boxed{=}$ key. Note that on many calculators the $\boxed{\sqrt{x}}$ key is the inverse (shift, or second function) of the $\boxed{x^2}$ key, and to activate the $\boxed{\sqrt{x}}$ key, one must first press the $\boxed{\text{INV}}$ ($\boxed{\text{Shift}}$ or $\boxed{\text{2dF}}$) key followed by the $\boxed{x^2}$ key.

**EXAMPLE 16.**   To find $\sqrt{529}$, enter 529, then press the $\boxed{\sqrt{x}}$ key to see immediately that $\pm23$ is the square root of 529.

If the $\boxed{\sqrt{x}}$ key is the inverse, shift, or second function of the $\boxed{x^2}$ key, enter 529, then press the $\boxed{\text{INV}}$ ($\boxed{\text{Shift}}$ or $\boxed{\text{2dF}}$ key) followed by the $\boxed{x^2}$ key to activate the $\boxed{\sqrt{x}}$ key, and you will see immediately that $\sqrt{529} = 23$ without having to press the $\boxed{=}$ key.

### 1.7.8   Finding the *n*th Root

To find the *n*th root of a number, enter the number, press the $\boxed{\sqrt[x]{y}}$ key, then enter the value of the root $n$ and hit the $\boxed{=}$ key to find the root. If the $\boxed{\sqrt[x]{y}}$ is the inverse, shift, or second function of the $\boxed{y^x}$ key, enter the number, press the $\boxed{\text{INV}}$ ($\boxed{\text{Shift}}$, or $\boxed{\text{2dF}}$) key followed by the $\boxed{y^x}$ key, then enter the value of the root $n$ and hit the $\boxed{=}$ key to find the answer.

**EXAMPLE 17.**

(a)   To find $\sqrt[3]{17,576}$, enter 17,576, press the $\boxed{\text{INV}}$ key followed by the $\boxed{y^x}$ key, enter 3, then hit the $\boxed{=}$ key to learn $26 = \sqrt[3]{17,576}$.

(b)   To find $\sqrt[5]{32,768}$, enter 32,768, hit the $\boxed{\text{INV}}$ key followed by the $\boxed{y^x}$ key, then enter 5 and hit the $\boxed{=}$ key to learn $8 = \sqrt[5]{32,768}$.

(c)   From Rule 8 in Section 1.1, $\sqrt[5]{32,768} = 32,768^{1/5} = 32,768^{0.2}$. To use this latter form, simply enter 32,768, press the $\boxed{y^x}$ key, enter 0.2, and hit the $\boxed{=}$ key to find $32,768^{0.2} = 8$.
To make use of similar conversions, recall that $\sqrt{x} = x^{1/2} = x^{0.5}$, $\sqrt[3]{x} = x^{1/3} \approx x^{0.33}$, $\sqrt[4]{x} = x^{1/4} = x^{0.25}$, and so forth. See Problems 1.25 to 1.27.

### 1.7.9   Logarithms

To find the value of the common logarithm $\log_{10} x$, enter the value of $x$ and simply press the $\boxed{\log}$ key. The answer will appear without the need to press the $\boxed{=}$ key.

**EXAMPLE 18.**

    (a)  To find the value of log 24, enter 24 and hit the $\boxed{\log}$ key. The screen will immediately display 1.3802112, indicating that $\log 24 = 1.3802112$.

    (b)  To find log 175, enter 175 and hit the $\boxed{\log}$ key. You will see 2.243038, which is the value of log 175.

### 1.7.10  Natural Logarithms

To find the value of the natural logarithm $\ln x$, enter the value of $x$ and press the $\boxed{\ln x}$ key. The answer will appear immediately without the need to press the $\boxed{=}$ key.

**EXAMPLE 19.**

    (a)  To find ln 20, enter 20 and hit the $\boxed{\ln x}$ key to see $2.9957323 = \ln 20$.

    (b)  For ln 0.75, enter 0.75 and hit the $\boxed{\ln x}$ key. You will find $\ln 0.75 = -0.2876821$.

### 1.7.11  Exponential Functions

To find the value of an exponential function $y = a^x$, enter the value of $a$ and press the $\boxed{y^x}$ key, then enter the value of $x$ and hit the $\boxed{=}$ key, similar to what was done in Section 1.7.6.

**EXAMPLE 20.**

    (a)  Given $y = 1.5^{3.2}$, enter 1.5, press the $\boxed{y^x}$ key, then enter 3.2, and hit the $\boxed{=}$ key to get $1.5^{3.2} = 3.6600922$.

    (b)  For $y = 256^{-1.25}$, enter 256, hit the $\boxed{y^x}$ key, then enter 1.25 followed immediately by the $\boxed{\pm}$ key to make it negative, and then press the $\boxed{=}$ key to learn $256^{-1.25} = 0.0009766$.

### 1.7.12  Natural Exponential Functions

To find the value of a natural exponential function $y = e^x$, enter the value of $x$, press the $\boxed{e^x}$ key, and the answer will appear immediately without the need of hitting the $\boxed{=}$ key. If the $\boxed{e^x}$ key is the inverse, shift, or second function of the $\boxed{\ln x}$ key, enter the value of $x$, hit the $\boxed{\text{INV}}$ key followed by the $\boxed{\ln x}$ key, and the answer will also appear immediately.

**EXAMPLE 21.**

    (a)  Given $y = e^{1.4}$, enter 1.4, press the $\boxed{\text{INV}}$ key followed by the $\boxed{\ln x}$ key, and you will see that $e^{1.4} = 4.0552$.

    (b)  For $e^{-0.65}$, enter 0.65, hit the $\boxed{\pm}$ key to make it negative, then press the $\boxed{\text{INV}}$ key followed by the $\boxed{\ln x}$ key to find $e^{-0.65} = 0.5220458$.

See also Problems 1.24 to 1.27.

# Solved Problems

**EXPONENTS**

**1.1.**    Simplify the following expressions using the rules of exponents from Section 1.1.

    (a)  $x^3 \cdot x^4$

$$x^3 \cdot x^4 = x^{3+4} = x^7 \hspace{3cm} \text{(Rule 1)}$$

(b)  $x^5 \cdot x^{-3}$

$$x^5 \cdot x^{-3} = x^{5+(-3)} = x^2 \qquad \text{(Rule 1)}$$

$$\left[ x^5 \cdot x^{-3} = x^5 \cdot \frac{1}{x^3} = x \cdot x \cdot x \cdot x \cdot x \cdot \frac{1}{x \cdot x \cdot x} = x^2 \right]$$

(c)  $x^{-2} \cdot x^{-4}$

$$x^{-2} \cdot x^{-4} = x^{-2+(-4)} = x^{-6} = \frac{1}{x^6} \qquad \text{(Rule 1)}$$

$$\left[ x^{-2} \cdot x^{-4} = \frac{1}{x \cdot x} \cdot \frac{1}{x \cdot x \cdot x \cdot x} = \frac{1}{x^6} \right]$$

(d)  $x^{1/2} \cdot x^3$

$$x^{1/2} \cdot x^3 = x^{(1/2)+3} = x^{3\frac{1}{2}} = x^{7/2} = \sqrt{x^7} \qquad \text{(Rules 7 and 1)}$$

$$[x^{1/2} \cdot x^3 = (\sqrt{x})(x \cdot x \cdot x) = (\sqrt{x})(\sqrt{x} \cdot \sqrt{x} \cdot \sqrt{x} \cdot \sqrt{x} \cdot \sqrt{x} \cdot \sqrt{x}) = (x^{1/2})^7 = x^{7/2}]$$

(e)  $\dfrac{x^{10}}{x^6}$

$$\frac{x^{10}}{x^6} = x^{10-6} = x^4 \qquad \text{(Rule 2)}$$

(f)  $\dfrac{x^4}{x^6}$

$$\frac{x^4}{x^6} = x^{4-6} = x^{-2} = \frac{1}{x^2} \qquad \text{(Rules 2 and 6)}$$

$$\left[ \frac{x^4}{x^6} = \frac{x \cdot x \cdot x \cdot x}{x \cdot x \cdot x \cdot x \cdot x \cdot x} = \frac{1}{x^2} \right]$$

(g)  $\dfrac{x^5}{x^{-6}}$

$$\frac{x^5}{x^{-6}} = x^{5-(-6)} = x^{5+6} = x^{11} \qquad \text{(Rule 2)}$$

$$\left[ \frac{x^5}{x^{-6}} = \frac{x^5}{(1/x^6)} = x^5 \cdot x^6 = x^{11} \right]$$

(h)  $\dfrac{x^5}{\sqrt{x}}$

$$\frac{x^5}{\sqrt{x}} = \frac{x^5}{x^{1/2}} = x^{5-(1/2)} = x^{4\frac{1}{2}} = x^{9/2} = \sqrt{x^9} \qquad \text{(Rules 2 and 7)}$$

(i)  $(x^4)^{-3}$

$$(x^4)^{-3} = x^{4(-3)} = x^{-12} = \frac{1}{x^{12}} \qquad \text{(Rules 3 and 6)}$$

(j)  $(\sqrt[3]{x})^4$

$$(\sqrt[3]{x})^4 = (x^{1/3})^4 = x^{(1/3)(4)} = x^{4/3} \qquad \text{(Rules 8 and 3)}$$

(k)  $\dfrac{1}{x^3} \cdot \dfrac{1}{y^3}$

$$\frac{1}{x^3} \cdot \frac{1}{y^3} = x^{-3} \cdot y^{-3} = (xy)^{-3} = \frac{1}{(xy)^3} \qquad \text{(Rules 6 and 4)}$$

(l)  $\left(\dfrac{x^5}{y^7}\right)^2$

$$\left(\frac{x^5}{y^7}\right)^2 = \frac{x^{5\cdot 2}}{y^{7\cdot 2}} = \frac{x^{10}}{y^{14}} \qquad\qquad (\text{Rules 3 and 5})$$

See also Problems 1.24, 1.26, and 1.27.

## POLYNOMIALS

**1.2.**  Perform the indicated arithmetic operations on the following polynomials:

(a)  $35xy + 52xy$
$35xy + 52xy = 87xy$

(b)  $22yz^2 - 46yz^2$
$22yz^2 - 46yz^2 = -24yz^2$

(c)  $79x^2y^3 - 46x^2y^3$
$79x^2y^3 - 46x^2y^3 = 33x^2y^3$

(d)  $16x_1x_2 + 62x_1x_2$
$16x_1x_2 + 62x_1x_2 = 78x_1x_2$

(e)  $57y_1y_2 - 70y_1y_2$
$57y_1y_2 - 70y_1y_2 = -13y_1y_2$

(f)  $0.5x^2y^3z^5 + 0.9x^2y^3z^5$
$0.5x^2y^3z^5 + 0.9x^2y^3z^5 = 1.4x^2y^3z^5$

**1.3.**  Add or subtract the following polynomials as indicated. Note that in subtraction the sign of every term within the parentheses must be changed before corresponding elements are added.

(a)  $(25x - 9y) + (32x + 16y)$

$$(25x - 9y) + (32x + 16y) = 57x + 7y$$

(b)  $(84x - 31y) - (76x + 43y)$
Multiplying each term in the second set of parentheses by $-1$, which in effect changes the sign of the said terms, and then simply adding, we have

$$(84x - 31y) - (76x + 43y) = 84x - 31y - 76x - 43y = 8x - 74y$$

(c)  $(9x^2 + 7x) - (3x^2 - 4x)$
Changing the signs of each of the terms in the second set of parentheses and then adding, we have

$$(9x^2 + 7x) - (3x^2 - 4x) = 9x^2 + 7x - 3x^2 + 4x = 6x^2 + 11x$$

(d)  $(42x^2 + 23x) - (5x^2 + 11x - 82)$

$$(42x^2 + 23x) - (5x^2 + 11x - 82) = 42x^2 + 23x - 5x^2 - 11x + 82$$

$$= 37x^2 + 12x + 82$$

**1.4.**  Perform the indicated operations, recalling that each term in the first polynomial must be multiplied by each term in the second and their products summed.

(a)  $(2x + 7)(4x - 5)$

$$(2x + 7)(4x - 5) = 8x^2 - 10x + 28x - 35 = 8x^2 + 18x - 35$$

(b)  $(5x - 6y)(4x - 3y)$

$$(5x - 6y)(4x - 3y) = 20x^2 - 15xy - 24xy + 18y^2 = 20x^2 - 39xy + 18y^2$$

(c)  $(2x - 9)^2$

$$(2x - 9)^2 = (2x - 9)(2x - 9) = 4x^2 - 18x - 18x + 81 = 4x^2 - 36x + 81$$

(d)  $(2x + 3y)(2x - 3y)$

$$(2x + 3y)(2x - 3y) = 4x^2 - 6xy + 6xy - 9y^2 = 4x^2 - 9y^2$$

(e)  $(4x + 3y)(5x^2 - 2xy + 6y^2)$

$$(4x + 3y)(5x^2 - 2xy + 6y^2) = 20x^3 - 8x^2y + 24xy^2 + 15x^2y - 6xy^2 + 18y^3$$

$$= 20x^3 + 7x^2y + 18xy^2 + 18y^3$$

(f)  $(3x^3 - 5x^2y^2 - 2y^3)(7x - 4y)$

$$(3x^3 - 5x^2y^2 - 2y^3)(7x - 4y) = 21x^4 - 12x^3y - 35x^3y^2 + 20x^2y^3 - 14xy^3 + 8y^4$$

## FACTORING

**1.5.**  Simplify each of the following polynomials by factoring out the greatest common factor:

(a)  $32x - 8$
$$32x - 8 = 8(4x - 1)$$

(b)  $18x^2 + 27x$
$$18x^2 + 27x = 9x(2x + 3)$$

(c)  $14x^5 - 35x^4$
$$14x^5 - 35x^4 = 7x^4(2x - 5)$$

(d)  $45x^2y^5 - 75x^4y^3$
$$45x^2y^5 - 75x^4y^3 = 15x^2y^3(3y^2 - 5x^2)$$

(e)  $55x^8y^9 - 22x^6y^4 - 99x^5y^7$

$$55x^8y^9 - 22x^6y^4 - 99x^5y^7 = 11x^5y^4(5x^3y^5 - 2x - 9y^3)$$

**1.6.**  Factor each of the following using integer coefficients:

(a)  $x^2 + 10x + 21$

Here, using the notation from Rule 1 in Section 1.3, $m = 1$, $n = 10$, and $p = 21$. For simplicity we limit our search to positive integers such that:

(1)  $a \cdot b = 1$    [1, 1] Henceforth, when $a = b = 1$, this step will be omitted and only one order will be considered.

(2)  $c \cdot d = 21$    [1, 21; 3, 7; 7, 3; 21, 1]

(3)  $ad + bc = 10$. With $a = 1 = b$, $(c + d)$ must equal 10. Since only $3 + 7$ and $7 + 3 = 10$,

$$x^2 + 10x + 21 = (x + 3)(x + 7) \quad \text{or} \quad (x + 7)(x + 3)$$

Whenever $a \cdot b = 1$, the order of factors is unimportant and only one ordering of a given pair will be mentioned.

(b)  $x^2 + 8x + 16$

(1)  $c \cdot d = 16$    [1, 16; 2, 8; 4, 4]

(2)  $c + d = 8$    [$1 + 16 \neq 8$; $2 + 8 \neq 8$; $4 + 4 = 8$]

$$x^2 + 8x + 16 = (x + 4)(x + 4)$$

(c)  $x^2 + 13x + 36$

(1)  $c \cdot d = 36$    [1, 36; 2, 18; 3, 12; 4, 9; 6, 6]

(2)  $c + d = 13$    [only $4 + 9 = 13$]

$$x^2 + 13x + 36 = (x + 4)(x + 9)$$

**1.7.**  Factor each of the following, noting that the coefficient of the $x$ term is negative while the constant term is positive.

(a)  $x^2 - 13x + 30$

With $(c \cdot d)$ positive and $(c + d)$ negative, the two integer factors must both be negative. We can no longer limit ourselves to positive factors for $c$ and $d$.

(1)  $c \cdot d = 30$    [$-1, -30$; $-2, -15$; $-3, -10$; $-5, -6$]

(2)  $c + d = -13$    [only $-3 + (-10) = -13$]

$$x^2 - 13x + 30 = (x - 3)(x - 10)$$

(b)  $x^2 - 15x + 36$

(1)  $c \cdot d = 36$    [$-1, -36$; $-2, -18$; $-3, -12$; $-4, -9$; $-6, -6$]

(2)  $c + d = -15$    [$-3 + (-12) = -15$]

$$x^2 - 15x + 36 = (x - 3)(x - 12)$$

**1.8.**  Factor each of the following, noting that the coefficient of the $x$ term is now positive and the constant term is negative.

(a)  $x^2 + 19x - 42$

With $(c \cdot d)$ negative, the two integer factors must be of opposite signs; for $(c + d)$ to be positive when one of the two factors is negative, the factor with the larger absolute value must be positive.
(1)  $c \cdot d = -42$    $[-1, 42; -2, 21; -3, 14; -6, 7]$
(2)  $c + d = 19$    [Only $-2 + 21 = 19$]

$$x^2 + 19x - 42 = (x - 2)(x + 21)$$

(b)  $x^2 + 18x - 63$
(1)  $c \cdot d = -63$    $[-1, 63; -3, 21; -7, 9]$
(2)  $c + d = 18$    $[-3 + 21 = 18]$

$$x^2 + 18x - 63 = (x - 3)(x + 21)$$

**1.9.**  Factor each of the following expressions, in which both the coefficient of the $x$ term and the constant term are now negative.

(a)  $x^2 - 8x - 48$

With $(c \cdot d)$ and $(c + d)$ both negative, the factors must be of different signs and the factor with the larger absolute value must be negative.
(1)  $c \cdot d = -48$    $[1, -48; 2, -24; 3, -16; 4, -12; 6, -8]$
(2)  $c + d = -8$    $[4 + (-12) = -8]$

$$x^2 - 8x - 48 = (x + 4)(x - 12)$$

(b)  $x^2 - 26x - 56$
(1)  $c \cdot d = -56$    $[1, -56; 2, -28; 4, -14; 7, -8]$
(2)  $c + d = -26$    $[2 + (-28) = -26]$

$$x^2 - 26x - 56 = (x + 2)(x - 28)$$

**1.10.**  Factor each of the following, noting that there is no $x$ term and the constant term is negative.

(a)  $x^2 - 81$

Here $(c \cdot d)$ is negative and $(c + d) = 0$. For this to be true, the factors must be of different signs and of the same absolute value.
(1)  $c \cdot d = -81$    $[9, -9]$
(2)  $c + d = 0$    $[9 + (-9) = 0]$

$$x^2 - 81 = (x + 9)(x - 9)$$

(b)  $x^2 - 169$
(1)  $c \cdot d = -169$    $[13, -13]$
(2)  $c + d = 0$    $[13 + (-13) = 0]$

$$x^2 - 169 = (x + 13)(x - 13)$$

**1.11.**  Use the techniques and procedures developed in Problems 1.5 to 1.10 to factor the following expressions in which the coefficient of the $x^2$ term is no longer limited to 1.

(a)  $5x^2 + 47x + 18$
(1)  $a \cdot b = 5$.    Factors are $[5, 1]$, giving $(5x + ?)(x + ?)$.
(2)  $c \cdot d = 18$    $[1, 18; 2, 9; 3, 6]$
(3)  $ad + bc = 47$. Here all the pairs of possible factors from step (2) must be tried in *both* orders in step (1), that is, $[5, 1]$ with $[1, 18; 18, 1; 2, 9; 9, 2; 3, 6; 6, 3]$

Of all the possible combinations of factors above, only $(5 \cdot 9) + (1 \cdot 2) = 47$. Carefully arranging the factors, therefore, to ensure that **5** multiplies 9 and **1** multiplies 2, we have

$$5x^2 + 47x + 18 = (5x + 2)(x + 9)$$

(b) $3x^2 + 22x + 24$

    (1) $a \cdot b = 3$. Factors are [**3, 1**], giving $(3x + ?)(x + ?)$.

    (2) $c \cdot d = 24$    [1, 24; 24, 1; 2, 12; 12, 2; 3, 8; 8, 3; 4, 6; 6, 4]

    (3) $ad + bc = 22$    [$(3 \cdot 6) + (1 \cdot 4) = 22$]. Then arranging the factors to ensure that **3** multiplies 6 and **1** multiplies 4, we have

$$3x^2 + 22x + 24 = (3x + 4)(x + 6)$$

(c) $3x^2 - 35x + 22$

    (1) $a \cdot b = 3$    [**3, 1**]

    (2) $c \cdot d = 22$    [$-1, -22; -22, -1; -2, -11; -11, -2$], as in Problem 1.7.

    (3) $ad + bc = -35$    [$(3 \cdot -11) + (1 \cdot -2) = -35$]. Here rearranging the factors so **3** multiplies $-11$ and **1** multiplies $-2$, we obtain

$$3x^2 - 35x + 22 = (3x - 2)(x - 11)$$

(d) $7x^2 - 32x + 16$

    (1) $a \cdot b = 7$    [**7, 1**]

    (2) $c \cdot d = 16$    [$-1, -16; -16, -1; -2, -8; -8, -2; -4, -4$]

    (3) $ad + bc = -32$    [$(7 \cdot -4) + (1 \cdot -4) = -32$]

$$7x^2 - 32x + 16 = (7x - 4)(x - 4)$$

(e) $5x^2 + 7x - 52$

    (1) $a \cdot b = 5$    [**5, 1**]

    (2) $c \cdot d = -52$    [1, 52; 2, 26; 4, 13; each combination of which must be considered in both orders and with alternating signs]

    (3) $ad + bc = 7$    [$(5 \cdot 4) + (1 \cdot -13) = 7$]

$$5x^2 + 7x - 52 = (5x - 13)(x + 4)$$

(f) $3x^2 - 13x - 56$

    (1) $a \cdot b = 3$    [**3, 1**]

    (2) $c \cdot d = -56$    [1, 56; 2, 28; 4, 14; 7, 8; considered as in (e)]

    (3) $ad + bc = -13$    [$(3 \cdot -7) + (1 \cdot 8) = -13$]

$$3x^2 - 13x - 56 = (3x + 8)(x - 7)$$

(g) $11x^2 + 12x - 20$

    (1) $a \cdot b = 11$    [**11, 1**]

    (2) $c \cdot d = -20$    [1, 20; 2, 10; 4, 5; considered as above]

    (3) $ad + bc = 12$    [$(11 \cdot 2) + (1 \cdot -10) = 12$]

$$11x^2 + 12x - 20 = (11x - 10)(x + 2)$$

(h) $7x^2 - 39x - 18$

    (1) $a \cdot b = 7$    [**7, 1**]

    (2) $c \cdot d = -18$    [1, 18; 2, 9; 3, 6; considered as above]

    (3) $ad + bc = -39$    [$(7 \cdot -6) + (1 \cdot 3) = -39$]

$$7x^2 - 39x - 18 = (7x + 3)(x - 6)$$

**1.12.** Redo Problem 1.11 to factor the following expressions in which the coefficient of the $x^2$ term now has multiple factors.

(a) $6x^2 + 23x + 20$

    (1) $a \cdot b = 6$. Factors are [**1, 6**; **2, 3**; **3, 2**; **6, 1**].

    (2) $c \cdot d = 20$    [1, 20; 2, 10; 4, 5; 5, 4; 10, 2; 20, 1]

    (3) $ad + bc = 23$. Here all the pairs of possible factors from step (2) must be tried with all the possible factors in step (1).

Of all the possible combinations of factors above, only $(2 \cdot 4) + (3 \cdot 5) = 23$. Carefully arranging the factors, therefore, to ensure that **2** multiplies 4 and **3** multiplies 5, we have

$$6x^2 + 23x + 20 = (2x + 5)(3x + 4)$$

(b)  $4x^2 + 15x + 14$
   (1)  $a \cdot b = 4$      [**1, 4; 2, 2; 4, 1**]
   (2)  $c \cdot d = 14$      [1, 14; 2, 7; 7, 2; 14, 1]
   (3)  $ad + bc = 15$      [$(\mathbf{4} \cdot 2) + (\mathbf{1} \cdot 7) = 15$] Arranging the factors to ensure that **4** multiplies 2 and **1** multiplies 7, we have

$$4x^2 + 15x + 14 = (4x + 7)(x + 2)$$

(c)  $8x^2 + 34x + 21$
   (1)  $a \cdot b = 8$      [**1, 8; 2, 4; 4, 2; 8, 1**]
   (2)  $c \cdot d = 21$      [1, 21; 3, 7; 7, 3; 21, 1]
   (3)  $ad + bc = 34$      [$(\mathbf{2} \cdot 3) + (\mathbf{4} \cdot 7) = 34$] Arranging the factors so that **2** multiplies 3 and **4** multiplies 7, we have

$$8x^2 + 34x + 21 = (2x + 7)(4x + 3)$$

(d)  $6x^2 - 17x + 10$
   Here with the $-17x$ term, if we wish to limit $a$ and $b$ to positive factors for convenience, $c$ and $d$ must both be negative.
   (1)  $a \cdot b = 6$      [**1, 6; 2, 3; 3, 2; 6, 1**]
   (2)  $c \cdot d = 10$      [$-1, -10; -2, -5; -5, -2; -10, -1$]
   (3)  $ad + bc = -17$      [$(\mathbf{6} \cdot -2) + (\mathbf{1} \cdot -5) = -17$] Arranging the factors so that **6** multiplies $-2$ and **1** multiplies $-5$,

$$6x^2 - 17x + 10 = (6x - 5)(x - 2)$$

(e)  $9x^2 - 30x + 16$
   (1)  $a \cdot b = 9$      [**1, 9; 3, 3; 9, 1**]
   (2)  $c \cdot d = 16$      [$-1, -16; -2, -8; -4, -4; -8, -2; -16, -1$]
   (3)  $ad + bc = -30$      [$(\mathbf{3} \cdot -8) + (\mathbf{3} \cdot -2) = -30$]

$$9x^2 - 30x + 16 = (3x - 2)(3x - 8)$$

*Note*: Quadratic equations more complicated than this are seldom factored. If factors are needed for solution, the quadratic formula is generally used. See Section 3.5.

**1.13.**  Using Rule 2 from Section 1.3 and the techniques developed in Problems 1.5 to 1.11, factor the following polynomials:

(a)  $x^2 + 11xy + 28y^2$
   (1)  $c \cdot d = 28$      [1, 28; 2, 14; 4, 7]
   (2)  $c + d = 11$      [$4 + 7 = 11$]

$$x^2 + 11xy + 28y^2 = (x + 4y)(x + 7y)$$

(b)  $x^2 - 19xy + 60y^2$
   (1)  $c \cdot d = 60$      [$-1, -60; -2, -30; -3, -20; -4, -15; -5, -12; -6, -10$], for reasons analogous to those in Problem 1.7.
   (2)  $c + d = -19$      [$-4 + (-15) = -19$]

$$x^2 - 19xy + 60y^2 = (x - 4y)(x - 15y)$$

(c)  $x^2 - 13xy - 48y^2$
   (1)  $c \cdot d = -48$      [$1, -48; 2, -24; 3, -16; 4, -12; 6, -8$], for reasons analogous to those in Problem 1.9.
   (2)  $c + d = -13$      [$3 + (-16) = -13$]

$$x^2 - 13xy - 48y^2 = (x + 3y)(x - 16y)$$

(d)  $x^2 + 11xy - 42y^2$
    (1)  $c \cdot d = -42$     $[-1, 42; -2, 21; -3, 14; -6, 7]$, for reasons similar to those in Problem 1.8.
    (2)  $c + d = 11$     $[-3 + 14 = 11]$

$$x^2 + 11xy - 42y^2 = (x - 3y)(x + 14y)$$

(e)  $3x^2 + 29xy + 18y^2$
    (1)  $a \cdot b = 3$     $[\mathbf{3, 1}]$
    (2)  $c \cdot d = 18$     $[1, 18; 18, 1; 2, 9; 9, 2; 3, 6; 6, 3]$, as in Problems 1.11 (a) and (b).
    (3)  $ad + bc = 29$     $[(\mathbf{3} \cdot 9) + (\mathbf{1} \cdot 2) = 29]$ Then rearranging the factors for proper multiplication, we obtain

$$3x^2 + 29xy + 18y^2 = (3x + 2y)(x + 9y)$$

(f)  $7x^2 - 36xy + 45y^2$
    (1)  $a \cdot b = 7$     $[\mathbf{7, 1}]$
    (2)  $c \cdot d = 45$     $[-1, -45; -45, -1; -3, -15; -15, -3; -5, -9; -9, -5]$, as in Problem 1.11 (c).
    (3)  $ad + bc = -36$     $[(\mathbf{7} \cdot -3) + (\mathbf{1} \cdot -15) = -36]$

$$7x^2 - 36xy + 45y^2 = (7x - 15y)(x - 3y)$$

(g)  $5x^2 + 12xy - 44y^2$
    (1)  $a \cdot b = 5$     $[\mathbf{5, 1}]$
    (2)  $c \cdot d = -44$     $[1, 44; 2, 22; 4, 11;$ each combination of which must be considered in both orders and with alternating signs as in Problem 1.11 (e)]
    (3)  $ad + bc = 12$     $[(\mathbf{5} \cdot -2) + (\mathbf{1} \cdot 22) = 12]$

$$5x^2 + 12xy - 44y^2 = (5x + 22y)(x - 2y)$$

(h)  $8x^2 + 46xy + 45y^2$
    (1)  $a \cdot b = 8$     $[\mathbf{1, 8}; \mathbf{2, 4}; \mathbf{4, 2}; \mathbf{1, 8}]$, as in Problem 1.12
    (2)  $c \cdot d = 45$     $[1, 45; 3, 15; 5, 9; 9, 5; 15, 3; 45, 1]$
    (3)  $ad + bc = 46$     $[(\mathbf{2} \cdot 5) + (\mathbf{4} \cdot 9) = 46]$

$$8x^2 + 46xy + 45y^2 = (2x + 9y)(4x + 5y)$$

(i)  $4x^2 - 25y^2$
    (1)  $a \cdot b = 4$     $[\mathbf{1, 4}; \mathbf{2, 2}]$
    (2)  $c \cdot d = -25$     $[1, -25; -25, 1; 5, -5]$
    (3)  $ad + bc = 0$     $[(\mathbf{2} \cdot 5) + (\mathbf{2} \cdot -5) = 0]$

$$4x^2 - 25y^2 = (2x + 5y)(2x - 5y)$$

## FRACTIONS

**1.14.**  According to the *fundamental principle of fractions* expressed in Rule 1 of Section 1.4, when the numerator and denominator of a rational number are multiplied or divided by the same nonzero polynomial, the result will be equivalent to the original expression. Use this principle to reduce the following fractions to their lowest terms by finding and canceling the greatest common factor:

(a)  $\dfrac{16}{96}$

        (b)  $\dfrac{21x^5}{3x^2}$

$$\frac{16}{96} = \frac{16 \cdot 1}{16 \cdot 6} = \frac{1}{6}$$

$$\frac{21x^5}{3x^2} = \frac{3x^2 \cdot 7x^3}{3x^2 \cdot 1} = 7x^3$$

(c)  $\dfrac{27z}{36z^2 - 63z}$

$$\frac{27z}{36z^2 - 63z} = \frac{9z \cdot 3}{9z(4z - 7)} = \frac{3}{4z - 7}$$

(d) $\dfrac{x^2 + 7x + 12}{x^2 + 9x + 20}$

By factoring, $\qquad \dfrac{x^2 + 7x + 12}{x^2 + 9x + 20} = \dfrac{(x+4)(x+3)}{(x+4)(x+5)} = \dfrac{x+3}{x+5}$

**1.15.** Use Rule 1 to raise the following fractions to higher terms with a common denominator of 48:

(a) $\dfrac{1}{3}$

$\qquad \dfrac{1}{3} = \dfrac{1}{3} \cdot \dfrac{16}{16} = \dfrac{16}{48}$

(b) $\dfrac{1}{4}$

$\qquad \dfrac{1}{4} = \dfrac{1}{4} \cdot \dfrac{12}{12} = \dfrac{12}{48}$

(c) $\dfrac{1}{2}$

$\qquad \dfrac{1}{2} = \dfrac{1}{2} \cdot \dfrac{24}{24} = \dfrac{24}{48}$

(d) $\dfrac{1}{6}$

$\qquad \dfrac{1}{6} = \dfrac{1}{6} \cdot \dfrac{8}{8} = \dfrac{8}{48}$

**1.16.** Use Rule 2 of Section 1.4 to multiply the following rational expressions and reduce all answers to the lowest terms.

(a) $\dfrac{7w^3 x^4}{3y^2 z^2} \cdot \dfrac{2x^2 y^5}{21w^6 z^3}$

   (a) When multiplying rational expressions involving quotients of monomials, multiply numerator and denominator separately, then reduce to lowest terms, recalling from Section 1.1 that exponents of the same variable are added in multiplication and subtracted in division.

$$\dfrac{7w^3 x^4}{3y^2 z^2} \cdot \dfrac{2x^2 y^5}{21w^6 z^3} = \dfrac{14w^3 x^6 y^5}{63w^6 y^2 z^5} = \dfrac{7w^3 y^2}{7w^3 y^2} \cdot \dfrac{2x^6 y^3}{9w^3 z^5} = \dfrac{2x^6 y^3}{9w^3 z^5}$$

(b) $\dfrac{5x^6 y^3}{4w^4 z^2} \cdot \dfrac{8wz^3}{15x^2 y}$

$$\dfrac{5x^6 y^3}{4w^4 z^2} \cdot \dfrac{8wz^3}{15x^2 y} = \dfrac{40wx^6 y^3 z^3}{60w^4 x^2 yz^2} = \dfrac{20wx^2 yz^2}{20wx^2 yz^2} \cdot \dfrac{2x^4 y^2 z}{3w^3} = \dfrac{2x^4 y^2 z}{3w^3}$$

**1.17.** Multiply the following rational expressions involving quotients of binomials and reduce to lowest terms.

(a) $\dfrac{x-5}{x+8} \cdot \dfrac{x+2}{x-9}$

   (a) When multiplying rational expressions involving quotients of binomials, multiply numerator and denominator separately as above, remembering from Example 5 that each term in the first polynomial must be multipled by each term in the second, and their products summed.

$$\dfrac{x-5}{x+8} \cdot \dfrac{x+2}{x-9} = \dfrac{x^2 - 3x - 10}{x^2 - x - 72}$$

(b) $\dfrac{x+7}{x^3} \cdot \dfrac{x^2}{x-4}$

$$\dfrac{x+7}{x^3} \cdot \dfrac{x^2}{x-4} = \dfrac{x^2(x+7)}{x^3(x-4)} = \dfrac{x^2 \cdot (x+7)}{x^2 \cdot x(x-4)} = \dfrac{x+7}{x^2 - 4x}$$

**1.18.** Divide the following expressions by inverting the divisor and multiplying as in Rule 3 of Section 1.4.

(a) $\dfrac{6xy^5}{65w^4 z^4} \div \dfrac{2x^3 y}{11w^4 z^2}$

$$\dfrac{6xy^5}{65w^4 z^4} \div \dfrac{2x^3 y}{11w^4 z^2} = \dfrac{6xy^5}{65w^4 z^4} \cdot \dfrac{11w^4 z^2}{2x^3 y} = \dfrac{66w^4 xy^5 z^2}{130w^4 x^3 yz^4} = \dfrac{33y^4}{65x^2 z^2}$$

(b) $\dfrac{7x - 14y}{12x - 2y} \div \dfrac{12x + 32y}{18x - 3y}$

$$\dfrac{7x - 14y}{12x - 2y} \div \dfrac{12x + 32y}{18x - 3y} = \dfrac{7x - 14y}{12x - 2y} \cdot \dfrac{18x - 3y}{12x + 32y}$$

$$= \dfrac{7(x - 2y)}{2(6x - y)} \cdot \dfrac{3(6x - y)}{4(3x + 8y)} = \dfrac{21(x - 2y)}{8(3x + 8y)} = \dfrac{21x - 42y}{24x + 64y}$$

**1.19.** Add or subtract the following fractions as in Rule 5:

(a) $\dfrac{a}{b} + \dfrac{c}{d}$

(a) To add or subtract two fractions, you must ensure that they have a common denominator. Since the product of the individual denominators will automatically be a common denominator, simply multiply each fraction in the manner indicated below. With $(d/d) = 1 = (b/b)$, the values of the original fractions remain unchanged.

$$\dfrac{a}{b} + \dfrac{c}{d} = \left( \dfrac{a}{b} \cdot \dfrac{d}{d} \right) + \left( \dfrac{c}{d} \cdot \dfrac{b}{b} \right) = \dfrac{ad}{bd} + \dfrac{bc}{bd} = \dfrac{ad + bc}{bd}$$

(b) $\dfrac{1}{5} + \dfrac{1}{7}$

$$\dfrac{1}{5} + \dfrac{1}{7} = \dfrac{1 \cdot 7}{5 \cdot 7} + \dfrac{1 \cdot 5}{7 \cdot 5} = \dfrac{7 + 5}{35} = \dfrac{12}{35}$$

(c) $\dfrac{1}{3} - \dfrac{2}{17}$

$$\dfrac{1}{3} - \dfrac{2}{17} = \dfrac{1 \cdot 17}{3 \cdot 17} - \dfrac{2 \cdot 3}{17 \cdot 3} = \dfrac{17 - 6}{51} = \dfrac{11}{51}$$

**1.20.** Add or subtract the following fractions by finding the *least common denominator*.

(a) $\dfrac{11}{12} - \dfrac{7}{18}$

(a) To find the least common denominator, use the *fundamental principle of arithmetic* to decompose the denominators into products of prime numbers. Cancel all prime numbers common to the different denominators, then multiply both numerator and denominator of each term by the product of the remaining primes in the other denominator, as is demonstrated below.

$$\dfrac{11}{12} - \dfrac{7}{18} = \dfrac{11}{\cancel{2} \cdot 2 \cdot \cancel{3}} - \dfrac{7}{\cancel{2} \cdot \cancel{3} \cdot 3}$$

Multiplying both numerator and denominator of the first term by 3, therefore, and of the second term by 2,

$$\dfrac{11}{12} - \dfrac{7}{18} = \dfrac{11 \cdot 3}{12 \cdot 3} - \dfrac{7 \cdot 2}{18 \cdot 2} = \dfrac{33 - 14}{36} = \dfrac{19}{36}$$

(b) $\dfrac{5}{6} + \dfrac{3}{8}$

$$\dfrac{5}{6} + \dfrac{3}{8} = \dfrac{5}{\cancel{2} \cdot 3} + \dfrac{3}{\cancel{2} \cdot 2 \cdot 2}$$

Multiplying the first term by $\frac{4}{4}$ and the second by $\frac{3}{3}$,

$$\dfrac{5}{6} + \dfrac{3}{8} = \dfrac{5 \cdot 4}{6 \cdot 4} + \dfrac{3 \cdot 3}{8 \cdot 3} = \dfrac{20 + 9}{24} = \dfrac{29}{24} = 1\dfrac{5}{24}$$

(c) $\dfrac{8}{21} - \dfrac{13}{28}$

$$\dfrac{8}{21} - \dfrac{13}{28} = \dfrac{8}{3 \cdot \cancel{7}} - \dfrac{13}{2 \cdot 2 \cdot \cancel{7}}$$

Multiplying the first term by $\frac{4}{4}$ and the second by $\frac{3}{3}$,

$$\frac{8}{21} - \frac{13}{28} = \frac{8 \cdot 4}{21 \cdot 4} - \frac{13 \cdot 3}{28 \cdot 3} = \frac{32 - 39}{84} = -\frac{7}{84} = -\frac{1}{12}$$

**1.21.** Add or subtract the following fractions by finding the least common denominators. Reduce all answers to the lowest terms.

(a)  $\dfrac{4}{9x} + \dfrac{2}{5x}$

$$\frac{4}{9x} + \frac{2}{5x} = \frac{4}{9 \cdot x} + \frac{2}{5 \cdot x}$$

Multiplying the first term by $\frac{5}{5}$ and the second by $\frac{9}{9}$,

$$\frac{4}{9x} + \frac{2}{5x} = \left(\frac{4}{9x} \cdot \frac{5}{5}\right) + \left(\frac{2}{5x} \cdot \frac{9}{9}\right) = \frac{20}{45x} + \frac{18}{45x} = \frac{38}{45x}$$

(b)  $\dfrac{5}{6x} - \dfrac{7}{9y}$

$$\frac{5}{6x} - \frac{7}{9y} = \frac{5}{2 \cdot 3 \cdot x} - \frac{7}{3 \cdot 3 \cdot y}$$

Multiplying the first term by $3y/3y$ and the second by $2x/2x$,

$$\frac{5}{6x} - \frac{7}{9y} = \left(\frac{5}{6x} \cdot \frac{3y}{3y}\right) - \left(\frac{7}{9y} \cdot \frac{2x}{2x}\right)$$

$$= \frac{15y}{18xy} - \frac{14x}{18xy} = \frac{15y - 14x}{18xy}$$

(c)  $\dfrac{6}{x+8} + \dfrac{7}{x}$

Multiplying the first term by $x/x$ and the second by $(x+8)/(x+8)$ since there are no common factors,

$$\frac{6}{x+8} + \frac{7}{x} = \left[\frac{6}{(x+8)} \cdot \left(\frac{x}{x}\right)\right] + \left[\frac{7}{x} \cdot \left(\frac{x+8}{x+8}\right)\right]$$

$$= \frac{6x}{x(x+8)} + \frac{7x+56}{x(x+8)} = \frac{13x+56}{x^2+8x}$$

(d)  $\dfrac{12}{x^2-81} + \dfrac{7x}{x+9}$

$$\frac{12}{x^2-81} + \frac{7x}{x+9} = \frac{12}{(x+9)(x-9)} + \frac{7x}{(x+9)}$$

Multiplying the second term by $(x-9)/(x-9)$, therefore,

$$\frac{12}{x^2-81} + \frac{7x}{x+9} = \frac{12}{(x+9)(x-9)} + \left[\frac{7x}{(x+9)} \cdot \left(\frac{x-9}{x-9}\right)\right]$$

$$= \frac{12 + 7x^2 - 63x}{(x+9)(x-9)} = \frac{7x^2 - 63x + 12}{x^2 - 81}$$

(e)  $\dfrac{9}{x-3} + \dfrac{6x}{x^2-8x+15}$

$$\frac{9}{x-3} + \frac{6x}{x^2-8x+15} = \frac{9}{(x-3)} + \frac{6x}{(x-3)(x-5)}$$

Multiplying the first term by $(x - 5)/(x - 5)$,

$$\frac{9}{x - 3} + \frac{6x}{x^2 - 8x + 15} = \left[\frac{9}{(x - 3)} \cdot \left(\frac{x - 5}{x - 5}\right)\right] + \frac{6x}{(x - 3)(x - 5)}$$

$$= \frac{9x - 45 + 6x}{(x - 3)(x - 5)} = \frac{15(x - 3)}{(x - 3)(x - 5)} = \frac{15}{x - 5}$$

## RADICALS

**1.22.** Using the properties of radicals set forth in Section 1.5, simplify the following radicals.

(a)  $\sqrt{3} \cdot \sqrt{27}$

$$\sqrt{3} \cdot \sqrt{27} = \sqrt{3 \cdot 27} = \sqrt{81} = \pm 9 \qquad \text{(Rule 3)}$$

(b)  $\sqrt{5} \cdot \sqrt{45}$

$$\sqrt{5} \cdot \sqrt{45} = \sqrt{5 \cdot 45} = \sqrt{225} = \pm 15 \qquad \text{(Rule 3)}$$

(c)  $\sqrt{245}$

$$\sqrt{245} = \sqrt{5 \cdot 49} = \sqrt{5} \cdot \sqrt{49} = \pm 7\sqrt{5} \qquad \text{(Rule 3)}$$

(d)  $\sqrt{108}$

$$\sqrt{108} = \sqrt{3 \cdot 36} = \pm 6\sqrt{3} \qquad \text{(Rule 3)}$$

(e)  $(\sqrt[3]{192})/(\sqrt[3]{3})$

$$\frac{\sqrt[3]{192}}{\sqrt[3]{3}} = \sqrt[3]{\frac{192}{3}} = \sqrt[3]{64} = 4 \qquad \text{(Rule 4)}$$

(f)  $(\sqrt{294})/(\sqrt{6})$

$$\frac{\sqrt{294}}{\sqrt{6}} = \sqrt{\frac{294}{6}} = \sqrt{49} = \pm 7 \qquad \text{(Rule 4)}$$

(g)  $\sqrt[5]{\sqrt[3]{x}}$

$$\sqrt[5]{\sqrt[3]{x}} = \sqrt[5 \cdot 3]{x} = \sqrt[15]{x} \qquad \text{(Rule 2)}$$

(h)  $\sqrt{\sqrt[4]{y}}$

$$\sqrt{\sqrt[4]{y}} = \sqrt[2 \cdot 4]{y} = \sqrt[8]{y} \qquad \text{(Rule 2)}$$

(i)  $\sqrt{36x^4 y^6}$

$$\sqrt{36x^4 y^6} = \pm 6x^2 y^3 \qquad \text{(Rule 2)}$$

(j)  $\sqrt{98x^7 y^5}$

$$\sqrt{98x^7 y^5} = \sqrt{49x^6 y^4} \cdot \sqrt{2xy} = \pm 7x^3 y^2 \sqrt{2xy} \qquad \text{(Rules 2 and 3)}$$

**1.23.** Use the properties of radicals to solve for $y$ in each of the following instances.

(a)  $\sqrt{3y} = 6x^5$

Squaring both sides of the equation, then using Rule 1,

$$(\sqrt{3y})^2 = (6x^5)^2$$

$$3y = 36x^{10}$$

$$y = 12x^{10}$$

(b)   $\sqrt{7y} = 42x^3$

      Squaring both sides, we obtain

$$(\sqrt{7y})^2 = (42x^3)^2$$

$$7y = 1764x^6$$

$$y = 252x^6$$

## USE OF A CALCULATOR

**1.24.**  Practice the use of a pocket calculator and the rules of exponents from Section 1.1 to solve for $y$. Round all answers to five decimal places.

(a)   $y = 4^5 \cdot 4^3$

$$y = 4^{5+3} = 4^8 \qquad\qquad\qquad \text{(Rule 1)}$$

To find $4^8$, enter 4 on your calculator, press the $\boxed{y^x}$ key, enter 8 for the power, and hit the $\boxed{=}$ key.

$$y = 65{,}536$$

(b)   $y = 13^7 \div 13^4$

$$y = 13^{(7-4)} = 13^3 \qquad\qquad\qquad \text{(Rule 2)}$$

Entering 13, pressing the $\boxed{y^x}$ key, then entering 3 and hitting the $\boxed{=}$ key,

$$y = 2197$$

(c)   $y = 17^6 \div 17^8$

$$y = 17^{(6-8)} = 17^{-2}$$

To find $17^{-2}$, enter 17, press the $\boxed{y^x}$ key, then enter 2 followed by the $\boxed{\pm}$ key to make it negative, and hit the $\boxed{=}$ key.

$$y = 0.00346$$

(d)   $y = (5^3)^2$

$$y = 5^{(3\cdot 2)} = 5^6 \qquad\qquad\qquad \text{(Rule 3)}$$

Entering 5, pressing the $\boxed{y^x}$ key, then entering 6 and hitting the $\boxed{=}$ key,

$$y = 15{,}625$$

(e)   $y = (2^{-4})^2$

$$y = 2^{(-4\cdot 2)} = 2^{-8}$$

Entering 2 and pressing the $\boxed{y^x}$ key, then entering 8 followed by the $\boxed{\pm}$ key, and hitting the $\boxed{=}$ key,

$$y = 0.00391$$

**1.25.**  Use a calculator and the rules of radicals from Section 1.5 to estimate the value of $y$. Round all answers to five decimal places.

(a)   $y = \sqrt{3} \cdot \sqrt{17}$

$$y = \sqrt{3 \cdot 17} = \sqrt{51} \qquad\qquad\qquad \text{(Rule 3)}$$

To find $\sqrt{51}$, enter 51, hit the $\boxed{\text{INV}}$ key followed by the $\boxed{x^2}$ key to activate the $\boxed{\sqrt{x}}$ key, and you will get 7.14143.

Recalling that an even root can be positive or negative,

$$y = \pm 7.14143$$

(b)   $y = \sqrt{63} \cdot \sqrt{37}$

$$y = \sqrt{63 \cdot 37} = \sqrt{2331}$$

Entering 2331, then pressing the $\boxed{\text{INV}}$ key followed by the $\boxed{x^2}$ key,

$$y = \pm 48.28043$$

(c)   $y = \sqrt[3]{48} \cdot \sqrt[3]{74}$

$$y = \sqrt[3]{48 \cdot 74} = \sqrt[3]{3552}$$

To find $\sqrt[3]{3552}$, enter 3552, press the $\boxed{\text{INV}}$ key followed by the $\boxed{y^x}$ key to activate the $\boxed{\sqrt[x]{y}}$ key, then enter 3 for the root, and hit the $\boxed{=}$ key.

$$y = 15.25777$$

(d)   $y = \sqrt[5]{22} \cdot \sqrt[5]{45}$

$$y = \sqrt[5]{22 \cdot 45} = \sqrt[5]{990}$$

Entering 990, pressing the $\boxed{\text{INV}}$ key followed by the $\boxed{y^x}$ key, then entering 5 for the root and hitting the $\boxed{=}$ key,

$$y = 3.97308$$

(e)   $y = \dfrac{\sqrt[3]{2668}}{\sqrt[3]{46}}$

$$y = \sqrt[3]{\dfrac{2668}{46}} = \sqrt[3]{58} \qquad\qquad \text{(Rule 4)}$$

Entering 58, pressing the $\boxed{\text{INV}}$ key followed by the $\boxed{y^x}$ key, then entering 3 and hitting the $\boxed{=}$ key,

$$y = 3.87088$$

(f)   $y = \dfrac{\sqrt{3293}}{\sqrt{89}}$

$$y = \sqrt{\dfrac{3293}{89}} = \sqrt{37}$$

Entering 37, pressing the $\boxed{\text{INV}}$ key followed by the $\boxed{x^2}$ key,

$$y = \pm 6.08276$$

(g)   $y = \sqrt{\sqrt[4]{6561}}$

$$y = \sqrt[8]{6561} \qquad\qquad \text{(Rule 2)}$$

Entering 6561, pressing the $\boxed{\text{INV}}$ key followed by the $\boxed{y^x}$ key, then entering 8 and hitting the $\boxed{=}$ key,

$$y = \pm 3$$

**1.26.**   (a) Express the following radicals in their exponential forms. (b) Then use your calculator and the rules of exponents from Section 1.1 to solve for $y$.

(1)   $y = \sqrt{3} \cdot \sqrt{17}$

(a)                                       $y = 3^{1/2} \cdot 17^{1/2}$

(b)                                       $y = (3 \cdot 17)^{1/2} = 51^{1/2} = 51^{0.5}$ \qquad\qquad (Rule 4)

To find $51^{0.5}$, enter 51, hit the $\boxed{y^x}$ key, then enter 0.5 and hit the $\boxed{=}$ key to learn $51^{0.5} = 7.14143$. Recalling that an even root can be positive or negative,

$$y = \pm 7.14143$$

Compare with Problem 1.25 (a).

(2)   $y = \sqrt[5]{22} \cdot \sqrt[5]{45}$

(a) $$y = 22^{1/5} \cdot 45^{1/5}$$

(b) $$y = (22 \cdot 45)^{1/5} = 990^{1/5} = 990^{0.2}$$

Entering 990, pressing the $\boxed{y^x}$ key, then entering 0.2 and hitting the $\boxed{=}$ key,

$$y = 3.97308$$

Compare with Problem 1.25 (d).

(3)   $$y = \dfrac{\sqrt[3]{2668}}{\sqrt[3]{46}}$$

(a) $$y = \frac{2668^{1/3}}{46^{1/3}}$$

(b) $$y = \left(\frac{2668}{46}\right)^{1/3} = 58^{1/3} = 58^{0.33} \qquad \text{(Rule 5)}$$

Entering 58, pressing the $\boxed{y^x}$ key, then entering 0.33 and hitting the $\boxed{=}$ key,

$$y = 3.81884$$

Compare with Problem 1.25 (e). The slight discrepancy here is due simply to limiting $\frac{1}{3}$ to a two-decimal equivalent.

**1.27.**   Use your calculator and the rules of exponents to solve for $y$. Note that all the radicals are expressed in exponential form.

(a)   $y = 522^{-(1/2)} \div 29^{-(1/2)}$

$$y = \left(\frac{522}{29}\right)^{-(1/2)} = 18^{-(1/2)} = 18^{-0.5} \qquad \text{(Rule 5)}$$

To find $18^{-0.5}$, enter 18, press the $\boxed{y^x}$ key, then enter 0.5 followed by the $\boxed{\pm}$ key to make it negative, and hit the $\boxed{=}$ key.

$$y = \pm 0.23570$$

(b)   $y = 15^{-(1/4)} \cdot 23^{-(1/4)}$

$$y = (15 \cdot 23)^{-(1/4)} = 345^{-(1/4)} = 345^{-0.25}$$

Entering 345, hitting the $\boxed{y^x}$ key, then entering 0.25 followed by the $\boxed{\pm}$ key to make it negative and hitting the $\boxed{=}$ key,

$$y = \pm 0.23203$$

(c)   $y = (6561^{1/4})^{1/2}$

$$y = (6561)^{1/4 \cdot 1/2} = 6561^{1/8} = 6561^{0.125} \qquad \text{(Rule 3)}$$

Entering 6561, pressing the $\boxed{y^x}$ key, entering 0.125, then hitting the $\boxed{=}$ key,

$$y = \pm 3$$

*Note*: The same answer can be obtained by entering 6561, pressing the $\boxed{\text{INV}}$ key followed by the $\boxed{y^x}$, then entering 8 for the root, and hitting the $\boxed{=}$ key.

(d)   $y = (34^{2/5})^2$

$$y = 34^{(2/5 \cdot 2)} = 34^{4/5} = 34^{0.8}$$

Entering 34, pressing the $\boxed{y^x}$ key, then entering 0.8 and hitting the $\boxed{=}$ key,

$$y = 16.79512$$

(e)   $y = [14^{-(3/2)}]^{4/5}$

$$y = [14^{-(3/2)}]^{4/5} = 14^{[-(3/2) \cdot 4/5]} = 14^{-(12/10)} = 14^{-1.2}$$

Entering 14, pressing the $\boxed{y^x}$ key, then entering 1.2 followed by the $\boxed{\pm}$ key to make it negative, and hitting the $\boxed{=}$ key,

$$y = 0.04214$$

## PROOFS

**1.28.** Given the polynomial $mx^2 + nx + p$, show that the factors are $(ax + c)(bx + d)$ where $ab = m$, $ad + bc = n$, and $cd = p$.

By reversing the process of multiplication, we can express the polynomial

$$mx^2 + nx + p = (ax + c)(bx + d)$$

Carrying out the multiplication on the right-hand side,

$$mx^2 + nx + p = abx^2 + (ad + bc)x + cd \qquad (1.1)$$

From $(1.1)$ it is evident that $ab = m$, $ad + bc = n$, and $cd = p$.

**1.29.** Given a polynomial in two variables, $mx^2 + nxy + py^2$, show that the factors $(ax + cy)(bx + dy)$ are such that $ab = m$, $ad + bc = n$, and $cd = p$.

$$mx^2 + nxy + py^2 = (ax + cy)(bx + dy)$$

Multiplying,

$$mx^2 + nxy + py^2 = abx^2 + (ad + bc)xy + cdy^2$$

Hence, $ab = m$, $ad + bc = n$, and $cd = p$.

**1.30.** Use the laws of exponents to prove Rule 1 for radicals, namely, $(\sqrt[n]{x})^n = x$.

From Section 1.1,

$$(\sqrt[n]{x})^n = (x^{1/n})^n$$

From Rule 3 of exponents,

$$(x^{1/n})^n = x^{(1/n)n} = x^{(1)} = x$$

# Supplementary Problems

## EXPONENTS

**1.31.** Use the rules of exponents to simplify the following expressions:

(a) $x^4 \cdot x^3$      (b) $x^5 \cdot x^{1/2}$      (c) $x^6 \cdot x^{-4}$      (d) $x^{-2} \cdot x^{-7}$

**1.32.** Simplify the following exponential expressions:

(a) $\dfrac{x^5}{x^2}$      (b) $\dfrac{x^3}{x^5}$      (c) $\dfrac{x^2}{x^{-4}}$      (d) $\dfrac{x^{-4}}{x^3}$

**1.33.** Simplify the following:

(a) $(x^3)^4$      (b) $(x^4)^{-2}$      (c) $(x^3 y^4)^2$      (d) $\left(\dfrac{x^2}{y^3}\right)^3$

**1.34.** Simplify the following:

(a) $\dfrac{x^2}{x^{1.5}}$      (b) $(x^3)^{1/2}$      (c) $(x^{1/3})^2$      (d) $(x^4)^{-(1/5)}$

## POLYNOMIALS

**1.35.**  Perform the indicated arithmetical operations for each of the following polynomial expressions:

    (a) $76xy - 22xy$              (b) $14x_1x_2 + 31x_1x_2$        (c) $32xy^2 - 48xy^2$

    (d) $50xyz - 43xyz$

**1.36.**  Add or subtract the following polynomials as indicated:

    (a) $(6x + 9y) + (3x - 8y)$           (b) $(20x + 13y) - (4x + 7y)$

    (c) $(12x - 17y) - (9x - 8y)$         (d) $(5x^2 - 12xy + 3y^2) - (2x^2 + 9xy + 7y^2)$

**1.37.**  Multiply each of the following polynomial expressions:

    (a) $(5x + 2y)(8x + 3y)$            (b) $(6x - 15y)(10x - 3y)$

    (c) $(3x^2 - 5xy + 4y^2)(6x + 2y)$        (d) $(20x - 5y)(8x^2 + 2xy + 3y^2)$

## FACTORING

**1.38.**  Simplify each of the following polynomials by factoring out the greatest common factor:

    (a) $12 - 15x$           (b) $42x^3 + 28x$        (c) $15x^3y^4 + 55x^2y^5$

    (d) $18x^2y^6z^4 + 22x^3y^4z^3 - 26x^5y^4z^6$

**1.39.**  Factor each of the following quadratic equations:

    (a) $x^2 + 10x + 24$       (b) $x^2 - 10x + 21$       (c) $x^2 + 9x - 22$

    (d) $x^2 - 7x - 60$        (e) $x^2 - 49$           (f) $x^2 - 23x + 120$

**1.40.**  Factor each of the following quadratic equations in which the coefficient of the $x^2$ is no longer 1:

    (a) $3x^2 + 17x + 10$       (b) $5x^2 - 19x + 12$       (c) $7x^2 - 22x - 24$

    (d) $4x^2 + 16x + 15$       (e) $6x^2 - 23x + 20$       (f) $8x^2 - 41x - 42$

## FRACTIONS

**1.41.**  Reduce the following fractions to their lowest terms:

    (a) $\dfrac{14}{35}$            (b) $\dfrac{22}{110}$          (c) $\dfrac{8x^5}{56x^3}$

    (d) $\dfrac{32x^3}{6x^4}$         (e) $\dfrac{x^2 + x - 30}{x^2 + 14x + 48}$       (f) $\dfrac{x^2 - 8x + 12}{x^2 - 11x + 18}$

**1.42.**  Multiply or divide the following fractions as indicated:

    (a) $\dfrac{7x}{9y} \cdot \dfrac{4x}{3y}$      (b) $\dfrac{5x^3y^2}{2wz^4} \cdot \dfrac{8w^2y^5}{7x^2z^3}$      (c) $\dfrac{12x}{25y} \div \dfrac{3x}{5y}$      (d) $\dfrac{11w^2x^4}{8yz^3} \div \dfrac{3wx^5}{2y^3z}$

**1.43.**  Multiply or divide the following rational expressions:

    (a) $\dfrac{x + 5}{x - 8} \cdot \dfrac{x - 3}{x + 9}$       (b) $\dfrac{x - 4}{x - 1} \cdot \dfrac{x + 7}{x - 6}$

    (c) $\dfrac{x + 12}{x + 9} \div \dfrac{x - 6}{x + 2}$       (d) $\dfrac{2x - 7}{x + 4} \div \dfrac{3x + 2}{x - 9}$

**1.44.**  Add or subtract the following functions, finding the least common denominator where necessary:

    (a) $\dfrac{2}{5} + \dfrac{3}{8}$          (b) $\dfrac{7}{12} - \dfrac{2}{11}$         (c) $\dfrac{x}{6} + \dfrac{3x}{7}$

    (d) $\dfrac{5}{18x} - \dfrac{3}{16x}$       (e) $\dfrac{3}{4} - \dfrac{1}{6}$         (f) $\dfrac{7}{9} - \dfrac{5}{12}$

**1.45.**    Add or subtract the following as indicated:

(a)   $\dfrac{13}{x+9} - \dfrac{6}{x}$          (b)   $\dfrac{8}{x-2} + \dfrac{3}{x+4}$          (c)   $\dfrac{41}{x-7} - \dfrac{15}{x^2 - 49}$

(d)   $\dfrac{17}{x-6} - \dfrac{15}{x^2 - 14x + 48}$

## RADICALS

**1.46.**    Simplify the following radicals:

(a)   $\sqrt{2} \cdot \sqrt{32}$          (b)   $\sqrt{7} \cdot \sqrt{63}$          (c)   $\sqrt{243}$

(d)   $\sqrt{605}$          (e)   $\sqrt{75} \div \sqrt{3}$          (f)   $\sqrt{288} \div \sqrt{6}$

**1.47.**    Simplify the following radicals:

(a)   $\sqrt{169x^6 y^8}$          (b)   $\sqrt{81x^2 y^4 z^6}$          (c)   $\sqrt{108x^3 y^5}$          (d)   $\sqrt{450x^4 y^5 z^6}$

## USE OF CALCULATOR

**1.48.**    Use a calculator to multiply or divide the following:

(a)   $5236 \cdot 0.015$          (b)   $0.065 \cdot 3.75$          (c)   $145 \div 0.25$          (d)   $0.675 \div 0.045$

**1.49.**    Estimate the following using a calculator:

(a)   $6^5$          (b)   $3^7$          (c)   $8^{-4}$          (d)   $12^{-6}$          (e)   $\sqrt{784}$

(f)   $\sqrt[3]{2197}$          (g)   $\sqrt[4]{50{,}625}$          (h)   $784^{0.5}$          (i)   $50{,}625^{0.25}$

# Answers to Supplementary Problems

**1.31.**    (a)   $x^7$          (b)   $x^{5.5} = x^{(11/2)} = \sqrt{x^{11}}$          (c)   $x^2$          (d)   $x^{-9} = \dfrac{1}{x^9}$

**1.32.**    (a)   $x^3$          (b)   $x^{-2} = \dfrac{1}{x^2}$          (c)   $x^6$          (d)   $x^{-7} = \dfrac{1}{x^7}$

**1.33.**    (a)   $x^{12}$          (b)   $x^{-8} = \dfrac{1}{x^8}$          (c)   $x^6 y^8$          (d)   $\dfrac{x^6}{y^9}$

**1.34.**    (a)   $x^{1/2} = \sqrt{x}$          (b)   $x^{3/2} = \sqrt{x^3}$          (c)   $x^{2/3} = \sqrt[3]{x^2}$          (d)   $x^{-(4/5)} = \dfrac{1}{\sqrt[5]{x^4}}$

**1.35.**    (a)   $54xy$          (b)   $45x_1 x_2$          (c)   $-16xy^2$          (d)   $7xyz$

**1.36.**    (a)   $9x + y$          (b)   $16x + 6y$          (c)   $3x - 9y$          (d)   $3x^2 - 21xy - 4y^2$

**1.37.**    (a)   $40x^2 + 31xy + 6y^2$          (b)   $60x^2 - 168xy + 45y^2$          (c)   $18x^3 - 24x^2 y + 14xy^2 + 8y^3$

(d)   $160x^3 + 50xy^2 - 15y^3$

**1.38.**    (a)   $3(4 - 5x)$          (b)   $14x(3x^2 + 2)$          (c)   $5x^2 y^4(3x + 11y)$          (d)   $2x^2 y^4 z^3(9y^2 z + 11x - 13x^3 z^3)$

**1.39.**    (a)   $(x + 4)(x + 6)$          (b)   $(x - 3)(x - 7)$          (c)   $(x - 2)(x + 11)$          (d)   $(x + 5)(x - 12)$

(e)   $(x + 7)(x - 7)$          (f)   $(x - 8)(x - 15)$

**1.40.**    (*a*)   $(3x + 2)(x + 5)$     (*b*)   $(5x - 4)(x - 3)$     (*c*)   $(7x + 6)(x - 4)$     (*d*)   $(2x + 3)(2x + 5)$
         (*e*)   $(2x - 5)(3x - 4)$     (*f*)   $(8x + 7)(x - 6)$

**1.41.**    (*a*)   $\dfrac{2}{5}$    (*b*)   $\dfrac{1}{5}$    (*c*)   $\dfrac{x^2}{7}$    (*d*)   $\dfrac{16}{3x}$    (*e*)   $\dfrac{x - 5}{x + 8}$    (*f*)   $\dfrac{x - 6}{x - 9}$

**1.42.**    (*a*)   $\dfrac{28x^2}{27y^2}$    (*b*)   $\dfrac{20wxy^7}{7z^7}$    (*c*)   $\dfrac{4}{5}$    (*d*)   $\dfrac{11wy^2}{12xz^2}$

**1.43.**    (*a*)   $\dfrac{x^2 + 2x - 15}{x^2 + x - 72}$    (*b*)   $\dfrac{x^2 + 3x - 28}{x^2 - 7x + 6}$    (*c*)   $\dfrac{x^2 + 14x + 24}{x^2 + 3x - 54}$    (*d*)   $\dfrac{2x^2 - 25x + 63}{3x^2 + 14x + 8}$

**1.44.**    (*a*)   $\dfrac{31}{40}$    (*b*)   $\dfrac{53}{132}$    (*c*)   $\dfrac{25x}{42}$    (*d*)   $\dfrac{13}{144x}$    (*e*)   $\dfrac{7}{12}$    (*f*)   $\dfrac{13}{36}$

**1.45.**    (*a*)   $\dfrac{7x - 54}{x^2 + 9x}$    (*b*)   $\dfrac{11x + 26}{x^2 + 2x - 8}$    (*c*)   $\dfrac{41x + 272}{x^2 - 49}$    (*d*)   $\dfrac{17x - 151}{x^2 - 14x + 48}$

**1.46.**    (*a*)   $\pm 8$    (*b*)   $\pm 21$    (*c*)   $\pm 9\sqrt{3}$    (*d*)   $\pm 11\sqrt{5}$    (*e*)   $\pm 5$    (*f*)   $\pm 4\sqrt{3}$

**1.47.**    (*a*)   $\pm 13x^3y^4$    (*b*)   $\pm 9xy^2z^3$    (*c*)   $\pm 6xy^2\sqrt{3xy}$    (*d*)   $\pm 15x^2y^2z^3\sqrt{2y}$

**1.48.**    (*a*)   78.54    (*b*)   0.24375    (*c*)   580    (*d*)   15

**1.49.**    (*a*)   7776    (*b*)   2187    (*c*)   0.0002441    (*d*)   0.0000003    (*e*)   $\pm 28$    (*f*)   13
         (*g*)   $\pm 15$    (*h*)   $\pm 28$    (*i*)   $\pm 15$

# Chapter 2

# Equations and Graphs

## 2.1 EQUATIONS

A mathematical statement setting two algebraic expressions equal to each other is called an *equation*. A *solution* of an equation is a number or numbers which, if substituted for each occurrence of the variable or variables, leads to the same value on each side of the equality sign.

The same quantity $c$ can be added, subtracted, multiplied, or divided on both sides of the equation without affecting the equality, assuming $c \neq 0$ for division or multiplication. For all real numbers $a$, $b$, and $c$, if $a = b$, the *properties of equality* can be summed up as follows:

(a) Addition property:          $a + c = b + c$
(b) Subtraction property:       $a - c = b - c$
(c) Multiplication property:    $ac = bc$
(d) Division property:          $a/c = b/c \quad (c \neq 0)$

The equivalent equations that result from the use of these steps can then be used to find a simplified equation in which the solution is obvious, that is, with the unknown variable by itself to the left of the equality sign and the solution set to the right of the equality sign.

**EXAMPLE 1.**    An equation, such as the one below, is solved in three easy steps, using the properties of equality set out above. Given

$$\frac{x}{4} - 2 = \frac{x}{5} + 1$$

1)   Move any term with the unknown variable to the left, here by subtracting $x/5$ from both sides of the equation:

$$\boxed{\frac{x}{4} - 2 - \frac{x}{5} = \frac{x}{5} + 1 - \frac{x}{5}}$$   (Subtraction Property)

$$\frac{x}{4} - \frac{x}{5} - 2 = 1$$

2)   Move any term without the unknown variable to the right, here by adding 2 to both sides of the equation:

$$\boxed{\frac{x}{4} - \frac{x}{5} - 2 + 2 = 1 + 2}$$   (Addition Property)

$$\frac{x}{4} - \frac{x}{5} = 3$$

3)   Simplify both sides of the equation until the unknown variable is by itself on the left and the solution set on the right, here by multiplying both sides of the equation by 20 and then subtracting:

$$\boxed{20 \cdot \left(\frac{x}{4} - \frac{x}{5}\right) = 3 \cdot 20}$$   (Multiplication Property)

$$5x - 4x = 60 \qquad x = 60$$

See Problems 2.1 and 2.2.

27

## 2.2 CARTESIAN COORDINATE SYSTEM

A *Cartesian coordinate system* is composed of a horizontal line and a vertical line set perpendicular to each other in a plane. The lines are called the *coordinate axes*; their point of intersection, the *origin*. The horizontal line is referred to as the *x axis*; the vertical line, the *y axis*. The four sections into which the plane is divided by the intersection of the axes are called *quadrants*.

Each point in the plane is uniquely associated with an ordered pair of numbers, known as *coordinates*, describing the location of the point in relation to the origin. The first coordinate, called the *x coordinate* or *abscissa*, gives the distance of the point from the vertical axis; the second, the *y coordinate* or *ordinate*, records the distance of the point from the horizontal axis. To the right of the *y* axis, *x* coordinates are positive; to the left, negative. Above the *x* axis, *y* coordinates are positive; below, negative. See Problem 2.3.

**EXAMPLE 2.**    The signs of the coordinates in each of the quadrants are illustrated in Fig. 2-1. Note that the quadrants are numbered counterclockwise.

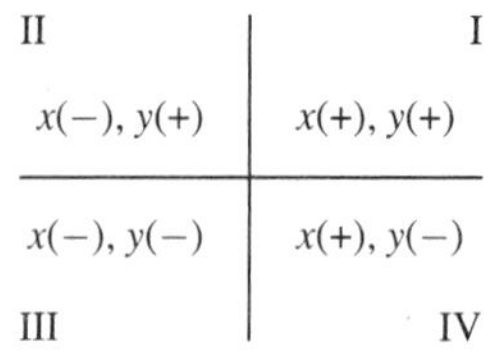

II     |     I

$x(-), y(+)$   |   $x(+), y(+)$

$x(-), y(-)$   |   $x(+), y(-)$

III     |     IV

**Fig. 2-1**

**EXAMPLE 3.**    The coordinates give the location of the point $P$ in relation to the origin. The point $(4, 2)$ is four units to the right of the *y* axis, two units above the *x* axis. Point $P(-4, 2)$ is four units to the left of the *y* axis, two units above the *x* axis. Point $P(-4, -2)$ is four units to the left of the *y* axis, two units below the *x* axis; $P(4, -2)$, four units to the right of the *y* axis, two units below the *x* axis. See Fig. 2-2 and Problem 2.3.

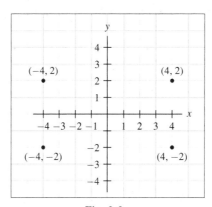

**Fig. 2-2**

## 2.3 LINEAR EQUATIONS AND GRAPHS

An equation in which all variables are raised to the first power and no cross product of variables occurs is called a *linear equation*. The *standard form* of a linear equation is

$$ax + cy = d$$

where $a$, $c$, and $d$ are real numbers and $a$ and $c$ do not both equal zero.

The graph of a linear equation is a *straight line*. To graph a linear equation on a Cartesian coordinate system, one need only (1) find two points that satisfy the equation, (2) connect them with a straight line, and (3) extend the straight line as far as needed. The coordinates of the points on a line are found by selecting different values of $x$ and solving the equation for each of the corresponding values of $y$ or by selecting different values of $y$ and finding the corresponding values of $x$. In graphing an equation, $x$ is usually placed on the horizontal axis and called the *independent variable*; $y$ is placed on the vertical axis and called the *dependent variable*. See Example 4 and Problems 2.3 to 2.4.

**EXAMPLE 4.**    In graphing an equation, such as

$$6x + 3y = 18$$

1)   Pick any two values for $x$ and solve the equation for each of the corresponding values of $y$. If $x = -1$ and $x = 1$ are selected,

At $x = -1$,     $6(-1) + 3y = 18$     $3y = 24$     $y = 8$     point$_a$ : $(-1, 8)$

At $x = 1$,         $6(1) + 3y = 18$     $3y = 12$     $y = 4$     point$_b$ : $(1, 4)$

2)   Plot the points and connect them with a straight line.

3)   Extend the straight line as far as needed. See Fig. 2-3.

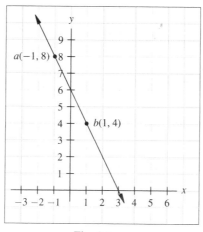

**Fig. 2-3**

## 2.4   SLOPES

The *slope* of a line measures the *change in the dependent variable* $y$, also known as the "*rise*," divided by a *change in the independent variable* $x$, also known as the "*run*." The slope indicates both the steepness and direction of a line. Going from left to right, a positively sloped line moves up, a negatively sloped line moves down. The greater the absolute value of the slope, the steeper the line. The slope of a horizontal line, $y = k_1$ (a constant), is 0; the slope of a vertical line, $x = k_2$ (a constant), is undefined. In mathematical terms, using the Greek capital letter delta ($\Delta$) to symbolize a change, the slope $m$ of a line can be written in four different but equally valid ways:

$$m = \frac{\Delta y}{\Delta x} = \frac{y_1 - y_2}{x_1 - x_2} = \frac{y_2 - y_1}{x_2 - x_1} = \frac{\text{rise}}{\text{run}}    \qquad (x_1 \neq x_2) \qquad\qquad (2.1)$$

**EXAMPLE 5.**    To find the slope of the equation

$$6x + 3y = 18$$

substitute the coordinates of any two points that satisfy the equation into the formula for the slope in equation ($2.1$).

$$\text{Slope} = m = \frac{\Delta y}{\Delta x} = \frac{y_1 - y_2}{x_1 - x_2}$$

Selecting from Example 4, $(-1, 8)$ as the first point, so that $x_1 = -1$ and $y_1 = 8$, and $(1, 4)$ as the second point, so that $x_2 = 1$ and $y_2 = 4$, and substituting above,

$$\text{Slope} = m = \frac{8 - 4}{-1 - (1)} = \frac{4}{-2} = -2$$

*Note*:

1) Reversing the order of coordinates, if done consistently in numerator and denominator, does not change the value or sign of the slope, as indicated by the formula in ($2.1$).

$$\text{Slope} = m = \frac{y_2 - y_1}{x_2 - x_1} = \frac{4 - 8}{1 - (-1)} = \frac{-4}{2} = -2$$

2) A negative slope indicates that the line moves down from left to right, as is evident from the graph of the equation in Fig. 2-3.

3) The absolute value of the slope $|2|$ indicates that the line changes vertically at a rate of two units for every one-unit movement horizontally. See Fig. 2-3 and Problems 2.5 to 2.8.

## 2.5   INTERCEPTS

The *x intercept* is the point where the graph crosses the *x* axis; the *y intercept*, the point where the line crosses the *y* axis. Since the line crosses the *x* axis where $y = 0$, the *x* coordinate of the *x* intercept is found by setting $y = 0$ and solving the equation for *x*. Similarly, since the line crosses the *y* axis where $x = 0$, the *y* coordinate of the *y* intercept is obtained by setting *x* equal to zero and solving the equation for *y*.

**EXAMPLE 6.**   To find the *x* intercept of the equation

$$6x + 3y = 18$$

set $y = 0$ and solve for *x*.

$$6x + 3(0) = 18 \qquad x = 3$$

The *x* intercept is $(3, 0)$, as seen in Fig. 2-3.
To find the *y* intercept, set $x = 0$ and solve for *y*.

$$6(0) + 3y = 18 \qquad y = 6$$

The *y* intercept is $(0, 6)$, as can also be seen in Fig. 2-3. See Problems 2.9 to 2.13.

## 2.6   THE SLOPE-INTERCEPT FORM

By starting with the standard form of a linear equation

$$ax + cy = d$$

and solving for *y* in terms of *x*, we get

$$cy = -ax + d$$

$$y = -\frac{a}{c}x + \frac{d}{c}$$

Then by letting $m = -a/c$ and $b = d/c$,

$$y = mx + b \qquad\qquad\qquad\qquad\qquad (2.2)$$

Equation *(2.2)* is the *slope-intercept form* of a linear equation where $m$ is the slope of the line; $(0, b)$, the coordinates of the $y$ intercept; and $(b/m, 0)$, the coordinates of the $x$ intercept, as is demonstrated in Examples 7 and 8 and illustrated in Problems 2.9 to 2.14.

**EXAMPLE 7.**    Deriving the slope-intercept form from the standard form of an equation is accomplished by solving the equation for $y$ in terms of $x$. Thus, given

$$6x + 3y = 18$$

$$3y = -6x + 18$$

$$y = -2x + 6$$

where in terms of equation *(2.2)* $m = -2$, the slope of the line as found in Example 5; $b = 6$, the $y$ coordinate of the $y$ intercept $(0, 6)$, as was found in Example 6; and the $x$ intercept, being $(-b/m, 0)$, is $(-6/-2, 0)$ or $(3, 0)$, as was also found in Example 6.

**EXAMPLE 8.**    Given the equations of four different lines

$$(a)\ y = -\tfrac{1}{4}x + 3 \qquad (b)\ y = 2x - 1 \qquad (c)\ y = 5 \qquad (d)\ y = -3x + 6$$

and knowing that in the slope-intercept form of a linear equation the slope is given by the coefficient of $x$, one knows immediately that the slopes of the lines are $(a)$ $-\tfrac{1}{4}$, $(b)$ 2, $(c)$ 0, and $(d)$ $-3$.

Knowing that the constant term in the slope-intercept form is the $y$ coordinate of the $y$ intercept, where $x = 0$, one also knows that the line crosses the $y$ axis at $(a)$ $(0, 3)$, $(b)$ $(0, -1)$, $(c)$ $(0, 5)$, and $(d)$ $(0, 6)$.

Because the $x$ intercept of a linear equation in the slope-intercept form is $(-b/m, 0)$, we know that the graph crosses the $x$ axis at $(a)$ $(12, 0)$, $(b)$ $(\tfrac{1}{2}, 0)$, and $(d)$ $(2, 0)$. With $m = 0$ in $(c)$ and division by 0 undefined, there is no $x$ intercept in $(c)$. See Fig. 2-4 and Problems 2.11 and 2.13.

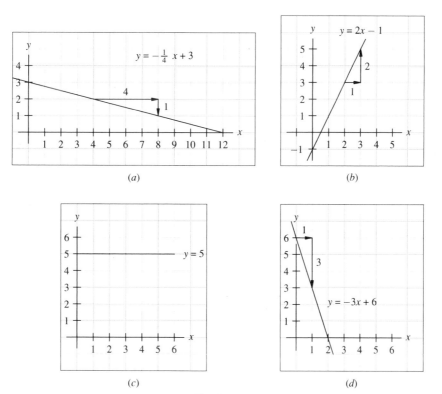

**Fig. 2-4**

## 2.7  DETERMINING THE EQUATION OF A STRAIGHT LINE

Determining the equation of a straight line depends on the amount of information available. If one knows both the slope and the $y$ intercept, one need only use the slope-intercept form of an equation, as demonstrated in Example 9. If one knows the slope $m$ and a single point $(x_1, y_1)$ on the line, as shown in Example 10, one can use the *point-slope formula*

$$y - y_1 = m(x - x_1) \qquad (2.3)$$

And if one simply knows a few points on the line, as demonstrated in Example 11, one can use $(2.3)$ after the *two-point formula*

$$m = \frac{y_2 - y_1}{x_2 - x_1} \qquad (2.4)$$

See also Problems 2.14 to 2.20.

**EXAMPLE 9.**    The equation of a line where the slope $m$ and the $y$ intercept $(0, b)$ are given, such as in Fig. 2-5, can be found by substituting $m = \frac{1}{4}$ and $b = 3$ in the slope-intercept form

$$y = mx + b$$

$$y = \tfrac{1}{4}x + 3$$

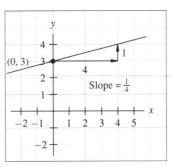

**Fig. 2-5**

**EXAMPLE 10.**    The equation of a line where the slope $m$ and a single point $(x_1, y_1)$ are known, such as in Fig. 2-6, can be found by substituting $m = -5$, $x_1 = 1$, and $y_1 = 25$ in the point-slope formula in equation $(2.3)$.

$$y - y_1 = m(x - x_1)$$

$$y - 25 = -5(x - 1)$$

$$y = -5x + 30$$

**EXAMPLE 11.**    The equation of a line where two points $(x_1, y_1)$ and $(x_2, y_2)$ are known, such as in Fig. 2-7, can be found by substituting $x_1 = 4$, $y_1 = 1$, $x_2 = 6$, and $y_2 = 2$ in the two-point formula in equation $(2.4)$ to find the slope

$$m = \frac{y_2 - y_1}{x_2 - x_1} = \frac{2 - 1}{6 - 4} = \frac{1}{2}$$

and then substituting the values of $m$, $x_1$, and $y_1$ in the point-slope formula

$$y - y_1 = m(x - x_1)$$

$$y - 1 = \tfrac{1}{2}(x - 4)$$

$$y = \tfrac{1}{2}x - 1$$

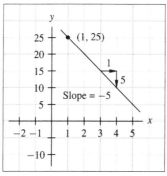

**Fig. 2-6**

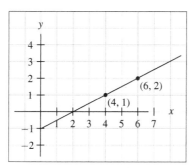

**Fig. 2-7**

## 2.8 APPLICATIONS OF LINEAR EQUATIONS IN BUSINESS AND ECONOMICS

Many areas in business and economics are handled effectively by linear equations, as is demonstrated in Examples 12 to 15 and Problems 2.21 to 2.34.

**EXAMPLE 12.**   Linear equations are often suitable for describing production constraints. Assume a firm has 240 hours of skilled labor available each week to produce two products. Each unit of the first product $x$ requires 3 hours of skilled labor. A unit of the second product $y$ requires 4 hours. (*a*) Express the firm's labor constraint in terms of an equation. (*b*) Draw a graph showing all the different possible ways labor can be allocated between the two goods.

    (*a*)   With 3 hours required for each unit of $x$ and 4 hours for each unit of $y$, and a total of 240 hours available, the constraint can be expressed in the standard form of a linear equation:

$$3x + 4y = 240 \qquad\qquad (2.5)$$

    (*b*)   To facilitate graphing, convert (*2.5*) to the slope-intercept form and graph, as in Fig. 2-8.

$$4y = -3x + 240$$

$$y = -\tfrac{3}{4}x + 60$$

Note that only points in the first quadrant are relevant to this problem. This will also be true of many other economic problems.

**EXAMPLE 13.**   Using a linear equation to express *straight-line* or *linear depreciation*, the current value $y$ of a moving firm's van after $x$ years is estimated to be

$$y = 68{,}000 - 8000x$$

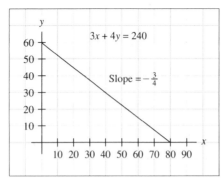

**Fig. 2-8**

Find (*a*) the initial value of the van, (*b*) the value after 3 years, and (*c*) the salvage value after 8 years. (*d*) Draw the graph.

(*a*)   Substituting $x = 0$, $y = 68,000 - 8000(0) = 68,000$

(*b*)   Substituting $x = 3$, $y = 68,000 - 8000(3) = 44,000$

(*c*)   Substituting $x = 8$, $y = 68,000 - 8000(8) = 4000$

(*d*)   See Fig. 2-9 and Problems 2.25 and 2.26.

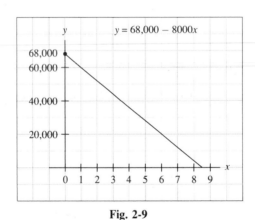

**Fig. 2-9**

## EXAMPLE 14.

(*a*)   A firm which has *fixed costs* FC of \$560, for rent and executives' salaries which must be met regardless of output level, and a *marginal cost* MC of \$9, which is the expense incurred for each additional unit of output $x$, faces a *total cost* C which can be expressed by a linear equation of the form $y = mx + b$, where $y = C$, $m = \text{MC} = 9$, and $b = \text{FC} = 560$. Thus,

$$C = 9x + 560$$

If 140 units are produced so that $x = 140$,

$$C = 9(140) + 560 = 1820$$

(*b*)   If the firm operates in *pure competition*, where it receives a constant price $p$ for each unit of output $x$, its *total revenue* R can be expressed by the linear equation

$$R = p \cdot x$$

If 140 units are sold at $p = 25$,

$$R = 25(140) = 3500$$

(c)   With *profit* $\pi$ being the difference between total revenue and total cost, the profit level can also be expressed as a linear equation

$$\pi = R - C$$

Substituting from above,

$$\pi = 25(x) - [9x + 560] = 16x - 560$$

If 140 units are produced and sold,

$$\pi = 16(140) - 560 = 1680$$

See also Problems 2.21 to 2.24.

**EXAMPLE 15.**    Economists are often called on to maximize utility subject to some budget constraint. The constraint is generally represented by a *budget line* depicting all the different possible combinations of goods a person can buy with a given budget $B$. Assume a woman has \$150 to spend on two goods $x$ and $y$ whose respective prices are $p_x = \$5$ and $p_y = \$2$. (a) Draw her budget line. Show what happens to the budget line (b) if her budget falls by 20 percent, (c) if $p_x$ is cut in half, and (d) if $p_y$ increases by 50¢.

(a)   The formula for a budget line is the standard form of a linear equation

$$p_x \cdot x + p_y \cdot y = B$$

Substituting $p_x = 5$, $p_y = 2$, and $B = \$150$,

$$5x + 2y = 150 \tag{2.6}$$

Converting (2.6) to the slope-intercept form and graphing,

$$y = -2.5x + 75$$

See the solid line in Fig. 2-10(a).

(b)   If the budget falls by 20 percent, the new budget is 120, that is, $150 - 0.2(150) = 120$. The equation for the new budget line is

$$5x + 2y = 120$$

$$y = -2.5x + 60$$

See the dashed line in Fig. 2-10(a). Lowering the budget causes the budget line to shift parallel to the left.

(c)   If $p_x$ is cut in half, the original equation in (2.6) becomes

$$2.5x + 2y = 150$$

$$y = -1.25x + 75$$

The vertical intercept remains the same, but the slope changes and becomes flatter. See the dashed line in Fig. 2-10(b). With the price for $x$ cut in half, twice as much $x$ can be bought with the given budget.

(d)   If $p_y$ increases by 50 cents,

$$5x + 2.5y = 150$$

$$y = -2x + 60$$

A change in $p_y$ causes both the vertical intercept and the slope to change, as shown in Fig. 2-10(c). With an increase in $p_y$, the vertical intercept drops because less $y$ can be bought with the given budget and the slope becomes flatter. Notice that the horizontal intercept does not change. With no changes in $p_x$ and $B$ the consumer can still buy the same maximum amount of $x$ if she chooses to spend her entire budget on $x$.

See also Problems 2.31 and 2.32.

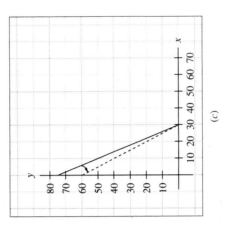

(a)

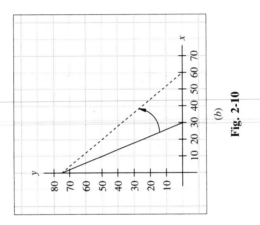

(b)

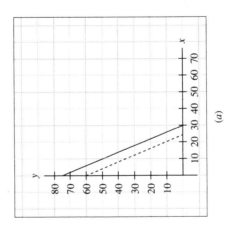

(c)

**Fig. 2-10**

# Solved Problems

## SOLVING LINEAR EQUATIONS

**2.1.** Use the properties of equality to solve the following linear equations by moving all terms with the unknown variable to the left, all other terms to the right, and then simplifying:

(a)  $3x + 8 = 5x - 6$

$$3x + 8 = 5x - 6$$

$$3x - 5x = -6 - 8$$

$$-2x = -14$$

$$x = 7$$

(b)  $32 - 3x = 9x - 40$

$$32 - 3x = 9x - 40$$

$$-3x - 9x = -40 - 32$$

$$-12x = -72$$

$$x = 6$$

(c)  $6(4x + 5) - 3x = 19 - 2(7x + 82)$

$$6(4x + 5) - 3x = 19 - 2(7x + 82)$$

$$24x + 30 - 3x = 19 - 14x - 164$$

$$24x - 3x + 14x = 19 - 164 - 30$$

$$35x = -175$$

$$x = -5$$

(d)  $6x - 13 = 3(2x - 11) + 20$

$$6x - 13 = 3(2x - 11) + 20$$

$$6x - 13 = 6x - 33 + 20$$

$$6x - 6x = -33 + 20 + 13$$

$$0 = 0$$

With the equation reduced to $0 = 0$, the equation is an identity. An *identity* is an equation in which any real number can be substituted for $x$ and the equation will remain valid.

**2.2.** Solve for $x$ by *clearing the denominator*, that is, by multiplying both sides of the equation by the least common denominator (LCD) as soon as is feasible.

(a)  $\dfrac{x}{4} - \dfrac{x}{5} = 6$

$$\frac{x}{4} - \frac{x}{5} = 6$$

Multiplying both sides of the equation by 20,

$$20 \cdot \left( \frac{x}{4} - \frac{x}{5} \right) = 6 \cdot 20$$

$$5x - 4x = 120$$

$$x = 120$$

(b) $\dfrac{x}{6} - 5 = \dfrac{x}{9} + 1$

$$\dfrac{x}{6} - 5 = \dfrac{x}{9} + 1$$

$$\dfrac{x}{6} - \dfrac{x}{9} = 1 + 5$$

$$18 \cdot \left( \dfrac{x}{6} - \dfrac{x}{9} \right) = 6 \cdot 18$$

$$3x - 2x = 108$$

$$x = 108$$

(c) $\dfrac{8}{x} + \dfrac{6}{x + 5} = \dfrac{12}{x}$      $(x \neq 0, -5)$

$$\dfrac{8}{x} + \dfrac{6}{x + 5} = \dfrac{12}{x}$$

$$x(x + 5) \cdot \left( \dfrac{8}{x} + \dfrac{6}{x + 5} \right) = \dfrac{12}{x} \cdot x(x + 5)$$

$$8(x + 5) + 6x = 12(x + 5)$$

$$14x + 40 = 12x + 60$$

$$2x = 20 \qquad x = 10$$

(d) $\dfrac{14}{x + 3} + \dfrac{20}{x - 4} = \dfrac{38}{x - 4}$      $(x \neq -3, 4)$

$$(x + 3)(x - 4) \left( \dfrac{14}{x + 3} + \dfrac{20}{x - 4} \right) = \left( \dfrac{38}{x - 4} \right)(x + 3)(x - 4)$$

$$14(x - 4) + 20(x + 3) = 38(x + 3)$$

$$14x - 56 + 20x + 60 = 38x + 114$$

$$-4x = 110 \qquad x = -27.5$$

(e) $\dfrac{36}{x - 5} - \dfrac{25}{2x} = \dfrac{26}{x - 5}$      $(x \neq 0, 5)$

$$2x(x - 5) \cdot \left( \dfrac{36}{x - 5} - \dfrac{25}{2x} \right) = \left( \dfrac{26}{x - 5} \right) \cdot 2x(x - 5)$$

$$72x - 25(x - 5) = 52x$$

$$-5x + 125 = 0 \qquad x = 25$$

## GRAPHING ON A CARTESIAN COORDINATE SYSTEM

**2.3.**    Plot the following points: (a) $(3, 4)$, (b) $(5, -2)$, (c) $(-6, 3)$, (d) $(0, -4)$, (e) $(-7, 0)$, (f) $(-8, -2)$, (g) $(2, -3)$, (h) $(-5, -4)$.

(a)    To plot $(3, 4)$, move 3 units to the right of the origin, then 4 units up, as in Fig. 2-11.

(b)    To locate $(5, -2)$, move 5 units to the right of the origin, then 2 units down as in Fig. 2-11.

(c)    To find $(-6, 3)$, move 6 units to the left, 3 units up.

(d)    To plot $(0, -4)$, make no movement horizontally along the $x$ axis, simply move 4 units down.

(e)    For $(-7, 0)$, move 7 units to the left and stop, making no movement vertically along the $y$ axis.

         For $(f)$, $(g)$, and $(h)$, see Fig. 2-11.

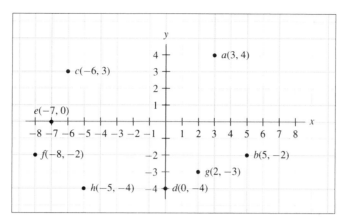

**Fig. 2-11**

**2.4.** Convert the following linear equations in standard form to the slope-intercept form by solving for $y$ in terms of $x$.

(a)   $56x + 7y = 91$

$$56x + 7y = 91$$

$$7y = -56x + 91$$

$$y = -8x + 13$$

(b)   $42x - 6y = 90$

$$42x - 6y = 90$$

$$-6y = -42x + 90$$

$$y = 7x - 15$$

(c)   $72x - 8y = 0$

$$72x - 8y = 0$$

$$-8y = -72x$$

$$y = 9x$$

(d)   $16y = 176$

$$16y = 176$$

$$y = 11$$

*Note*:

1)   If there is no constant term in the standard form of a linear equation as in (c) above, $b = 0$ in the slope-intercept form and the $y$ coordinate of the $y$ intercept $(0, b)$ is zero. This means that the graph crosses the $y$ axis at the origin. See also Problem 2.13 (g).

2)   If there is no $x$ term in the standard form of a linear equation as in (d), the slope $m = 0$ in the slope-intercept form, $y$ equals a constant, and the graph of the equation is a horizontal line. See Problem 2.13 (h).

## SLOPES

**2.5.** Find the slopes of the lines of the following equations:

(a)   $y = -5x + 16$      (b)   $y = \frac{1}{6}x - 5$      (c)   $y = 4.3x - 8$      (d)   $y = -7x$      (e)   $y = 18$

(f)   $y = -\frac{1}{2}$      (g)   $x = 6$

In the slope-intercept form of a linear equation, the coefficient of $x$ is the slope of the line. The slopes of the lines of the first four equations are immediately evident, therefore, namely, (a) $-5$, (b) $\frac{1}{6}$, (c) 4.3, and (d) $-7$.

The slopes of (e) and (f) are 0 because there are no $x$'s in the equations. The equations could also be written $y = (0)x + 18$ and $y = (0)x - \frac{1}{2}$; the graphs would be horizontal lines.

The slope of (g), a vertical line at $x = 6$, is undefined because vertical lines have the same $x$ coordinate (here 6) for all values of $y$. If $x_1 = x_2$, $x_2 - x_1 = 0$ and division by zero is not permissible. Hence the earlier warning in parentheses:

$$m = \frac{y_2 - y_1}{x_2 - x_1} \quad (x_1 \neq x_2)$$

**2.6.** Find the slopes of the lines of the following equations by first converting the equations to the slope-intercept form:

(a)   $24x + 6y = 30$
     $y = -4x + 5$    $m = -4$

(b)   $15x - 3y = 48$
     $y = 5x - 16$    $m = 5$

(c)   $9x + \frac{1}{2}y = -14$
     $y = -18x - 28$    $m = -18$

(d)   $-36x + 16y = 80$
     $y = 2\frac{1}{4}x + 5$    $m = 2\frac{1}{4}$

(e)   $14x - 4y = 1$
     $y = 3\frac{1}{2}x - \frac{1}{4}$    $m = 3\frac{1}{2}$

(f)   $14x + 56y = 280$
     $y = -\frac{1}{4}x + 5$    $m = -\frac{1}{4}$

**2.7.** To illustrate the steepness and direction conveyed by the slope, (a) draw five separate lines each with y intercept (0, 0) and having slopes of $\frac{1}{4}$, $\frac{1}{2}$, 1, 2, and 4 respectively; (b) draw five separate lines each with y intercept (0, 4) and having slopes of $-4$, $-2$, $-1$, $-\frac{2}{3}$ and $-\frac{1}{2}$, respectively.

(a)   See Fig. 2-12(a); (b) see Fig. 2-12(b).

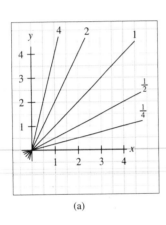   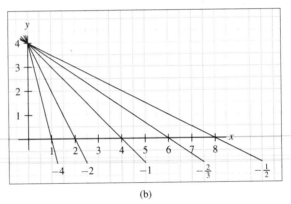

(a)                              (b)

**Fig. 2-12**

**2.8.** Use the different variations of the formula in Section 2.4 to find the slopes of the lines passing through the following points. Recall that the order in which you treat the points does not matter provided you are consistent within the problem.

(a)   $(5, 8)$, $(7, 14)$

$$m = \frac{\Delta y}{\Delta x} = \frac{y_2 - y_1}{x_2 - x_1} = \frac{14 - 8}{7 - 5} = \frac{6}{2} = 3$$

(b)   $(6, 10)$, $(9, 4)$

$$m = \frac{\text{rise}}{\text{run}} = \frac{y_2 - y_1}{x_2 - x_1} = \frac{4 - 10}{9 - 6} = \frac{-6}{3} = -2$$

(c)   $(-2, 5)$, $(1, -7)$

$$m = \frac{y_2 - y_1}{x_2 - x_1} = \frac{\text{rise}}{\text{run}} = \frac{-7 - 5}{1 - (-2)} = \frac{-12}{3} = -4$$

(d)   $(4, -19)$, $(6, -9)$

$$m = \frac{y_2 - y_1}{x_2 - x_1} = \frac{\Delta y}{\Delta x} = \frac{-9 - (-19)}{6 - 4} = \frac{10}{2} = 5$$

For the remaining sections of the problem deliberately reverse the order in which the points are taken.

(e)   $(8, 6)$, $(12, 16)$

$$m = \frac{\text{rise}}{\text{run}} = \frac{y_1 - y_2}{x_1 - x_2} = \frac{6 - 16}{8 - 12} = \frac{-10}{-4} = 2\frac{1}{2}$$

(f)   $(-1, 3), (8, 15)$

$$m = \frac{\Delta y}{\Delta x} = \frac{y_1 - y_2}{x_1 - x_2} = \frac{3 - 15}{-1 - 8} = \frac{-12}{-9} = 1\frac{1}{3}$$

(g)   $(-2, -13), (-6, -5)$

$$m = \frac{y_1 - y_2}{x_1 - x_2} = \frac{\Delta y}{\Delta x} = \frac{-13 - (-5)}{-2 - (-6)} = \frac{-8}{4} = -2$$

(h)   $(5, -7), (12, -8)$

$$m = \frac{y_1 - y_2}{x_1 - x_2} = \frac{\text{rise}}{\text{run}} = \frac{-7 - (-8)}{5 - 12} = \frac{1}{-7} = -\frac{1}{7}$$

## INTERCEPTS

**2.9.**   Find the $y$ intercept for each of the following equations:

(a)   $5x + y = 9$

The $y$ intercept occurs where the line crosses the $y$ axis, which is the point where $x = 0$. Setting $x = 0$ in (a) above and solving for $y$, we have

$$5(0) + y = 9$$

$$y = 9$$

Uniting the two pieces of information, the $y$ intercept is $(0, 9)$.

(b)   $7x - 4y = 56$
Setting $x = 0$,

$$7(0) - 4y = 56$$

$$-4y = 56$$

$$y = -14 \qquad y \text{ intercept: } (0, -14)$$

(c)   $y = 5x - 17$
Setting $x = 0$,

$$y = 5(0) - 17$$

$$y = -17 \qquad y \text{ intercept: } (0, -17)$$

If the equation is in the slope-intercept form $y = mx + b$ as in (c), the $y$ intercept $(0, b)$ can be inferred directly from the equation, as was explained in Example 7.

(d)   $y = 18x - 33$

$$y \text{ intercept: } (0, -33)$$

(e)   $y = 3x + \frac{1}{4}$

$$y \text{ intercept: } (0, \tfrac{1}{4})$$

(f)   $y = \frac{1}{5}x - 2$

$$y \text{ intercept: } (0, -2)$$

**2.10.**   Find the $x$ intercept for the following equations:

(a)   $y = 9x - 72$

The $x$ intercept is the point where the line crosses the $x$ axis. Since the line crosses the $x$ axis where $y = 0$, zero is the $y$ coordinate of the $x$ intercept. To find the $x$ coordinate, simply set $y = 0$ and solve the equation for $x$:

$$0 = 9x - 72$$
$$-9x = -72$$
$$x = 8 \qquad x \text{ intercept: } (8, 0)$$

(b)   $y = 13x + 169$
       Setting $y = 0$,

$$0 = 13x + 169$$
$$-13x = 169$$
$$x = -13 \qquad x \text{ intercept: } (-13, 0)$$

(c)   $y = 32x - 8$

$$0 = 32x - 8$$
$$-32x = -8$$
$$x = \tfrac{1}{4} \qquad x \text{ intercept: } (\tfrac{1}{4}, 0)$$

(d)   $y = 7x$

$$0 = 7x$$
$$x = 0 \qquad x \text{ intercept: } (0, 0)$$

**2.11.**   Find the $x$ intercept in terms of the parameters of the slope-intercept form of a linear equation $y = mx + b$.

Setting $y = 0$,

$$0 = mx + b$$
$$mx = -b$$
$$x = -\frac{b}{m} \qquad\qquad\qquad\qquad\qquad\qquad\qquad (2.7)$$

The $x$ intercept of the slope-intercept form is $(-b/m, 0)$.

**2.12.**   Use the information in Problem 2.11 to speed the process of finding the $x$ intercepts for the following equations:

(a)   $y = 16x + 64$
       Here $m = 16$, $b = 64$. Substituting in (2.7),

$$x = -\frac{64}{16}$$
$$x = -4 \qquad x \text{ intercept: } (-4, 0)$$

(b)   $y = 18x - 9$
       With $m = 18$, $b = -9$, from (2.7),

$$x = -\left(\frac{-9}{18}\right)$$
$$x = \frac{1}{2} \qquad x \text{ intercept: } (\tfrac{1}{2}, 0)$$

(c)   $y = 15x + 120$
       Here $m = 15$, $b = 120$, and

$$x = -\frac{120}{15}$$
$$x = -8 \qquad x \text{ intercept: } (-8, 0)$$

(d)   $y = 5x - 125$
    With $m = 5, b = -125$,

$$x = -\left(\frac{-125}{5}\right)$$

$$x = 25 \qquad\qquad x \text{ intercept: } (25, 0)$$

(e)   $18x + 5y = 54$
    If the equation is not in the slope-intercept form, find the slope-intercept form or simply substitute 0
    for $y$.

$$18x + 5(0) = 54$$

$$x = 3 \qquad x \text{ intercept: } (3, 0)$$

**2.13.**   Find the $y$ intercepts and the $x$ intercepts and use them as the two points needed to graph the following
linear equations:

(a)   $y = 5x + 10$
    $y$ intercept: $(0, 10)$, $x$ intercept: $(-2, 0)$. See Fig. 2-13.

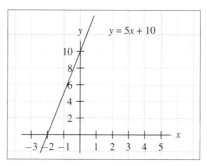

**Fig. 2-13**

(b)   $y = -3x + 9$
    $y$ intercept: $(0, 9)$, $x$ intercept: $(3, 0)$. See Fig. 2-14.

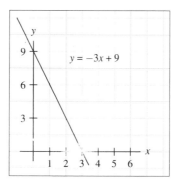

**Fig. 2-14**

(c)   $y = 4x - 8$
    $y$ intercept: $(0, -8)$, $x$ intercept: $(2, 0)$. See Fig. 2-15.

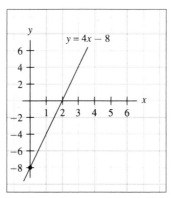

**Fig. 2-15**

(*d*)  $y = -x + 6$

    $y$ intercept: $(0, 6)$, $x$ intercept: $(6, 0)$. See Fig. 2-16.

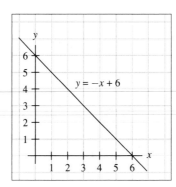

**Fig. 2-16**

(*e*)  $y = \frac{1}{3}x + 4$

    $y$ intercept: $(0, 4)$, $x$ intercept: $(-12, 0)$. See Fig. 2-17.

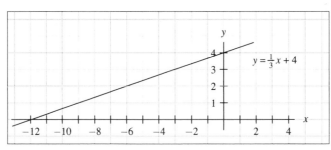

**Fig. 2-17**

(*f*)  $y = -\frac{1}{5}x + 2$

    $y$ intercept: $(0, 2)$, $x$ intercept: $(10, 0)$. See Fig. 2-18.

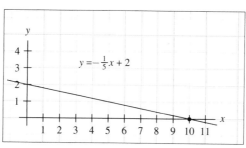

**Fig. 2-18**

(g)   $y = 2x$

   $y$ intercept: $(0, 0)$, $x$ intercept $(0, 0)$. When the $x$ and $y$ intercepts coincide, as occurs when the graph passes through the origin, a second distinct point must be found. Letting $x = 1$ for ease of computation, $y = 2(1) = 2$. The graph can now be constructed from the points $(0, 0)$ and $(1, 2)$. See Fig. 2-19.

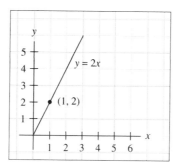

**Fig. 2-19**

(h)   $y = 3$

   $y$ intercept: $(0, 3)$; the $x$ intercept does not exist because $y$ cannot be set equal to zero without involving a contradiction. Since $y = 3$ independently of $x$, $y$ will equal 3 for any value of $x$. The graph is simply a horizontal line 3 units above the $x$ axis. Paralleling the $x$ axis, it never crosses it. See Fig. 2-20.

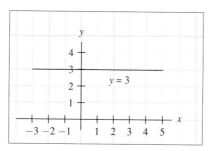

**Fig. 2-20**

## DETERMINING THE EQUATION FOR A STRAIGHT LINE

**2.14.**   Find the equation for the following straight lines with:

   (a)   Slope $= 7$, $y$ intercept: $(0, 16)$

Using the slope-intercept form $y = mx + b$ throughout, and substituting $m = 7$, $b = 16$ here,

$$y = 7x + 16$$

(b)   Slope $= -6$, $y$ intercept: $(0, 45)$
        Substituting $m = -6$, $b = 45$,

$$y = -6x + 45$$

(c)   Slope $= 0.35$, $y$ intercept: $(0, -5.5)$

$$y = 0.35x - 5.5$$

**2.15.**   Use the point-slope formula from equation $(2.3)$ to derive the equation for each of the following:

(a)   Line passing through $(3, 11)$, slope $-4$.
        Taking the point-slope formula,

$$y - y_1 = m(x - x_1)$$

and substituting $x_1 = 3$, $y_1 = 11$, and $m = -4$, we have

$$y - 11 = -4(x - 3)$$
$$y = -4x + 12 + 11$$
$$y = -4x + 23$$

(b)   Line passing through $(-7, 4)$, slope $5$.

$$y - 4 = 5[x - (-7)]$$
$$y = 5x + 35 + 4$$
$$y = 5x + 39$$

(c)   Line passing through $(8, -2)$, slope $\frac{1}{4}$.

$$y - (-2) = \frac{1}{4}(x - 8)$$
$$y + 2 = \frac{1}{4}x - 2$$
$$y = \frac{1}{4}x - 4$$

(d)   Line passing through $(-5, -3)$, slope $-7$.

$$y - (-3) = -7[x - (-5)]$$
$$y + 3 = -7x - 35$$
$$y = -7x - 38$$

**2.16.**   Find the equation for the line passing through $(-3, 6)$ and parallel to the line having the equation $y = 5x + 8$.

*Parallel lines* have the same slope. The slope of the line we seek therefore is 5. Combining this knowledge with the given coordinates in the point-slope formula,

$$y - 6 = 5[x - (-3)]$$
$$y = 5x + 15 + 6$$
$$y = 5x + 21$$

**2.17.**   Determine the equation for the line passing through $(6, 4)$ and perpendicular to the line having the equation $y = 2x + 15$.

*Perpendicular lines* have slopes that are negative reciprocals of one another. Given a line whose slope is 2, the slope of a line perpendicular to it must be $-\frac{1}{2}$.

$$y - 4 = -\tfrac{1}{2}(x - 6)$$

$$y = -\tfrac{1}{2}x + 3 + 4$$

$$y = -\tfrac{1}{2}x + 7$$

**2.18.** Find the equations for the lines passing through the following points:

(*a*)   (3, 13) and (7, 45)

Using the two-point formula and the procedure demonstrated in Example 11,

$$m = \frac{y_2 - y_1}{x_2 - x_1} = \frac{45 - 13}{7 - 3} = \frac{32}{4} = 8$$

then substituting $m = 8$, $x_1 = 3$, and $y_1 = 13$ in the point-slope formula

$$y - y_1 = m(x - x_1)$$

$$y - 13 = 8(x - 3)$$

$$y = 8x - 11$$

(*b*)   (2, 18), (5, −3)

$$m = \frac{-3 - 18}{5 - 2} = \frac{-21}{3} = -7$$

Using $m = -7$, $x_1 = 2$, and $y_1 = 18$ in the point-slope formula,

$$y - 18 = -7(x - 2)$$

$$y = -7x + 32$$

(*c*)   (3, −17), (0, 19)

$$m = \frac{19 - (-17)}{0 - 3} = \frac{36}{-3} = -12$$

With one of the points (0, 19) the vertical intercept, the equation can be written immediately.

$$y = -12x + 19$$

(*d*)   (0, −2), (8, 0)

$$m = \frac{0 - (-2)}{8 - 0} = \frac{2}{8} = \frac{1}{4}$$

With (0, −2) the *y* intercept,

$$y = \frac{1}{4}x - 2$$

**2.19.** Prove the formula for the slope given in Section 2.4:

$$m = \frac{y_1 - y_2}{x_1 - x_2}$$

Given two points $(x_1, y_1)$ and $(x_2, y_2)$ on the same line, both must satisfy the slope-intercept form of the equation:

$$y_1 = mx_1 + b$$

$$y_2 = mx_2 + b$$

Subtracting $y_2$ from $y_1$ and recalling from Example 5 that the order is irrelevant, as long as it remains consistent,

$$y_1 - y_2 = mx_1 + b - mx_2 - b$$

$$y_1 - y_2 = m(x_1 - x_2)$$

Then dividing by $(x_1 - x_2)$ and rearranging terms,

$$m = \frac{y_1 - y_2}{x_1 - x_2}$$

**2.20.** Verify the point-slope formula:

$$y - y_1 = m(x - x_1)$$

For any point $(x, y)$ to be on a line which passes through the point $(x_1, y_1)$ and has slope $m$, it must be true that

$$m = \frac{y - y_1}{x - x_1}$$

Multiplying both sides of the equation by $(x - x_1)$ and rearranging,

$$y - y_1 = m(x - x_1)$$

## LINEAR EQUATIONS IN BUSINESS AND ECONOMICS

**2.21.** A firm has a fixed cost of $7000 for plant and equipment and a variable cost of $600 for each unit produced. What is the total cost $C$ of producing (*a*) 15 and (*b*) 30 units of output?

$$C = 600x + 7000$$

(*a*) If $x = 15$,                     $C = 600(15) + 7000 = 16{,}000$

(*b*) If $x = 30$                      $C = 600(30) + 7000 = 25{,}000$

**2.22.** Find the total cost of producing (*a*) 20 units and (*b*) 35 units of output for a firm that has fixed costs of $3500 and a marginal cost of $400 per unit.

$$C = 400x + 3500$$

(*a*) If $x = 20$,                     $C = 400(20) + 3500 = 11{,}500$

(*b*) If $x = 35$,                     $C = 400(35) + 3500 = 17{,}500$

**2.23.** A firm operating in pure competition receives $45 for each unit of output sold. It has a variable cost of $25 per item and a fixed cost of $1600. What is its profit level $\pi$ if it sells (*a*) 150 items, (*b*) 200 items, and (*c*) 75 items?

$$\pi = \text{revenue } (R) - \text{ cost } (C)$$

where $R = 45x$ and $C = 25x + 1600$. Substituting,

$$\pi = 45x - (25x + 1600)$$

$$\pi = 20x - 1600$$

(*a*) At $x = 150$,                    $\pi = 20(150) - 1600 = 1400$

(*b*) At $x = 200$,                    $\pi = 20(200) - 1600 = 2400$

(*c*) At $x = 75$,                     $\pi = 20(75) - 1600 = -100$        (a loss)

**2.24.** Find the profit level of a firm in pure competition that has a fixed cost of $950, a variable cost of $70, and a selling price of $85 when it sells (*a*) 50 units and (*b*) 80 units.

$$\pi = R - C$$

where $R = 85x$ and $C = 70x + 950$. Substituting,

$$\pi = 85x - (70x + 950)$$

$$\pi = 15x - 950$$

(*a*) At $x = 50$,                     $\pi = 15(50) - 950 = -200$        (a loss)

(*b*) At $x = 80$,                     $\pi = 15(80) - 950 = 250$

**2.25.** Find (*a*) the value after 6 years and (*b*) the salvage value after 8 years of a combine whose current value *y* after *x* years is

$$y = 67{,}500 - 7750x$$

(*a*) $$y = 67{,}500 - 7750(6) = 21{,}000$$

(*b*) $$y = 67{,}500 - 7750(8) = 5500$$

**2.26.** For tax purposes the value *y* of a computer after *x* years is

$$y = 3{,}000{,}000 - 450{,}000x$$

Find (*a*) the value of the computer after 3 years, and (*b*) the salvage value after 5 years.

(*a*) $$y = 3{,}000{,}000 - 450{,}000(3) = 1{,}650{,}000$$

(*b*) $$y = 3{,}000{,}000 - 450{,}000(5) = 750{,}000$$

**2.27.** A retiree receives $5120 a year interest from $40,000 placed in two bonds, one paying 14 percent and the other 12 percent. How much is invested in each bond?

Let $x =$ the amount invested at 14 percent; then $(40{,}000 - x)$ is the amount invested at 12 percent, and the total annual interest *y* from the two amounts is given by

$$y = .14x + .12(40{,}000 - x)$$

Substituting $y = 5120$,

$$5120 = 0.14x + 4800 - 0.12x$$

$$320 = .02x$$

$$x = 16{,}000 \text{ at } 14\%$$

And $40{,}000 - 16{,}000 = 24{,}000$ at 12%.

**2.28.** With $60,000 to invest, how much should a broker invest at 11 percent and how much at 15 percent to earn 14 percent on the total investment?

If $x =$ the amount invested at 11 percent, then $(60{,}000 - x)$ is the amount invested at 15 percent, and the total annual interest *y* is

$$y = .11x + .15(60{,}000 - x)$$

Substituting the desired level of *y*: $0.14(60{,}000) = 8400$,

$$8400 = .11x + 9000 - .15x$$

$$-600 = -0.04x$$

$$x = 15{,}000 \text{ at } 11\%$$

And $60{,}000 - 15{,}000 = 45{,}000$ at 15%.

**2.29.** A candy maker wants to blend candy worth 45 cents a pound with candy worth 85 cents a pound to obtain 200 pounds of a mixture worth 70 cents a pound. How much of each type should go into the mixture?

Let $x =$ the amount of candy worth 45 cents; then $(200 - x)$ is the amount of candy worth 85 cents, and the value *y* of the mixture is

$$y = .45x + .85(200 - x)$$

Substituting the value of the desired mixture for *y*,

$$.70(200) = .45x + .85(200 - x)$$

$$140 = .45x + 170 - .85x$$

$$-30 = -.4x$$

$$x = 75 \text{ pounds at } 45¢$$

And $200 - 75 = 125$ pounds at 85¢.

**2.30.** How much wine with an 18% alcohol content must be mixed with 6000 gallons of wine with a 12% alcohol content to obtain wine with a 16% alcohol content?

Letting $x =$ the amount of wine with 18% alcohol to be mixed and $y =$ the alcohol content of the mixture, then

$$y = .18x + .12(6000)$$

Then substituting the desired mixture of $(6000 + x)$ gallons with 16% alcohol for $y$,

$$.16(6000 + x) = .18x + .12(6000)$$

$$960 + .16x = .18x + 720$$

$$-0.02x = -240$$

$$x = 12,000 \text{ gallons}$$

**2.31.** An *isocost curve* shows the different combinations of two goods that can be purchased with a given budget $B$. Blast furnaces can be heated with either gas $(x)$ or coal $(y)$. Given $p_x = 100$, $p_y = 400$, and $B = 8000$, $(a)$ draw the isocost curve. Always starting from the original data, draw a new isocost curve $(b)$ if $B$ increases by 50 percent, $(c)$ if $p_x$ doubles, and $(d)$ if $p_y$ decreases by 37.5 percent.

$(a)$   With $p_x = 100$, $p_y = 400$, and $B = 8000$, the standard form of the equation for the isocost curve is

$$100x + 400y = 8000$$

$$y = -0.25x + 20$$

The graph is the solid line in Fig. 2-21$(a)$.

$(b)$   With a 50 percent increase in expenditures, the new budget is $8000+0.5(8000) = 12,000$. The new equation is

$$100x + 400y = 12,000$$

$$y = -0.25x + 30$$

The graph is the dashed line in Fig. 2-21$(a)$.

$(c)$   If $p_x$ doubles, the new price is 200, and the new equation is

$$200x + 400y = 8000$$

$$y = -0.5x + 20$$

The graph is the dashed line in Fig. 2-21$(b)$.

$(d)$   If $p_y$ drops by 37.5 percent, the new $p_y$ is $400 - 0.375(400) = 250$.

$$100x + 250y = 8000$$

$$y = -0.4x + 32$$

The graph appears as a dashed line in Fig. 2-21$(c)$.

**2.32.** Given two goods $x$ and $y$ with prices $p_x$ and $p_y$ and a budget $B$, $(a)$ determine the slope-intercept form of the equation for the isocost curve. Using the information from $(a)$, indicate what will happen to the graph of the curve $(b)$ if the budget changes, $(c)$ if $p_x$ changes, and $(d)$ if $p_y$ changes.

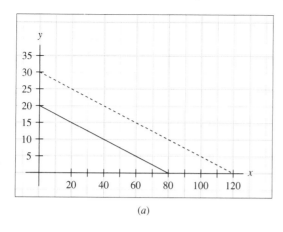

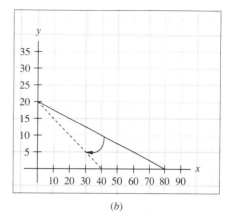

(a)

(b)

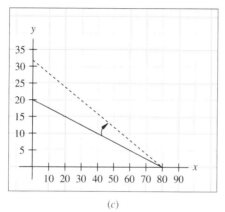

(c)

**Fig. 2-21**

(a)  Starting with the standard form and solving for $y$,

$$p_x \cdot x + p_y \cdot y = B$$

$$p_y \cdot y = -(p_x \cdot x) + B$$

$$y = -\left(\frac{p_x}{p_y}\right)x + \frac{B}{p_y} \tag{2.8}$$

This is the familiar slope-intercept form of a linear equation $y = mx + b$, where $m = -(p_x/p_y) =$ the slope and $b = B/p_y =$ the $y$ coordinate of the $y$ intercept. The $y$ intercept of the isocost line, therefore, is $(0, B/p_y)$, and as the $x$ intercept of the standard form is $(-b/m, 0)$, the $x$ intercept of the isocost line is $[-(B/p_y)/(-p_x/p_y), 0] = (B/p_x, 0)$.

(b)  From (2.8) it is clear that a change in the budget $B$ will affect the $x$ and $y$ intercepts but not the slope. This means that the isocost curve will shift parallel to the original curve; to the right for an increase in $B$ and to the left, for a decrease in $B$.

(c)  If $p_x$ changes, the slope $-p_x/p_y$ and the $x$ intercept $(B/p_x, 0)$ will change, but not the $y$ intercept $(0, B/p_y)$. The graph will get steeper for an increase in $p_x$ and flatter for a decrease in $p_x$, pivoting around the same $y$ intercept.

(d)  For a change in $p_y$, the slope $-p_x/p_y$ and the $y$ intercept $(0, B/p_y)$ will change, but not the $x$ intercept $(B/p_x, 0)$. An increase in $p_y$ makes the $y$ intercept lower and the slope flatter. A decrease in $p_y$ makes the $y$ intercept higher and the slope steeper. In both instances the $x$ intercept remains the same.

**2.33.** Increasing at a constant rate, a company's profits $y$ have gone from \$535 million in 1985 to \$570 million in 1990. Find the expected level of profit for 1995 if the trend continues.

Letting $x = 0$ for 1985 and $x = 5$ for 1990, we have two points on a graph: $(0, 535)$ and $(5, 570)$. Since profits $y$ are changing at a constant rate, the slope $(\Delta y / \Delta x)$ is constant and the two points can be connected by a straight line, as in Fig. 2-22.

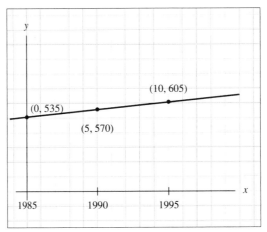

**Fig. 2-22**

Then using the two-point formula to find the slope, we have

$$m = \frac{570 - 535}{5 - 0} = \frac{35}{5} = 7$$

With $(0, 535)$ the vertical intercept, we can now write

$$y = 7x + 535$$

Then substituting $x = 10$ for 1995, we have

$$y = 7(10) + 535 = 605$$

**2.34.** Given that $0°$ (Celsius) is equal to $32°F$ (Fahrenheit) and $100°C$ is equal to $212°F$ and that there is a straight-line relationship between temperatures measured on the different scales, express the relationship as a linear equation.

Starting with the two given points: $(0, 32)$ and $(100, 212)$ in Fig. 2-23, and using the two-point formula to find the slope,

$$m = \frac{212 - 32}{100 - 0} = \frac{180}{100} = 1.8$$

With $(0, 32)$ the vertical intercept,

$$F = 1.8C + 32$$

# Supplementary Problems

**SOLVING EQUATIONS**

**2.35.** Solve the following linear equations by using the properties of equality:
   (a)  $6x - 7 = 3x + 2$          (b)  $60 - 8x = 3x + 5$
   (c)  $7x + 5 = 4(3x - 8) + 7$          (d)  $3(2x + 9) = 4(5x - 21) - 1$

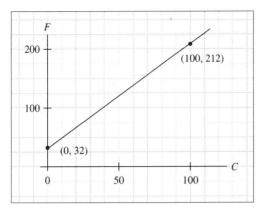

**Fig. 2-23**

**2.36.** Solve the following equations by clearing the denominator:

(a) $\dfrac{x}{3} + \dfrac{x}{8} = 22$

(b) $\dfrac{x}{5} - 4 = \dfrac{x}{9} + 8$

(c) $\dfrac{240}{x-2} = \dfrac{576}{x+5}$

(d) $\dfrac{675}{x+3} = \dfrac{120}{x-6} + \dfrac{375}{x+3}$

## SLOPES

**2.37.** Convert the following linear equations in standard form to the slope-intercept form:

(a) $3x + 12y = 96$

(b) $12x - 16y = 240$

(c) $22x + 2y = 86$

(d) $28x - 4y = -104$

**2.38.** Find the slopes of the following linear equations:

(a) $y = -16x + 23$

(b) $y = \frac{1}{2}x - 9$

(c) $y = 15.5x$

(d) $y = -\frac{1}{8}x - 3$

(e) $y = 19$

(f) $x = 27$

**2.39.** Find the slopes of the lines passing through the following points:

(a) $(2,\ 7),\ (5,\ 19)$

(b) $(1,\ 9),\ (3,\ 3)$

(c) $(3,\ 56),\ (7,\ 8)$

(d) $(9,\ 19),\ (15,\ 22)$

(e) $(-4,\ 0),\ (1,\ -3)$

(f) $(-7,\ 0),\ (-1,\ 0)$

## INTERCEPTS

**2.40.** Find the $y$ intercepts of each of the following equations in standard form:

(a) $8x - 3y = 15$

(b) $13x + \frac{1}{2}y = 4.5$

(c) $-26x + 7y = -56$

(d) $47x - 18y = -9$

**2.41.** Find the $y$ intercepts of each of the following equations in slope-intercept form:

(a) $y = -6.5x - 23$

(b) $y = 108x + 14$

(c) $y = 79.5x - 250$

(d) $y = -3x + 0.65$

**2.42.** Find the $x$ intercepts of the standard form equations below:

(a) $6x - 13y = 126$

(b) $\frac{1}{4}x + 17y = -8$

(c) $-5x - \frac{1}{2}y = 55$

(d) $-13x + 19y = -39$

**2.43.** Find the $x$ intercepts of each of the following equations in slope-intercept form:

(a) $y = 7x + 35$

(b) $y = -8x + 64$

(c) $y = \frac{2}{5}x - 16$

(d) $y = -1.5x - 9$

## FINDING THE EQUATION FOR A STRAIGHT LINE

**2.44.**    Derive the equation for a straight line having
(*a*)  Slope: $-13$,  *y* intercept: $(0, 22)$
(*b*)  Slope: $\frac{4}{9}$,  *y* intercept: $(0, -6)$
(*c*)  Slope: $-\frac{2}{3}$,  *y* intercept: $(0, 49)$
(*d*)  Slope: $23.5$,  *y* intercept: $(0, -70)$

**2.45.**    Use the point-slope formula to derive the equation for each of the following straight lines:
(*a*)  Passing through $(2, 6)$, slope $= 7$        (*b*)  Passing through $(8, -5)$, slope $= 3$
(*c*)  Passing through $(-10, 12)$, slope $= \frac{1}{2}$
(*d*)  Passing through $(3, -11)$, slope $= -5$

**2.46.**    Use the two-point formula to derive the equation for a straight line passing through:
(*a*)  $(2, 6)$, $(5, 18)$
(*b*)  $(-1, 10)$, $(4, -5)$
(*c*)  $(-6, -3)$, $(9, 7)$
(*d*)  $(4, -5)$, $(10, -8)$

## BUSINESS AND ECONOMICS APPLICATIONS

**2.47.**    Find the value after (*a*) 2 years and (*b*) 4 years of a photocopier which was purchased for $12,250 initially and is depreciating at a constant rate of $1995 a year.

**2.48.**    Estimate the current value after (*a*) 3 years and (*b*) 7 years of a printing press purchased for $265,000 and depreciating linearly by $32,000 a year.

**2.49.**    A firm has fixed costs of $122,000 and variable costs of $750 a unit. Determine the firm's total cost of producing (*a*) 25 units and (*b*) 50 units.

**2.50.**    Estimate the total cost of producing (*a*) 10 units and (*b*) 100 units for a company with fixed costs of $85,000 and variable costs of $225 a unit.

**2.51.**    Determine the level of profit on sales of (*a*) 75 units and (*b*) 125 units for a company operating in a purely competitive market which receives $120 for each item sold and has a fixed cost of $1800 and variable costs of $55 a unit.

**2.52.**    How much of a $50,000 account should a broker invest at 8 percent and how much at 12 percent to average a 9.5 percent yearly rate of return?

**2.53.**    How many tons of coal valued at $140 a ton should be mixed with coal worth $190 a ton to obtain a mixture of 32,000 tons of coal worth $175 a ton?

**2.54.**    Growing at a constant rate, a firm's costs have increased from $265 million in 1988 to $279 million in 1992. Estimate costs in 1996 if the trend continues.

**2.55.**    Sales in millions of dollars have declined linearly from $905 in 1989 to $887.75 in 1992. On the basis of the trend, project the level of sales in 1995.

# Answers to Supplementary Problems

**2.35.**    (*a*)  $x = 3$       (*b*)  $x = 5$       (*c*)  $x = 6$       (*d*)  $x = 8$

**2.36.**    (*a*)  $x = 48$     (*b*)  $x = 135$    (*c*)  $x = 7$    (*d*)  $x = 12$

**2.37.**    (*a*)  $y = -\frac{1}{4}x + 8$   (*b*)  $y = \frac{3}{4}x - 15$   (*c*)  $y = -11x + 43$   (*d*)  $y = 7x + 26$

**2.38.**    (*a*)  $-16$   (*b*)  $\frac{1}{2}$   (*c*)  15.5   (*d*)  $-\frac{1}{8}$   (*e*)  0   (*f*)  undefined

**2.39.**    (*a*)  4   (*b*)  $-3$   (*c*)  $-12$   (*d*)  $\frac{1}{2}$   (*e*)  $-\frac{3}{5}$   (*f*)  0

**2.40.**    (*a*)  $(0, -5)$   (*b*)  $(0, 9)$   (*c*)  $(0, -8)$   (*d*)  $(0, \frac{1}{2})$

**2.41.**    (*a*)  $(0, -23)$   (*b*)  $(0, 14)$   (*c*)  $(0, -250)$   (*d*)  $(0, 0.65)$

**2.42.**    (*a*)  $(21, 0)$   (*b*)  $(-32, 0)$   (*c*)  $(-11, 0)$   (*d*)  $(3, 0)$

**2.43.**    (*a*)  $(-5, 0)$   (*b*)  $(8, 0)$   (*c*)  $(40, 0)$   (*d*)  $(-6, 0)$

**2.44.**    (*a*)  $y = -13x + 22$   (*b*)  $y = \frac{4}{9}x - 6$   (*c*)  $y = -\frac{2}{3}x + 49$   (*d*)  $y = 23.5x - 70$

**2.45.**    (*a*)  $y = 7x - 8$   (*b*)  $y = 3x - 29$   (*c*)  $y = \frac{1}{2}x + 17$   (*d*)  $y = -5x + 4$

**2.46.**    (*a*)  $y = 4x - 2$   (*b*)  $y = -3x + 7$   (*c*)  $y = \frac{2}{3}x + 1$   (*d*)  $y = -\frac{1}{2}x - 3$

**2.47.**    (*a*)  $8260   (*b*)  $4270

**2.48.**    (*a*)  $169,000   (*b*)  $41,000

**2.49.**    (*a*)  $140,750   (*b*)  $159,500

**2.50.**    (*a*)  $87,250   (*b*)  $107,500

**2.51.**    (*a*)  $3075   (*b*)  $6325

**2.52.**    $31,250 at 8 percent, $18,750 at 12 percent

**2.53.**    22,400 tons worth $190 a ton, 9600 tons worth $140 a ton

**2.54.**    $293 million

**2.55.**    $870.5 million

# Chapter 3

# Functions

## 3.1 CONCEPTS AND DEFINITIONS

A *function* is a *rule* ($f$) which assigns to each value of a variable ($x$), called the *argument* of the function, one and only one value [$f(x)$], called the *value* of the function. A function is written $y = f(x)$, which is read "$y$ is a function of $x$." The *domain* of a function is the set of all possible values of $x$; the *range* of a function is the set of all possible values of $f(x)$.

Functions are frequently defined by algebraic formulas as is illustrated in Example 1. Other letters such as $g$ or $h$ can also be used to express a function. If there is more than one function, different letters must be used to distinguish between them. Frequently encountered functions are listed below.

*Constant function*:     $f(x) = a_0$

*Linear function*:     $f(x) = a_1 x + a_0$

*Quadratic function*:     $f(x) = a_2 x^2 + a_1 x + a_0$     $(a_2 \neq 0)$

*Cubic function*:     $f(x) = a_3 x^3 + a_2 x^2 + a_1 x + a_0$     $(a_3 \neq 0)$

*Polynomial function of degree n*:     $f(x) = a_n x^n + a_{n-1} x^{n-1} + \cdots + a_0$

$(n = \text{nonnegative integer}; a_n \neq 0)$

Note that constant, linear, quadratic, and cubic functions are polynomial functions in which $n = 0, 1, 2,$ and 3, respectively.

*Rational function*:     $f(x) = \dfrac{g(x)}{h(x)}$

where $g(x)$ and $h(x)$ are both polynomials and $h(x) \neq 0$. Note that a rational function derives its name from the fact that it expresses a *ratio* of functions.

*Power function*:     $f(x) = ax^n$     $(n = \text{any real number})$

**EXAMPLE 1.**     The function $f(x) = 7x - 6$ is the rule that takes a number, multiplies it by 7, and then subtracts 6 from the product. If a value is given for $x$, the value is substituted for $x$ in the formula and the equation solved for $f(x)$. For example,

$$\text{If } x = 3, f(3) = 7(3) - 6 = 15$$

$$\text{If } x = 4, f(4) = 7(4) - 6 = 22$$

See Problems 3.1 to 3.4.

**EXAMPLE 2.**    Given below are examples of different functions:

Constant:        $f(x) = 14, \qquad g(x) = -9$

Linear:         $f(x) = 5x - 3, \qquad g(x) = -8x, \qquad h(x) = 22$

Quadratic:      $f(x) = 3x^2 + 8x - 7, \qquad g(x) = x^2 - 4x, \qquad h(x) = 6x^2$

Cubic:        $f(x) = 4x^3 - 2x^2 + 9x + 5, \qquad g(x) = 7x^3 + 4, \qquad h(x) = 2x^3$

Polynomial:       $f(x) = 8x^4 + 3x^2 - 5x + 9, \qquad g(x) = 2x^5 - x^3 + 7$

Rational:       $f(x) = \dfrac{x - 9}{x^2 - 4} \quad (x \neq \pm 2), \qquad g(x) = \dfrac{5x}{x - 3} \quad (x \neq 3)$

Power:        $f(x) = 4x^5, \qquad g(x) = x^{2/3}, \qquad h(x) = 9x^{-2} \qquad (x \neq 0)$

The domain of polynomial functions, including constant, linear, quadratic, and cubic functions, is the set of all real numbers; the domain of rational and power functions excludes any value of $x$ involving an undefined operation, such as division by zero.

## 3.2   GRAPHING FUNCTIONS

The graph of a linear function is a straight line. All linear equations are functions except those of the form $x = c$, a constant, the graph of which is a vertical line [see Problem 3.4 $(f)$]. In graphing, $y$ generally replaces $f(x)$ as the symbol for a function and, as the dependent variable, is placed on the vertical axis. Linear graphs are the topic of Chapter 2.

The graph of a nonlinear function on a Cartesian coordinate system is accomplished by determining a number of ordered pairs of values which satisfy the function. The ordered pairs are found by picking different values of $x$ and computing for each the corresponding value of $y$. Each ordered pair specifies a single point lying on the graph of the function. Connecting the points by a smooth curve then completes the graph of the function. The graph of a quadratic function is illustrated in Example 3; the graph of a rational function, in Example 4.

**EXAMPLE 3.**    To graph a nonlinear function, pick some representative values of $x$; solve for $y$; plot the resulting ordered pairs $(x, y)$; and connect them with a smooth line. The procedure is illustrated in Fig. 3-1 for the quadratic function $y = 2x^2$.

| $x$ | $f(x)$ | $= 2x^2$ | $= y$ | Points |
|---|---|---|---|---|
| $-2$ | $f(-2)$ | $= 2(-2)^2$ | $= 8$ | $(-2, 8)$ |
| $-1$ | $f(-1)$ | $= 2(-1)^2$ | $= 2$ | $(-1, 2)$ |
| $0$ | $f(0)$ | $= 2(0)^2$ | $= 0$ | $(0, 0)$ |
| $1$ | $f(1)$ | $= 2(1)^2$ | $= 2$ | $(1, 2)$ |
| $2$ | $f(2)$ | $= 2(2)^2$ | $= 8$ | $(2, 8)$ |

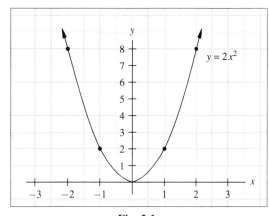

**Fig. 3-1**

The graph of a *quadratic function* is a *parabola*. The graph of a parabola is symmetric about a line called the *axis of symmetry*. The point of intersection of the parabola and its axis is called the *vertex*. In Fig. 3-1, the axis of symmetry coincides with the $y$ axis; the vertex is $(0, 0)$. See Problems 3.18 and 3.22.

**EXAMPLE 4.**    The procedure for graphing in Example 3 is repeated in Fig. 3-2 for the rational function $y = 2/x$ $(x \neq 0)$.

| $x$ | $f(x)$ | $= 2/x$ | $= y$ | Points |
|---|---|---|---|---|
| $-4$ | $f(-4) = 2/-4$ | | $= -\frac{1}{2}$ | $(-4, -\frac{1}{2})$ |
| $-2$ | $f(-2) = 2/-2$ | | $= -1$ | $(-2, -1)$ |
| $-1$ | $f(-1) = 2/-1$ | | $= -2$ | $(-1, -2)$ |
| $-\frac{1}{2}$ | $f(-\frac{1}{2}) = 2/-\frac{1}{2}$ | | $= -4$ | $(-\frac{1}{2}, -4)$ |
| $\frac{1}{2}$ | $f(\frac{1}{2}) = 2/\frac{1}{2}$ | | $= 4$ | $(\frac{1}{2}, 4)$ |
| $1$ | $f(1) = 2/1$ | | $= 2$ | $(1, 2)$ |
| $2$ | $f(2) = 2/2$ | | $= 1$ | $(2, 1)$ |
| $4$ | $f(4) = 2/4$ | | $= \frac{1}{2}$ | $(4, \frac{1}{2})$ |

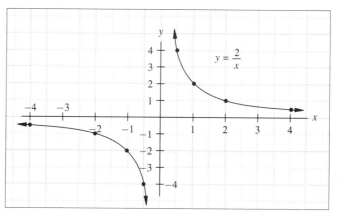

**Fig. 3-2**

The graph of $y = 2/x$, a *rational function*, is plotted in two diagonally opposite quadrants. As $x \to 0$, the graph approaches the $y$ axis. The $y$ axis in this instance is called the *vertical asymptote*. As $x \to \infty$, the graph approaches the $x$ axis, in this case called the *horizontal asymptote*. See Problems 3.19 and 3.23.

## 3.3   THE ALGEBRA OF FUNCTIONS

Two or more functions can be combined to obtain a new function by addition, subtraction, multiplication, or division of the original functions. Given two functions $f$ and $g$, with $x$ in the domain of both $f$ and $g$,

$$(f + g)(x) = f(x) + g(x)$$

$$(f - g)(x) = f(x) - g(x)$$

$$(f \cdot g)(x) = f(x) \cdot g(x)$$

$$(f \div g)(x) = f(x) \div g(x) \qquad [g(x) \neq 0]$$

Functions can also be combined by substituting one function $y = f(x)$ for every occurrence of $y$ in another function $z = g(y)$. Written $z = g[f(x)]$, it is called a *composition of functions*. See Examples 5 and 6 and Problems 3.5 to 3.17.

**EXAMPLE 5.**    If $f(x) = 3x + 8$ and $g(x) = 5x - 4$, then from the properties of the algebra of functions given above,

$$(f + g)(x) = (3x + 8) + (5x - 4) = 8x + 4$$

$$(f - g)(x) = (3x + 8) - (5x - 4) = -2x + 12$$

$$(f \cdot g)(x) = (3x + 8)(5x - 4) = 15x^2 + 28x - 32$$

$$(f \div g)(x) = \frac{3x + 8}{5x - 4} \qquad (x \neq 0.8)$$

**EXAMPLE 6.**   Given $z = g(y) = y^2 + 3y - 8$ and $y = f(x) = x + 2$, the composite function $g[f(x)]$ is found by substituting $f(x)$ for each occurrence of $y$ in $g(y)$.

$$g(y) = y^2 \qquad + 3y \qquad\qquad - 8$$

$$g[f(x)] = [f(x)]^2 \quad + 3[f(x)] \qquad - 8$$

$$= (x + 2)^2 \ + 3(x + 2) \qquad - 8$$

$$= (x^2 + 4x + 4) + (3x + 6) - 8$$

$$= x^2 + 7x \ + 2$$

See also Problems 3.7 and 3.30 to 3.32.

## 3.4   APPLICATIONS OF LINEAR FUNCTIONS FOR BUSINESS AND ECONOMICS

Linear functions are used often in business and economics and frequently combined to form new functions. Instead of $f(x)$, $g(x)$, or $h(x)$, for example, the functional notation $C(x)$ is commonly used to represent a cost function; $R(x)$ is used to represent a revenue function; and $\pi(x)$ is used to represent a profit function. Assume a firm operating in a purely competitive market which has a constant marginal revenue or selling price of \$60, a fixed cost of \$450, and a variable cost of \$35 an item. By letting $x$ represent the number of items produced and sold, the firm's total revenue $R$ and total cost $C$ can be expressed by the following functions of $x$:

$$R(x) = 60x$$

$$C(x) = 35(x) + 450$$

Functions are also frequently combined for business and economic purposes, using the algebra of functions. For instance, the profit function of the firm mentioned above is easily obtained by subtracting the total cost function from the total revenue function:

$$\pi(x) = R(x) - C(x)$$

$$\pi(x) = 60x - [35(x) + 450]$$

$$\pi(x) = 25x - 450$$

**EXAMPLE 7.**   The total revenue $R$, total cost $C$, and profit level $\pi$ for the firm mentioned in Section 3.4 at output levels $(x)$ of 70 and 90 units, respectively, assuming $R = 60x$, $C = 35x + 450$, and $\pi = 25x - 450$, are found as follows:

At $x = 70$,

$$R(70) = 60(70) = 4200$$

$$C(70) = 35(70) + 450 = 2900$$

$$\pi(70) = 25(70) - 450 = 1300$$

At $x = 90$

$$R(90) = 60(90) = 5400$$

$$C(90) = 35(90) + 450 = 3600$$

$$\pi(90) = 25(90) - 450 = 1800$$

See also Problems 3.8 to 3.17.

## 3.5 SOLVING QUADRATIC EQUATIONS

By setting $y = 0$, a quadratic function $y = ax^2 + bx + c$, can be expressed as a *quadratic equation* $ax^2 + bx + c = 0$, where $a$, $b$, and $c$ are constants and $a \neq 0$. Quadratic equations in this form can be solved by factoring or use of the *quadratic formula*

$$x = \frac{-b \pm \sqrt{b^2 - 4ac}}{2a}$$ (3.1)

as is illustrated in Example 8.

**EXAMPLE 8.** The quadratic equation

$$5x^2 + 47x + 18 = 0$$

is solved below by means of (*a*) factoring and (*b*) the quadratic formula.

(*a*) Factoring the equation, as in Problem 1.11 (*a*),

$$5x^2 + 47x + 18 = (5x + 2)(x + 9) = 0$$

For $(5x + 2)(x + 9)$ to equal 0, $5x + 2 = 0$ or $x + 9 = 0$. Setting each in turn equal to 0 and solving for $x$, we have

$$5x + 2 = 0 \qquad\qquad x + 9 = 0$$

$$x = -0.4 \qquad\qquad x = -9$$

(*b*) Using the quadratic formula where in terms of the present equation $a = 5$, $b = 47$, and $c = 18$,

$$x = \frac{-47 \pm \sqrt{(47)^2 - 4(5)(18)}}{2(5)}$$

$$x = \frac{-47 \pm \sqrt{2209 - 360}}{10} = \frac{-47 \pm \sqrt{1849}}{10}$$

Using a calculator, as explained in Section 1.7.7,

$$x = \frac{-47 \pm 43}{10}$$

Then adding and subtracting 43 in turn in the numerator,

$$x = \frac{-47 + 43}{10} \qquad\qquad x = \frac{-47 - 43}{10}$$

$$x = -\frac{4}{10} = -0.4 \qquad\qquad x = -\frac{90}{10} = -9$$

See also Problems 3.20 and 3.21. A third method of solving quadratic equations, called *completing the square*, is not discussed here but can be found in Dowling, *Schaum's Outline of Calculus for Business, Economics, and the Social Sciences*, Section 2.6.

## 3.6 FACILITATING NONLINEAR GRAPHING

The graph of a *quadratic function* $y = ax^2 + bx + c$ is simplified by three helpful hints:

1. If $a < 0$, the parabola opens down; if $a > 0$, the parabola opens up.
2. As proved in Problem 10.27, the coordinates of the vertex are the ordered pair $(x, y)$, where $x = -b/2a$, and $y = (4ac - b^2)/4a$.
3. The $x$ intercepts can be found by setting the function equal to zero and using the quadratic formula or factoring to solve for $x$, as is demonstrated in Example 9 and Problem 3.22.

The graph of a *rational function* is made easier by finding the asymptotes. The *vertical asymptote* is the line $x = k$ where $k$ is found after all cancellation is completed by solving the denominator, when set equal to zero, for $x$; the *horizontal asymptote* is the line $y = m$, where $m$ is found by first solving the original equation for $x$ and then solving the denominator of that equation, when set equal to zero, for $y$. See Example 11 and Problem 3.23.

**EXAMPLE 9.**    Applying the hints given in Section 3.6 to graph the quadratic function,

$$y = -x^2 + 6x + 7$$

note that $a = -1$, $b = 6$, and $c = 7$. Thus

   1.  With $a = -1 < 0$, the parabola opens down.

   2.  $$x = \frac{-b}{2a} = \frac{-6}{2(-1)} \qquad\qquad y = \frac{4ac - b^2}{4a} = \frac{4(-1)(7) - 6^2}{4(-1)}$$

$$x = 3 \qquad\qquad y = \frac{-28 - 36}{-4} = 16$$

       The coordinates of the vertex are (3, 16).

   3.  Setting $y = 0$ and factoring,

$$-x^2 + 6x + 7 = 0$$

$$(-x - 1)(x - 7) = 0$$

$$-x - 1 = 0 \qquad\qquad x - 7 = 0$$

$$x = -1 \qquad\qquad x = 7$$

       The $x$ intercepts are $(-1, 0)$ and $(7, 0)$. See Fig. 3-3 and Problem 3.22.

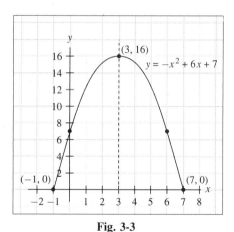

**Fig. 3-3**

## 3.7    APPLICATIONS OF NONLINEAR FUNCTIONS IN BUSINESS AND ECONOMICS

Many of the problems encountered in economics and business do not lend themselves to analysis in terms of simple linear functions. They require different types of functions. Revenue and profit functions for monopolistic competition, for instance, are frequently expressed in terms of quadratic functions; cost-benefit analysis is usually handled with rational functions. See Examples 10 and 11 and Problems 3.24 to 3.29.

**EXAMPLE 10.**    A sneaker company's profits $\pi$ for each unit $x$ sold has been estimated as

$$\pi(x) = -x^2 + 300x - 9000$$

With the coefficient of the $x^2$ term negative, the parabola opens down. The coordinates of the vertex are the ordered pair $[-b/2a, (4ac - b^2)/4a]$, where in terms of the given equation, $a = -1$, $b = 300$, $c = -9000$. Substituting,

$$x = \frac{-b}{2a} = \frac{-(300)}{2(-1)} \qquad \pi = \frac{4ac - b^2}{4a} = \frac{4(-1)(-9000) - (300)^2}{4(-1)}$$

$$x = 150 \qquad \pi = \frac{36,000 - 90,000}{-4} = \frac{-54,000}{-4} = 13,500$$

Thus the vertex is at (150, 13,500), which in terms of Fig. 3-4 indicates that profit is maximized at \$13,500 when 150 units are sold. To find the $x$ intercepts to polish up the graph, set $\pi = 0$ and factor or use the quadratic formula to find at $\pi = 0$, $x = 33.81$ and $x = 266.19$. Hence the $x$ intercepts are (33.81, 0) and (266.19, 0). See Fig. 3-4 and Problems 3.25 and 3.26.

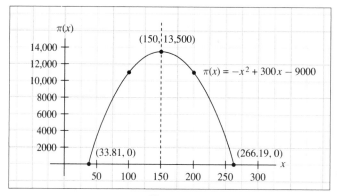

**Fig. 3-4**

**EXAMPLE 11.** Assume that the cost $C$ in thousands of dollars of removing $x$ percent of the sulfur dioxide from the exhaust of a copper smelting plant is given by the rational function

$$C(x) = \frac{15x}{105 - x} \qquad (0 \le x \le 100)$$

By finding a select number of ordered pairs that satisfy the function and graphing, one can see the escalating costs of cleaning up the last percentage points of the pollutant. Although outside the restricted domain here, the vertical asymptote of a rational function can always be found, after all cancellation is completed, by solving the denominator, set equal to zero, for $x$. With the denominator $105 - x = 0$, $x = 105$. Thus, the vertical asymptote occurs at $x = 105$. See Fig. 3-5 and Problem 3.27.

# Solved Problems

## FUNCTIONS

**3.1.** Evaluate the following functions at the given values of $x$.

(a) $f(x) = x^2 - 4x + 7$ at (1) $x = 5$, (2) $x = -4$

(1) Substituting 5 for each occurrence of $x$ in the function,

$$f(5) = (5)^2 - 4(5) + 7 = 12$$

(2) Now substituting $-4$ for each occurrence of $x$,

$$f(-4) = (-4)^2 - 4(-4) + 7 = 39$$

| $x$ | $C$ |
|-----|-----|
| 70  | 30  |
| 80  | 48  |
| 90  | 90  |
| 100 | 300 |

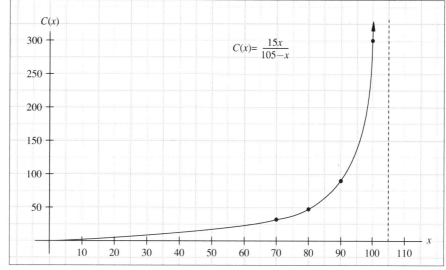

$$C(x) = \frac{15x}{105 - x}$$

**Fig. 3-5**

(b)   $f(x) = 2x^3 - 5x^2 + 8x - 11$ at (1) $x = 3$, (2) $x = -2$

   (1)   $f(3) = 2(3)^3 - 5(3)^2 + 8(3) - 11 = 22$
   (2)   $f(-2) = 2(-2)^3 - 5(-2)^2 + 8(-2) - 11 = -63$

(c)   $f(x) = \dfrac{3x^2 - 8x + 18}{x - 4}$ at (1) $x = 6$, (2) $x = -2$

   (1)   $f(6) = \dfrac{3(6)^2 - 8(6) + 18}{(6) - 4} = \dfrac{78}{2} = 39$

   (2)   $f(-2) = \dfrac{3(-2)^2 - 8(-2) + 18}{(-2) - 4} = \dfrac{46}{-6} = -7\dfrac{2}{3}$

**3.2.**   Parameters and other expressions can also be substituted into functions as the independent variable. Evaluate the following functions at the given values of $x$ by substituting the various parameters for $x$ and simplifying.

(a)   $f(x) = 2x^2 + 5x + 9$ at (1) $x = a$, (2) $x = a - 3$

   (1)   $f(a) = 2(a)^2 + 5(a) + 9$
       $= 2a^2 + 5a + 9$
   (2)   $f(a - 3) = 2(a - 3)^2 + 5(a - 3) + 9$
       $= 2(a - 3)(a - 3) + 5(a - 3) + 9$
       $= 2a^2 - 7a + 12$

(b)   $f(x) = x^2 + 7x + 4$ at (1) $x = 3a$, (2) $x = a + 5$

   (1)   $f(3a) = (3a)^2 + 7(3a) + 4$
       $= 9a^2 + 21a + 4$
   (2)   $f(a + 5) = (a + 5)^2 + 7(a + 5) + 4$
       $= (a + 5)(a + 5) + 7(a + 5) + 4$
       $= a^2 + 17a + 64$

(c)   $f(x) = \dfrac{x^2 - 13}{x + 4}$ at (1) $x = a$, (2) $x = a - 2$

   (1)   $f(a) = \dfrac{a^2 - 13}{a + 4}$

$$(2) \quad f(a-2) = \frac{(a-2)^2 - 13}{(a-2)+4} = \frac{a^2 - 4a - 9}{a+2}$$

**3.3.** In the graphs in Fig. 3-6, where $y$ replaces $f(x)$ as the dependent variable in functions, as is common in graphing, indicate which graphs are graphs of functions and which are not.

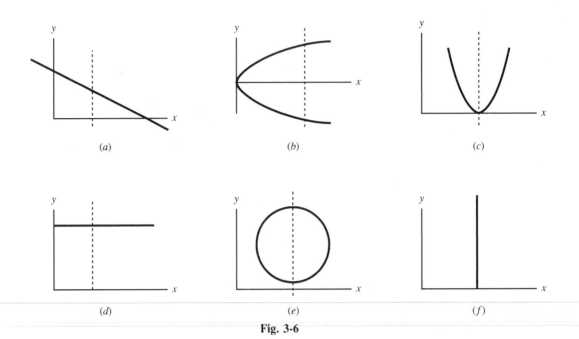

Fig. 3-6

From the definition of a function, we know that for each value of $x$, there can be one and only one value of $y$. If, for any value of $x$ in the domain, therefore, a vertical line can be drawn which intersects the graph at more than one point, the graph is not the graph of a function. Applying this criterion, which is known as the *vertical line test*, we see that $(a)$, $(c)$, and $(d)$ are functions; $(b)$, $(e)$ and $(f)$ are not.

**3.4.** Which of the following equations are functions and why?

($a$)  $y = -3x + 8$

The equation $y = -3x + 8$ is a function because for each value of the independent variable $x$ there is one and only one value of the dependent variable $y$. For example, if $x = 1$, $y = -3(1) + 8 = 5$. The graph would be similar to $(a)$ in Fig. 3-6.

($b$)  $y^2 = x$

The equation $y^2 = x$, which is equivalent to $y = \pm\sqrt{x}$, is not a function because for each positive value of $x$, there are two values of $y$. For example, if $y^2 = 4$, $y = \pm 2$. The graph would be similar to that of $(b)$ in Fig. 3-6, illustrating that a parabola whose axis is parallel to the $x$ axis cannot be a function.

($c$)  $y = x^2$

The equation $y = x^2$ is a function. For each value of $x$ there is only one value of $y$. For instance, if $x = -6$, $y = 36$. While it is also true that $y = 36$ when $x = 6$, it is irrelevant. The definition of a function simply demands that for each value of $x$ there be one and only one value of $y$, *not* that for each value of $y$ there be one and only one value of $x$. The graph would look like $(c)$ in Fig. 3-6, demonstrating that a parabola with axis parallel to the $y$ axis is a function, regardless of whether it opens up or down.

(*d*)   $y = 6$

>   The equation $y = 6$ is a function. For each value of $x$ there is one and only one value of $y$. The graph would look like (*d*) in Fig. 3-6.

(*e*)   $x^2 + y^2 = 81$

>   The equation $x^2 + y^2 = 81$ is not a function. If $x = 0$, $y^2 = 81$, and $y = \pm 9$. The graph would be a circle, similar to (*e*) in Fig. 3-6. A circle does not pass the vertical line test.

(*f*)   $x = 8$

>   The equation $x = 8$ is not a function. The graph of $x = 8$ is a vertical line. This means that at $x = 8$, $y$ has an infinite number of values. The graph would look like (*f*) in Fig. 3-6.

## THE ALGEBRA OF FUNCTIONS

**3.5.**   Use the rules of the algebra of functions to combine the following functions by finding (1) $(f + g)(x)$, (2) $(f - g)(x)$, (3) $(f \cdot g)(x)$, and (4) $(f \div g)(x)$.

(*a*)   $f(x) = 4x - 5$, $g(x) = 7x - 3$

   (1)   $(f + g)(x) = f(x) + g(x) = (4x - 5) + (7x - 3) = 11x - 8$
   (2)   $(f - g)(x) = f(x) - g(x) = (4x - 5) - (7x - 3) = -3x - 2$
   (3)   $(f \cdot g)(x)\ = f(x) \cdot g(x)$
         $= (4x - 5) \cdot (7x - 3) = 28x^2 - 47x + 15$
   (4)   $(f \div g)(x) = f(x) \div g(x) = \dfrac{4x - 5}{7x - 3} \qquad \left(x \neq \dfrac{3}{7}\right)$

(*b*)   $f(x) = x^2 + 3$, $g(x) = 4x - 7$

   (1)   $(f + g)(x) = (x^2 + 3) + (4x - 7) = x^2 + 4x - 4$
   (2)   $(f - g)(x) = (x^2 + 3) - (4x - 7) = x^2 - 4x + 10$
   (3)   $(f \cdot g)(x) = (x^2 + 3) \cdot (4x - 7) = 4x^3 - 7x^2 + 12x - 21$
   (4)   $(f \div g)(x) = \dfrac{x^2 + 3}{4x - 7} \qquad \left(x \neq \dfrac{7}{4}\right)$

(*c*)   $f(x) = \dfrac{3}{x}$, $g(x) = \dfrac{4}{x + 5} \qquad (x \neq 0, -5)$

   (1)   $(f + g)(x) = \dfrac{3}{x} + \dfrac{4}{x + 5} = \dfrac{3(x + 5) + 4x}{x(x + 5)} = \dfrac{7x + 15}{x^2 + 5x}$

   (2)   $(f - g)(x) = \dfrac{3}{x} - \dfrac{4}{x + 5} = \dfrac{3(x + 5) - 4x}{x(x + 5)} = \dfrac{15 - x}{x^2 + 5x}$

   (3)   $(f \cdot g)(x) = \dfrac{3}{x} \cdot \dfrac{4}{x + 5} = \dfrac{12}{x(x + 5)} = \dfrac{12}{x^2 + 5x}$

   (4)   $(f \div g)(x) = \dfrac{3}{x} \div \dfrac{4}{x + 5} = \dfrac{3}{x} \cdot \dfrac{x + 5}{4} = \dfrac{3x + 15}{4x}$

**3.6.**   Given

$$f(x) = \frac{3x}{x + 5} \qquad g(x) = \frac{x - 4}{x + 1} \qquad h(x) = \frac{x + 6}{x - 2} \qquad (x \neq -5, -1, 2)$$

use the algebra of functions to find the following:

(*a*)   $(f + h)(a)$

$$(f + h)(a) = f(a) + h(a)$$

Substituting $a$ for each occurrence of $x$,

$$(f + h)(a) = \frac{3a}{a+5} + \frac{a+6}{a-2} = \frac{3a(a-2) + (a+6)(a+5)}{(a+5)(a-2)}$$

$$= \frac{(3a^2 - 6a) + (a^2 + 11a + 30)}{a^2 + 3a - 10} = \frac{4a^2 + 5a + 30}{a^2 + 3a - 10}$$

(b) $(g \cdot h)(t)$

$$(g \cdot h)(t) = g(t) \cdot h(t) = \frac{(t-4)}{(t+1)} \cdot \frac{(t+6)}{(t-2)} = \frac{t^2 + 2t - 24}{t^2 - t - 2}$$

(c) $(h - f)(x + 1)$

$$(h - f)(x + 1) = h(x + 1) - f(x + 1)$$

Substituting $x + 1$ for each occurrence of $x$,

$$(h - f)(x + 1) = \frac{(x+1)+6}{(x+1)-2} - \frac{3(x+1)}{(x+1)+5} \qquad (x \neq 1, -6)$$

$$= \frac{x+7}{x-1} - \frac{3x+3}{x+6} = \frac{(x+7)(x+6) - (3x+3)(x-1)}{(x-1)(x+6)}$$

$$= \frac{(x^2 + 13x + 42) - (3x^2 - 3)}{x^2 + 5x - 6} = \frac{-2x^2 + 13x + 45}{x^2 + 5x - 6}$$

(d) $(g \div h)(t - 3)$

$$(g \div h)(t - 3) = g(t - 3) \div h(t - 3)$$

Substituting $t - 3$ for each occurrence of $x$ and inverting $h$,

$$(g \div h)(t - 3) = \frac{(t-3)-4}{(t-3)+1} \cdot \frac{(t-3)-2}{(t-3)+6} \qquad (t \neq 2, -3, 5)$$

$$= \frac{t-7}{t-2} \cdot \frac{t-5}{t+3} = \frac{t^2 - 12t + 35}{t^2 + t - 6}$$

**3.7.** Given $f(x) = x^4$, $g(x) = x^2 - 3x + 4$, and

$$h(x) = \frac{x}{x-5} \qquad (x \neq 5)$$

find the following composite functions, as in Example 6.

(a) $g[f(x)]$

Substituting $f(x)$ for every occurrence of $x$ in $g(x)$,

$$g[f(x)] = (x^4)^2 - 3(x^4) + 4 = x^8 - 3x^4 + 4$$

(b) $f[h(x)]$

Substituting $h(x)$ for each occurrence of $x$ in $f(x)$

$$f[h(x)] = \left(\frac{x}{x-5}\right)^4$$

(c) $h[f(x)]$

$$h[f(x)] = \frac{x^4}{x^4 - 5} \qquad (x \neq \pm\sqrt[4]{5})$$

(d) $h[g(x)]$

$$h[g(x)] = \frac{(x^2 - 3x + 4)}{(x^2 - 3x + 4) - 5} = \frac{x^2 - 3x + 4}{x^2 - 3x - 1} \qquad (x^2 - 3x - 1 \neq 0)$$

## LINEAR FUNCTIONS IN BUSINESS AND ECONOMICS

**3.8.**   A plumber charges $50 for a house visit plus $35 an hour for each extra hour of work. Express the cost $C$ of a plumber as a function of the number of hours $x$ the job involves.

$$C(x) = 35x + 50$$

**3.9.**   A recording artist receives a fee of $9000 plus $2.75 for every album sold. Express her revenue $R$ as a function of the number of albums $x$ sold.

$$R(x) = 2.75x + 9000$$

**3.10.**   An apple orchard charges $3.25 to enter and 60 cents a pound for whatever is picked. Express the cost $C$ as a function of the number of pounds $x$ of apples picked.

$$C(x) = 0.60x + 3.25$$

**3.11.**   A potter exhibiting at a fair receives $24 for each ceramic sold minus a flat exhibition fee of $85. Express the revenue $R$ he receives as a function of the number $x$ of ceramics sold.

$$R(x) = 24x - 85$$

**3.12.**   An office machine worth $12,000 depreciates in value by $1500 a year. Using linear or straight-line depreciation, express the value $V$ of the machine as a function of years $t$.

$$V(t) = 12,000 - 1500t$$

**3.13.**   A farmer in a purely competitive market receives $35 a bushel for each bushel of wheat sold. ($a$) Express his revenue $R$ as a function of the number $x$ of bushels sold and evaluate the function at ($b$) $x = 10,000$ and ($c$) $x = 25,000$.

($a$)   $$R(x) = 35x$$

($b$)   $$R(10,000) = 35(10,000) = 350,000$$

($c$)   $$R(25,000) = 35(25,000) = 875,000$$

**3.14.**   The farmer in Problem 3.13 has a fixed cost of $425,000 and a marginal cost of $15 a bushel. ($a$) Express his cost $C$ as a function of the number $x$ of bushels produced and estimate the function at ($b$) $x = 5000$ and ($c$) $x = 20,000$.

($a$)   $$C(x) = 15x + 425,000$$

($b$)   $$C(5000) = 15(5000) + 425,000 = 500,000$$

($c$)   $$C(20,000) = 15(20,000) + 425,000 = 725,000$$

**3.15.**   ($a$) Use the algebra of functions to express the same farmer's profit $\pi$ as a function of the number $x$ of bushels sold, and evaluate the function at ($b$) $x = 20,000$ and ($c$) $x = 30,000$.

(a)
$$\pi(x) = R(x) - C(x)$$
$$\pi(x) = 35(x) - [15(x) + 425{,}000]$$
$$\pi(x) = 20x - 425{,}000$$

(b)
$$\pi(20{,}000) = 20(20{,}000) - 425{,}000 = -25{,}000$$

(c)
$$\pi(30{,}000) = 20(30{,}000) - 425{,}000 = 175{,}000$$

**3.16.** A company has a fixed cost of \$8250 and a marginal cost of \$450 for each item produced. (a) Express the cost $C$ as a function of the number $x$ of items produced and evaluate the function at (b) $x = 20$ and (c) $x = 50$.

(a)
$$C(x) = 450x + 8250$$

(b)
$$C(20) = 450(20) + 8250 = 17{,}250$$

(c)
$$C(50) = 450(50) + 8250 = 30{,}750$$

**3.17.** Assume that the company in Problem 3.16 receives \$800 for each item sold. (a) Express the company's profit $\pi$ as a function of the number $x$ of items sold and evaluate the function at (b) $x = 25$ and (c) $x = 40$.

(a) With total revenue, $R(x) = 800x$, the profit function is
$$\pi(x) = R(x) - C(x)$$
$$\pi(x) = 800x - [450x + 8250] = 350x - 8250$$

(b)
$$\pi(25) = 350(25) - 8250 = 500$$

(c)
$$\pi(40) = 350(40) - 8250 = 5750$$

For earlier practical applications that could have also been expressed as functions, see Problems 2.21 to 2.34.

## GRAPHING NONLINEAR FUNCTIONS

**3.18.** Graph the following quadratic functions and identify the vertex and axis of each:

(a)  $f(x) = x^2 - 2$

First select some representative values of $x$ and solve for $f(x)$. Then, using $y$ for $f(x)$, plot the resulting ordered pairs and connect them with a smooth line. The closer the points, the more accurate the graph. See Fig. 3-7.

| $x$ | $f(x)$ | $= x^2$ | $- 2$ | $= y$ |
|-----|--------|---------|-------|-------|
| $-3$ | $f(-3)$ | $= (-3)^2$ | $- 2$ | $= 7$ |
| $-2$ | $f(-2)$ | $= (-2)^2$ | $- 2$ | $= 2$ |
| $-1$ | $f(-1)$ | $= (-1)^2$ | $- 2$ | $= -1$ |
| $0$ | $f(0)$ | $= (0)^2$ | $- 2$ | $= -2$ |
| $1$ | $f(1)$ | $= (1)^2$ | $- 2$ | $= -1$ |
| $2$ | $f(2)$ | $= (2)^2$ | $- 2$ | $= 2$ |
| $3$ | $f(3)$ | $= (3)^2$ | $- 2$ | $= 7$ |

Vertex: $(0, -2)$.
Axis: $x = 0$, which is the $y$ axis.

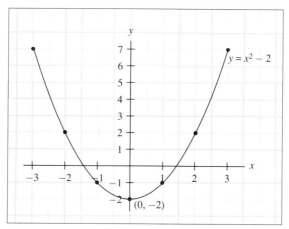

**Fig. 3-7**

(b)   $f(x) = -x^2 + 4x + 32$

| $x$ | $f(x)$ | $= -x^2$ | $+ 4x$ | $+ 32$ | $= y$ |
|-----|--------|----------|--------|--------|-------|
| $-4$ | $f(-4) =$ | $-(-4)^2$ | $+ 4(-4)$ | $+ 32$ | $= 0$ |
| $-2$ | $f(-2) =$ | $-(-2)^2$ | $+ 4(-2)$ | $+ 32$ | $= 20$ |
| $0$ | $f(0) =$ | $-(0)^2$ | $+ 4(0)$ | $+ 32$ | $= 32$ |
| $2$ | $f(2) =$ | $-(2)^2$ | $+ 4(2)$ | $+ 32$ | $= 36$ |
| $4$ | $f(4) =$ | $-(4)^2$ | $+ 4(4)$ | $+ 32$ | $= 32$ |
| $6$ | $f(6) =$ | $-(6)^2$ | $+ 4(6)$ | $+ 32$ | $= 20$ |
| $8$ | $f(8) =$ | $-(8)^2$ | $+ 4(8)$ | $+ 32$ | $= 0$ |

Vertex: (2, 36).
Axis: $x = 2$.
See Fig. 3-8.

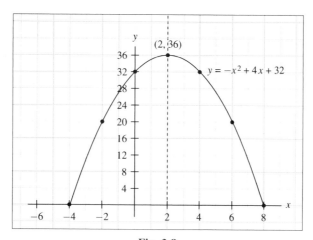

**Fig. 3-8**

Notice the symmetry about the axis: $x = 2$. For $x = 2 \pm 2$, $y = 32$; for $x = 2 \pm 4$, $y = 20$; for $x = 2 \pm 6$, $y = 0$.

(c)   $f(x) = x^2 - 8x + 17$

| $x$ | $f(x) = x^2 \quad - 8x \quad + 17$ | $= y$ |
|---|---|---|
| 2 | $f(2) = (2)^2 - 8(2) + 17$ | $= 5$ |
| 3 | $f(3) = (3)^2 - 8(3) + 17$ | $= 2$ |
| 4 | $f(4) = (4)^2 - 8(4) + 17$ | $= 1$ |
| 5 | $f(5) = (5)^2 - 8(5) + 17$ | $= 2$ |
| 6 | $f(6) = (6)^2 - 8(6) + 17$ | $= 5$ |

Vertex: $(4, 1)$.
Axis: $x = 4$.
See Fig. 3-9.

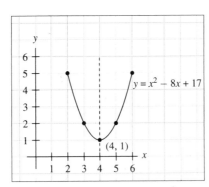

**Fig. 3-9**

Notice the symmetry about the axis: $x = 4$. For $x = 4 \pm 1$, $y = 2$; for $x = 4 \pm 2$, $y = 5$.

(d)   $f(x) = -x^2 + 4x + 9$

| $x$ | $f(x) = -x^2 \quad + 4x \quad + 9$ | $= y$ |
|---|---|---|
| 0 | $f(0) = -(0)^2 + 4(0) + 9$ | $= 9$ |
| 1 | $f(1) = -(1)^2 + 4(1) + 9$ | $= 12$ |
| 2 | $f(2) = -(2)^2 + 4(2) + 9$ | $= 13$ |
| 3 | $f(3) = -(3)^2 + 4(3) + 9$ | $= 12$ |
| 4 | $f(4) = -(4)^2 + 4(4) + 9$ | $= 9$ |

Vertex: $(2, 13)$.
Axis: $x = 2$.
See Fig. 3-10.

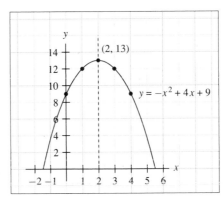

**Fig. 3-10**

**3.19.** Determine a number of representative points on the graphs of the following rational functions and draw a sketch of the graph.

(a)  $y = \dfrac{8}{x-4}$

| x | y |
|---|---|
| −4 | −1 |
| 0 | −2 |
| 2 | −4 |
| 3 | −8 |
| - - - | - - - |
| 5 | 8 |
| 6 | 4 |
| 8 | 2 |
| 12 | 1 |

Vertical asymptote: $x = 4$.
Horizontal asymptote: $y = 0$.
See Fig. 3-11.

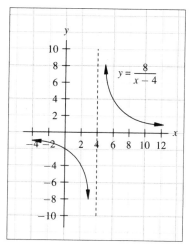

**Fig. 3-11**

(b)  $y = \dfrac{-5}{x+2}$

| x | y |
|---|---|
| −7 | 1 |
| −6 | 1.25 |
| −4 | 2.5 |
| −3 | 5 |
| - - - | - - - |
| −1 | −5 |
| 0 | −2.5 |
| 2 | −1.25 |
| 3 | −1 |

Vertical asymptote: $x = -2$.
Horizontal asymptote: $y = 0$.
See Fig. 3-12.

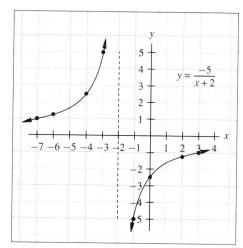

**Fig. 3-12**

## SOLVING QUADRATIC EQUATIONS

**3.20.** Solve the following quadratic equations by factoring.

(a)  $x^2 + 10x + 21 = 0$

From Problem 1.6 (a),

$$x^2 + 10x + 21 = (x+3)(x+7) = 0$$

For $(x + 3)(x + 7)$ to equal 0, $x + 3 = 0$ or $x + 7 = 0$. Setting each in turn equal to 0 and solving for $x$, we have

$$x + 3 = 0 \qquad x + 7 = 0$$

$$x = -3 \qquad x = -7$$

(b) $x^2 - 13x + 30 = 0$

From Problem 1.7 (a),

$$x^2 - 13x + 30 = (x - 3)(x - 10) = 0$$

$$x - 3 = 0 \qquad x - 10 = 0$$

$$x = 3 \qquad x = 10$$

(c) $x^2 + 19x - 42 = 0$

From Problem 1.8 (a),

$$x^2 + 19x - 42 = (x - 2)(x + 21) = 0$$

$$x - 2 = 0 \qquad x + 21 = 0$$

$$x = 2 \qquad x = -21$$

(d) $x^2 - 8x - 48 = 0$

From Problem 1.9 (a),

$$x^2 - 8x - 48 = (x + 4)(x - 12) = 0$$

$$x + 4 = 0 \qquad x - 12 = 0$$

$$x = -4 \qquad x = 12$$

(e) $3x^2 + 22x + 24 = 0$

From Problem 1.11 (b),

$$3x^2 + 22x + 24 = (3x + 4)(x + 6)$$

$$3x + 4 = 0 \qquad x + 6 = 0$$

$$x = -\tfrac{4}{3} \qquad x = -6$$

**3.21.** Solve the following quadratic equations using the quadratic formula:

$$x = \frac{-b \pm \sqrt{b^2 - 4ac}}{2a} \tag{3.1}$$

(a) $3x^2 - 35x + 22 = 0$

Substituting $a = 3$, $b = -35$, and $c = 22$ in the formula

$$x = \frac{-(-35) \pm \sqrt{(-35)^2 - 4(3)(22)}}{2(3)}$$

Using a calculator, as explained in Sections 1.7.6 and 1.7.7,

$$x = \frac{35 \pm \sqrt{1225 - 264}}{6} = \frac{35 \pm \sqrt{961}}{6} = \frac{35 \pm 31}{6}$$

Adding and subtracting 31 in turn in the numerator,

$$x = \frac{35 + 31}{6} = 11 \qquad\qquad x = \frac{35 - 31}{6} = \frac{2}{3}$$

(b)   $7x^2 - 32x + 16 = 0$

$$x = \frac{-(-32) \pm \sqrt{(-32)^2 - 4(7)(16)}}{2(7)}$$

$$x = \frac{32 \pm \sqrt{1024 - 448}}{14} = \frac{32 \pm \sqrt{576}}{14} = \frac{32 \pm 24}{14}$$

$$x = \frac{32 + 24}{14} = 4 \qquad x = \frac{32 - 24}{14} = \frac{4}{7}$$

(c)   $5x^2 + 7x - 52 = 0$

$$x = \frac{-(7) \pm \sqrt{(7)^2 - 4(5)(-52)}}{2(5)}$$

$$x = \frac{-7 \pm \sqrt{49 + 1040}}{10} = \frac{-7 \pm \sqrt{1089}}{10} = \frac{-7 \pm 33}{10}$$

$$x = \frac{-7 + 33}{10} = 2.6 \qquad x = \frac{-7 - 33}{10} = -4$$

(d)   $3x^2 - 13x - 56 = 0$

$$x = \frac{-(-13) \pm \sqrt{(-13)^2 - 4(3)(-56)}}{2(3)}$$

$$x = \frac{13 \pm \sqrt{169 + 672}}{6} = \frac{13 \pm \sqrt{841}}{6} = \frac{13 \pm 29}{6}$$

$$x = \frac{13 + 29}{6} = 7 \qquad x = \frac{13 - 29}{6} = -2\frac{2}{3}$$

To see how the same solutions for (a) through (d) can also be obtained by factoring, see Problem 1.11 (c) to ( f ). For practice with the method of solving quadratic equations by completing the square, see Dowling, *Schaum's Outline of Calculus for Business, Economics, and the Social Sciences*, Section 2.6.

## FACILITATING NONLINEAR GRAPHING

**3.22.**   Use the three steps outlined in Section 3.6 for graphing $y = ax^2 + bx + c$, to sketch the following quadratic functions:

(a)   $y = -x^2 + 10x - 16$

(1)   Here $a = -1$, $b = 10$, and $c = -16$. With $a < 0$, the parabola opens down.

(2)
$$x = \frac{-b}{2a} = \frac{-(10)}{2(-1)} \qquad\qquad y = \frac{4ac - b^2}{4a} = \frac{4(-1)(-16) - (10)^2}{4(-1)}$$

$$x = 5 \qquad\qquad\qquad y = \frac{64 - 100}{-4} = \frac{-36}{-4} = 9$$

The coordinates of the vertex are (5, 9).

(3)   Setting the equation equal to zero in preparation to find the $x$ coordinates of the horizontal intercepts,

$$-x^2 + 10x - 16 = 0$$

Dividing both sides by $-1$ before factoring,

$$x^2 - 10x + 16 = 0$$

$$(x - 2)(x - 8) = 0$$

$$x = 2 \qquad x = 8$$

Horizontal intercepts: $(2, 0)$, $(8, 0)$. See Fig. 3-13.

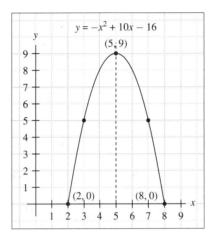

**Fig. 3-13**

(b)   $y = x^2 - 8x + 18$

(1)   With $a = 1 > 0$, the parabola opens upward.

(2)
$$x = \frac{-b}{2a} = \frac{-(-8)}{2(1)} \qquad y = \frac{4ac - b^2}{4a} = \frac{4(1)(18) - (-8)^2}{4(1)}$$

$$x = 4 \qquad\qquad y = \frac{72 - 64}{4} = 2$$

Coordinates of the vertex: $(4, 2)$

(3)   With the vertex at $(4, 2)$ and the parabola opening upward, there are no $x$ intercepts. See Fig. 3-14.

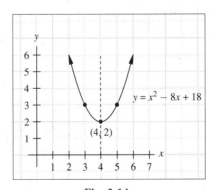

**Fig. 3-14**

(c)   $y = -2x^2 + 10x - 8$

(1) With $a = -2 < 0$, the parabola opens downward.

(2)
$$x = \frac{-b}{2a} = \frac{-(10)}{2(-2)} \qquad y = \frac{4ac - b^2}{4a} = \frac{4(-2)(-8) - (10)^2}{4(-2)}$$

$$x = 2.5 \qquad\qquad y = \frac{64 - 100}{-8} = \frac{-36}{-8} = 4.5$$

Coordinates of the vertex: $(2.5, 4.5)$

(3)  $-2x^2 + 10x - 8 = 0$

Dividing by $-2$, then factoring,

$$x^2 - 5x + 4 = 0$$

$$(x - 1)(x - 4) = 0 \qquad (x = 1, x = 4)$$

The coordinates of the $x$ intercepts are $(1, 0)$ and $(4, 0)$. See Fig. 3-15.

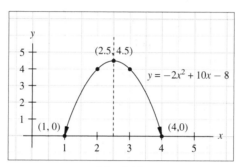

**Fig. 3-15**

(d)  $y = \frac{1}{2}x^2 + 6x + 21$

(1) With $a = \frac{1}{2} > 0$, the parabola opens up.

(2)
$$x = \frac{-(6)}{2(1/2)} \qquad\qquad y = \frac{4(1/2)(21) - (6)^2}{4(1/2)}$$

$$x = -6 \qquad\qquad y = \frac{42 - 36}{2} = 3$$

Coordinates of the vertex: $(-6, 3)$

(3) The parabola does not cross the $x$ axis. See Fig. 3-16.

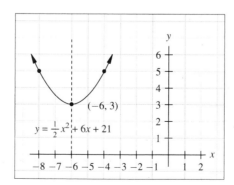

**Fig. 3-16**

(e)   $y = -x^2 + 8x - 7$

    (1)  The parabola opens down.

    (2)  $x = \dfrac{-(8)}{2(-1)} = 4$          $y = \dfrac{4(-1)(-7) - (8)^2}{4(-1)} = 9$

    Vertex coordinates: $(4, 9)$

    (3)  $-x^2 + 8x - 7 = 0$

    Dividing by $-1$ and factoring,

$$x^2 - 8x + 7 = 0$$

$$(x - 1)(x - 7) = 0 \qquad (x = 1, x = 7)$$

    $x$ intercepts: $(1, 0), (7, 0)$.

See Fig. 3-17.

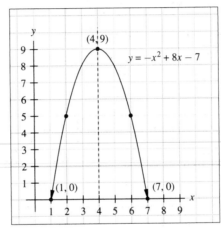

**Fig. 3-17**

(f)   $y = x^2 + 8x + 12$

    (1)  The parabola opens up.

    (2)  $x = \dfrac{-(8)}{2(1)} = -4$          $y = \dfrac{4(1)(12) - (8)^2}{4(1)} = -4$

    Coordinates of the vertex: $(-4, -4)$

    (3)              $x^2 + 8x + 12 = 0$

$$(x + 2)(x + 6) = 0$$

$$x = -2 \qquad x = -6$$

    $x$ intercepts: $(-2, 0), (-6, 0)$.

See Fig. 3-18.

To see how graphing similar functions can also be simplified by the process of completing the square, see Dowling, *Schaum's Outline of Calculus for Business, Economics, and the Social Sciences*, Problem 3.25.

**3.23.**  Do a rough sketch of the graphs of the following rational functions by finding (1) the vertical asymptote, (2) the horizontal asymptote, and (3) selecting a number of representative points and graphing, as was suggested in Section 3.6.

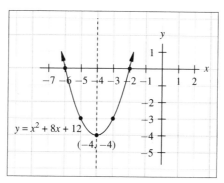

**Fig. 3-18**

(a)   $y = \dfrac{-4}{x + 6}$

   (1)   Setting the denominator equal to zero and solving for $x$ to find the vertical asymptote,

$$x + 6 = 0 \qquad x = -6 \qquad \text{vertical asymptote}$$

   (2)   Solving the original equation for $x$ to find the horizontal asymptote,

$$y = \frac{-4}{x + 6}$$

$$y(x + 6) = -4$$

$$yx = -6y - 4$$

$$x = -\frac{(6y + 4)}{y}$$

   Then setting the denominator equal to zero and solving for $y$,

$$y = 0 \qquad \text{horizontal asymptote}$$

   (3)   Finding representative points and graphing in Fig. 3-19,

| $x$ | $y$ |
|-----|-----|
| $-10$ | $1$ |
| $-8$ | $2$ |
| $-7$ | $4$ |
| - - - | - - - |
| $-5$ | $-4$ |
| $-4$ | $-2$ |
| $-2$ | $-1$ |

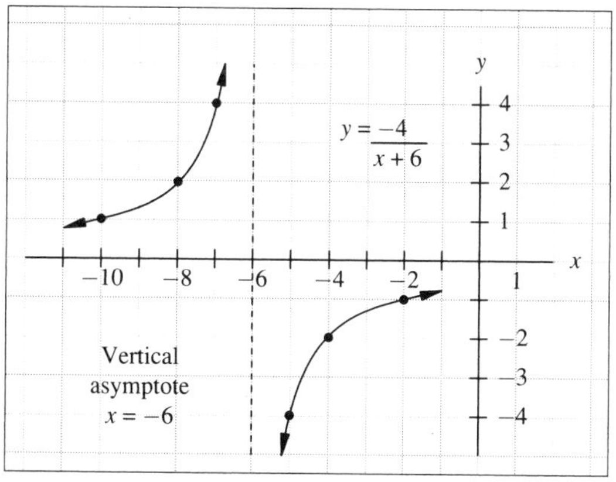

**Fig. 3-19**

(b)   $y = \dfrac{-3}{x - 5}$

(1)   $x - 5 = 0$

$x = 5$     vertical asymptote

(2)   $y(x - 5) = -3$

$yx = -3 + 5y$

$x = \dfrac{-3 + 5y}{y}$

$y = 0$     horizontal asymptote

(3)

| $x$ | $y$ |
|-----|------|
| 8 | −1 |
| 7 | −1.5 |
| 6 | −3 |
| - - - | - - - |
| 4 | 3 |
| 3 | 1.5 |
| 2 | 1 |

See Fig. 3-20.

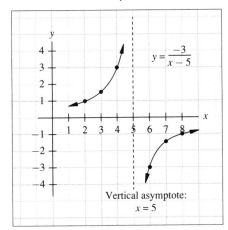

Fig. 3-20

(c)   $y = \dfrac{2x + 3}{x - 4}$

(1)   Vertical asymptote: $x = 4$

(2)   Multiplying both sides of the equation by $(x - 4)$ and solving for $x$ to find the horizontal asymptote,

$$y(x - 4) = 2x + 3$$

$$yx - 4y = 2x + 3$$

$$yx - 2x = 4y + 3$$

$$x(y - 2) = 4y + 3$$

$$x = \dfrac{4y + 3}{y - 2}$$

Then setting the denominator equal to zero and solving for $y$,

$$y - 2 = 0 \qquad y = 2 \qquad \text{horizontal asymptote}$$

For step (3), see top of next page.

(d)   $y = \dfrac{3x + 5}{4x - 9}$

(1)   $4x - 9 = 0$

$4x = 9$

$x = 2.25$

vertical asymptote

(2)   $y(4x - 9) = 3x + 5$

$4xy - 9y = 3x + 5$

$4xy - 3x = 9y + 5$

$x(4y - 3) = 9y + 5$

$x = \dfrac{9y + 5}{4y - 3}$

Setting the denominator equal to zero and solving for $y$,

$$y = \tfrac{3}{4} \qquad \text{horizontal asymptote}$$

(3)

| $x$ | $y$ |
|---|---|
| 0 | $-0.75$ |
| 1 | $-1.67$ |
| 2 | $-3.5$ |
| 3 | $-9.0$ |
| - - - | - - - - |
| 5 | 13.0 |
| 6 | 7.5 |
| 7 | 5.67 |
| 8 | 4.75 |

See Fig. 3-21.

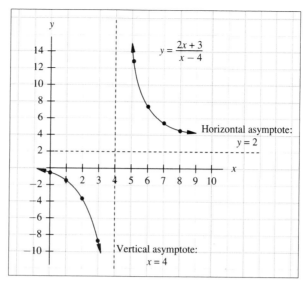

**Fig. 3-21**

(3)

| $x$ | $y$ |
|---|---|
| 0 | $-0.56$ |
| 1.0 | $-1.6$ |
| 2.0 | $-11.0$ |
| - - - | - - - - - |
| 2.5 | 12.5 |
| 3.5 | 3.1 |
| 4.5 | 2.06 |

See Fig. 3-22.

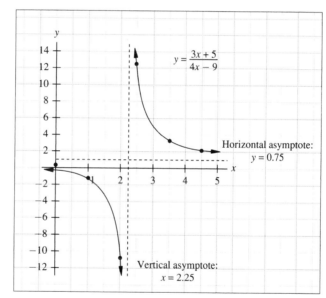

**Fig. 3-22**

## NONLINEAR FUNCTIONS IN BUSINESS AND ECONOMICS

**3.24.** Find the total revenue $R$ functions for producers facing the following linear demand functions:

(a) $Q = -8P + 425$  (b) $Q = -\frac{1}{2}P + 235$

(a) By definition total revenue is equal to price x quantity, or

$$R = P \cdot Q$$

Substituting for $Q$ from above,

$$R = P(-8P + 425)$$
$$R = -8P^2 + 425P$$

(b)
$$R = P(-\tfrac{1}{2}P + 235)$$
$$R = -\tfrac{1}{2}P^2 + 235P$$

Notice how a total revenue function derived from a linear demand function is always a quadratic function.

**3.25.** Given the following total revenue $R(x)$ and total cost $C(x)$ functions, (1) express profit $\pi$ as a function of $x$, (2) determine the maximum level of profit by finding the vertex of $\pi(x)$, and (3) find the $x$ intercepts and draw a rough sketch of the graph.

(a)   $R(x) = 600x - 5x^2$          $C(x) = 100x + 10,500$

(1)                                $\pi(x) = R(x) - C(x)$

$$\pi(x) = 600x - 5x^2 - (100x + 10,500)$$

$$\pi(x) = -5x^2 + 500x - 10,500$$

(2)   Using the method of Problem 3.22, but substituting $\pi$ for $y$,

$$x = \frac{-(500)}{2(-5)} \qquad\qquad \pi = \frac{4(-5)(-10,500) - (500)^2}{4(-5)}$$

$$x = 50 \qquad\qquad \pi = \frac{210,000 - 250,000}{-20} = 2000$$

Vertex: $(50, 2000)$               Maximum: $\pi(50) = 2000$

(3)   Factoring $\pi(x) = 0$ to find the $x$ intercepts,

$$-5x^2 + 500x - 10,500 = 0$$

$$-5(x^2 - 100x + 2100) = 0$$

$$(x - 30)(x - 70) = 0$$

$$x = 30 \qquad\qquad x = 70$$

$x$ intercepts: $(30, 0), (70, 0)$

See Fig. 3-23.

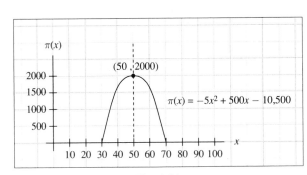

**Fig. 3-23**

(b)   $R(x) = 280x - 2x^2$          $C(x) = 60x + 5600$

(1)   $\pi(x) = -2x^2 + 220x - 5600$

(2)                        $x = \dfrac{-(220)}{2(-2)} = 55$          $\pi = \dfrac{4(-2)(-5600) - (220)^2}{4(-2)} = 450$

Vertex: $(55, 450)$               Maximum: $\pi(55) = 450$

(3)   Factoring $\pi(x) = 0$ to find the $x$ intercepts,

$$-2(x^2 - 110x + 2800) = 0$$

$$(x - 40)(x - 70) = 0$$

$$x = 40 \qquad x = 70$$

$x$ intercepts: $(40, 0), (70, 0)$

See Fig. 3-24.

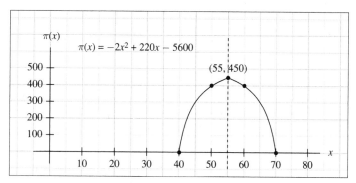

**Fig. 3-24**

(c)   $R(x) = 540x - 3x^2 \qquad C(x) = 90x + 16{,}200$

(1)   $\pi(x) = -3x^2 + 450x - 16{,}200$

(2)   $\qquad x = \dfrac{-(450)}{2(-3)} = 75 \qquad\qquad \pi = \dfrac{4(-3)(-16{,}200) - (450)^2}{4(-3)} = 675$

Vertex: $(75, 675)$ \qquad\qquad Maximum: $\pi(75) = 675$

(3)   Factoring $\pi(x) = 0$ to find the $x$ intercepts,

$$-3(x^2 - 150x + 5400) = 0$$

$$(x - 60)(x - 90) = 0$$

$$x = 60 \qquad x = 90$$

$x$ intercepts: $(60, 0), (90, 0)$

See Fig. 3-25.

(d)   $R(x) = 48x - 3x^2 \qquad C(x) = 6x + 120$

(1)   $\pi(x) = -3x^2 + 42x - 120$

(2)   $\qquad x = \dfrac{-(42)}{2(-3)} = 7 \qquad\qquad \pi = \dfrac{4(-3)(-120) - (42)^2}{4(-3)} = 27$

Vertex: $(7, 27)$ \qquad Maximum: $\pi(7) = 27$

(3)   Solving $\pi(x) = 0$ to find the $x$ intercepts,

$$-3(x^2 - 14x + 40) = 0$$

$$(x - 4)(x - 10) = 0$$

$$x = 4 \qquad x = 10$$

$x$ intercepts: $(4, 0), (10, 0)$

See Fig. 3-26.

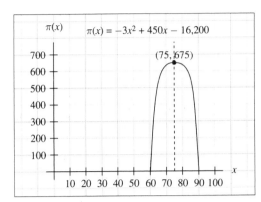

**Fig. 3-25**

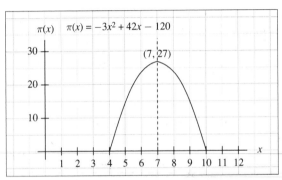

**Fig. 3-26**

**3.26.** The long-run average cost AC($x$) can be approximated by a quadratic function. Find the minimum long-run average cost (LRAC) by finding the vertex of AC($x$) and sketch the graph, given

$$AC(x) = x^2 - 120x + 4100$$

(1)    Here, with $a > 0$, the parabola opens up.

(2)          $x = \dfrac{-(-120)}{2(1)} = 60$          $AC = \dfrac{4(1)(4100) - (-120)^2}{4(1)} = 500$

Vertex: $(60, 500)$          Minimum: $AC(60) = 500$

(3)    With the parabola opening up from the vertex $(60, 500)$, the graph does not cross the $x$ axis. See Fig. 3-27. To see how similar problems can also be solved by the method of completing the square, see Dowling, *Schaum's Outline of Calculus for Business, Economics, and the Social Sciences*, Problems 3.33 to 3.35.

**3.27.** The cost of scrubbers to clean carbon monoxide from the exhaust of a blast furnace is estimated by the rational function

$$C(x) = \frac{150x}{110 - x} \qquad (0 \le x \le 100)$$

where $C$ is the cost in thousands of dollars of removing $x$ percent of the carbon monoxide. Graph the equation to show the sharply rising costs of cleaning up the final percentages of the toxic.

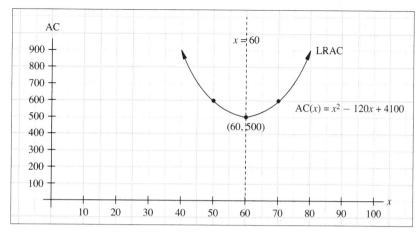

**Fig. 3-27**

| $x$ | $y$ |
|-----|------|
| 60  | 180  |
| 80  | 400  |
| 90  | 675  |
| 100 | 1500 |

Vertical asymptote:

$$110 - x = 0$$

$$x = 110$$

See Fig. 3-28.

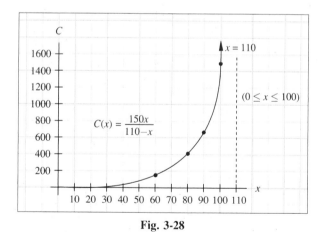

**Fig. 3-28**

**3.28.** Items such as automobiles are subject to *accelerated depreciation* whereby they lose more of their value faster than they do under linear depreciation. Assume a car worth \$10,000 with a lifetime of 10 years with no salvage value. (*a*) Using a solid line, draw a graph of the value $V(t)$ of a car under accelerated depreciation and (*b*) using a dotted line show the value of the same car under straight-line depreciation, given

(*a*)   $V(t) = 100t^2 - 2000t + 10{,}000$          (*b*)       $V(t) = 10{,}000 - 1000t$

(a)

| t | V |
|---|---|
| 0 | 10,000 |
| 2 | 6,400 |
| 5 | 2,500 |
| 8 | 400 |

See Fig. 3-29.

(b)

| t | V |
|---|---|
| 0 | 10,000 |
| 2 | 8,000 |
| 5 | 5,000 |
| 8 | 2,000 |

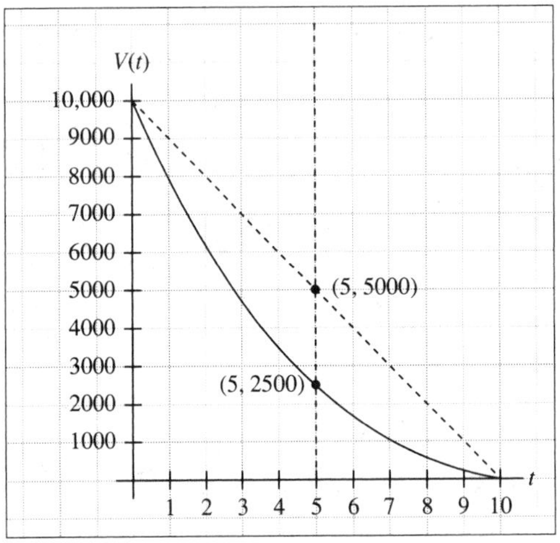

**Fig. 3-29**

Note how at $t = 5$, the value of the car is \$5000 under linear depreciation but only \$2500 under accelerated depreciation.

**3.29.** Draw a similar set of graphs for a car worth \$12,000 with a lifespan of 8 years, given (a) $V(t) = 187.5t^2 - 3000t + 12,000$ under accelerated depreciation and (b) $V(t) = 12,000 - 1500t$ under linear depreciation.

(a)

| t | V |
|---|---|
| 0 | 12,000 |
| 2 | 6,750 |
| 4 | 3,000 |
| 6 | 750 |

See Fig. 3-30.

(b)

| t | V |
|---|---|
| 0 | 12,000 |
| 2 | 9,000 |
| 4 | 6,000 |
| 6 | 3,000 |

## COMPOSITION OF FUNCTIONS IN BUSINESS AND ECONOMICS

**3.30.** A factory's cost $C(q)$ is a function of the number of units produced; its level of output $q(t)$ is a function of time. Express the factory's cost as a function of the time given each of the following circumstances:

(a) $C(q) = 1500 + 40q$ $\qquad$ $q(t) = 16t - \frac{1}{4}t^2$

Substituting $q(t)$ for each occurrence of $q$ in $C(q)$,

$$C[q(t)] = 1500 + 40(16t - \tfrac{1}{4}t^2)$$

$$= 1500 + 640t - 10t^2$$

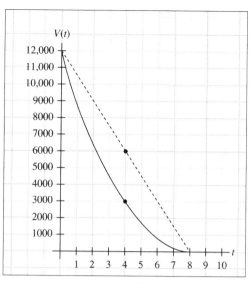

**Fig. 3-30**

(b)                     $C(q) = q^2 + 3q + 75$            $q(t) = 8(t - \frac{1}{4})$

$$q(t) = 8(t - \tfrac{1}{4}) = 8t - 2$$

Substituting for $q$ in $C(q)$,

$$C[q(t)] = (8t - 2)^2 + 3(8t - 2) + 75$$

$$= (64t^2 - 32t + 4) + (24t - 6) + 75$$

$$= 64t^2 - 8t + 73$$

**3.31.** Environmentalists have estimated that the average level of carbon monoxide in the air is $L(n) = (1 + 0.6n)$ parts per million (ppm) when the number of people is $n$ thousand. Assuming that the population in thousands at time $t$ is $n(t) = 400 + 30t + 0.15t^2$, (a) express the level of carbon monoxide in the air as a function of time and (b) estimate the level of carbon monoxide at $t = 5$.

(a)  Setting up the composite function $L[n(t)]$ by substituting $n(t)$ for each occurrence of $n$ in $L(n)$, we have

$$L[n(t)] = 1 + 0.6(400 + 30t + 0.15t^2)$$

$$= 241 + 18t + 0.09t^2$$

(b)                     $L[n(5)] = 241 + 18(5) + 0.09(5)^2 = 333.25$ ppm

**3.32.** The frog population $F$ measured in hundreds in a given region depends on the insect population $m$ in thousands: $F(m) = 65 + \sqrt{m/8}$. The insect population in turn varies with the amount of rainfall $r$ given in inches: $m(r) = 43r + 7.5$.

(a)  Express the frog population as a function of the rainfall and (b) estimate the frog population when the rainfall is 1.5 inches.

(a)   $F[m(r)] = 65 + \sqrt{\dfrac{43r + 7.5}{8}}$

(b)   $F[m(1.5)] = 65 + \sqrt{\dfrac{43(1.5) + 7.5}{8}} = 65 + \sqrt{9} = 68$ or 6800 frogs

# Supplementary Problems

## FUNCTIONS

**3.33.** Evaluate the following functions at the given numerical values of $x$:
   (a)  $f(x) = x^2 - 9x + 42$ at $x = 3$
   (b)  $g(x) = 2x^2 + 5x - 9$ at $x = -4$
   (c)  $h(x) = x^3 - 3x^2 + 6x - 7$ at $x = 2$
   (d)  $F(x) = (5x^2 - 4x + 7)/(3x - 7)$ at $x = 5$

**3.34.** Evaluate the following functions at the given parametric values of $x$:
   (a)  $f(x) = 7x^3 - 12x^2 - 38x + 115$ at $x = a$
   (b)  $g(x) = 8x^2 + 5x - 13$ at $x = a + 2$
   (c)  $h(x) = 4x^2 - 6x + 7$ at $x = b - 4$
   (d)  $G(x) = (12x^2 - 9x + 15)/(4x - 3)$ at $x = b + 5$

## ALGEBRA OF FUNCTIONS

**3.35.** Given $f(x) = 7x - 2$ and $g(x) = 3x + 8$, find (a) $(f + g)(x)$ and (b) $(f \cdot g)(x)$.

**3.36.** Given $g(x) = 4x - 9$ and $h(x) = 12 - 5x$, find (a) $(g - h)(x)$ and (b) $(g \div h)(x)$.

**3.37.** Given $F(x) = 3x^2 - 7x + 8$ and $G(x) = 9x - 4$, find (a) $(F + G)(x)$ and (b) $(F \cdot G)(x)$.

**3.38.** Given $f(x) = 30x^2 - x - 99$ and $h(x) = 5x + 9$, find (a) $(f - h)(x)$ and (b) $(f \div h)(x)$.

**3.39.** Given $G(x) = 6/(x + 3)$ and $H(x) = 11/x^2$, find (a) $(G + H)(x)$ and (b) $(G \cdot H)(x)$.

**3.40.** Given $f(x) = (x - 7)/(x + 2)$ and $g(x) = (x + 3)/(x - 8)$, find (a) $(f - g)(x)$ and (b) $(f \div g)(x)$.

**3.41.** Given $f(x) = x^3 - 3x + 4$ and $g(x) = 5x^2$, find the composite functions (a) $f[g(x)]$ and (b) $g[f(x)]$.

**3.42.** Given $F(x) = (9x - 2)/4x$ and $G(x) = x^5$, find (a) $F[G(x)]$ and (b) $G[F(x)]$, also called *functions of functions*.

## LINEAR FUNCTIONS IN BUSINESS AND ECONOMICS

**3.43.** A firm has a fixed cost of $125,000 and variable costs per item manufactured of $685. Express the firm's total cost **TC** as a function of output $x$.

**3.44.** A new car bought today for $13,500 depreciates by $2250 at the end of each calendar year. Express the value $V$ of the car as a function of years $t$.

**3.45.** A company in a purely competitive market receives $95 in revenue for each item sold. If the company has fixed costs of $8800 and a marginal cost of $67.50 per item, express the company's profit $\pi$ as a function of the number of items $x$ sold.

## SOLVING QUADRATIC EQUATIONS

**3.46.** Solve the following quadratic equations by factoring:
   (a)  $x^2 - 11x + 28$          (b)  $x^2 + 5x - 24$              (c)  $x^2 + 11x + 18$
   (d)  $x^2 - 8x - 48$

**3.47.**     Use the quadratic formula to solve each of the following quadratic equations:

(a)   $6x^2 + 31x + 40$                    (b)   $8x^2 - 50x + 33$                         (c)   $4x^2 - 31x - 45$

(d)   $9x^2 - 67.5x + 126$

## GRAPHING NONLINEAR FUNCTIONS

**3.48.**     For each of the following quadratic functions, indicate the shape of the parabola by determining (1) whether the parabola opens up or down, (2) the coordinates of the vertex, and (3) the $x$ intercepts, if any:

(a)   $y = -x^2 + 12x - 27$                              (b)   $y = x^2 - 8x - 48$

(c)   $y = -x^2 - 16x - 28$                              (d)   $y = x^2 - 5x + 30$

**3.49.**     Indicate the general shape of the following rational functions by finding (1) the vertical asymptote VA, (2) the horizontal asymptote HA, and (3) a small number of representative points on the graph:

(a)   $y = \dfrac{6}{x - 8}$            (b)   $y = \dfrac{-7}{x + 3}$            (c)   $y = \dfrac{x + 9}{x - 6}$            (d)   $y = \dfrac{4 - x}{5 + x}$

## NONLINEAR BUSINESS AND ECONOMICS FUNCTIONS

**3.50.**     Determine the total revenue **TR** function for each of the firms confronted with the following linear demand functions:

(a)   $P = -17.5Q + 2675$                              (b)   $P = -9.8Q + 860$

**3.51.**     For each set of the following total revenue and total cost functions, first express profit $\pi$ as a function of output $x$ and then determine the maximum level of profit by finding the vertex of the parabola:

(a)   TR $= -6x^2 + 1200x$, TC $= 180x + 3350$

(b)   TR $= -4x^2 + 900x$, TC $= 140x + 6100$

(c)   TR $= -7x^2 + 4100x$, TC $= 600x + 12{,}500$

(d)   TR $= -9x^2 + 6600x$, TC $= 660x + 19{,}600$

# Answers to Supplementary Problems

**3.33.**     (a)   $f(3) = 24$          (b)   $g(-4) = 3$          (c)   $h(2) = 1$          (d)   $F(5) = 14$

**3.34.**     (a)   $f(a) = 7a^3 - 12a^2 - 38a + 115$

(b)   $g(a + 2) = 8a^2 + 37a + 29,$          (c) $h(b - 4) = 4b^2 - 38b + 95$

(d)   $G(b + 5) = \dfrac{12b^2 + 111b + 270}{4b + 17}$

**3.35.**     (a)   $(f + g)(x) = 10x + 6$          (b)   $(f \cdot g)(x) = 21x^2 + 50x - 16$

**3.36.**     (a)   $(g - h)(x) = 9x - 21$          (b)   $(g \div h)(x) = \dfrac{4x - 9}{12 - 5x}$

**3.37.**     (a)   $(F + G)(x) = 3x^2 + 2x + 4$          (b)   $(F \cdot G)(x) = 27x^3 - 75x^2 + 100x - 32$

**3.38.**     (a)   $(f - h)(x) = 30x^2 - 6x - 108$          (b)   $(f \div h)(x) = 6x - 11$

**3.39.**     (a)   $(G + H)(x) = \dfrac{6x^2 + 11x + 33}{x^3 + 3x^2}$          (b)   $(G \cdot H)(x) = \dfrac{66}{x^3 + 3x^2}$

**3.40.**     (a)   $(f - g)(x) = \dfrac{-20x + 50}{x^2 - 6x - 16}$          (b)   $(f \div g)(x) = \dfrac{x^2 - 15x + 56}{x^2 + 5x + 6}$

**3.41.**    (a)   $f[g(x)] = 125x^6 - 15x^2 + 4$
            (b)   $g[f(x)] = 5(x^3 - 3x + 4)^2 = 5x^6 - 30x^4 + 40x^3 + 45x^2 - 120x + 80$

**3.42.**    (a)   $F[G(x)] = \dfrac{9x^5 - 2}{4x^5}$          (b)   $G[F(x)] = \left(\dfrac{9x - 2}{4x}\right)^5$

**3.43.**    $TC = 685x + 125,000$

**3.44.**    $V = -2250t + 13,500$

**3.45.**    $\pi = 27.50x - 8800$

**3.46.**    (a)   $x = 4, 7$        (b)   $x = 3, -8$        (c)   $x = -2, -9$        (d)   $x = -4, 12$

**3.47.**    (a)   $x = -2.5, -2.67$        (b)   $x = 0.75, 5.5$        (c)   $x = -1.25, 9$        (d)   $x = 3.5, 4$

**3.48.**    (a)   Opens down;  vertex: (6, 9);  $x$ intercepts:  (3, 0), (9, 0).        (b)   Opens up;  vertex: (4, −64);
            $x$ intercepts: (−4, 0), (12, 0).        (c)   Opens down;  vertex: (−8, 36);  $x$ intercepts: (−2, 0), (−14, 0).
            (d)   Opens up;  vertex: (2.5, 23.75);  no $x$ intercepts.

**3.49.**    (a)   VA: $x = 8$                                      (b)   VA: $x = -3$
            HA: $y = 0$                                             HA: $y = 0$

| $x$ | $y$ |
|-----|-----|
| 2   | −1  |
| 7   | −6  |
| - - - | - - - |
| 9   | 6   |
| 14  | 1   |

| $x$ | $y$ |
|-----|-----|
| −10 | 1   |
| −4  | 7   |
| - - - | - - - |
| −2  | −7  |
| 4   | −1  |

(c)   VA: $x = 6$                                      (d)   VA: $x = -5$
            HA: $y = 1$                                             HA: $y = -1$

| $x$ | $y$ |
|-----|-----|
| 3   | −4  |
| 5   | −14 |
| - - - | - - - |
| 7   | 16  |
| 9   | 6   |

| $x$ | $y$ |
|-----|-----|
| −10 | −2.8 |
| −6  | −10 |
| - - - | - - - - |
| −4  | 8   |
| 0   | 0.8 |

**3.50.**    (a)   $TR = -17.5Q^2 + 2675Q$          (b)   $TR = -9.8Q^2 + 860Q$

**3.51.**    (a)   $\pi = -6x^2 + 1020x - 3350$. Vertex: (85, 40,000). Maximum: $\pi(85) = 40,000$.
            (b)   $\pi = -4x^2 + 760x - 6100$. Vertex: (95, 30,000). Maximum: $\pi(95) = 30,000$.
            (c)   $\pi = -7x^2 + 3500x - 12,500$. Vertex: (250, 425,000). Maximum: $\pi(250) = 425,000$.
            (d)   $\pi = -9x^2 + 5940x - 19,600$. Vertex: (330, 960,500). Maximum: $\pi(330) = 960,500$.

# Chapter 4

# Systems of Equations

## 4.1 INTRODUCTION

The focus of presentation has thus far concentrated on the analysis of individual equations. Many mathematical models, however, involve more than one equation. The equations of such models are said to form *a system of equations*. The solution to a system of equations is a set of values which concurrently satisfy all the equations of the system. A system of equations consisting of $n$ equations and $v$ variables is described as an $n \times v$ *system* or a system with *dimensions* $n \times v$, read "$n$ by $v$."

For a *unique solution* to a system of equations to exist, there must be at least as many equations as variables. A system with an equal number of equations and variables ($n = v$) is an *exactly constrained system*. An *underconstrained system* has fewer equations than variables ($n < v$); an *overconstrained system* has more equations than variables ($n > v$). An underconstrained system will have an unlimited number of solutions or no solution, never a unique solution. An exactly constrained or overconstrained system may have a unique solution, an infinite number of solutions, or no solution. This chapter deals with different methods of solving systems of equations and the many varied applications of such systems to business and economics.

**EXAMPLE 1.**    The dimensions of the following systems of equations are specified below.

$(a)$    $y = 5x + 7$       $(b)$    $y = 3x_1 + 7x_2 - 4x_3$       $(c)$    $y = 6x - 12$

$\quad\quad y = 8x - 3$       $\quad\quad\quad y = 9x_1 - 5x_2 + 2x_3$       $\quad\quad\quad y = 9x + 15$

$\quad\quad\quad\quad\quad\quad\quad\quad\quad\quad\quad\quad\quad\quad\quad\quad\quad\quad\quad\quad\quad\quad y = 7x - 11$

System $(a)$ has two equations and two variables: $x$ and $y$. It is a $(2 \times 2)$ system, and, therefore, exactly constrained. System $(b)$ has two equations and four variables: $x_1$, $x_2$, $x_3$, and $y$. It is a $(2 \times 4)$ system which is underconstrained. System $(c)$ has three equations and two variables: $x$ and $y$. It is an overconstrained system with dimensions $(3 \times 2)$.

## 4.2 GRAPHICAL SOLUTIONS

The graph of a $(2 \times 2)$ linear system of equations consists of two straight lines. If the two lines intersect at some point $(x_1, y_1)$, then the point $(x_1, y_1)$ satisfies both equations and the coordinates of the point represent a unique solution to the system. If the two lines do not intersect, there is no point in common that satisfies both equations and hence no solution exists. The two equations are *inconsistent*. Finally, if the two equations have identical graphs, they have an unlimited number of points in common and the system has an infinite number of solutions. Such equations are called *equivalent* or *dependent equations*. The conditions for the three different solution possibilities are summarized below and illustrated in Example 2.

For a $(2 \times 2)$ system of linear equations in the slope-intercept form presented in equation $(2.2)$:

$$y = m_1 x + b_1$$

$$y = m_2 x + b_2$$

1. If $m_1 \neq m_2$, there is a unique solution to the system.
2. If $m_1 = m_2$, and $b_1 \neq b_2$, the equations are inconsistent and there is no solution to the system.
3. If $m_1 = m_2$, and $b_1 = b_2$, the equations are equivalent and there is an infinite number of solutions to the system.

**EXAMPLE 2.** The different systems of equations given below are solved by graphing as follows:

(a) $x - 2y = -2$  (b) $10x + 2y = 30$  (c) $21x - 7y = 14$

$6x + 3y = 33$  $15x + 3y = 75$  $-15x + 5y = -10$

Converting the equations to the slope-intercept form, graphing in Fig. 4-1, then following up with observations,

(a) $y = \frac{1}{2}x + 1$  (b) $y = -5x + 15$  (c) $y = 3x - 2$

$y = -2x + 11$  $y = -5x + 25$  $y = 3x - 2$

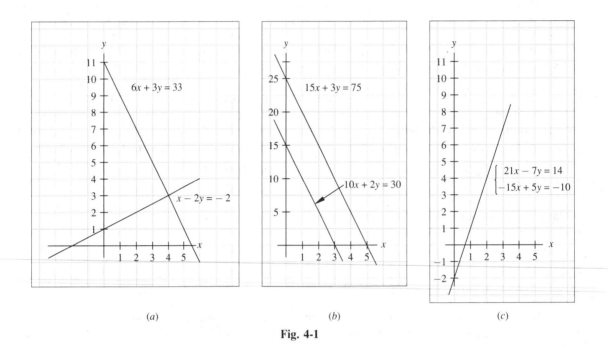

Fig. 4-1

(a) The solution is $x = 4$, $y = 3$. It is a unique solution because the system is exactly determined ($n = v$) and the equations are consistent and independent ($m_1 \neq m_2$).

(b) There is no unique solution. The system is exactly determined ($n = v$), but the equations are inconsistent ($m_1 = m_2$ but $b_1 \neq b_2$).

(c) There is an infinite number of solutions satisfying both equations simultaneously. The system is exactly determined but the equations are dependent or equivalent ($m_1 = m_2$ and $b_1 = b_2$).

## 4.3 SUPPLY-AND-DEMAND ANALYSIS

Supply-and-demand analysis in introductory economics courses is commonly expressed in terms of graphs. By longstanding custom, price $P$ is treated as the dependent variable and graphed on the $y$ axis. Quantity $Q$ is regarded as the independent variable and is placed on the $x$ axis. The point of intersection of the supply curve $S$ and the demand curve $D$ represents *equilibrium* in the model. Equilibrium occurs at a single price $P_e$, called the *equilibrium price*, at which quantity supplied $Q_s$ equals quantity demanded $Q_d$. The point at which $Q_s = Q_d$ is called the *equilibrium quantity* $Q_e$. An illustration is provided in Example 3.

Supply-and-demand analysis can also be handled with equations. The object is to find $P_e$ and $Q_e$. Since $P_e$ occurs when $Q_s = Q_d = Q_e$, one need only equate the supply and demand equations in whatever form they are presented. A demonstration of the method is offered in Example 4 and Problems 4.4 and 4.5.

**EXAMPLE 3.**    The graphical solution to the following system of supply and demand equations is presented in Fig. 4-2, followed by observations:

$$\text{Supply: } P = 1 + \tfrac{1}{2}Q$$

$$\text{Demand: } P = 10 - \tfrac{5}{8}Q$$

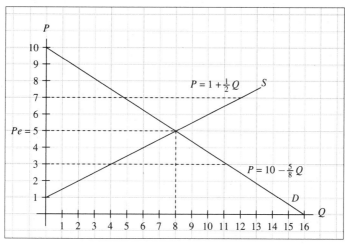

**Fig. 4-2**

The solution to the system of supply and demand equations occurs at $P = 5 = P_e$ and $Q = 8 = Q_e$. Notice that:

1)   At $P_e = 5$, $Q_s = Q_d = 8 = Q_e$; and at $Q_e = 8$, $P_s = P_d = 5 = P_e$.

2)   At $P > 5$, $Q_s > Q_d$. At $P = 7$, for example, $Q_s = 12$, $Q_d = 4.8$. When $Q_s > Q_d$, it is called a *surplus*. In a purely competitive market, a surplus tends to drive the price down toward $P_e$.

3)   At $P < 5$, $Q_d > Q_s$. At $P = 3$, for instance, $Q_d = 11.2$, $Q_s = 4$. When $Q_d > Q_s$, it is called a *shortage*. In a free market, a shortage drives prices upward toward $P_e$.

4)   Since a purely competitive market tends toward $P_e$, as explained in items 3 and 4 above, it is said to be *self-equilibrating*.

**EXAMPLE 4.**    Sometimes it is more convenient to solve a system of supply and demand equations algebraically. Consider a different set of equations in which $Q$ is now expressed as the dependent variable, as typically happens in mathematical economics.

$$\text{Supply: } Q = -50 + 6P$$

$$\text{Demand: } Q = 230 - 8P$$

Since we seek the price that will bring equilibrium to the market, and equilibrium occurs when $Q_s = Q_d$, we set the equations equal to each other and solve for $P$.

$$Q_s = Q_d$$

$$-50 + 6P = 230 - 8P$$

$$14P = 280$$

$$P = 20 = P_e$$

We then find $Q_e$ by substituting $P_e = 20$ in either (a) the supply or (b) the demand equation.

(a)   $Q_s = -50 + 6(20)$                                 (b)   $Q_d = 230 - 8(20)$

$\quad\quad Q_s = -50 + 120 = 70 = Q_e$                     $\quad\quad Q_d = 230 - 160 = 70 = Q_e$

## 4.4  BREAK-EVEN ANALYSIS

In planning a new business, it is important to know the minimum volume of sales needed to turn a profit. Estimating the point at which a firm's balance sheet moves from a loss to a profit is called *break-even analysis*. The break-even point can be defined as the level of sales at which (1) profit equals zero or (2) total revenue equals total cost. The break-even point is found both algebraically and graphically for linear functions in Example 5 and for a quadratic function in Example 6.

**EXAMPLE 5.**    Given total revenue $R(x) = 80x$, total cost $C(x) = 30x + 2000$, then profit $\pi(x) = R(x) - C(x) = 50x - 2000$, and the break-even point can be found in any of the three following ways:

(a)   Setting $\pi = 0$ and solving for $x$,

$$50x - 2000 = 0$$

$$x = 40 \quad\quad \text{Break-even level of output}$$

(b)   Setting $R(x) = C(x)$ and solving for $x$,

$$80x = 30x + 2000$$

$$50x = 2000$$

$$x = 40 \quad\quad \text{Break-even level of output}$$

(c)   Graphing $R(x)$ and $C(x)$ and finding the point of intersection, as in Fig. 4-3.

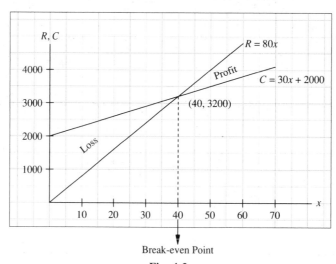

**Fig. 4-3**

Note that for $x < 40$, $C(x) > R(x)$, which is a loss; for $x > 40$, $R(x) > C(x)$ and there is a profit. See also Problem 4.6.

**EXAMPLE 6.**    Assume $R(x) = -3x^2 + 48x$ and $C(x) = 6x + 120$, then $\pi(x) = R(x) - C(x) = -3x^2 + 42x - 120$, and the break-even point can be found as in Example 5 in any of the following ways.

(a)  Setting $\pi(x) = 0$, dividing by $-3$, and factoring to find $x$,

$$-3x^2 + 42x - 120 = 0$$

$$x^2 - 14x + 40 = 0$$

$$(x - 4)(x - 10) = 0$$

$$x = 4, \qquad x = 10 \qquad \text{Break-even levels of output}$$

(b)  Setting $R(x) = C(x)$ and simplifying,

$$-3x^2 + 48x = 6x + 120$$

$$-3x^2 + 42x - 120 = 0$$

then factoring, as in (a), to find $x$,

$$x = 4, \qquad x = 10 \qquad \text{Break-even levels of output}$$

(c)  Graphing $R(x)$ and $C(x)$ and finding the points of intersection. Using the techniques of Section 3.6 to graph the quadratic function $R(x) = -3x^2 + 48x$,

1) With $a < 0$, the parabola opens down.

2)     
$$x = \frac{-b}{2a} \qquad\qquad y = \frac{4ac - b^2}{4a}$$

$$x = \frac{-(48)}{2(-3)} = 8 \qquad\qquad y = \frac{4(-3)(0) - (48)^2}{4(-3)} = \frac{-2304}{-12} = 192$$

Coordinates of the vertex: $(8, 192)$

3) Setting $R(x) = 0$ and factoring to find the $x$ intercepts,

$$-3x^2 + 48x = 0$$

$$-3x(x - 16) = 0$$

$$-3x = 0 \qquad\qquad x - 16 = 0$$

$$x = 0 \qquad\qquad x = 16$$

The $x$ intercepts are $(0, 0)$ and $(16, 0)$. See Fig. 4-4. From the points of intersection in Fig. 4-4, the break-even points are $(4, 144)$ and $(10, 180)$. Note that for $x < 4$, $C(x) > R(x)$ and there is a loss. For $4 < x < 10$, $R(x) > C(x)$ and there is a profit; and for $x > 10$, there is again a loss. See also Problem 4.7.

## 4.5  ELIMINATION AND SUBSTITUTION METHODS

For a unique solution to a system of equations to exist, (1) the equations must be consistent (noncontradictory), (2) they must be independent (not transformations such as multiples, squares, or square roots of each other), and (3) there must be as many equations as variables (an exactly constrained system). Two popular means of solving such systems are (a) the elimination method and (b) the substitution method.

The *elimination method* is so named because the procedure calls for one variable at a time to be removed or eliminated until a single variable is left set equal to a constant. To use the elimination method to solve a $2 \times 2$ system in $x$ and $y$, for instance, select one variable (say, $x$) to be removed from both equations and determine a common denominator or least common denominator (LCD). Multiplying the first equation by the coefficient of $x$ in the second equation, then multiplying the second equation by the coefficient of $x$ in the first equation, will always provide a common denominator, although not necessarily a LCD. The coefficients

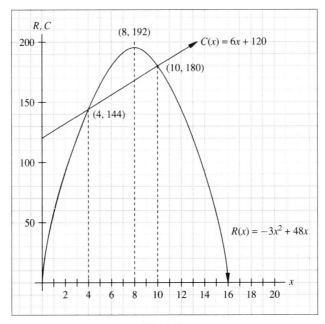

**Fig. 4-4**

of $x$ will then be identical in absolute value, and one need only add or subtract the two equations to eliminate $x$ and find the solution for $y$. See Example 7 (*a*). For solution of a $(3 \times 3)$ system, see Problem 4.14.

The *substitution method* in a $2 \times 2$ system in $x$ and $y$ involves solving one of the equations for one of the variables in terms of the other (say, $y$ in terms of $x$), then substituting the newly found value of that variable $y$ into the other equation and solving for $x$. See Example 7 (*b*).

**EXAMPLE 7.**    The system of simultaneous equations

$$2x + 5y = 52 \tag{4.1}$$

$$3x - 4y = -14 \tag{4.2}$$

is solved below by the two different methods.

(*a*)  Elimination method:

1)  Selecting $x$ to be eliminated and multiplying (*4.1*) by 3 and (*4.2*) by 2,

$$3(2x + 5y = 52) = 6x + 15y = 156 \tag{4.3}$$

$$2(3x - 4y = -14) = 6x - 8y = -28 \tag{4.4}$$

2)  Subtracting (*4.4*) from (*4.3*) by changing the signs of all the terms in (*4.4*) and then adding it to (*4.3*) to eliminate $x$,

$$23y = 184$$

$$y = 8$$

3)  Then substituting $y = 8$ in (*4.1*) or (*4.2*) to solve for $x$,

$$2x + 5(8) = 52$$

$$x = 6$$

(b)   Substitution method:

   1)   Solving one of the original equations for $x$ or $y$, here solving (4.1) for $x$,

$$2x = 52 - 5y$$

$$x = 26 - 2.5y \qquad\qquad (4.5)$$

   2)   Substituting the value of $x$ in (4.5) into the other original equation (4.2),

$$3(26 - 2.5y) - 4y = -14$$

$$78 - 7.5y - 4y = -14$$

$$-11.5y = -92$$

$$y = 8$$

   3)   Then substituting $y = 8$ in (4.1) or (4.2), here (4.2),

$$3x - 4(8) = -14$$

$$x = 6$$

See also Problems 4.8 to 4.13. For solution of a $(3 \times 3)$ system, see Problem 4.14. Still other methods of solution can be found in the chapters on linear algebra. See Section 5.10 (Gaussian elimination method), Section 6.3 (Cramer's rule), and Section 6.6 (matrix inversion).

## 4.6   INCOME DETERMINATION MODELS

Income determination models generally express the equilibrium level of income in a four-sector economy as

$$Y = C + I + G + (X - Z)$$

where $Y$ = income, $C$ = consumption, $I$ = investment, $G$ = government expenditures, $X$ = exports, and $Z$ = imports. By substituting the information supplied in a model, it is an easy matter to solve for the equilibrium level of income. *Aggregation* (summation) of the variables on the right-hand side of the equation also allows the equation to be graphed in two-dimensional space. See Example 8 and Problems 4.15 to 4.20.

**EXAMPLE 8.**    Assume a simple two-sector economic model in which $Y = C + I$, $C = C_0 + bY$, and $I = I_0$. Assume further that $C_0 = 95$, $b = 0.8$, and $I_0 = 75$. The equilibrium level of income can be calculated in terms of (a) the general parameters and (b) the specific values assigned to these parameters.

   (a)   The *equilibrium equation* is

$$Y = C + I$$

Substituting for $C$ and $I$,

$$Y = C_0 + bY + I_0$$

Solving for $Y$,

$$Y - bY = C_0 + I_0$$

$$(1 - b)Y = C_0 + I_0$$

$$Y = \frac{1}{1 - b}(C_0 + I_0) = Y_e$$

The solution in this form is called the reduced form. The *reduced-form equation* expresses the *endogenous* or dependent variable (here $Y$) as an explicit function of the *exogenous* or independent variables ($C_0$, $I_0$) and the parameters ($b$). An *explicit function* is one in which the endogenous variable is alone on the left-hand side of the equation and all the exogenous variables and parameters are on the right.

(*b*) The specific equilibrium level of income can be calculated by substituting the numerical values for the parameters in either (1) the original equation or (2) the reduced form equation:

(1) $\qquad Y = C_0 + bY + I_0$

$\qquad\qquad Y = 95 + 0.8Y + 75$

$\qquad\quad Y - 0.8Y = 170$

$\qquad\qquad 0.2Y = 170$

$\qquad\qquad\quad Y_e = 850$

(2) $\quad Y = \dfrac{1}{1-b}(C_0 + I_0)$

$\qquad Y = \dfrac{1}{1-0.8}(95 + 75)$

$\qquad Y = \dfrac{1}{0.2}(170)$

$\qquad Y = 5(170)$

$\qquad Y_e = 850$

The term $1/(1-b)$ is called the *autonomous-expenditure multiplier* in economics. It measures the multiple effect each dollar of autonomous spending has on the equilibrium level of income. Since $b = $ the marginal propensity to consume (MPC) in the income determination model, the multiplier is $1/(1 - \text{MPC})$. See Problems 4.15 to 4.20.

## 4.7  IS-LM ANALYSIS

The *IS schedule* is a locus of points representing all the different combinations of interest rates and income levels consistent with equilibrium in the commodity or goods market. The *LM schedule* is a locus of points representing all the different combinations of interest rates and income levels consistent with equilibrium in the money market. *IS-LM analysis* seeks to find the level of income and the rate of interest at which both the commodity market and the money market will be in equilibrium. This can be accomplished with the techniques used for solving a system of simultaneous equations. Unlike the simple income determination model in Section 4.6, IS-LM analysis deals explicitly with the interest rate and incorporates its effect into the model. See Example 9 and Problems 4.21 to 4.24.

**EXAMPLE 9.**    The commodity market for a simple two-sector economy is in equilibrium when $Y = C + I$. The money market is in equilibrium when the supply of money $M_s$ equals the demand for money $M_d$, which in turn is composed of the *transaction-precautionary demand* for money $M_t$ and the *speculative demand* for money $M_w$. Assume a two-sector economy where $C = 34 + 0.9Y$, $I = 41 - 80i$, $M_s = 325$, $M_t = 0.4Y$, and $M_w = 65 - 124i$.

Commodity-market equilibrium (IS) exists when $Y = C + I$. Substituting the given values into the equation,

$$Y = 34 + 0.9Y + 41 - 80i$$

$$Y - 0.9Y = 75 - 80i$$

$$0.1Y + 80i - 75 = 0 \qquad\qquad\qquad\qquad (4.6)$$

Monetary equilibrium (LM) exists when $M_s = M_t + M_w$. Substituting the specific values into the equation,

$$325 = 0.4Y + 65 - 124i$$

$$0.4Y - 124i - 260 = 0 \qquad\qquad\qquad\qquad (4.7)$$

A condition of simultaneous equilibrium in both markets can now be found by solving (*4.6*) and (*4.7*) simultaneously:

$$0.1Y + 80i - 75 = 0 \qquad\qquad\qquad\qquad (4.6)$$

$$0.4Y - 124i - 260 = 0 \qquad\qquad\qquad\qquad (4.7)$$

Using the elimination method from Section 4.5 and simplifying the procedure by multiplying (4.6) by $-4$ to obtain (4.8) and then simply adding to eliminate $Y$ and solve for $i$, we obtain

$$-0.4Y - 320i + 300 = 0 \qquad (4.8)$$
$$\underline{0.4Y - 124i - 260 = 0}$$
$$-444i + 40 = 0$$

$$i = \frac{-40}{-444} \approx 0.09 = 9\% = i_e$$

Then substituting $i = 0.09$ in (4.6) or (4.7) to find $Y$,

$$0.1Y + 80(0.09) - 75 = 0$$

$$0.1Y = 67.8$$

$$Y_e = 678$$

The commodity and money markets will be in simultaneous equilibrium when $Y = 678$ and $i = 0.09$. At that point, $C = 34 + 0.9(678) = 644.2$, $I = 41 - 80(0.09) = 33.8$, $M_t = 0.4(678) = 271.2$, and $M_w = 65 - 124(0.09) = 53.8$. Checking the answer, $Y = C + I = 644.2 + 33.8 = 678$, and $M_d = M_t + M_w = 271.2 + 53.8 = 325 = M_s$.

## 4.8  ECONOMIC AND MATHEMATICAL MODELING (OPTIONAL)

Economic modeling and mathematical modeling differ in a way that is at times perplexing. Assume the following system of linear supply and demand equations presented in standard form:

$$\text{Supply: } 6P - 3Q = 36 \qquad (4.9a)$$

$$\text{Demand: } 8P + 2Q = 192 \qquad (4.9b)$$

In economics, where price is treated as the dependent variable in the tradition of the great English economist Alfred Marshall, the slope-intercept forms of the equations are found by solving each for $P$ in terms of $Q$, in effect making $P$ a function of $Q$:

$$\text{Supply: } P = \tfrac{1}{2}Q + 6 \qquad (4.10a)$$

$$\text{Demand: } P = -\tfrac{1}{4}Q + 24 \qquad (4.10b)$$

In mathematical models, where $Q$ is regarded as the dependent variable, as originally taught by the famous French economist Leon Walras, the slope-intercept forms of the equations are found by solving for $Q$ in terms of $P$, making $Q$ a function of $P$:

$$\text{Supply: } Q = 2P - 12 \qquad (4.11a)$$

$$\text{Demand: } Q = -4P + 96 \qquad (4.11b)$$

When graphed, the different dependent variables go on the $y$ axis. While confusing at first, this merely reverses the position of the graphs and does not affect the solutions. See Fig. 4-5.

Today, while most economists think like Walras and agree that price determines quantity demanded and quantity supplied, they continue to graph like Marshall and place $Q$ on the horizontal axis and $P$ on the vertical axis, which implies that quantity supplied and demanded determines price.

## 4.9  IMPLICIT FUNCTIONS AND INVERSE FUNCTIONS (OPTIONAL)

Consider an alternate form of the supply equation in (4.9a):

$$6P - 3Q - 36 = 0 \qquad (4.12)$$

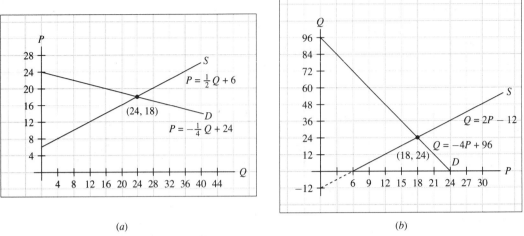

Fig. 4-5

The equation implies a relationship between $P$ and $Q$ but does not indicate which variable depends on which. Rewritten slightly as

$$f(P, Q) = 6P - 3Q - 36 = 0 \qquad (4.13)$$

it is termed an *implicit function* because with both variables to one side of the equality sign, it suggests possible dependence between the variables but does not specify the nature of the dependence. If an implicit function can be solved algebraically for one variable in terms of the other, as in (4.10a) or (4.11a), it can be rewritten as an explicit function:

$$P = g(Q) = \tfrac{1}{2}Q + 6 \qquad (4.14)$$

or

$$Q = h(P) = 2P - 12 \qquad (4.15)$$

An *explicit function* has the dependent variable on the left of the equality sign and the independent variable on the right and thus clearly demonstrates the order and nature of dependence.

Whenever an implicit function, such as (4.13), can be converted into two distinct explicit functions, such as (4.14) and (4.15), the explicit functions are called *inverse functions of each other*. Inverse functions are important, in part because of the different ways economists and mathematicians deal with supply-and-demand analysis. While most functions encountered in elementary economics have inverse functions, not all functions do. Put formally, given a function $y = f(x)$, an *inverse function*, written $x = f^{-1}(y)$, exists if and only if each value of $y$ yields one and only one value of $x$. From a practical point of view, however, all one need do to find the inverse of an ordinary economics function is to solve the given equation algebraically for the independent variable in terms of the dependent variable, as is illustrated in Example 10.

**EXAMPLE 10.**    The functions (4.14) and (4.15) are inverse functions of each other:

$$g(Q) = h^{-1}(Q) \qquad (4.16)$$

$$h(P) = g^{-1}(P) \qquad (4.17)$$

The inverse relationship in (4.16) can be tested by solving (4.15) algebraically for $P$ in terms of $Q$:

$$h(P): \qquad Q = 2P - 12$$

$$-2P = -Q - 12$$

$$P = \tfrac{1}{2}Q + 6 = h^{-1}(Q) = g(Q)$$

Similarly, the inverse relationship in (*4.17*) can be tested by solving (*4.14*) algebraically for $Q$ in terms of $P$:

$$g(Q): \quad P = \tfrac{1}{2}Q + 6$$

$$-\tfrac{1}{2}Q = -P + 6$$

$$Q = 2P - 12 = g^{-1}(P) = h(P)$$

# Solved Problems

## SYSTEMS OF EQUATIONS

**4.1.** Give the dimensions of each of the following systems of equations and indicate whether the system is exactly constrained, underconstrained, or overconstrained.

    (*a*)   $y = 4x_1 - 7x_2 - 6x_3$     (*b*)   $5x + 2y - 8z = 42$     (*c*)   $2x + 5y = 9$

           $y = 9x_1 + 8x_2 - 4x_3$           $9x - 7y + 4z = 37$            $7x - 4y = 2$

           $y = 7x_1 - 3x_2 + 2x_3$           $6x - 3y - 5z = -9$            $3x + 8y = 7$

    (*a*)   With three equations (*n*) and four variables (*v*)— $x_1$, $x_2$, $x_3$, and $y$— it is a $(3 \times 4)$ system. Since $n < v$, it is an underconstrained system.

    (*b*)   Since there are three equations and three variables, it is a $(3 \times 3)$ system. With $n = v$, it is exactly constrained.

    (*c*)   With $n = 3$ and $v = 2$, it is a $(3 \times 2)$ system. Because $n > v$, it is overconstrained.

**4.2.** Indicate what types of solutions are possible for the different systems of equations in Problem 4.1.

    (*a*)   An underconstrained system may have an unlimited number of possible solutions or no solution, but it will not have a unique solution for lack of sufficient equations as constraints.

    (*b*)   An exactly constrained system may have a unique solution, multiple solutions, or no solution. It will have multiple solutions if two or more of the equations are equivalent or dependent equations. It will have no solution if two or more of the equations are inconsistent.

    (*c*)   An overconstrained system may have a unique solution, no solution, or multiple solutions. It typically has no solution; if the equations are not dependent, they must be inconsistent.

## GRAPHICAL SOLUTIONS

**4.3.** Solve each of the following systems of equations graphically by (1) finding the slope-intercept forms and (2) graphing:

    (*a*)   $12x + 3y = 21$          $-2x + 4y = -8$

         (1)   Finding the slope-intercept forms by solving each equation for $y$ in terms of $x$,

$$12x + 3y = 21 \qquad -2x + 4y = -8$$

$$y = -4x + 7 \qquad y = \tfrac{1}{2}x - 2$$

         (2)   See Fig. 4-6.

    As seen by the intersection in Fig. 4-6, the solution is $x = 2$, $y = -1$.

    (*b*)   $\tfrac{1}{2}x + 2y = 6$          $-1\tfrac{1}{2}x + \tfrac{1}{2}y = -5$

         (1)   $y = -\tfrac{1}{4}x + 3$        $y = 3x - 10$

         (2)   See Fig. 4-7.

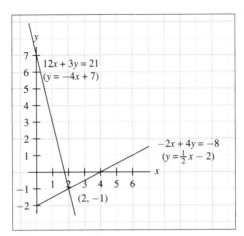

**Fig. 4-6**

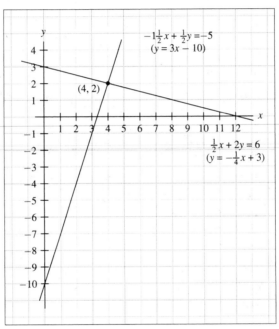

**Fig. 4-7**

As seen in Fig. 4-7, the solution is $x = 4$, $y = 2$.

    (c)   $-6x + 2y = -16$        $x + 5y = 40$

        (1)   $y = 3x - 8$        $y = -\frac{1}{5}x + 8$

        (2)   See Fig. 4-8.

    From Fig. 4-8, the solution is $x = 5$, $y = 7$.

    (d)   $-8x + 4y = -12$        $2x + 6y = 24$

        (1)   $y = 2x - 3$        $y = -\frac{1}{3}x + 4$

        (2)   See Fig. 4-9.

    As seen in Fig. 4-9, the solution is $x = 3$, $y = 3$.

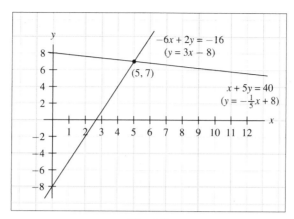

**Fig. 4-8**

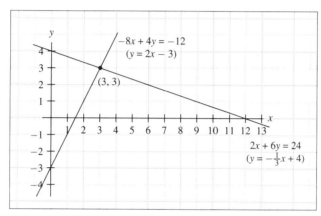

**Fig. 4-9**

(e)　$-3x + 2y = -6$　　　　　$x + \frac{1}{4}y = 2$

　　(1)　$y = 1\frac{1}{2}x - 3$　　　　$y = -4x + 8$

　　(2)　See Fig. 4-10.

As seen in Fig. 4-10, the solution is $x = 2$, $y = 0$.

(f)　$5x - 10y = -30$　　　　$-7x + 14y = 70$

　　(1)　$y = \frac{1}{2}x + 3$　　　　$y = \frac{1}{2}x + 5$

　　(2)　See Fig. 4-11.

As seen in Fig. 4-11, there is no intersection and thus no solution. The equations are inconsistent because in terms of Section 4.2, $m_1 = m_2$ but $b_1 \neq b_2$. Inconsistent equations graph as parallel lines.

(g)　$2x + 8y = 48$　　　　　$3x + 12y = 72$

　　(1)　$y = -\frac{1}{4}x + 6$　　　　$y = -\frac{1}{4}x + 6$

　　(2)　See Fig. 4-12.

As seen in Fig. 4-12, there is an infinite number of solutions and no unique solution because the two equations are equivalent or dependent. In terms of Sections 4.2, $m_1 = m_2$ and $b_1 = b_2$. With the two equations equivalent, there is, in effect, only one independent equation and the system is underconstrained. An underconstrained system has multiple solutions or no solution, never a unique solution.

(h)　$6x - 3y = -3$　　　　　$7x - 7y = -14$　　　　　$5x + 5y = 50$

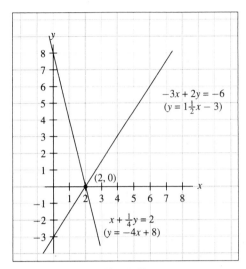

Fig. 4-10

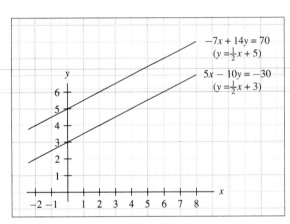

Fig. 4-11

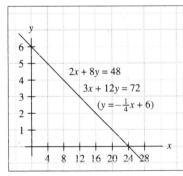

Fig. 4-12

(1)   $y = 2x + 1$          $y = x + 2$          $y = -x + 10$
(2)   See Fig. 4-13.

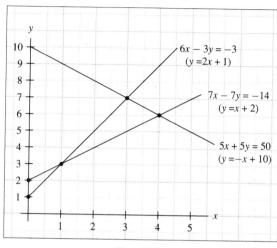

**Fig. 4-13**

As seen in Fig. 4-13, there is no unique solution because there is no one point at which all the graphs intersect simultaneously. An overconstrained system in which $n > v$ seldom has a unique solution.

## SUPPLY AND DEMAND

**4.4.**   Find the equilibrium price $P_e$ and quantity $Q_e$ for each of the following markets using (1) equations and (2) graphs:

(a)   Supply:   $P = 3Q + 10$

Demand: $P = -\frac{1}{2}Q + 80$

(1)   Setting $P_s = P_d$ for equilibrium and solving for $Q$,

$$3Q + 10 = -\tfrac{1}{2}Q + 80$$

$$3\tfrac{1}{2}Q = 70$$

$$Q = 20 = Q_e$$

Then substituting $Q_e = 20$ in the supply (or demand) equation,

$$P = 3(20) + 10$$

$$P = 70 = P_e$$

(2)   See Fig. 4-14.

(b)   Supply:   $P = \frac{1}{4}Q + 200$

Demand: $P = -\frac{1}{2}Q + 800$

(1)   $\frac{1}{4}Q + 200 = -\frac{1}{2}Q + 800$

$$\tfrac{3}{4}Q = 600$$

$$Q = 800 = Q_e$$

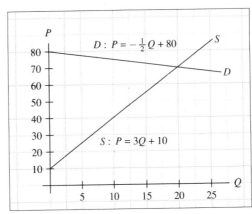

**Fig. 4-14**

Substituting $Q_e = 800$ in the supply (or demand) equation,

$$P = \tfrac{1}{4}(800) + 200 = 400 = P_e$$

(2)   See Fig. 4-15.

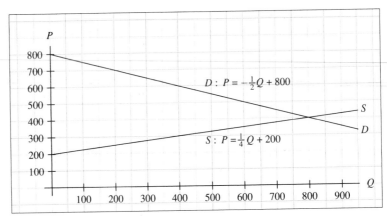

**Fig. 4-15**

**4.5.**   Find the equilibrium price and quantity for the following mathematical models of supply and demand, using (1) equations and (2) graphs.

(a)   Supply:   $Q = 5P + 10$

Demand: $Q = -3P + 50$

(1)   Setting $Q_s = Q_d$ for equilibrium and solving for $P$,

$$5P + 10 = -3P + 50$$

$$8P = 40$$

$$P = 5 = P_e$$

Then substituting $P_e = 5$ in the supply (or demand) equation,

$$Q = 5(5) + 10 = 35 = Q_e$$

(2)    See Fig. 4-16. Note that $Q$ is placed on the $y$ axis.

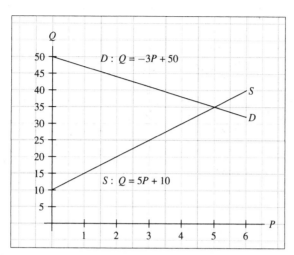

**Fig. 4-16**

(b)    Supply:   $Q = \frac{2}{3}P + 150$

Demand: $Q = -\frac{1}{3}P + 450$

(1)   $\frac{2}{3}P + 150 = -\frac{1}{3}P + 450$

$P = 300 = P_e$

Substituting $P_e = 300$ in the supply (or demand) equation,

$$Q = \frac{2}{3}(300) + 150 = 350 = Q_e$$

(2)    See Fig. 4-17.

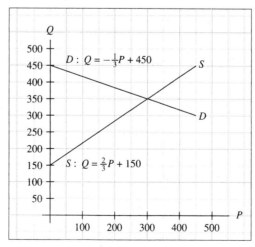

**Fig. 4-17**

**BREAK-EVEN ANALYSIS**

**4.6.** Find the break-even point for the firms with the following linear total revenue $R(x)$ and total cost $C(x)$ functions by (1) finding the profit $\pi(x)$ function, setting it equal to zero, and solving for $x$, (2) setting $R(x) = C(x)$ and solving for $x$, and (3) graphing $R(x)$ and $C(x)$ and finding the point of intersection.

(a)   $R(x) = 55x$

$C(x) = 30x + 250$

(1)   Finding the profit function $\pi(x)$,

$$\pi(x) = R(x) - C(x) = 55x - (30x + 250) = 25x - 250$$

Setting it equal to zero and solving for $x$,

$$25x - 250 = 0$$

$$x = 10 \qquad \text{Break-even level of output}$$

(2)   Setting $R(x) = C(x)$ and solving for $x$,

$$55x = 30x + 250$$

$$25x = 250$$

$$x = 10 \qquad \text{Break-even level of output}$$

(3)   See Fig. 4-18.

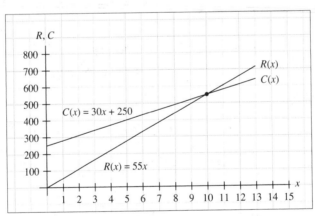

**Fig. 4-18**

(b)   $R(x) = 50x$

$C(x) = 35x + 90$

(1)   Setting $\pi(x) = 0$ and solving for $x$,

$$\pi(x) = 50x - (35x + 90) = 0$$

$$15x - 90 = 0$$

$$x = 6 \qquad \text{Break-even level of output}$$

(2)   Setting $R(x) = C(x)$ and solving for $x$,

$$50x = 35x + 90$$

$$15x = 90$$

$$x = 6 \qquad \text{Break-even level of output}$$

(3)   See Fig. 4-19.

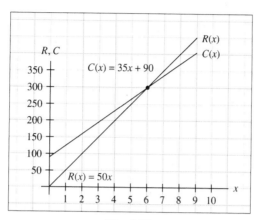

**Fig. 4-19**

**4.7.**   Find the break-even point for the firms with the following quadratic total revenue $R(x)$ functions by (1) finding the profit $\pi(x)$ function, setting it equal to zero, and solving for the $x$'s; (2) setting $R(x) = C(x)$ and solving for the $x$'s; and (3) graphing $R(x)$ and $C(x)$ and finding the point of intersection.

(a)   $R(x) = -4x^2 + 72x \qquad C(x) = 16x + 180$

(1)   $\pi(x) = -4x^2 + 72x - (16x + 180) = -4x^2 + 56x - 180$
Setting $\pi(x) = 0$ and factoring,

$$-4x^2 + 56x - 180 = 0$$

$$-4(x^2 - 14x + 45) = 0$$

$$(x - 5)(x - 9) = 0$$

$$x = 5 \qquad x = 9 \qquad \text{Break-even levels of output}$$

(2)   Setting $R(x) = C(x)$,

$$-4x^2 + 72x = 16x + 180$$

$$-4x^2 + 56x - 180 = 0$$

Then factoring as in (1),

$$x = 5 \qquad x = 9 \qquad \text{Break-even levels of output}$$

(3)   Using the techniques of Section 3.6 to graph the quadratic function $R(x) = -4x^2 + 72x$,

(i)   With $a = -4 < 0$, the parabola opens down.

(ii)

$$x = \frac{-b}{2a} \qquad\qquad y = \frac{4ac - b^2}{4a}$$

$$x = \frac{-(72)}{2(-4)} = 9 \qquad\qquad y = \frac{4(-4)(0) - (72)^2}{4(-4)} = \frac{-5184}{-16} = 324$$

Vertex: (9, 324)

(iii)   Setting $R(x) = 0$ and factoring to find the $x$ intercepts,

$$-4x^2 + 72x = 0$$

$$-4x(x - 18) = 0$$

$$-4x = 0 \qquad\qquad x - 18 = 0$$

$$x = 0 \qquad\qquad x = 18$$

The $x$ intercepts are (0, 0) and (18, 0). See Fig. 4-20 for the graphs of $R(x)$ and $C(x)$ and the points of intersection.

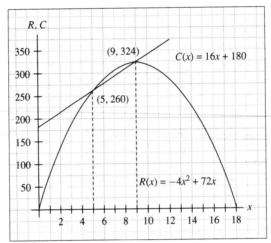

**Fig. 4-20**

(b)   $R(x) = -5x^2 + 750x \quad C(x) = 100x + 20,000$

(1)   Setting $\pi = 0$ and factoring,

$$\pi = -5x^2 + 650x - 20,000 = 0$$

$$-5(x^2 - 130x + 4000) = 0$$

$$(x - 50)(x - 80) = 0$$

$$x = 50 \qquad x = 80 \qquad \text{Break-even levels of output}$$

(2)   Setting $R(x) = C(x)$,

$$-5x^2 + 750x = 100x + 20,000$$

and factoring as in (1),

$$x = 50 \qquad x = 80 \qquad \text{Break-even levels of output}$$

(3)   Graphing $R(x) = -5x^2 + 750x$, as above,

(i)   With $a = -5 < 0$, the parabola opens down.

(ii)   $x = \dfrac{-(750)}{2(-5)} = 75 \qquad y = \dfrac{4(-5)(0) - (750)^2}{4(-5)} = \dfrac{-562{,}500}{-20} = 28{,}125$

$$\text{Vertex: } (75,\ 28{,}125)$$

(iii)   Setting $R(x) = 0$ and factoring for the $x$ intercepts,

$$-5x^2 + 750x = 0$$

$$-5x(x - 150) = 0$$

$$-5x = 0 \qquad\qquad x - 150 = 0$$

$$x = 0 \qquad\qquad x = 150$$

The $x$ intercepts are $(0, 0)$ and $(150, 0)$. See Fig. 4-21.

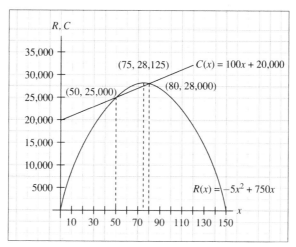

**Fig. 4-21**

## ELIMINATION AND SUBSTITUTION METHODS

**4.8.**   Solve the following systems of simultaneous equations, using the elimination method.

(a)

$$3x + 4y = 37 \tag{4.18}$$

$$8x + 5y = 76 \tag{4.19}$$

Multiplying (*4.18*) by 8 and (*4.19*) by 3 in preparation to eliminate $x$,

$$24x + 32y = 296 \tag{4.20}$$

$$24x + 15y = 228 \tag{4.21}$$

Subtracting (*4.21*) from (*4.20*) to eliminate $x$,

$$17y = 68$$

$$y = 4$$

Then substituting $y = 4$ in (4.18) or (4.19) to solve for $x$,

$$3x + 4(4) = 37$$

$$x = 7$$

(b)
$$5x + 2y = 46 \qquad (4.22)$$

$$9x - 7y = -2 \qquad (4.23)$$

Here picking $y$ for elimination and multiplying (4.22) by 7 and (4.23) by 2,

$$35x + 14y = 322 \qquad (4.24)$$

$$18x - 14y = -4 \qquad (4.25)$$

Then *adding* (4.24) and (4.25) to eliminate $y$,

$$53x = 318$$

$$x = 6$$

Substituting $x = 6$ in (4.22) or (4.23),

$$5(6) + 2y = 46$$

$$y = 8$$

**4.9.**   Solve the following systems of simultaneous equations, using the substitution method.

(a)
$$12x - 7y = 106 \qquad (4.26)$$

$$8x + y = 82 \qquad (4.27)$$

Solving (4.27) for $y$,

$$y = 82 - 8x \qquad (4.28)$$

Substituting the value of $y$ from (4.28) into the *other* equation (4.26),

$$12x - 7(82 - 8x) = 106$$

$$12x - 574 + 56x = 106$$

$$x = 10$$

Then substituting $x = 10$ in (4.26) or (4.27),

$$12(10) - 7y = 106$$

$$y = 2$$

(b)
$$5x + 24y = 100 \qquad (4.29)$$

$$7x - 15y = -103 \qquad (4.30)$$

Solving (4.29) for $x$,

$$5x = 100 - 24y$$

$$x = 20 - 4.8y \qquad (4.31)$$

Substituting (*4.31*) in the *other* equation (*4.30*),

$$7(20 - 4.8y) - 15y = -103$$

$$140 - 33.6y - 15y = -103$$

$$-48.6y = -243$$

$$y = 5$$

Finally, substituting $y = 5$ in (*4.29*) or (*4.30*),

$$5x + 24(5) = 100$$

$$x = -4$$

**4.10.** Given the following system of simultaneous equations for two substitute goods, beef $b$ and pork $p$, find the equilibrium price and quantity for each market, using the substitution method.

$$Q_b^s = 15P_b - 5 \qquad\qquad Q_p^s = 32P_p - 6$$

$$Q_b^d = -3P_b + P_p + 82 \qquad\qquad Q_p^d = 2P_b - 4P_p + 92$$

Setting $Q^s = Q^d$ for the equilibrium in each market,

$$Q_b^s = Q_b^d \qquad\qquad\qquad Q_p^s = Q_p^d$$

$$15P_b - 5 = -3P_b + P_p + 82 \qquad\qquad 32P_p - 6 = 2P_b - 4P_p + 92$$

$$18P_b - P_p = 87 \qquad\qquad\qquad -2P_b + 36P_p = 98$$

This reduces the problem to two equations and two unknowns:

$$18P_b - P_p = 87 \tag{4.32}$$

$$-2P_b + 36P_p = 98 \tag{4.33}$$

Solving for $P_p$ in (*4.32*),

$$P_p = 18P_b - 87 \tag{4.34}$$

Then substituting (*4.34*) in (*4.33*), the *other* equation, and solving for $P_b$,

$$-2P_b + 36(18P_b - 87) = 98$$

$$-2P_b + 648P_b - 3132 = 98$$

$$P_b = 5$$

Now substituting $P_b = 5$ in (*4.32*) or (*4.33*) to find $P_p$,

$$18(5) - P_p = 87$$

$$P_p = 3$$

Finally, substituting $P_b = 5$ and $P_p = 3$ in either the supply or demand functions for each market to find $Q_b$ and $Q_p$,

$$Q_b^d = -3(5) + (3) + 82 \qquad\qquad Q_p^d = 2(5) - 4(3) + 92$$

$$Q_b^d = 70 = Q_b^s \qquad\qquad\qquad Q_p^d = 90 = Q_p^s$$

**4.11.** Given the following system of equations for two complementary goods, trousers $t$ and jackets $j$, find the equilibrium price and quantity, using the elimination method.

$$Q_t^s = 3P_t - 60 \qquad\qquad Q_j^s = 2P_j - 120$$

$$Q_t^d = -5P_t - 2P_j + 410 \qquad\qquad Q_j^d = -P_t - 3P_j + 295$$

Setting $Q_s = Q_d$ for the equilibrium in each market,

$$Q_t^s = Q_t^d \qquad\qquad\qquad\qquad Q_j^s = Q_j^d$$

$$3P_t - 60 = -5P_t - 2P_j + 410 \qquad\qquad 2P_j - 120 = -P_t - 3P_j + 295$$

$$8P_t + 2P_j = 470 \qquad\qquad\qquad\qquad P_t + 5P_j = 415$$

This leaves two equations and two unknowns:

$$8P_t + 2P_j = 470 \tag{4.35}$$

$$P_t + 5P_j = 415 \tag{4.36}$$

Multiplying (4.36) by 8,

$$8P_t + 40P_j = 3320 \tag{4.37}$$

Subtracting (4.37) from the *other* equation (4.35), which here involves changing the sign of every term in (4.37) before, in effect, adding them to (4.35),

$$-38P_j = -2850$$

$$P_j = 75$$

Substituting $P_j = 75$ in (4.35) or (4.36),

$$8P_t + 2(75) = 470$$

$$P_t = 40$$

Then substituting $P_j = 75$, $P_t = 40$ in $Q_s$ or $Q_d$ for each market,

$$Q_t^d = -5(40) - 2(75) + 410 \qquad Q_j^d = -(40) - 3(75) + 295$$

$$Q_t^d = 60 = Q_t^s \qquad\qquad\qquad Q_j^d = 30 = Q_j^s$$

**4.12.** The elimination and substitution methods can also be used to solve systems of quadratic equations. Solve the following system of quadratic supply and demand equations, using the substitution method.

$$\text{Supply: } P - 3Q^2 + 10Q - 5 = 0 \tag{4.38}$$

$$\text{Demand: } P + Q^2 + 3Q - 20 = 0 \tag{4.39}$$

Solving (4.38) for $P$ in terms of $Q$,

$$P = 3Q^2 - 10Q + 5 \tag{4.40}$$

Substituting (4.40) for $P$ in the other equation (4.39),

$$(3Q^2 - 10Q + 5) + Q^2 + 3Q - 20 = 0$$

$$4Q^2 - 7Q - 15 = 0$$

Using the quadratic formula from Section 3.5 to solve for $Q$,

$$Q = 3 \qquad\qquad Q = -1.25$$

Since both $P$ and $Q$ are restricted to positive values in economics, $Q = 3$. Substituting $Q = 3$ in (4.38), (4.39), or (4.40),

$$P - 3(3)^2 + 10(3) - 5 = 0$$

$$P = 2$$

**4.13.** Use the elimination method to find $P_e$ and $Q_e$, given

$$\text{Supply: } P - 2Q^2 + 3Q + 71 = 0 \tag{4.41}$$

$$\text{Demand: } 3P + Q^2 + 5Q - 102 = 0 \tag{4.42}$$

Multiplying ($4.41$) by 3,

$$3P - 6Q^2 + 9Q + 213 = 0 \qquad (4.43)$$

Subtracting ($4.43$) from ($4.42$), be careful to change the sign of every term in ($4.43$) before adding:

$$7Q^2 - 4Q - 315 = 0$$

Using the quadratic formula,

$$Q = 7 \qquad\qquad Q \approx -6.43$$

Substituting $Q = 7$ in ($4.41$) or ($4.42$),

$$P - 2(7)^2 + 3(7) + 71 = 0$$

$$P = 6$$

**4.14.** Supply-and-demand analysis can also involve more than two interrelated markets. Use the elimination method to find $P_e$ and $Q_e$ for each of the three markets, given the system of equations:

$$Q_{s_1} = 6P_1 - 8 \qquad\qquad Q_{d_1} = -5P_1 + P_2 + P_3 + 23$$

$$Q_{s_2} = 3P_2 - 11 \qquad\qquad Q_{d_2} = P_1 - 3P_2 + 2P_3 + 15$$

$$Q_{s_3} = 3P_3 - 5 \qquad\qquad Q_{d_3} = P_1 + 2P_2 - 4P_3 + 19$$

For equilibrium in each market,

$$Q_{s_1} = Q_{d_1}$$

$$6P_1 - 8 = -5P_1 + P_2 + P_3 + 23$$

$$11P_1 - P_2 - P_3 = 31 \qquad (4.44)$$

$$Q_{s_2} = Q_{d_2}$$

$$3P_2 - 11 = P_1 - 3P_2 + 2P_3 + 15$$

$$-P_1 + 6P_2 - 2P_3 = 26 \qquad (4.45)$$

$$Q_{s_3} = Q_{d_3}$$

$$3P_3 - 5 = P_1 + 2P_2 - 4P_3 + 19$$

$$-P_1 - 2P_2 + 7P_3 = 24 \qquad (4.46)$$

This leaves three equations with three unknowns:

$$11P_1 - P_2 - P_3 = 31 \qquad (4.44)$$

$$-P_1 + 6P_2 - 2P_3 = 26 \qquad (4.45)$$

$$-P_1 - 2P_2 + 7P_3 = 24 \qquad (4.46)$$

To solve a $3 \times 3$ system, the following hints generally prove helpful: (1) using any two of the equations, eliminate one of the three variables; (2) then, using the third equation and either of the others already used, eliminate the same variable as before; (3) this will leave two equations with two unknowns, which can then be solved by now familiar methods, as shown below.

(1)   Selecting $P_2$ for elimination and multiplying ($4.44$) by 2,

$$22P_1 - 2P_2 - 2P_3 = 62 \qquad (4.47)$$

Subtracting ($4.47$) from ($4.46$),

$$-23P_1 + 9P_3 = -38 \qquad (4.48)$$

(2)   Now multiplying ($4.46$) by 3,

$$-3P_1 - 6P_2 + 21P_3 = 72 \qquad (4.49)$$

and adding it to (4.45), the previously unused equation,

$$-4P_1 + 19P_3 = 98 \tag{4.50}$$

(3)    This leaves two equations with two unknowns:

$$-23P_1 + 9P_3 = -38 \tag{4.48}$$

$$-4P_1 + 19P_3 = 98 \tag{4.50}$$

which can be solved with earlier methods, giving

$$P_1 = 4 \qquad P_3 = 6$$

Then substituting $P_1 = 4$, $P_3 = 6$ in (4.44), (4.45), or (4.46),

$$11(4) - P_2 - (6) = 31$$

$$P_2 = 7$$

## INCOME DETERMINATION MODELS

**4.15.** (a) Find the reduced-form equation and (b) solve for the equilibrium level of income (1) directly and (2) with the reduced-form equation, given

$$Y = C + I + G \qquad C = C_0 + bY \qquad I = I_0 \qquad G = G_0$$

where $C_0 = 135$, $b = 0.8$, $I_0 = 75$, and $G_0 = 30$.

(a)    From Section 4.6,

$$Y = C + I + G$$

Substituting the given values and solving for $Y$ in terms of the parameters (b) and exogenous variables ($C_0$, $I_0$, and $G_0$),

$$Y = C_0 + bY + I_0 + G_0$$

$$Y - bY = C_0 + I_0 + G_0$$

$$(1 - b)Y = C_0 + I_0 + G_0$$

$$Y_e = \frac{1}{1 - b}(C_0 + I_0 + G_0)$$

(b)    (1)    $Y = C + I + G$

$$Y = 135 + 0.8Y + 75 + 30$$

$$Y - 0.8Y = 240$$

$$0.2Y = 240$$

$$Y_e = 1200$$

(2)    $Y = \dfrac{1}{1 - b}(C_0 + I_0 + G_0)$

$$Y = \frac{1}{1 - 0.8}(135 + 75 + 30)$$

$$Y = \frac{1}{0.2}(240)$$

$$Y = 5(240)$$

$$Y_e = 1200$$

**4.16.** Find the equilibrium level of income ($Y_e$), given

$$Y = C + I + G$$

when $C = 125 + 0.8Y$, $I = 45$, and $G = 90$.

From Problem 4.15, we know that the reduced form equation is

$$Y_e = \frac{1}{1 - b}(C_0 + I_0 + G_0)$$

and when $b = \text{MPC} = 0.8$, the multiplier $1/(1 - b) = 5$. Using this knowledge, we can write immediately,

$$Y_e = 5(125 + 45 + 90) = 1300$$

**4.17.** Find (a) the reduced form, (b) the value of the equilibrium level of income $Y_e$, and (c) the effect on the multiplier, given the following model in which investment is not autonomous but a function of income:

$$Y = C + I \qquad C = C_0 + bY \qquad I = I_0 + aY$$

and $C_0 = 65$, $I_0 = 70$, $b = 0.6$, and $a = 0.2$.

(a)
$$Y = C + I$$

$$Y = C_0 + bY + I_0 + aY$$

$$Y - bY - aY = C_0 + I_0$$

$$(1 - b - a)Y = C_0 + I_0$$

$$Y_e = \frac{1}{1 - b - a}(C_0 + I_0)$$

(b)	Solving directly, although one may also use the reduced form,

$$Y = C + I$$

$$Y = 65 + 0.6Y + 70 + 0.2Y$$

$$Y - 0.6Y - 0.2Y = 135$$

$$0.2Y = 135$$

$$Y_e = 675$$

(c)	When investment is a function of income, and not autonomous, the multiplier changes from $1/(1 - b)$ to $1/(1 - b - a)$. This increases the value of the multiplier because it reduces the denominator of the fraction and makes the quotient larger, as substitution of the values of the parameters in the problem shows

$$\frac{1}{1 - b} = \frac{1}{1 - 0.6} = \frac{1}{0.4} = 2.5$$

while
$$\frac{1}{1 - b - a} = \frac{1}{1 - 0.6 - 0.2} = \frac{1}{0.2} = 5$$

**4.18.** Find (a) the reduced form, (b) the numerical value of $Y_e$, and (c) the effect on the multiplier when a lump-sum tax is added to the model and consumption becomes a function of disposable income ($Y_d$), given

$$Y = C + I \qquad C = C_0 + bY_d \qquad I = I_0 \qquad Y_d = Y - T$$

where $C_0 = 100$, $b = 0.6$, $I_0 = 40$, and $T = 50$.

(a)
$$Y = C + I = C_0 + bY_d + I_0$$

$$Y = C_0 + b(Y - T) + I_0 = C_0 + bY - bT + I_0$$

$$Y - bY = C_0 + I_0 - bT$$

$$Y_e = \frac{1}{1 - b}(C_0 + I_0 - bT) \qquad\qquad\qquad (4.51)$$

(b)
$$Y = 100 + 0.6Y_d + 40 = 140 + 0.6(Y - T)$$
$$Y = 140 + 0.6(Y - 50) = 140 + 0.6Y - 30$$
$$Y - 0.6Y = 110$$
$$0.4Y = 110$$
$$Y_e = 275$$

Or substituting in (4.51),

$$Y = \frac{1}{1 - 0.6}[100 + 40 - 0.6(50)]$$

$$Y = \frac{1}{0.4}(110)$$

$$Y_e = 275$$

(c)  As seen in part (a), incorporation of a lump-sum tax into the model does not change the value of the multiplier. It remains $1/(1 - b)$. Only the aggregate value of the exogenous variables is changed by an amount equal to $-bT$. Incorporation of other autonomous variables such as $G_0$, $X_0$, or $Z_0$ will not affect the value of the multiplier.

**4.19.**  Find (a) the reduced form, (b) the numerical value of $Y_e$, and (c) the effect on the multiplier if an income tax $T$ with a proportional component $t$ is incorporated into the model, given

$$Y = C + I \qquad C = C_0 + bY_d \qquad T = T_0 + tY \qquad Y_d = Y - T \qquad I = I_0$$

where $C_0 = 85$, $I_0 = 30$, $T_0 = 20$, $b = 0.75$, and $t = 0.2$.

(a)
$$Y = C + I = C_0 + bY_d + I_0$$
$$Y = C_0 + b(Y - T) + I_0 = C_0 + b(Y - T_0 - tY) + I_0$$
$$Y = C_0 + bY - bT_0 - btY + I_0$$
$$Y - bY + btY = C_0 + I_0 - bT_0$$
$$(1 - b + bt)Y = C_0 + I_0 - bT_0$$
$$Y_e = \frac{1}{1 - b + bt}(C_0 + I_0 - bT_0) \qquad\qquad (4.52)$$

(b)
$$Y = C + I = 85 + 0.75Y_d + 30 = 115 + 0.75(Y - T)$$
$$Y = 115 + 0.75(Y - 20 - 0.2Y)$$
$$Y = 115 + 0.75Y - 15 - 0.15Y$$
$$Y - 0.75Y + 0.15Y = 100$$
$$0.4Y = 100$$
$$Y_e = 250$$

or substituting in (4.52),

$$Y = \frac{1}{1 - 0.75 + (0.75)(0.2)}[85 + 30 - 0.75(20)]$$

$$Y_e = \frac{1}{0.4}(100) = 250$$

(c)  The multiplier is changed from $1/(1 - b)$ to $1/(1 - b + bt)$. This reduces the size of the multiplier because it makes the denominator larger and the fraction smaller:

$$\frac{1}{1 - b} = \frac{1}{1 - 0.75} = \frac{1}{0.25} = 4$$

$$\frac{1}{1 - b + bt} = \frac{1}{1 - 0.75 + 0.75(0.2)} = \frac{1}{1 - 0.75 + 0.15} = \frac{1}{0.4} = 2.5$$

**4.20.** If the foreign sector is added to the model and there is a positive marginal propensity to import $z$, find (a) the reduced form, (b) the equilibrium level of income, and (c) the effect on the multiplier, given

$$Y = C + I + G + (X - Z) \qquad C = C_0 + bY \qquad Z = Z_0 + zY$$

$$I = I_0 \qquad\qquad G = G_0 \qquad\qquad X = X_0$$

where $C_0 = 70$, $I_0 = 90$, $G_0 = 65$, $X_0 = 80$, $Z_0 = 40$, $b = 0.9$, and $z = 0.15$.

(a)
$$Y = C + I + G + (X - Z)$$

$$Y = C_0 + bY + I_0 + G_0 + X_0 - Z_0 - zY$$

$$Y - bY + zY = C_0 + I_0 + G_0 + X_0 - Z_0$$

$$(1 - b + z)Y = C_0 + I_0 + G_0 + X_0 - Z_0$$

$$Y_e = \frac{1}{1 - b + z}(C_0 + I_0 + G_0 + X_0 - Z_0) \qquad (4.53)$$

(b)  Using the reduced form in (4.53),

$$Y = \frac{1}{1 - 0.9 + 0.15}(70 + 90 + 65 + 80 - 40)$$

$$Y_e = \frac{1}{0.25}(265) = 1060$$

(c)  Introduction of the marginal propensity to import $z$ into the model reduces the size of the multiplier by making the denominator in the multiplier larger:

$$\frac{1}{1 - b} = \frac{1}{1 - 0.9} = \frac{1}{0.1} = 10$$

$$\frac{1}{1 - b + z} = \frac{1}{1 - 0.9 + 0.15} = \frac{1}{0.25} = 4$$

See also Problems 13.25 and 13.26. For a graphical analysis of the same type of issues, see Dowling, *Schaum's Outline of Introduction to Mathematical Economics*, Problems 2.7 to 2.10.

## IS-LM ANALYSIS

**4.21.** Find (a) the level of income and the rate of interest that concurrently bring equilibrium to the economy and (b) estimate the level of consumption $C$, investment $I$, the transaction-precautionary demand for money $M_t$, and the speculative demand for money $M_w$, when the money supply $M_s = 300$, and

$$C = 102 + 0.7Y \qquad I = 150 - 100i \qquad M_t = 0.25Y \qquad M_w = 124 - 200i$$

(a)  Equilibrium in the commodity market (IS) exists when

$$Y = C + I = 102 + 0.7Y + 150 - 100i$$

Moving everything to the left,

$$Y - 0.7Y - 102 - 150 + 100i = 0$$

$$0.3Y + 100i - 252 = 0 \qquad (4.54)$$

Equilibrium in the money market (LM) occurs when

$$M_s = M_t + M_w$$

$$300 = 0.25Y + 124 - 200i$$

$$0.25Y - 200i - 176 = 0 \qquad\qquad (4.55)$$

Simultaneous equilibrium in both markets requires that

$$\text{IS:} \quad 0.3Y + 100i - 252 = 0 \qquad\qquad (4.54)$$

$$\text{LM:} \quad 0.25Y - 200i - 176 = 0 \qquad\qquad (4.55)$$

Multiplying (4.54) by 2 as explained in Section 4.5,

$$0.6Y + 200i - 504 = 0 \qquad\qquad (4.56)$$

Adding (4.56) to (4.55) to eliminate $i$,

$$0.85Y - 680 = 0$$

$$Y_e = 800$$

Then substituting $Y = 800$ in (4.54) or (4.55),

$$0.3(800) + 100i - 252 = 0$$

$$100i = 12$$

$$i_e = 0.12$$

(b)   At $Y_e = 800$ and $i_e = 0.12$,

$$C = 102 + 0.7(800) = 662 \qquad M_t = 0.25(800) = 200$$

$$I = 150 - 100(0.12) = 138 \qquad M_w = 124 - 200(0.12) = 100$$

and

$$C + I = Y \qquad\qquad M_t + M_w = M_s$$

$$662 + 138 = 800 \qquad\qquad 200 + 100 = 300$$

**4.22.**  If the money supply in Problem 4.21 increases by 17, (a) what happens to the equilibrium level of income and rate of interest? (b) What are $C$, $I$, $M_t$, and $M_w$ at the new equilibrium?

(a)   If the money supply increases by 17, the new LM equation is

$$M_s = M_t + M_w$$

$$317 = 0.25Y + 124 - 200i$$

$$0.25Y - 200i - 193 = 0 \qquad\qquad (4.57)$$

The IS equation remains unchanged from (4.54):

$$0.3Y + 100i - 252 = 0 \qquad\qquad (4.58)$$

Multiplying (4.58) by 2,

$$0.6Y + 200i - 504 = 0 \qquad\qquad (4.59)$$

Adding (4.59) to (4.57),

$$0.85Y - 697 = 0$$

$$Y_e = 820$$

Substituting $Y = 820$ in (4.57) or (4.58),

$$0.25(820) - 200i - 193 = 0$$

$$200i = 12$$

$$i_e = 0.06$$

An increase in the money supply, ceteris paribus, or other things remaining the same, leads to an increase in the equilibrium level of income and a decrease in the interest rate.

(b)    At $Y_e = 820$ and $i_e = 0.06$,

$$C = 102 + 0.7(820) = 676 \qquad M_t = 0.25(820) = 205$$

$$I = 150 - 100(0.06) = 144 \qquad M_w = 124 - 200(0.06) = 112$$

and

$$C + I = Y \qquad M_s = M_t + M_w$$

$$676 + 144 = 820 \qquad 317 = 205 + 112$$

**4.23.** Find (a) the equilibrium income level and interest rate, and (b) the levels of $C$, $I$, $M_t$, and $M_w$, when $M_s = 275$, and

$$C = 89 + 0.6Y \qquad I = 120 - 150i \qquad M_t = 0.1Y \qquad M_w = 240 - 250i$$

(a) For IS:
$$Y = 89 + 0.6Y + 120 - 150i$$

$$0.4Y + 150i - 209 = 0 \tag{4.60}$$

For LM:
$$M_s = M_t + M_w$$

$$275 = 0.1Y + 240 - 250i$$

$$0.1Y - 250i - 35 = 0 \tag{4.61}$$

In equilibrium,
$$0.4Y + 150i - 209 = 0 \tag{4.60}$$

$$0.1Y - 250i - 35 = 0 \tag{4.61}$$

Multiplying (4.61) by 4,

$$0.4Y - 1000i - 140 = 0 \tag{4.62}$$

Subtracting (4.62) from (4.60) to eliminate $Y$,

$$1150i - 69 = 0$$

$$i_e = 0.06$$

Substituting $i_e = 0.06$ into (4.60) or (4.61),

$$0.4Y + 150(0.06) - 209 = 0$$

$$0.4Y = 200$$

$$Y_e = 500$$

(b)    At $Y_e = 500$ and $i_e = 0.06$,

$$C = 89 + 0.6(500) = 389 \qquad M_t = 0.1(500) = 50$$

$$I = 120 - 150(0.06) = 111 \qquad M_w = 240 - 250(0.06) = 225$$

and

$$C + I = Y \qquad M_t + M_w = M_s$$

$$389 + 111 = 500 \qquad 50 + 225 = 275$$

**4.24.** Show what happens to the equilibrium conditions in Problem 4.23 if autonomous investment drops to 97.

If $I_0 = 97$, the IS equation becomes:

$$Y = 89 + 0.6Y + 97 - 150i$$

$$0.4Y + 150i - 186 = 0 \qquad (4.63)$$

The LM equation remains the same as (4.61),

$$0.1Y - 250i - 35 = 0 \qquad (4.64)$$

In equilibrium,

$$0.4Y + 150i - 186 = 0 \qquad (4.63)$$

$$0.1Y - 250i - 35 = 0 \qquad (4.64)$$

Multiplying (4.64) by 4,

$$0.4Y - 1000i - 140 = 0 \qquad (4.65)$$

Subtracting (4.65) from (4.63),

$$1150i - 46 = 0$$

$$i_e = 0.04$$

Substituting $i_e = 0.04$ in (4.63) or (4.64),

$$0.4Y + 150(0.04) - 186 = 0$$

$$0.4Y - 180 = 0$$

$$Y_e = 450$$

A fall in autonomous investment, *ceteris paribus*, leads to a decrease in the equilibrium level of income and a drop in the interest rate.

## MODELING AND INVERSE FUNCTIONS (OPTIONAL)

**4.25.** Given the following linear supply and demand equations in standard form

$$\text{Supply: } -7P + 14Q = -42 \qquad (4.66)$$

$$\text{Demand: } 3P + 12Q = 90 \qquad (4.67)$$

(*a*) convert them to the slope-intercept form in a manner conformable to (1) economic modeling and (2) mathematical modeling; (*b*) find the equilibrium price and quantity $P_e$ and $Q_e$ for each model algebraically; and (*c*) graph both sets of equations to find $P_e$ and $Q_e$.

(*a*)   (1)   Since economists consider $P = f(Q)$, the slope-intercept forms are found by solving each equation for $P$ in terms of $Q$. Solving (4.66) for $P$ in terms of $Q$,

$$\text{Supply: } P = 2Q + 6 \qquad (4.68)$$

Solving (4.67) for $P$ in terms of $Q$,

$$\text{Demand: } P = -4Q + 30 \qquad (4.69)$$

(2)   Since mathematicians consider $Q = F(P)$, the slope-intercept forms are found by solving each equation for $Q$ in terms of $P$. Solving (4.66) for $Q$ in terms of $P$,

$$\text{Supply: } Q = \tfrac{1}{2}P - 3 \qquad (4.70)$$

Solving (4.67) for $Q$ in terms of $P$,

$$\text{Demand: } Q = -\tfrac{1}{4}P + 7\tfrac{1}{2} \qquad (4.71)$$

(b)  (1)   In equilibrium $P_s = P_d = P_e$. Setting $P$ in (4.68) equal to $P$ in (4.69) and solving algebraically,

$$2Q + 6 = -4Q + 30$$

$$6Q = 24 \qquad Q = 4 = Q_e$$

Then substituting $Q = 4$ in (4.68) or (4.69)

$$P = 2(4) + 6 = 14 = P_e$$

(2)   In equilibrium $Q_s = Q_d = Q_e$. Setting $Q$ in (4.70) equal to $Q$ in (4.71) and solving algebraically,

$$\tfrac{1}{2}P - 3 = -\tfrac{1}{4}P + 7\tfrac{1}{2}$$

$$\tfrac{3}{4}P = 10\tfrac{1}{2} \qquad P = 14 = P_e$$

Substituting $P = 14$ in (4.70) or (4.71),

$$Q = \tfrac{1}{2}(14) - 3 = 4 = Q_e$$

(c)   In graphing an economic model, $Q$ is placed on the $x$ axis and $P$ on the $y$ axis, as shown in Fig. 4-22(a). In graphing a mathematical model, $P$ is placed on the $x$ axis and $Q$ on the $y$ axis, as shown in Fig. 4-22(b).

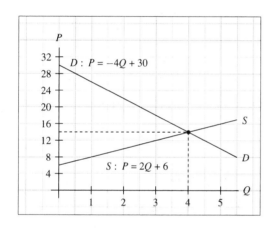

(a)

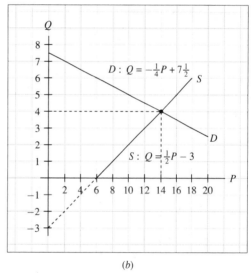

(b)

**Fig. 4-22**

**4.26.** Given the following linear supply and demand equations in standard form,

$$\text{Supply:}\quad 10P - 8Q = 50 \qquad\qquad (4.72)$$

$$\text{Demand: } 20P + 8Q = 700 \qquad\qquad (4.73)$$

(a) convert them to the slope-intercept form in a manner conformable to (1) economic modeling and (2) mathematical modeling; (b) find the equilibrium price $P_e$ and quantity $Q_e$ for each model algebraically; and (c) graph both sets of equations to find $P_e$ and $Q_e$.

(a)  (1)   For the economic model,

From (4.72),  $\qquad\qquad$  Supply:  $P = \tfrac{4}{5}Q + 5$  $\qquad\qquad\qquad$ (4.74)

From (4.73),  $\qquad\qquad$  Demand: $P = -\tfrac{2}{5}Q + 35$  $\qquad\qquad\qquad$ (4.75)

(2)   For the mathematical model,

From (4.72),                           Supply: $Q = 1\frac{1}{4}P - 6\frac{1}{4}$                             (4.76)

From (4.73),                           Demand: $Q = -2\frac{1}{2}P + 87\frac{1}{2}$                          (4.77)

(b)   (1) Setting $P$ in (4.74) equal to $P$ in (4.75),

$$\tfrac{4}{5}Q + 5 = -\tfrac{2}{5}Q + 35$$

$$\tfrac{6}{5}Q = 30 \qquad Q = 25 = Q_e$$

Substituting $Q = 25$ in (4.74) or (4.75),

$$P = \tfrac{4}{5}(25) + 5 = 25 = P_e$$

(2)   Setting $Q$ in (4.76) equal to $Q$ in (4.77),

$$1\tfrac{1}{4}P - 6\tfrac{1}{4} = -2\tfrac{1}{2}P + 87\tfrac{1}{2}$$

$$3\tfrac{3}{4}P = 93\tfrac{3}{4} \qquad P = 25 = P_e$$

Substituting $P = 25$ in (4.76) or (4.77),

$$Q = 1\tfrac{1}{4}(25) - 6\tfrac{1}{4} = 25 = Q_e$$

(c)   The economic graph with $Q$ on the $x$ axis and $P$ on the $y$ axis is found in Fig. 4-23(a). The mathematical graph with $P$ on the $x$ axis and $Q$ on the $y$ axis is found in Fig. 4-23(b).

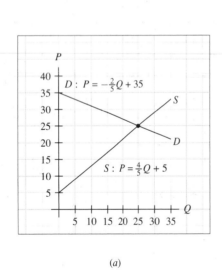

(a)

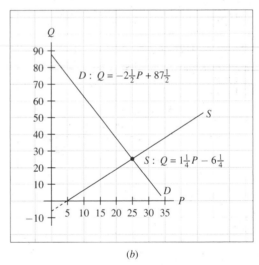

(b)

Fig. 4-23

**4.27.**   Given the implicit function

$$f(x, y) = 5x + 12y - 180 = 0 \tag{4.78}$$

find the explicit functions (a) $y = g(x)$ and (b) $x = h(y)$, if they exist.

(a)   To find the explicit function $y = g(x)$, if it exists, solve the implicit function $f(x, y)$ algebraically for $y$ in terms of $x$. If for each value of $x$, there is one and only one value of $y$, the explicit function $y = g(x)$ exists. Thus, from (4.78),

$$12y = -5x + 180$$

$$y = -\frac{5}{12}x + 15$$

Since there is only one value of $y$ for each value of $x$,

$$y = g(x) = -\frac{5}{12}x + 15 \qquad (4.79)$$

(b)  To find the explicit function $x = h(y)$, if it exists, solve the implicit function $f(x, y)$ algebraically for $x$ in terms of $y$. If for each value of $y$, there is one and only one value of $x$, the explicit function $x = h(y)$ exists. Thus, from (4.78),

$$5x = -12y + 180$$

$$x = -\frac{12}{5}y + 36$$

Since there is only one value of $x$ for each value of $y$,

$$x = h(y) = -\frac{12}{5}y + 36 \qquad (4.80)$$

Note that when two explicit functions can be derived from the same implicit function in two variables, the explicit functions are inverse functions of each other, as shown in Problem 4.28.

**4.28.**  Given $y = f(x)$, the inverse function is written $x = f^{-1}(y)$, which reads "$x$ is an inverse function of $y$." Find the inverse functions for the functions in Problem 4.27:

(a)                                    $y = g(x) = -\frac{5}{12}x + 15$                                    (4.79)

(b)                                    $x = h(y) = -\frac{12}{5}y + 36$                                    (4.80)

To find an inverse function, solve algebraically for the dependent variable in terms of the independent variable, then check to be sure it is a function. Thus

(a)  Solving $y = g(x) = -\frac{5}{12}x + 15$ in (4.79) for $x$ in terms of $y$,

$$\tfrac{5}{12}x = -y + 15$$

$$x = -\tfrac{12}{5}y + 36$$

Since there is only one value of the dependent variable, here $x$, for each value of the independent variable, here $y$, the inverse function $g^{-1}(y)$ exists and is written

$$x = g^{-1}(y) = -\tfrac{12}{5}y + 36 \qquad (4.81)$$

(b)  Solving $x = h(y) = -\tfrac{12}{5}y + 36$ in (4.80) for $y$ in terms of $x$,

$$\tfrac{12}{5}y = -x + 36$$

$$y = -\tfrac{5}{12}x + 15$$

Since there is only one value of the dependent variable $y$ for each value of the independent variable $x$, the inverse function $h^{-1}(x)$ exists and is written

$$y = h^{-1}(x) = -\tfrac{5}{12}x + 15 \qquad (4.82)$$

Note that the inverse function $g^{-1}(y)$ in (4.81) is equal to $h(y)$ in (4.80) and the inverse function $h^{-1}(x)$ in (4.82) is equal to $g(x)$ in (4.79), confirming the earlier observation that whenever an implicit function in two variables such as (4.78) yields two distinct explicit functions such as (4.79) and (4.80), the explicit functions are inverse functions of each other: From (4.79) and (4.82),

$$g(x) = h^{-1}(x) = -\tfrac{5}{12}x + 15$$

and from (4.80) and (4.81)

$$h(y) = g^{-1}(y) = -\tfrac{12}{5}y + 36$$

Note, too, that the slope of the inverse function is the reciprocal of the slope of the original or primitive function.

**4.29.**  Test (a) algebraically and (b) graphically to see if the inverse function exists for

$$y = x^2 \qquad (4.83)$$

(a)    Solving (*4.83*) for $x$ in terms of $y$,

$$x = \pm\sqrt{y}$$

Since there are two values of $x$ for each value of $y$, one positive and one negative, $x = \pm\sqrt{y}$ is not the inverse function of $y = x^2$ because $x = \pm\sqrt{y}$ is not a function.

(b)    Drawing up respective schedules, then graphing $y = x^2$ in Fig. 4-24(*a*) and $x = \pm\sqrt{y}$ in Fig. 4-24(*b*), we see that the graph of $x = \pm\sqrt{y}$ does not pass the vertical line test.

$$y = x^2 \qquad\qquad\qquad x = \pm\sqrt{y}$$

| $x$ | $y$ |
|-----|-----|
| $-3$ | 9 |
| $-2$ | 4 |
| $-1$ | 1 |
| 0 | 0 |
| 1 | 1 |
| 2 | 4 |
| 3 | 9 |

| $y$ | $x$ |
|-----|-----|
| 0 | 0 |
| 1 | $\pm1$ |
| 4 | $\pm2$ |
| 9 | $\pm3$ |

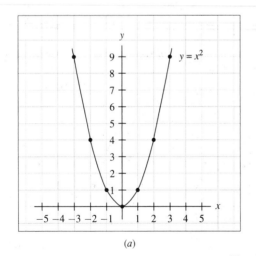

(a)

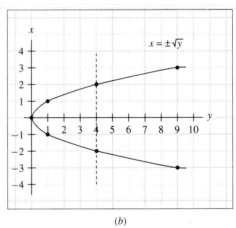

(b)

**Fig. 4-24**

**4.30.**   Given the domain restriction $x \geq 0$, show (*a*) algebraically and (*b*) graphically that the inverse function exists for $y = x^2$.

(a)    Given $y = x^2$ and the restricted domain $x \geq 0$, the inverse function is simply

$$x = \sqrt{y}$$

(b)    Constructing schedules and drawing the relevant graphs in Fig. 4-25, it is clear that the graph of the inverse function passes the vertical line test. For all $x \geq 0$,

(a) $y = x^2$

| $x$ | $y$ |
|---|---|
| 0 | 0 |
| 1 | 1 |
| 2 | 4 |
| 3 | 9 |

(b) $x = \sqrt{y}$

| $y$ | $x$ |
|---|---|
| 0 | 0 |
| 1 | 1 |
| 4 | 2 |
| 9 | 3 |

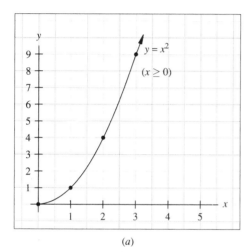

(a)

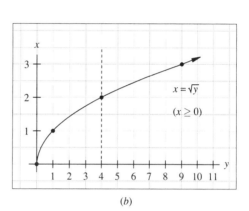

(b)

**Fig. 4-25**

# Supplementary Problems

## SYSTEMS OF EQUATIONS AND THEIR SOLUTIONS

**4.31.**    Give the dimensions of the following systems of equations:

(a)    $z = 4w + 3x + 6y$
         $z = 7w - 2x - 8y$
         $z = 9w - 4x + 5y$

(b)    $y = 6x_1 + 5x_2$
         $y = 4x_1 + 9x_2$
         $y = 3x_1 + 8x_2$

(c)    $y = 8x + 5$
         $y = 2x + 9$
         $y = 3x + 7$

**4.32.**    Use graphs on your own to solve each of the following systems of linear equations:

(a)    $7x + 2y = 33$
         $4x - 9y = -42$

(b)    $6x - 8y = 10$
         $5x + 3y = 47$

(c)    $3x + 4y = 26$
         $-x + 9y = 74$

(d)    $6x - 2y = 26$
         $15x - 5y = 85$

### SUPPLY AND DEMAND

**4.33.**    Using graphs on your own, with $P$ on the vertical axis as in economics, find the equilibrium quantity and price $(Q_e, P_e)$ in each of the following markets:

    (a)   Supply:  $P = \frac{1}{4}Q + 2$                           (b)   Supply:  $P = \frac{2}{5}Q + 3$

          Demand: $P = -\frac{3}{4}Q + 22$                      Demand: $P = -\frac{3}{5}Q + 33$

    (c)   Supply:  $P = \frac{3}{8}Q + 1$                           (d)   Supply:  $P = \frac{2}{3}Q - 4$

          Demand: $P = -\frac{1}{2}Q + 22$                      Demand: $P = -\frac{1}{6}Q + 11$

**4.34.**    Solve algebraically each of the following systems of linear supply and demand equations expressed in the mathematical format $Q = f(P)$:

    (a)   Supply:  $Q = 25P - 185$                     (b)   Supply:  $Q = 18P + 60$

          Demand: $Q = -32P + 1240$                 Demand: $Q = -22P + 1260$

    (c)   Supply:  $Q = 42P + 58$                       (d)   Supply:  $Q = 19P + 8$

          Demand: $Q = -37P + 2112$                 Demand: $Q = -13P + 2184$

### BREAK-EVEN ANALYSIS

**4.35.**    Find the break-even point for each of the following firms operating in a purely competitive market, given total revenue $R(x)$ and total cost $C(x)$ functions for each:

    (a)   $R(x) = 125x$                                 (b)   $R(x) = 37.5x$

          $C(x) = 85x + 5200$                          $C(x) = 22.25x + 1586$

    (c)   $R(x) = 495x$                                 (d)   $R(x) = 85.5x$

          $C(x) = 350x + 40,600$                       $C(x) = 79.75x + 5888$

**4.36.**    Find the break-even point for each of the following monopolistic firms, given

    (a)   $R(x) = -x^2 + 22x$                        (b)   $R(x) = -5x^2 + 163x$

          $C(x) = 7x + 36$                              $C(x) = 23x + 800$

    (c)   $R(x) = -2x^2 + 85x$                      (d)   $R(x) = -3x^2 + 201x$

          $C(x) = 11x + 420$                            $C(x) = 6x + 2250$

### ELIMINATION AND SUBSTITUTION METHODS

**4.37.**    Use the substitution method to solve each of the following systems of simultaneous equations:

    (a)   $3x - y = 1$           (b)   $7x + 2y = 62$           (c)   $5x + y = 26$

          $4x + 6y = 38$               $x + 6y = 26$               $8x - 3y = 60$

    (d)   $18x - 2y = 32$

          $12x + 5y = -23$

**4.38.**    Use the elimination method to solve each of the following systems of simultaneous equations:

    (a)   $4x - 3y = 22$           (b)   $-5x + 8y = 42$           (c)   $24x - 7y = 37$

          $7x - 6y = 34$               $15x + 9y = 6$              $-6x + 9y = 27$

    (d)   $11x + 3y = 53$

          $4x - 18y = 172$

**INCOME DETERMINATION MODELS**

**4.39.** Find the equilibrium level of income $Y_e$, given
   (a)  $Y = C + I$, $C = 150 + 0.75Y$, $I = 35$
   (b)  $Y = C + I$, $C = 275 + 0.75Y$, $I = 40 + 0.15Y$
   (c)  $Y = C + I + G$, $C = 320 + 0.65Y$, $I = 65 + 0.25Y$, $G = 150$
   (d)  $Y = C + I + G$, $C = 240 + 0.8Y_d$, $Y_d = Y - T$, $I = 70$, $G = 120$, $T = 50$
   (e)  $Y = C + I + G$, $C = 160 + 0.8Y_d$, $Y_d = Y - T$, $I = 80$, $G = 120$, $T = 40 + 0.25Y$
   (f)  $Y = C + I + G + (X - Z)$, $C = 420 + 0.85Y$, $I = 130$, $G = 310$, $X = 90$, $Z = 30 + 0.25Y$

**IS-LM CURVES**

**4.40.** Find the level of income $Y$ and the rate of interest $i$ that simultaneously bring equilibrium to the economy and estimate the level of consumption $C$, investment $I$, the speculative demand for money $M_w$, and the transaction-precautionary demand for money $M_t$ when
   (a)  the money supply $M_s = 1000$, $C = 950 + 0.75Y$, $I = 310 - 125i$, $M_w = 264 - 175i$, and $M_t = 0.15Y$,
   and (b)  $M_s = 800$, $C = 1200 + 0.6Y$, $I = 227 - 180i$, $M_w = 127 - 180i$, and $M_t = 0.2Y$.

# Answers to Supplementary Problems

**4.31.**    (a)  $3 \times 4$; underconstrained     (b)  $3 \times 3$; exactly constrained     (c)  $3 \times 2$; overconstrained

**4.32.**    (a)  (3, 6)     (b)  (7, 4)     (c)  (−2, 8)     (d)  none

**4.33.**    (a)  (20, 7)     (b)  (30, 15)     (c)  (24, 10)     (d)  (18, 8)

**4.34.**    (a)  $P_e = 25$, $Q_e = 440$          (b)  $P_e = 30$, $Q_e = 600$
        (c)  $P_e = 26$, $Q_e = 1150$       (d)  $P_e = 68$, $Q_e = 1300$

**4.35.**    (a)  $x = 130$     (b)  $x = 104$     (c)  $x = 280$     (d)  $x = 1024$

**4.36.**    (a)  $x = 3$, $x = 12$     (b)  $x = 8$, $x = 20$     (c)  $x = 7$, $x = 30$     (d)  $x = 15$, $x = 50$

**4.37.**    (a)  $x = 2$, $y = 5$     (b)  $x = 8$, $y = 3$     (c)  $x = 6$, $y = -4$     (d)  $x = 1$, $y = -7$

**4.38.**    (a)  $x = 10$, $y = 6$     (b)  $x = -2$, $y = 4$     (c)  $x = 3$, $y = 5$     (d)  $x = 7$, $y = -8$

**4.39.**    (a)  $Y_e = 740$     (b)  $Y_e = 3150$     (c)  $Y_e = 5350$     (d)  $Y_e = 1950$     (e)  $Y_e = 820$
       (f)  $Y_e = 2300$

**4.40.**    (a)  $Y = 5000$, $i = 0.08$, $C = 4700$, $I = 300$, $M_w = 250$, $M_t = 750$
       (b)  $Y = 3500$, $i = 0.15$, $C = 3300$, $I = 200$, $M_w = 100$, $M_t = 700$

# Chapter 5

# Linear (or Matrix) Algebra

## 5.1 INTRODUCTION

Linear (or matrix) algebra (1) permits expression of a system of linear equations in a succinct, simplified way; (2) provides a shorthand method to determine whether a solution exists before it is attempted; and (3) furnishes the means of solving the equation system. Although applicable only to linear equations, linear algebra is of immense help in economics and business because many economic relationships can be approximated by linear functions or converted to linear relationships. See Section 11.7.

**EXAMPLE 1.** For a company with several different retail stores selling the same line of products, a matrix provides a concise way of keeping track of stock.

| Store | PCs | Printers | Monitors | Modems |
|-------|-----|----------|----------|--------|
| 1 | 120 | 145 | 130 | 85 |
| 2 | 165 | 105 | 155 | 90 |
| 3 | 110 | 115 | 95 | 80 |
| 4 | 185 | 170 | 165 | 105 |

By reading across a row of the matrix, the firm can determine the level of stock in any of its outlets. By reading down a column of the matrix, the firm can determine the stock of any line of its products.

## 5.2 DEFINITIONS AND TERMS

A *matrix* is a rectangular array of numbers [7  2  6], parameters [*a  b  c*], or variables [*x  y  z*], arranged in meaningful order. The numbers (parameters, or variables) are referred to as *elements* of the matrix. The elements in a horizontal line constitute a *row* of the matrix; the elements in a vertical line constitute a *column* of the matrix. The number of rows $r$ and columns $c$ in a matrix defines the dimensions of the matrix ($r \times c$), which is read "$r$ by $c$." It is important to note that the row number always precedes the column number. In a *square matrix*, the number of rows equals the number of columns (that is, $r = c$). A matrix composed of several rows and a single column, such that its dimensions are ($r \times 1$), is called a *column vector*. A matrix consisting of a single row and several columns, with dimensions ($1 \times c$), is a *row vector*. A *transpose matrix*, designated by $A'$ or $A^T$, is a matrix in which the rows of $A$ are converted to columns and the columns of $A$ changed to rows.

**EXAMPLE 2.** Assume

$$A = \begin{bmatrix} a_{11} & a_{12} & a_{13} \\ a_{21} & a_{22} & a_{23} \\ a_{31} & a_{32} & a_{33} \end{bmatrix} \quad B = \begin{bmatrix} 3 & 8 & 6 \\ 5 & 2 & 9 \end{bmatrix} \quad C = [4 \quad 7 \quad 1] \quad X = \begin{bmatrix} x_1 \\ x_2 \\ x_3 \end{bmatrix}$$

Here A is a ($3 \times 3$) *general matrix* in which each element is represented by the same parameter $a$ distinguished one from another by a pair of subscripts. The subscripts are used to indicate the *address* or placement of the element in the matrix. The first subscript denotes the row in which the element appears; the second identifies the column. Positioning is precise within a matrix. Recalling that row is always listed before column (think of RC cola), note that $a_{13}$ is the element that appears in the first row, third column, while $a_{31}$ is the element located in the third row, first column.

The $B$ matrix is $2 \times 3$. To determine the number of rows in a matrix, always count down; to find the number of columns, simply count across. In $B$, the $b_{12}$ element is 8; the $b_{21}$ element is 5. The $C$ matrix is a row vector of numbers

128

with dimensions $1 \times 3$. The $X$ matrix is a column vector of variables; its dimensions are $3 \times 1$. The single subscripts in $X$ are used solely to distinguish between variables; single subscripts never denote address.

By taking the rows of $A$ and making them columns (or, equally effective, by taking the columns of $A$ and making them rows), the transpose of $A$ is found:

$$A' = \begin{bmatrix} a_{11} & a_{21} & a_{31} \\ a_{12} & a_{22} & a_{32} \\ a_{13} & a_{23} & a_{33} \end{bmatrix}$$

Similarly, the transpose of $X$ is

$$X' = [x_1 \quad x_2 \quad x_3]$$

## 5.3  ADDITION AND SUBTRACTION OF MATRICES

Addition (and subtraction) of two matrices $A + B$ (and $A - B$) requires that the two matrices be of the same dimensions. Each element of one matrix is then added to (subtracted from) the corresponding element of the other matrix. Thus $b_{11}$ in $B$ will be added to (subtracted from) $a_{11}$ in $A$, $b_{12}$ to $a_{12}$, and so on. See Examples 3 and 4 and Problems 5.4 to 5.7.

**EXAMPLE 3.**    The sum $A + B$ is calculated below, given the matrices

$$A = \begin{bmatrix} 7 & 4 \\ 6 & 5 \end{bmatrix} \qquad B = \begin{bmatrix} 3 & 9 \\ 8 & 2 \end{bmatrix}$$

$$A + B = \begin{bmatrix} 7+3 & 4+9 \\ 6+8 & 5+2 \end{bmatrix} = \begin{bmatrix} 10 & 13 \\ 14 & 7 \end{bmatrix}$$

The difference $C - D$, given matrices $C$ and $D$, is found as follows

$$C = \begin{bmatrix} 8 & 3 \\ 2 & 9 \end{bmatrix} \qquad D = \begin{bmatrix} 5 & 7 \\ 8 & 4 \end{bmatrix}$$

$$C - D = \begin{bmatrix} 8-5 & 3-7 \\ 2-8 & 9-4 \end{bmatrix} = \begin{bmatrix} 3 & -4 \\ -6 & 5 \end{bmatrix}$$

**EXAMPLE 4.**    Given

$$D = \begin{bmatrix} 20 & 15 & 35 & 10 \\ 25 & 5 & 30 & 15 \\ 10 & 40 & 25 & 20 \\ 15 & 20 & 35 & 10 \end{bmatrix}$$

where $D$ represents deliveries made to the different stores in Example 1, find the new level of stock.

To find the new level of stock, label $S$ the initial matrix given in Example 1, and solve for $S + D$. Adding the corresponding elements of each matrix as follows,

$$S + D = \begin{bmatrix} 120+20 & 145+15 & 130+35 & 85+10 \\ 165+25 & 105+5 & 155+30 & 90+15 \\ 110+10 & 115+40 & 95+25 & 80+20 \\ 185+15 & 170+20 & 165+35 & 105+10 \end{bmatrix}$$

$$= \begin{bmatrix} 140 & 160 & 165 & 95 \\ 190 & 110 & 185 & 105 \\ 120 & 155 & 120 & 100 \\ 200 & 190 & 200 & 115 \end{bmatrix}$$

## 5.4  SCALAR MULTIPLICATION

In linear algebra a real number such as 7, $-5$, or 0.08 is called a *scalar.* Multiplication of a matrix by a scalar involves multiplication of every element of the matrix by the scalar. The process is called *scalar multiplication* because it scales the matrix up or down according to the value of the scalar. See Example 5 and Problems 5.9 to 5.11.

**EXAMPLE 5.**  The result of scalar multiplication $Ak$, given $k = 5$ and

$$A = \begin{bmatrix} 4 & 9 \\ 2 & 6 \\ 3 & 7 \end{bmatrix}_{3 \times 2}$$

is

$$Ak = \begin{bmatrix} 4(5) & 9(5) \\ 2(5) & 6(5) \\ 3(5) & 7(5) \end{bmatrix}_{3 \times 2} = \begin{bmatrix} 20 & 45 \\ 10 & 30 \\ 15 & 35 \end{bmatrix}_{3 \times 2}$$

## 5.5  VECTOR MULTIPLICATION

Multiplication of a row vector $A$ by a column vector $B$ requires as a precondition that each vector have exactly the same number of elements. The product is then found by multiplying the individual elements of the row vector by their corresponding elements in the column vector and summing up the products:

$$AB = (a_{11} \cdot b_{11}) + (a_{12} \cdot b_{21}) + (a_{13} \cdot b_{31}) + \cdots$$

The product of a row-column multiplication will thus be a single element. Row-column vector multiplication is of paramount importance. It is the basis for all matrix multiplication. See Example 6 and Problems 5.12 to 5.16.

**EXAMPLE 6.**  The product $AB$ of the row vector $A$ and the column vector $B$, given

$$A = [2 \quad 6 \quad 5 \quad 8]_{1 \times 4} \qquad B = \begin{bmatrix} 7 \\ 3 \\ 9 \\ 4 \end{bmatrix}_{4 \times 1}$$

where $A$ and $B$ both have the same number of elements, is calculated as follows:

$$AB = [2(7) + 6(3) + 5(9) + 8(4)] = [14 + 18 + 45 + 32] = [109]$$

The product of vectors

$$C = [6 \quad 1 \quad 9]_{1 \times 3} \qquad D = \begin{bmatrix} 8 \\ 5 \\ 3 \end{bmatrix}_{3 \times 1}$$

where both again must have the same number of elements is

$$CD = [6(8) + 1(5) + 9(3)] = [48 + 5 + 27] = [80]$$

Be warned that reversing the order of multiplication in either of the above cases and having column-row multiplication ($BA$ or $DC$) would give totally different answers. See Problem 5.27.

## 5.6  MULTIPLICATION OF MATRICES

Multiplication of two matrices $AB$ with dimensions $(r_1 \times c_1)$ and $(r_2 \times c_2)$ requires that the matrices be *conformable*, that is, that $c_1 = r_2$ or the number of columns in the first matrix, called the "lead" matrix, equal the number of rows in the second matrix, known as the "lag" matrix. Each row vector in the lead matrix is then multiplied by each column vector in the lag matrix, using the rules for multiplying row and column vectors discussed in Section 5.5. Each row-column product, called an *inner product*, in turn forms an element of the product matrix, such that each element $c_{ij}$ of the product matrix $C$ is a scalar derived from the multiplication of the $i$th row of the lead matrix and the $j$th column of the lag matrix. See Examples 7 and 8 and Problems 5.17 to 5.31. For the commutative, associative, and distributive laws in matrix algebra, see Dowling, *Introduction to Mathematical Economics*, Section 10.7 and Problems 10.34 to 10.48.

**EXAMPLE 7.**    Given

$$A = \begin{bmatrix} 4 & 9 & 11 \\ 7 & 12 & 3 \end{bmatrix}_{2\times3} \qquad B = \begin{bmatrix} 8 & 2 \\ 6 & 20 \\ 7 & 12 \end{bmatrix}_{3\times2} \qquad C = \begin{bmatrix} 5 & 16 & 2 \\ 4 & 3 & 9 \end{bmatrix}_{2\times3}$$

An easy way to check conformability, which must be tested before any matrix multiplication, is to place the two sets of dimensions in the order in which they are to be multiplied, then mentally circle the last number of the first set and the first number of the second set. If the two numbers are equal, the number of columns in the lead matrix equals the number of rows in the lag matrix and the matrices are conformable for multiplication in the given order. Moreover, the numbers outside the circle will provide, in proper order, the dimensions of the product matrix. Thus, for $AB$,

$$2 \times \overbrace{(3 = 3)} \times 2$$
$$2 \times 2$$

The number of columns in the lead matrix equals the number of rows of the lag matrix, $3 = 3$; the matrices are conformable for multiplication in the given order; and the dimensions of the product matrix $AB$ will be $2 \times 2$. When two matrices such as $AB$ are conformable for multiplication, the product $AB$ is said to be *defined*.

For $BC$,

$$3 \times \overbrace{(2 = 2)} \times 3$$
$$3 \times 3$$

The number of columns in $B$ equals the number of rows in $C$, $2 = 2$; $B$ and $C$ are conformable in the given order. The product $BC$ is defined and will be a $3 \times 3$ matrix.

For $AC$,

$$2 \times \overbrace{(3 \neq 2)} \times 3$$

$A$ and $C$ are not conformable for multiplication. $AC$ is undefined.

**EXAMPLE 8.**    Having determined that $A$ and $B$ are conformable in Example 7, we can now find the product $AB$. Always remembering to use only rows $R$ from the lead matrix and only columns $C$ from the lag matrix, multiply the first row $R_1$ of the lead matrix by the first column $C_1$ of the lag matrix to find the first element $d_{11}$ ($= R_1C_1$) of the product matrix $D$. Then multiply the first row $R_1$ of the lead matrix by the second column $C_2$ of the lag matrix to find $d_{12}$ ($= R_1C_2$). Since there are no more columns left in the lag matrix to be multiplied by the first row of the lead matrix, move to the second row of the lead matrix. Multiply the second row $R_2$ of the lead matrix by the first column $C_1$ of the lag matrix to get $d_{21}$ ($= R_2C_1$). Finally, multiply the second row $R_2$ of the lead matrix by the second column $C_2$ of the lag matrix to get $d_{22}$ ($= R_2C_2$). Thus

$$AB = D = \begin{bmatrix} R_1C_1 & R_1C_2 \\ R_2C_1 & R_2C_2 \end{bmatrix}$$

$$= \begin{bmatrix} 4(8) + 9(6) + 11(7) & 4(2) + 9(20) + 11(12) \\ 7(8) + 12(6) + 3(7) & 7(2) + 12(20) + 3(12) \end{bmatrix} = \begin{bmatrix} 163 & 320 \\ 149 & 290 \end{bmatrix}$$

The product of $BC$ is calculated below, using the same procedure.

$$BC = E = \begin{bmatrix} R_1C_1 & R_1C_2 & R_1C_3 \\ R_2C_1 & R_2C_2 & R_2C_3 \end{bmatrix}$$

$$= \begin{bmatrix} 8(5) + 2(4) & 8(16) + 2(3) & 8(2) + 2(9) \\ 6(5) + 20(4) & 6(16) + 20(3) & 6(2) + 20(9) \\ 7(5) + 12(4) & 7(16) + 12(3) & 7(2) + 12(9) \end{bmatrix} = \begin{bmatrix} 48 & 134 & 34 \\ 110 & 156 & 192 \\ 83 & 148 & 122 \end{bmatrix}$$

See also Problems 5.17 to 5.31.

**EXAMPLE 9.** Referring to Example 1, assume that the price of PCs is \$900, printers \$500, monitors \$350, and modems \$200. To find the value $V$ of the stock in the different stores, express the prices as a column vector $P$, and multiply $S$ by $P$:

$$V = SP = \begin{bmatrix} 120 & 145 & 130 & 85 \\ 165 & 105 & 155 & 90 \\ 110 & 115 & 95 & 80 \\ 185 & 170 & 165 & 105 \end{bmatrix}_{4 \times 4} \begin{bmatrix} 900 \\ 500 \\ 350 \\ 200 \end{bmatrix}_{4 \times 1}$$

The matrices are conformable and the product matrix will be $4 \times 1$:

$$4 \times 4 = 4 \times 1$$
$$4 \times 1$$

Thus

$$V = \begin{bmatrix} R_1C_1 \\ R_2C_1 \\ R_3C_1 \\ R_4C_1 \end{bmatrix} = \begin{bmatrix} 120(900) + 145(500) + 130(350) + 85(200) \\ 165(900) + 105(500) + 155(350) + 90(200) \\ 110(900) + 115(500) + 95(350) + 80(200) \\ 185(900) + 170(500) + 165(350) + 105(200) \end{bmatrix}$$

$$= \begin{bmatrix} 243{,}000 \\ 273{,}250 \\ 205{,}750 \\ 330{,}250 \end{bmatrix}$$

Note that the product $SP$ is defined above when $P$ *post*multiplies $S$; unlike ordinary algebra, however, when $P$ *pre*multiplies $S$, the product $PS$ is not defined:

$$4 \times 1 \neq 4 \times 4$$

## 5.7  MATRIX EXPRESSION OF A SYSTEM OF LINEAR EQUATIONS

Matrix algebra permits concise expression of a system of linear equations. Given the system of linear equations

$$5x_1 + 12x_2 = 32 \qquad\qquad (5.1a)$$

$$7x_1 - 3x_2 = 25 \qquad\qquad (5.1b)$$

in which like terms are all aligned in the same column, as is typically done in ordinary algebra,

(a)  Create a *coefficient matrix* $A$ by copying the coefficients in the exact order as given above

$$A = \begin{bmatrix} 5 & 12 \\ 7 & -3 \end{bmatrix} \qquad\qquad (5.2)$$

(b)  Create a *column vector of variables* $X$ in which the variables read down in the same order as they appear from left to right above

$$X = \begin{bmatrix} x_1 \\ x_2 \end{bmatrix} \tag{5.3}$$

(c)  Create a *column vector of constants* $B$

$$B = \begin{bmatrix} 32 \\ 25 \end{bmatrix} \tag{5.4}$$

Then the system of equations can be expressed simply as

$$AX = B \tag{5.5}$$

as is illustrated in Examples 10 and 11 and Problems 5.32 and 5.33.

**EXAMPLE 10.**   To show that $AX = B$ accurately represents the system of equations in (*5.1a*) and (*5.1b*), find the product $AX$ and substitute in (*5.5*).
From (*5.2*) and (*5.3*),

$$AX = \begin{bmatrix} 5 & 12 \\ 7 & -3 \end{bmatrix}_{2\times2} \begin{bmatrix} x_1 \\ x_2 \end{bmatrix}_{2\times1} = \begin{bmatrix} 5x_1 + 12x_2 \\ 7x_1 - 3x_2 \end{bmatrix}_{2\times1}$$

Substituting in (*5.5*),

$$AX = B: \quad \begin{bmatrix} 5x_1 + 12x_2 \\ 7x_1 - 3x_2 \end{bmatrix}_{2\times1} = \begin{bmatrix} 32 \\ 25 \end{bmatrix}_{2\times1} \qquad \text{Q.E.D.}$$

Here, despite appearances, $AX$ is a $2 \times 1$ column vector because each row is composed of a single element which cannot be simplified further through addition or subtraction.

**EXAMPLE 11.**   By following the procedure outlined in Section 5.7 and mentally reversing the process of matrix multiplication, the system of equations

$$9w + 4x - 5y + 2z = 10$$

$$3w - 7x + 11y - 6z = -2$$

can be expressed in matrix notation as follows:

$$\begin{bmatrix} 9 & 4 & -5 & 2 \\ 3 & -7 & 11 & -6 \end{bmatrix}_{2\times4} \begin{bmatrix} w \\ x \\ y \\ z \end{bmatrix}_{4\times1} = \begin{bmatrix} 10 \\ -2 \end{bmatrix}_{2\times1}$$

Then by letting $A =$ the matrix of coefficients, $W =$ the column vector of variables, and $B =$ the column vector of constants, the given system of equations can be expressed in matrix form

$$A_{2\times4} W_{4\times1} = B_{2\times1}$$

and checked by multiplication.

## 5.8  AUGMENTED MATRIX

Given a system of equations in matrix form $AX = B$, the *augmented matrix* $A|B$ is the coefficient matrix $A$ with the column vector of constants $B$ set alongside it, separated by a line or bar. Thus, for the system of equations in (*5.1a*) and (*5.1b*),

$$A|B = \begin{bmatrix} 5 & 12 & \big| & 32 \\ 7 & -3 & \big| & 25 \end{bmatrix}$$

**EXAMPLE 12.**    The augmented matrix $A|B$ for

$$3x_1 + 8x_2 + 4x_3 = 58$$

$$6x_1 + 5x_2 - 2x_3 = 22$$

$$x_1 - 7x_2 + 9x_3 = 18$$

is

$$A|B = \begin{bmatrix} 3 & 8 & 4 & | & 58 \\ 6 & 5 & -2 & | & 22 \\ 1 & -7 & 9 & | & 18 \end{bmatrix}$$

## 5.9  ROW OPERATIONS

*Row operations* involve the application of simple algebraic operations to the rows of a matrix. With no change in the linear relationship, the three basic row operations allow (1) any two rows of a matrix to be interchanged, (2) any row or rows to be multiplied by a constant, provided the constant does not equal zero, and (3) any multiple of a row to be added to or subtracted from any other row, as is demonstrated in Example 13.

**EXAMPLE 13.**    Row operations are demonstrated below in terms of ordinary algebra for simplicity. Given

$$6x + 5y = 43 \qquad\qquad (5.6)$$

$$8x + 12y = 84 \qquad\qquad (5.7)$$

Without any change in the linear relationship, one may

1.   Interchange the two rows:

$$8x + 12y = 84$$

$$6x + 5y = 43$$

2.   Multiply any row by a constant, here $8x + 12y = 84$ by $\frac{1}{4}$, leaving

$$2x + 3y = 21$$

$$6x + 5y = 43$$

3.   Subtract a multiple of one row from another, here $3(2x + 3y = 21)$ from $6x + 5y = 43$, leaving

$$6x + 5y = \quad 43$$

$$\underline{-6x - 9y = -63}$$

$$-4y = -20$$

$$y = 5$$

4.   Substitute $y = 5$ in (5.6) or (5.7), getting

$$6x + 5(5) = 43$$

$$x = 3$$

## 5.10  GAUSSIAN METHOD OF SOLVING LINEAR EQUATIONS

To use the Gaussian elimination method of solving a system of linear equations, simply express the system of equations as an augmented matrix and apply row operations to the augmented matrix until the original coefficient matrix to the left of the bar is reduced to an identity matrix. An *identity matrix* is a square matrix which has 1 for every element on the *principal diagonal* from left to right (i.e., for every $a_{ij}$ element where $i = j$) and 0 everywhere else (i.e., for every $a_{ij}$ element where $i \neq j$). The solution to the original system of equations can then be found in the column vector to the right of the bar (see Example 14).

To transform the coefficient matrix to an identity matrix, work along the principal diagonal. First obtain a 1 in the $a_{11}$ position of the coefficient matrix; then use row operations to obtain 0s everywhere else in the first column. Next obtain a 1 in the $a_{22}$ position, and use row operations to obtain 0s everywhere else in the second column. Continue getting 1s along the principal diagonal and then clearing the columns until the identity matrix is completed. See Example 14 and Problems 5.34 to 5.39.

**EXAMPLE 14.**	The Gaussian elimination method is used below to solve for $x_1$ and $x_2$, given the system of linear equations:

$$3x_1 + 12x_2 = 102$$

$$4x_1 + 5x_2 = 48$$

First express the equations in an augmented matrix:

$$A|B = \begin{bmatrix} 3 & 12 & | & 102 \\ 4 & 5 & | & 48 \end{bmatrix}$$

Then

1a.	Multiply the first row by $\frac{1}{3}$ to obtain 1 in the $a_{11}$ position.

$$\begin{bmatrix} 1 & 4 & | & 34 \\ 4 & 5 & | & 48 \end{bmatrix}$$

1b.	Subtract 4 times row 1 from row 2 to clear the first column.

$$\begin{bmatrix} 1 & 4 & | & 34 \\ 0 & -11 & | & -88 \end{bmatrix}$$

2a.	Multiply row 2 by $-\frac{1}{11}$ to obtain 1 in the $a_{22}$ position.

$$\begin{bmatrix} 1 & 4 & | & 34 \\ 0 & 1 & | & 8 \end{bmatrix}$$

2b.	Subtract 4 times row 2 from row 1 to clear the second column.

$$\begin{bmatrix} 1 & 0 & | & 2 \\ 0 & 1 & | & 8 \end{bmatrix}$$

The solution is $x_1 = 2$, $x_2 = 8$ since

$$\begin{bmatrix} 1 & 0 \\ 0 & 1 \end{bmatrix} \begin{bmatrix} x_1 \\ x_2 \end{bmatrix} = \begin{bmatrix} 2 \\ 8 \end{bmatrix}$$

$$x_1 + 0 = 2$$

$$0 + x_2 = 8$$

# Solved Problems

## MATRIX FORMAT

**5.1.**	(a) Give the dimensions of each of the following matrices. (b) Find their transposes and indicate the new dimensions.

$$A = \begin{bmatrix} 4 & 8 \\ 7 & 2 \\ 3 & 5 \end{bmatrix} \qquad B = \begin{bmatrix} 3 & 4 & 4 \\ 1 & 7 & 9 \\ 9 & 2 & 6 \\ 8 & 5 & 4 \end{bmatrix} \qquad C = \begin{bmatrix} 6 \\ 13 \\ 9 \end{bmatrix}$$

$$D = \begin{bmatrix} 5 & 2 & 3 & 4 \\ 9 & 8 & 1 & 9 \end{bmatrix} \qquad E = [-9 \quad 5 \quad 2 \quad -3 \quad 7] \qquad F = \begin{bmatrix} 6 & 8 & 3 & 4 \\ 4 & 5 & 9 & 3 \\ 7 & 1 & 2 & 9 \end{bmatrix}$$

(a)  Recalling that dimensions are always listed row by column or $rc$, $A = 3 \times 2$, $B = 4 \times 3$, $C = 3 \times 1$, $D = 2 \times 4$, $E = 1 \times 5$, and $F = 3 \times 4$. $C$ is also called a column vector; $E$, a row vector. (b) The transpose of $A$ converts the rows of $A$ to columns and the columns of $A$ to rows.

$$A' = \begin{bmatrix} 4 & 7 & 3 \\ 8 & 2 & 5 \end{bmatrix}_{2 \times 3} \qquad B' = \begin{bmatrix} 3 & 1 & 9 & 8 \\ 4 & 7 & 2 & 5 \\ 4 & 9 & 6 & 4 \end{bmatrix}_{3 \times 4} \qquad C' = [6 \quad 13 \quad 9]_{1 \times 3}$$

$$D' = \begin{bmatrix} 5 & 9 \\ 2 & 8 \\ 3 & 1 \\ 4 & 9 \end{bmatrix}_{4 \times 2} \qquad E' = \begin{bmatrix} -9 \\ 5 \\ 2 \\ -3 \\ 7 \end{bmatrix}_{5 \times 1} \qquad F' = \begin{bmatrix} 6 & 4 & 7 \\ 8 & 5 & 1 \\ 3 & 9 & 2 \\ 4 & 3 & 9 \end{bmatrix}_{4 \times 3}$$

**5.2.**  Use your knowledge of subscripts and addresses to complete the following matrix, given $a_{12} = 9$, $a_{21} = -4$, $a_{13} = -5$, $a_{31} = 2$, $a_{23} = 7$, and $a_{32} = 3$.

$$A = \begin{bmatrix} 8 & \_ & \_ \\ \_ & 1 & \_ \\ \_ & \_ & 6 \end{bmatrix}$$

Since the subscripts are always given in row-column order, $a_{12} = 9$ means that 9 is situated in the first row, second column; $a_{21} = -4$ means that $-4$ appears in the second row, first column; and so forth. Thus,

$$A = \begin{bmatrix} 8 & 9 & -5 \\ -4 & 1 & 7 \\ 2 & 3 & 6 \end{bmatrix}$$

**5.3.**  A company with four retail stores has 35 TVs $t$, 60 stereos $s$, 55 VCRs (videocassette recorders) $v$, and 45 camcorders $c$ in store 1; $80t$, $65s$, $50v$, and $38c$ in store 2; $29t$, $36s$, $24v$, and $32c$ in store 3; and $62t$, $49s$, $54v$, and $33c$ in store 4. Express present inventory in matrix form.

| Retail store | $t$ | $s$ | $v$ | $c$ |
|---|---|---|---|---|
| 1 | 35 | 60 | 55 | 45 |
| 2 | 80 | 65 | 50 | 38 |
| 3 | 29 | 36 | 24 | 32 |
| 4 | 62 | 49 | 54 | 33 |

## MATRIX ADDITION AND SUBTRACTION

**5.4.**  Find the sum $A + B$ for the following matrices:

(a)  $A = \begin{bmatrix} 6 & 5 \\ 8 & 2 \end{bmatrix} \qquad B = \begin{bmatrix} 3 & 7 \\ 9 & 4 \end{bmatrix}$

$$A + B = \begin{bmatrix} 6+3 & 5+7 \\ 8+9 & 2+4 \end{bmatrix} = \begin{bmatrix} 9 & 12 \\ 17 & 6 \end{bmatrix}$$

(b)　$A = \begin{bmatrix} -4 & 7 \\ 1 & -8 \end{bmatrix}$　　$B = \begin{bmatrix} 5 & -2 \\ -6 & -9 \end{bmatrix}$

$$A + B = \begin{bmatrix} -4+5 & 7+(-2) \\ 1+(-6) & -8+(-9) \end{bmatrix} = \begin{bmatrix} 1 & 5 \\ -5 & -17 \end{bmatrix}$$

(c)　$A = [15 \quad 4 \quad 3 \quad 11]$　　$B = [8 \quad -6 \quad -9 \quad 5]$

$$A + B = [23 \quad -2 \quad -6 \quad 16]$$

(d)　$A = \begin{bmatrix} 12 & 5 \\ -4 & 18 \\ 7 & -6 \\ 24 & -2 \end{bmatrix}$　　$B = \begin{bmatrix} 9 & -8 \\ 7 & -4 \\ 6 & 32 \\ -5 & -4 \end{bmatrix}$

$$A + B = \begin{bmatrix} 21 & -3 \\ 3 & 14 \\ 13 & 26 \\ 19 & -6 \end{bmatrix}$$

**5.5.**　The parent company in Problem 5.3 sends out deliveries to its stores:

$$D = \begin{bmatrix} 8 & 6 & 9 & 5 \\ 4 & 7 & 5 & 2 \\ 6 & 3 & 0 & 8 \\ 5 & 9 & 7 & 4 \end{bmatrix}$$

What is the new level of inventory?

$$I_2 = I_1 + D = \begin{bmatrix} 35 & 60 & 55 & 45 \\ 80 & 65 & 50 & 38 \\ 29 & 36 & 24 & 32 \\ 62 & 49 & 54 & 33 \end{bmatrix} + \begin{bmatrix} 8 & 6 & 9 & 5 \\ 4 & 7 & 5 & 2 \\ 6 & 3 & 0 & 8 \\ 5 & 9 & 7 & 4 \end{bmatrix} = \begin{bmatrix} 43 & 66 & 64 & 50 \\ 84 & 72 & 55 & 40 \\ 35 & 39 & 24 & 40 \\ 67 & 58 & 61 & 37 \end{bmatrix}$$

**5.6.**　Find the difference $A - B$ for each of the following:

(a)　$A = \begin{bmatrix} 6 & 3 & 7 \\ 2 & 9 & 1 \end{bmatrix}$　　$B = \begin{bmatrix} 5 & 8 & 9 \\ 4 & 1 & 2 \end{bmatrix}$

$$A - B = \begin{bmatrix} 6-5 & 3-8 & 7-9 \\ 2-4 & 9-1 & 1-2 \end{bmatrix} = \begin{bmatrix} 1 & -5 & -2 \\ -2 & 8 & -1 \end{bmatrix}$$

(b)　$A = \begin{bmatrix} 13 \\ 9 \\ 18 \\ 5 \end{bmatrix}$　　$B = \begin{bmatrix} 8 \\ 3 \\ 15 \\ 7 \end{bmatrix}$

$$A - B = \begin{bmatrix} 13 - 8 \\ 9 - 3 \\ 18 - 15 \\ 5 - 7 \end{bmatrix} = \begin{bmatrix} 5 \\ 6 \\ 3 \\ -2 \end{bmatrix}$$

(c)   $A = \begin{bmatrix} 14 & 8 & -3 \\ 6 & 19 & 11 \\ 20 & -1 & 18 \end{bmatrix}$     $B = \begin{bmatrix} 9 & 4 & 5 \\ -8 & 16 & 7 \\ 13 & 2 & 12 \end{bmatrix}$

$$A - B = \begin{bmatrix} 5 & 4 & -8 \\ 14 & 3 & 4 \\ 7 & -3 & 6 \end{bmatrix}$$

**5.7.** A monthly report $R$ on sales for the company in Problem 5.5 indicates:

$$R = \begin{bmatrix} 21 & 16 & 36 & 18 \\ 44 & 26 & 21 & 19 \\ 11 & 17 & 13 & 20 \\ 33 & 28 & 34 & 12 \end{bmatrix}$$

What is the inventory level $I_3$ at the end of the month?

$$I_3 = I_2 - R = \begin{bmatrix} 43 & 66 & 64 & 50 \\ 84 & 72 & 55 & 40 \\ 35 & 39 & 24 & 40 \\ 67 & 58 & 61 & 37 \end{bmatrix} - \begin{bmatrix} 21 & 16 & 36 & 18 \\ 44 & 26 & 21 & 19 \\ 11 & 17 & 13 & 20 \\ 33 & 28 & 34 & 12 \end{bmatrix} = \begin{bmatrix} 22 & 50 & 28 & 32 \\ 40 & 46 & 34 & 21 \\ 24 & 22 & 11 & 20 \\ 34 & 30 & 27 & 25 \end{bmatrix}$$

**5.8.** Given

$$A = [59 \quad 24] \qquad B = \begin{bmatrix} 39 \\ 44 \end{bmatrix} \qquad C = [7 \quad 4 \quad 8]$$

$$D = \begin{bmatrix} 4 & 7 & 2 \\ 9 & 5 & 8 \\ 3 & 1 & 6 \end{bmatrix} \qquad E = \begin{bmatrix} 9 & 4 \\ 2 & 5 \end{bmatrix} \qquad F = \begin{bmatrix} 14 \\ 25 \\ 19 \end{bmatrix}$$

Determine for each of the following whether the products are defined, that is, conformable for multiplication in the order given. If so, indicate the dimensions of the product matrix (a) $AB$, (b) $AE$, (c) $EB$, (d) $BE$, (e) $CD$, (f) $CF$, (g) $DF$, (h) $FD$, (i) $FC$.

(a)   The dimensions of $AB$, in that order of multiplication, are
      $1 \times (2 = 2) \times 1$. The product $AB$ is defined because the numbers within the dashed circle indicate that the number of columns in the lead matrix $A$ equals the number of rows in the lag matrix $B$. The numbers outside the circle indicate that the product matrix will be $1 \times 1$.

(b)   The dimensions of $AE$ are $1 \times (2 = 2) \times 2$. The product $AE$ is defined; the product matrix will be $1 \times 2$.

(c)   The dimensions of $EB$ are $2 \times (2 = 2) \times 1$. The product $EB$ is defined; the product matrix will be $2 \times 1$.

(d)   The dimensions of $BE$ are $2 \times (1 \neq 2) \times 2$. The product $BE$ is undefined. The matrices are not conformable for multiplication in the order given. [Note that $EB$ in part (c) is defined. This illustrates matrix multiplication is not commutative: $EB \neq BE$.]

(e)    The dimensions of $CD$ are $1 \times (3 = 3) \times 3$. The product $CD$ is defined; the product matrix will be $1 \times 3$.

(f)    The dimensions of $CF$ are $1 \times (3 = 3) \times 1$. The product $CF$ is defined; the product matrix will be $1 \times 1$.

(g)    The dimensions of $DF$ are $3 \times (3 = 3) \times 1$. The product $DF$ is defined; the product matrix will be $3 \times 1$.

(h)    The dimensions of $FD$ are $3 \times (1 \neq 3) \times 3$. The matrices are not conformable for multiplication in the given order. Although $DF$ in part (g) is defined, the product $FD$ is not.

(i)    The dimensions of $FC$ are $3 \times (1 = 1) \times 3$. The product $FC$ is defined; the product matrix will be $3 \times 3$. Compare this answer to the $1 \times 1$ product for $CF$ in part (f).

## SCALAR AND VECTOR MULTIPLICATION

**5.9.** Determine $Ak$, given

$$A = \begin{bmatrix} 6 & 1 \\ 9 & 2 \\ 7 & 4 \end{bmatrix} \qquad k = 5$$

Here $k$ is a scalar, and scalar multiplication is possible with a matrix of any dimension.

$$Ak = \begin{bmatrix} 6(5) & 1(5) \\ 9(5) & 2(5) \\ 7(5) & 4(5) \end{bmatrix} = \begin{bmatrix} 30 & 5 \\ 45 & 10 \\ 35 & 20 \end{bmatrix}$$

**5.10.** Find $kA$, given

$$k = -3 \qquad A = \begin{bmatrix} 7 & -4 & 2 \\ -9 & 5 & -6 \\ 1 & -8 & 1 \end{bmatrix}$$

$$kA = \begin{bmatrix} -3(7) & -3(-4) & -3(2) \\ -3(-9) & -3(5) & -3(-6) \\ -3(1) & -3(-8) & -3(1) \end{bmatrix} = \begin{bmatrix} -21 & 12 & -6 \\ 27 & -15 & 18 \\ -3 & 24 & -3 \end{bmatrix}$$

**5.11.** A ski shop discounts all its skis, poles, and bindings by 25 percent at the end of the season. Assuming that $V_1$ is the value of stock in its three branches prior to the discount, find the value $V_2$ after the discount when

$$V_1 = \begin{bmatrix} 8400 & 7200 & 6800 \\ 4600 & 5400 & 5600 \\ 6200 & 7800 & 7400 \end{bmatrix}$$

A 25 percent reduction means that the equipment is selling for 75 percent of its original value. Hence $V_2 = 0.75V_1$, and

$$V_2 = 0.75 \begin{bmatrix} 8400 & 7200 & 6800 \\ 4600 & 5400 & 5600 \\ 6200 & 7800 & 7400 \end{bmatrix} = \begin{bmatrix} 6300 & 5400 & 5100 \\ 3450 & 4050 & 4200 \\ 4650 & 5850 & 5550 \end{bmatrix}$$

**5.12.** Find $AB$, given

$$A = [8 \quad 3 \quad 6] \qquad B = \begin{bmatrix} 2 \\ 5 \\ 7 \end{bmatrix}$$

The product $AB$ is defined: $1 \times (3 = 3) \times 1$. The product will be a $1 \times 1$ matrix, derived by multiplying each element of the row vector $A$ by its corresponding element in the column vector $B$, and then summing the products.

$$AB = [8(2) + 3(5) + 6(7)] = [16 + 15 + 42] = [73]$$

**5.13.** Find $AB$, given

$$A = [5 \quad 12] \qquad B = \begin{bmatrix} 21 \\ 10 \end{bmatrix}$$

The product $AB$ is defined: $1 \times \undergroup{2 = 2} \times 1$.

$$AB = [5(21) + 12(10)] = [225]$$

**5.14.** Find $AB$, given

$$A = [6 \quad 3 \quad 5 \quad 8] \qquad B = \begin{bmatrix} 11 \\ 15 \\ 18 \\ 9 \end{bmatrix}$$

The product is defined: $1 \times \undergroup{4 = 4} \times 1$.

$$AB = [6(11) + 3(15) + 5(18) + 8(9)] = [273]$$

**5.15.** If the price of a TV is \$400, the price of a stereo is \$300, the price of a VCR is \$250, and the price of a camcorder is \$500, use vector multiplication to find the value of stock for outlet 2 in Problem 5.3.

The value of stock is $V = QP$. The physical volume of stock in outlet 2 in vector form is $Q = [80 \quad 65 \quad 50 \quad 38]$. The price vector $P$ can be written

$$P = \begin{bmatrix} 400 \\ 300 \\ 250 \\ 500 \end{bmatrix}$$

The product $QP$ is defined: $1 \times \undergroup{4 = 4} \times 1$. Thus,

$$V = QP = 80(400) + 65(300) + 50(250) + 38(500) = 83,000$$

**5.16.** Redo Problem 5.15 for outlet 3 in Problem 5.3.

Here $Q = [29 \quad 36 \quad 24 \quad 32]$, $P$ remains the same, and

$$V = QP = 29(400) + 36(300) + 24(250) + 32(500) = 44,400$$

## MATRIX MULTIPLICATION

**5.17.** Determine whether the product $AB$ is defined, indicate what the dimensions of the product matrix will be, and find the product matrix, given

$$A = \begin{bmatrix} 8 & 5 \\ 6 & 3 \end{bmatrix} \qquad B = \begin{bmatrix} 2 & 7 \\ 4 & 1 \end{bmatrix}$$

The product $AB$ is defined: $2 \times \undergroup{2 = 2} \times 2$; the product matrix will be $2 \times 2$. Matrix multiplication is then accomplished by a simple series of row-column vector multiplications to determine the various elements of the product matrix. The $a_{11}$ element of the product matrix is determined by the product of the first row $R_1$ of the lead matrix and the first column $C_1$ of the lag matrix; the $a_{12}$ element of the product matrix is determined by the product of the first row $R_1$ of the lead matrix and the second column $C_2$ of the lag matrix. In general, the $a_{ij}$ element of the product matrix is determined by the product of the $i$th row $R_i$ of the lead matrix and the $j$th column $C_j$ of the lag matrix. Thus

$$AB = \begin{bmatrix} R_1 C_1 & R_1 C_2 \\ R_2 C_1 & R_2 C_2 \end{bmatrix}$$

$$AB = \begin{bmatrix} 8(2) + 5(4) & 8(7) + 5(1) \\ 6(2) + 3(4) & 6(7) + 3(1) \end{bmatrix} = \begin{bmatrix} 36 & 61 \\ 24 & 45 \end{bmatrix}$$

**5.18.** Redo Problem 5.17, given

$$A = \begin{bmatrix} 4 & 7 \\ 2 & 5 \end{bmatrix} \qquad B = \begin{bmatrix} 3 & 1 & 6 \\ 8 & 2 & 9 \end{bmatrix}$$

The product $AB$ is defined: $2 \times (2 = 2) \times 3$. The product $AB$ will be $2 \times 3$.

$$AB = \begin{bmatrix} R_1 C_1 & R_1 C_2 & R_1 C_3 \\ R_2 C_1 & R_2 C_2 & R_2 C_3 \end{bmatrix}$$

$$AB = \begin{bmatrix} 4(3) + 7(8) & 4(1) + 7(2) & 4(6) + 7(9) \\ 2(3) + 5(8) & 2(1) + 5(2) & 2(6) + 5(9) \end{bmatrix} = \begin{bmatrix} 68 & 18 & 87 \\ 46 & 12 & 57 \end{bmatrix}$$

**5.19.** Redo Problem 5.17, given

$$A = \begin{bmatrix} 3 & 5 \\ 9 & 2 \end{bmatrix} \qquad B = \begin{bmatrix} 8 & 1 \\ 6 & 7 \\ 4 & 2 \end{bmatrix}$$

The product $AB$ is not defined: $2 \times (2 \neq 3) \times 2$. The matrices are not conformable in the given order. The number of columns (2) in $A$ does not equal the number of rows (3) in $B$. Hence the matrices cannot be multiplied in the order presently given.

**5.20.** Redo Problem 5.17 for $BA$ in Problem 5.19.

The product $BA$ is defined: $3 \times (2 = 2) \times 2$. $BA$ will be $3 \times 2$.

$$BA = \begin{bmatrix} 8 & 1 \\ 6 & 7 \\ 4 & 2 \end{bmatrix} \begin{bmatrix} 3 & 5 \\ 9 & 2 \end{bmatrix} = \begin{bmatrix} R_1 C_1 & R_1 C_2 \\ R_2 C_1 & R_2 C_2 \\ R_3 C_1 & R_3 C_2 \end{bmatrix}$$

$$= \begin{bmatrix} 8(3) + 1(9) & 8(5) + 1(2) \\ 6(3) + 7(9) & 6(5) + 7(2) \\ 4(3) + 2(9) & 4(5) + 2(2) \end{bmatrix} = \begin{bmatrix} 33 & 42 \\ 81 & 42 \\ 30 & 24 \end{bmatrix}$$

**5.21.** Redo Problem 5.17 for $AB'$ in Problem 5.19, where $B'$ is the transpose of $B$:

$$B' = \begin{bmatrix} 8 & 6 & 4 \\ 1 & 7 & 2 \end{bmatrix}$$

The product $AB'$ is defined: $2 \times (2 = 2) \times 3$. $AB'$ will be $2 \times 3$.

$$AB' = \begin{bmatrix} 3 & 5 \\ 9 & 2 \end{bmatrix} \begin{bmatrix} 8 & 6 & 4 \\ 1 & 7 & 2 \end{bmatrix} = \begin{bmatrix} R_1 C_1 & R_1 C_2 & R_1 C_3 \\ R_2 C_1 & R_2 C_2 & R_2 C_3 \end{bmatrix}$$

$$= \begin{bmatrix} 3(8) + 5(1) & 3(6) + 5(7) & 3(4) + 5(2) \\ 9(8) + 2(1) & 9(6) + 2(7) & 9(4) + 2(2) \end{bmatrix} = \begin{bmatrix} 29 & 53 & 22 \\ 74 & 68 & 40 \end{bmatrix}$$

Note from Problems 5.19 to 5.21 that $AB \neq BA \neq AB'$. This further reflects the fact that matrix multiplication is not commutative.

**5.22.** Find the product $CD$, given

$$C = \begin{bmatrix} 4 & 1 \\ 6 & 9 \\ 7 & 2 \end{bmatrix} \qquad D = \begin{bmatrix} 5 & 8 & 6 \\ 4 & 3 & 1 \end{bmatrix}$$

$CD$ is defined: $3 \times (2 = 2) \times 3$. $CD$ will be $3 \times 3$.

$$CD = \begin{bmatrix} R_1 C_1 & R_1 C_2 & R_1 C_3 \\ R_2 C_1 & R_2 C_2 & R_2 C_3 \\ R_3 C_1 & R_3 C_2 & R_3 C_3 \end{bmatrix}$$

$$CD = \begin{bmatrix} 4(5)+1(4) & 4(8)+1(3) & 4(6)+1(1) \\ 6(5)+9(4) & 6(8)+9(3) & 6(6)+9(1) \\ 7(5)+2(4) & 7(8)+2(3) & 7(6)+2(1) \end{bmatrix} = \begin{bmatrix} 24 & 35 & 25 \\ 66 & 75 & 45 \\ 43 & 62 & 44 \end{bmatrix}$$

**5.23.** Find $EF$, given

$$E = \begin{bmatrix} 9 & 2 & 4 \\ 6 & 5 & 1 \end{bmatrix} \qquad F = \begin{bmatrix} 8 & 3 \\ 2 & 7 \\ 5 & 9 \end{bmatrix}$$

$EF$ is defined: $2 \times (3 = 3) \times 2$. $EF$ will be $2 \times 2$.

$$EF = \begin{bmatrix} R_1C_1 & R_1C_2 \\ R_2C_1 & R_2C_2 \end{bmatrix}$$

$$= \begin{bmatrix} 9(8)+2(2)+4(5) & 9(3)+2(7)+4(9) \\ 6(8)+5(2)+1(5) & 6(3)+5(7)+1(9) \end{bmatrix} = \begin{bmatrix} 96 & 77 \\ 63 & 62 \end{bmatrix}$$

**5.24.** Find $AB$, given

$$A = [2 \quad 6 \quad 5] \qquad B = \begin{bmatrix} 7 & 1 & 9 \\ 4 & 3 & 6 \\ 5 & 8 & 2 \end{bmatrix}$$

$AB$ is defined: $1 \times (3 = 3) \times 3$. $AB$ will be $1 \times 3$.

$$AB = [R_1C_1 \quad R_1C_2 \quad R_1C_3]$$

$$= [2(7)+6(4)+5(5) \quad 2(1)+6(3)+5(8) \quad 2(9)+6(6)+5(2)]$$

$$= [63 \quad 60 \quad 64]$$

**5.25.** Find $CD$, given

$$C = \begin{bmatrix} 3 \\ 7 \\ 5 \end{bmatrix} \qquad D = \begin{bmatrix} 2 & 7 & 4 \\ 5 & 9 & 1 \\ 3 & 6 & 2 \end{bmatrix}$$

$CD$ is not defined: $3 \times (1 \neq 3) \times 3$. Multiplication is impossible in the given order.

**5.26.** Find $DC$ from Problem 5.25.

$DC$ is defined: $3 \times (3 = 3) \times 1$. $DC$ will be $3 \times 1$.

$$DC = \begin{bmatrix} 2 & 7 & 4 \\ 5 & 9 & 1 \\ 3 & 6 & 2 \end{bmatrix} \begin{bmatrix} 3 \\ 7 \\ 5 \end{bmatrix}$$

$$= \begin{bmatrix} R_1C_1 \\ R_2C_1 \\ R_3C_1 \end{bmatrix} = \begin{bmatrix} 2(3)+7(7)+4(5) \\ 5(3)+9(7)+1(5) \\ 3(3)+6(7)+2(5) \end{bmatrix} = \begin{bmatrix} 75 \\ 83 \\ 61 \end{bmatrix}$$

**5.27.** Find $EF$, given

$$E = \begin{bmatrix} 8 \\ 2 \\ 5 \end{bmatrix} \qquad F = [3 \quad 6 \quad 4]$$

$EF$ is defined: $3 \times (1 = 1) \times 3$. $EF$ will be $3 \times 3$.

$$EF = \begin{bmatrix} R_1C_1 & R_1C_2 & R_1C_3 \\ R_2C_1 & R_2C_2 & R_2C_3 \\ R_3C_1 & R_3C_2 & R_3C_3 \end{bmatrix}$$

$$= \begin{bmatrix} 8(3) & 8(6) & 8(4) \\ 2(3) & 2(6) & 2(4) \\ 5(3) & 5(6) & 5(4) \end{bmatrix} = \begin{bmatrix} 24 & 48 & 32 \\ 6 & 12 & 8 \\ 15 & 30 & 20 \end{bmatrix}$$

**5.28.** Find $AB$, given

$$A = [6 \quad 1 \quad 9] \qquad B = \begin{bmatrix} 5 \\ 2 \end{bmatrix}$$

$AB$ is not defined: $1 \times 3 \neq 2 \times 1$.

**5.29.** Find $BA$ from Problem 5.28.

$BA$ is defined: $2 \times 1 = 1 \times 3$. $BA$ will be $2 \times 3$.

$$BA = \begin{bmatrix} 5 \\ 2 \end{bmatrix} [6 \quad 1 \quad 9]$$

$$= \begin{bmatrix} R_1C_1 & R_1C_2 & R_1C_3 \\ R_2C_1 & R_2C_2 & R_2C_3 \end{bmatrix} = \begin{bmatrix} 5(6) & 5(1) & 5(9) \\ 2(6) & 2(1) & 2(9) \end{bmatrix} = \begin{bmatrix} 30 & 5 & 45 \\ 12 & 2 & 18 \end{bmatrix}$$

**5.30.** Use the inventory matrix for the company in Problem 5.3 and the price vector from Problem 5.15 to determine the value of the inventory in all four of the company's outlets.

$V = QP$.　$QP$ is defined: $4 \times 4 = 4 \times 1$. $V$ will be $4 \times 1$.

$$V = \begin{bmatrix} 35 & 60 & 55 & 45 \\ 80 & 65 & 50 & 38 \\ 29 & 36 & 24 & 32 \\ 62 & 49 & 54 & 33 \end{bmatrix} \begin{bmatrix} 400 \\ 300 \\ 250 \\ 500 \end{bmatrix}$$

$$= \begin{bmatrix} R_1C_1 \\ R_2C_1 \\ R_3C_1 \\ R_4C_1 \end{bmatrix} = \begin{bmatrix} 35(400) + 60(300) + 55(250) + 45(500) \\ 80(400) + 65(300) + 50(250) + 38(500) \\ 29(400) + 36(300) + 24(250) + 32(500) \\ 62(400) + 49(300) + 54(250) + 33(500) \end{bmatrix} = \begin{bmatrix} 68,250 \\ 83,000 \\ 44,400 \\ 69,500 \end{bmatrix}$$

**5.31.** Find the product of the following matrices and their corresponding identity matrices, given

(a)　$A = \begin{bmatrix} 4 & 9 \\ 3 & 7 \end{bmatrix}$　　(b)　$B = \begin{bmatrix} 2 & 9 & 4 \\ 5 & 8 & 1 \\ 3 & 7 & 6 \end{bmatrix}$

(a)
$$AI = \begin{bmatrix} 4 & 9 \\ 3 & 7 \end{bmatrix} \begin{bmatrix} 1 & 0 \\ 0 & 1 \end{bmatrix}$$

$$= \begin{bmatrix} R_1C_1 & R_1C_2 \\ R_2C_1 & R_2C_2 \end{bmatrix} = \begin{bmatrix} 4(1) + 9(0) & 4(0) + 9(1) \\ 3(1) + 7(0) & 3(0) + 7(1) \end{bmatrix} = \begin{bmatrix} 4 & 9 \\ 3 & 7 \end{bmatrix}$$

(b)
$$BI = \begin{bmatrix} 2 & 9 & 4 \\ 5 & 8 & 1 \\ 3 & 7 & 6 \end{bmatrix} \begin{bmatrix} 1 & 0 & 0 \\ 0 & 1 & 0 \\ 0 & 0 & 1 \end{bmatrix} = \begin{bmatrix} R_1C_1 & R_1C_2 & R_1C_3 \\ R_2C_1 & R_2C_2 & R_2C_3 \\ R_3C_1 & R_3C_2 & R_3C_3 \end{bmatrix}$$

$$= \begin{bmatrix} 2(1) + 9(0) + 4(0) & 2(0) + 9(1) + 4(0) & 2(0) + 9(0) + 4(1) \\ 5(1) + 8(0) + 1(0) & 5(0) + 8(1) + 1(0) & 5(0) + 8(0) + 1(1) \\ 3(1) + 7(0) + 6(0) & 3(0) + 7(1) + 6(0) & 3(0) + 7(0) + 6(1) \end{bmatrix}$$

$$= \begin{bmatrix} 2 & 9 & 4 \\ 5 & 8 & 1 \\ 3 & 7 & 6 \end{bmatrix}$$

Multiplication of a matrix by a conformable identity matrix, regardless of the order of multiplication, leaves the original matrix unchanged: $AI = A = IA$, $BI = B = IB$. It is equivalent to multiplying by 1 in ordinary algebra.

## GAUSSIAN METHOD OF SOLVING LINEAR EQUATIONS

**5.32.** Express the following system of linear equations ($a$) in matrix form and ($b$) as an augmented matrix, letting $A$ = the coefficient matrix, $X$ = the column vector of variables, and $B$ = the column vector of constants.

$$8x_1 + 3x_2 = 28$$
$$5x_1 + 9x_2 = 46$$

($a$)
$$A \cdot X = B$$
$$\begin{bmatrix} 8 & 3 \\ 5 & 9 \end{bmatrix} \begin{bmatrix} x_1 \\ x_2 \end{bmatrix} = \begin{bmatrix} 28 \\ 46 \end{bmatrix}$$

($b$)
$$[A|B] = \begin{bmatrix} 8 & 3 & | & 28 \\ 5 & 9 & | & 46 \end{bmatrix}$$

**5.33.** Redo Problem 5.32, given

$$5x_1 + 9x_2 + 2x_3 = 35$$
$$4x_1 + 7x_2 + 6x_3 = 32$$
$$x_1 + 3x_2 + 8x_3 = 17$$

($a$)
$$A \cdot X = B$$
$$\begin{bmatrix} 5 & 9 & 2 \\ 4 & 7 & 6 \\ 1 & 3 & 8 \end{bmatrix} \begin{bmatrix} x_1 \\ x_2 \\ x_3 \end{bmatrix} = \begin{bmatrix} 35 \\ 32 \\ 17 \end{bmatrix}$$

($b$)
$$[A|B] = \begin{bmatrix} 5 & 9 & 2 & | & 35 \\ 4 & 7 & 6 & | & 32 \\ 1 & 3 & 8 & | & 17 \end{bmatrix}$$

Note from Problems 5.32 and 5.33 that if the equations are arranged so that in each successive equation the same variables are always placed directly under each other, as typically occurs in ordinary algebra, the coefficients of the first variable will always appear in the first column, the coefficients of the second variable in the second column, and so on. The coefficient matrix can then be formed by simply reading into it, in the order that they appear, the coefficients of the system of equations. Since the coefficients of each equation form a separate row and matrix multiplication always involves row-column operations, always express the variables as a column vector, following the same order in which they appear in the equations. If a particular variable does not appear in a given equation, its equivalent coefficient is 0, which must be included. See Problem 5.39.

**5.34.** Use the Gaussian elimination method to solve the following system of linear equations:

$$3x_1 + 8x_2 = 53$$
$$6x_1 + 2x_2 = 50$$

First express the equations in an augmented matrix:

$$[A|B] = \begin{bmatrix} 3 & 8 & | & 53 \\ 6 & 2 & | & 50 \end{bmatrix}$$

Then apply row operations to convert the coefficient matrix on the left to an identity matrix. The easiest way to do this is to convert the $a_{11}$ element to 1 and clear column 1, then convert the $a_{22}$ element to 1 and clear column 2, and so forth, as follows:

1a.　Multiply row 1 by $\frac{1}{3}$:

$$\begin{bmatrix} 1 & \frac{8}{3} & \Big| & \frac{53}{3} \\ 6 & 2 & \Big| & 50 \end{bmatrix}$$

1b.　Subtract 6 times row 1 from row 2:

$$\begin{bmatrix} 1 & \frac{8}{3} & \Big| & \frac{53}{3} \\ 0 & -14 & \Big| & -56 \end{bmatrix}$$

2a.　Multiply row 2 by $-\frac{1}{14}$:

$$\begin{bmatrix} 1 & \frac{8}{3} & \Big| & \frac{53}{3} \\ 0 & 1 & \Big| & 4 \end{bmatrix}$$

2b.　Subtract $\frac{8}{3}$ times row 2 from row 1:

$$\begin{bmatrix} 1 & 0 & \Big| & 7 \\ 0 & 1 & \Big| & 4 \end{bmatrix}$$

Thus, $\bar{x}_1 = 7$ and $\bar{x}_2 = 4$, since

$$\begin{bmatrix} 1 & 0 \\ 0 & 1 \end{bmatrix} \begin{bmatrix} x_1 \\ x_2 \end{bmatrix} = \begin{bmatrix} 7 \\ 4 \end{bmatrix}$$

**5.35.**　Redo Problem 5.34, given

$$4x_1 + 9x_2 = 62$$
$$5x_1 + 8x_2 = 58$$

The augmented matrix is

$$[A|B] = \begin{bmatrix} 4 & 9 & \Big| & 62 \\ 5 & 8 & \Big| & 58 \end{bmatrix}$$

1a.　Multiply row 1 by $\frac{1}{4}$:

$$\begin{bmatrix} 1 & \frac{9}{4} & \Big| & \frac{31}{2} \\ 5 & 8 & \Big| & 58 \end{bmatrix}$$

1b.　Subtract 5 times row 1 from row 2:

$$\begin{bmatrix} 1 & \frac{9}{4} & \Big| & \frac{31}{2} \\ 0 & -\frac{13}{4} & \Big| & -\frac{39}{2} \end{bmatrix}$$

2a.　Multiply row 2 by $-\frac{4}{13}$:

$$\begin{bmatrix} 1 & \frac{9}{4} & \Big| & \frac{31}{2} \\ 0 & 1 & \Big| & 6 \end{bmatrix}$$

2b.　Subtract $\frac{9}{4}$ times row 2 from row 1:

$$\begin{bmatrix} 1 & 0 & \Big| & 2 \\ 0 & 1 & \Big| & 6 \end{bmatrix}$$

Thus $\bar{x}_1 = 2$ and $\bar{x}_2 = 6$.

**5.36.**　Redo Problem 5.34, given

$$6x_1 + 4x_2 = 47$$
$$2x_1 + 9x_2 = 77$$

The augmented matrix is

$$[A|B] = \begin{bmatrix} 6 & 4 & | & 47 \\ 2 & 9 & | & 77 \end{bmatrix}$$

1a. Multiply row 1 by $\frac{1}{6}$:

$$\begin{bmatrix} 1 & \frac{2}{3} & | & \frac{47}{6} \\ 2 & 9 & | & 77 \end{bmatrix}$$

1b. Subtract 2 times row 1 from row 2:

$$\begin{bmatrix} 1 & \frac{2}{3} & | & \frac{47}{6} \\ 0 & \frac{23}{3} & | & \frac{184}{3} \end{bmatrix}$$

2a. Multiply row 2 by $\frac{3}{23}$:

$$\begin{bmatrix} 1 & \frac{2}{3} & | & \frac{47}{6} \\ 0 & 1 & | & 8 \end{bmatrix}$$

2b. Subtract $\frac{2}{3}$ times row 2 from row 1:

$$\begin{bmatrix} 1 & 0 & | & 2.5 \\ 0 & 1 & | & 8 \end{bmatrix}$$

Thus $\bar{x}_1 = 2.5$ and $\bar{x}_2 = 8$.

**5.37.** Redo Problem 5.34, given

$$4x_1 + 2x_2 + 5x_3 = 21$$
$$3x_1 + 6x_2 + x_3 = 31$$
$$x_1 + 8x_2 + 3x_3 = 37$$

The augmented matrix is

$$[A|B] = \begin{bmatrix} 4 & 2 & 5 & | & 21 \\ 3 & 6 & 1 & | & 31 \\ 1 & 8 & 3 & | & 37 \end{bmatrix}$$

1a. Multiply row 1 by $\frac{1}{4}$:

$$\begin{bmatrix} 1 & \frac{1}{2} & \frac{5}{4} & | & \frac{21}{4} \\ 3 & 6 & 1 & | & 31 \\ 1 & 8 & 3 & | & 37 \end{bmatrix}$$

1b. Subtract 3 times row 1 from row 2 and row 1 from row 3:

$$\begin{bmatrix} 1 & \frac{1}{2} & \frac{5}{4} & | & \frac{21}{4} \\ 0 & \frac{9}{2} & -\frac{11}{4} & | & \frac{61}{4} \\ 0 & \frac{15}{2} & \frac{7}{4} & | & \frac{127}{4} \end{bmatrix}$$

2a. Multiply row 2 by $\frac{2}{9}$:

$$\begin{bmatrix} 1 & \frac{1}{2} & \frac{5}{4} & | & \frac{21}{4} \\ 0 & 1 & -\frac{11}{18} & | & \frac{61}{18} \\ 0 & \frac{15}{2} & \frac{7}{4} & | & \frac{127}{4} \end{bmatrix}$$

2b. Subtract $\frac{1}{2}$ times row 2 from row 1 and $\frac{15}{2}$ times row 2 from row 3:

$$\begin{bmatrix} 1 & 0 & \frac{14}{9} & | & \frac{32}{9} \\ 0 & 1 & -\frac{11}{18} & | & \frac{61}{18} \\ 0 & 0 & \frac{57}{9} & | & \frac{57}{9} \end{bmatrix}$$

3a.   Multiply row 3 by $\frac{9}{57}$:

$$\begin{bmatrix} 1 & 0 & \frac{14}{9} & \Big| & \frac{32}{9} \\ 0 & 1 & -\frac{11}{18} & \Big| & \frac{61}{18} \\ 0 & 0 & 1 & \Big| & 1 \end{bmatrix}$$

3b.   Subtract $\frac{14}{9}$ times row 3 from row 1 and add $\frac{11}{18}$ times row 3 to row 2:

$$\begin{bmatrix} 1 & 0 & 0 & \big| & 2 \\ 0 & 1 & 0 & \big| & 4 \\ 0 & 0 & 1 & \big| & 1 \end{bmatrix}$$

Thus, $\bar{x}_1 = 2$, $\bar{x}_2 = 4$, and $\bar{x}_3 = 1$.

**5.38.**   Redo Problem 5.34, given

$$2x_1 + 4x_2 + 7x_3 = 82$$

$$6x_1 - 3x_2 + x_3 = 11$$

$$x_1 + 2x_2 - 5x_3 = -27$$

The augmented matrix is

$$[A|B] = \begin{bmatrix} 2 & 4 & 7 & \big| & 82 \\ 6 & -3 & 1 & \big| & 11 \\ 1 & 2 & -5 & \big| & -27 \end{bmatrix}$$

1a.   Multiply row 1 by $\frac{1}{2}$:

$$\begin{bmatrix} 1 & 2 & \frac{7}{2} & \Big| & 41 \\ 6 & -3 & 1 & \Big| & 11 \\ 1 & 2 & -5 & \Big| & -27 \end{bmatrix}$$

1b.   Subtract 6 times row 1 from row 2 and row 1 from row 3:

$$\begin{bmatrix} 1 & 2 & \frac{7}{2} & \Big| & 41 \\ 0 & -15 & -20 & \Big| & -235 \\ 0 & 0 & -\frac{17}{2} & \Big| & -68 \end{bmatrix}$$

2a.   Multiply row 2 by $-\frac{1}{15}$:

$$\begin{bmatrix} 1 & 2 & \frac{7}{2} & \Big| & 41 \\ 0 & 1 & \frac{4}{3} & \Big| & \frac{47}{3} \\ 0 & 0 & -\frac{17}{2} & \Big| & -68 \end{bmatrix}$$

2b.   Subtract 2 times row 2 from row 1 and leave row 3 as is:

$$\begin{bmatrix} 1 & 0 & \frac{5}{6} & \Big| & \frac{29}{3} \\ 0 & 1 & \frac{4}{3} & \Big| & \frac{47}{3} \\ 0 & 0 & -\frac{17}{2} & \Big| & -68 \end{bmatrix}$$

3a.   Multiply row 3 by $-\frac{2}{17}$:

$$\begin{bmatrix} 1 & 0 & \frac{5}{6} & \Big| & \frac{29}{3} \\ 0 & 1 & \frac{4}{3} & \Big| & \frac{47}{3} \\ 0 & 0 & 1 & \Big| & 8 \end{bmatrix}$$

3b.   Subtract $\frac{5}{6}$ times row 3 from row 1 and $\frac{4}{3}$ times row 3 from row 2:

$$\begin{bmatrix} 1 & 0 & 0 & \big| & 3 \\ 0 & 1 & 0 & \big| & 5 \\ 0 & 0 & 1 & \big| & 8 \end{bmatrix}$$

Thus $\bar{x}_1 = 3$, $\bar{x}_2 = 5$, and $\bar{x}_3 = 8$.

**5.39.** Redo Problem 5.34, given

$$5x_1 \qquad - 2x_3 = -3$$
$$4x_1 + 9x_2 \qquad = 51$$
$$- 6x_2 - x_3 = -36$$

The augmented matrix is

$$[A|B] = \begin{bmatrix} 5 & 0 & -2 & | & -3 \\ 4 & 9 & 0 & | & 51 \\ 0 & -6 & -1 & | & -36 \end{bmatrix}$$

1a.  Multiply row 1 by $\frac{1}{5}$:

$$\begin{bmatrix} 1 & 0 & -\frac{2}{5} & | & -\frac{3}{5} \\ 4 & 9 & 0 & | & 51 \\ 0 & -6 & -1 & | & -36 \end{bmatrix}$$

1b.  Subtract 4 times row 1 from row 2 and leave row 3 alone since the $a_{31}$ element is already 0.

$$\begin{bmatrix} 1 & 0 & -\frac{2}{5} & | & -\frac{3}{5} \\ 0 & 9 & \frac{8}{5} & | & \frac{267}{5} \\ 0 & -6 & -1 & | & -36 \end{bmatrix}$$

2a.  Multiply row 2 by $\frac{1}{9}$:

$$\begin{bmatrix} 1 & 0 & -\frac{2}{5} & | & -\frac{3}{5} \\ 0 & 1 & \frac{8}{45} & | & \frac{267}{45} \\ 0 & -6 & -1 & | & -36 \end{bmatrix}$$

2b.  Add 6 times row 2 to row 3 and ignore row 1 since $a_{12} = 0$.

$$\begin{bmatrix} 1 & 0 & -\frac{2}{5} & | & -\frac{3}{5} \\ 0 & 1 & \frac{8}{45} & | & \frac{267}{45} \\ 0 & 0 & \frac{1}{15} & | & -\frac{18}{45} \end{bmatrix}$$

3a.  Multiply row 3 by 15.

$$\begin{bmatrix} 1 & 0 & -\frac{2}{5} & | & -\frac{3}{5} \\ 0 & 1 & \frac{8}{45} & | & \frac{267}{45} \\ 0 & 0 & 1 & | & -6 \end{bmatrix}$$

3b.  Add $\frac{2}{5}$ times row 3 to row 1 and $-\frac{8}{45}$ times row 3 to row 2.

$$\begin{bmatrix} 1 & 0 & 0 & | & -3 \\ 0 & 1 & 0 & | & 7 \\ 0 & 0 & 1 & | & -6 \end{bmatrix}$$

Thus $\overline{x}_1 = -3$, $\overline{x}_2 = 7$, and $\overline{x}_3 = -6$.

# Supplementary Problems

Refer to the following matrices in answering the questions below:

$$A = \begin{bmatrix} 8 & -5 \\ 3 & 4 \end{bmatrix} \qquad B = \begin{bmatrix} 7 & 6 \\ -2 & 1 \end{bmatrix} \qquad C = \begin{bmatrix} 3 & -8 & 13 \\ 6 & 11 & -5 \end{bmatrix} \qquad D = \begin{bmatrix} 9 & 15 & 4 \\ 2 & -8 & 7 \end{bmatrix}$$

$$E = \begin{bmatrix} 5 & -7 \\ 9 & 22 \\ 4 & -5 \end{bmatrix} \quad F = \begin{bmatrix} 4 & -9 \\ 8 & -2 \\ 3 & 16 \end{bmatrix} \quad G = \begin{bmatrix} -4 & 9 & 15 \\ 5 & 12 & -9 \\ 18 & -6 & 20 \end{bmatrix} \quad H = \begin{bmatrix} 6 & 8 & -1 \\ -2 & 14 & 5 \\ 13 & -9 & 3 \end{bmatrix}$$

## MATRIX FORMAT

**5.40.** Give the dimensions of (*a*) $A$, (*b*) $C$, (*c*) $E$, and (*d*) $G$.

**5.41.** In the matrices above, identify the following elements: (*a*) $a_{21}$ (*b*) $b_{12}$ (*c*) $c_{23}$ (*d*) $d_{12}$ (*e*) $e_{31}$ (*f*) $f_{12}$ (*g*) $g_{13}$ (*h*) $g_{32}$ (*i*) $h_{23}$ (*j*) $h_{31}$.

**5.42.** Find the following transpose matrices: (*a*) $B'$, (*b*) $D'$, (*c*) $F'$, and (*d*) $H'$.

**5.43.** Give the dimensions of the following transpose matrices: (*a*) $B'$, (*b*) $D'$, (*c*) $F'$, and (*d*) $H'$.

## ADDITION AND SUBTRACTION OF MATRICES

**5.44.** Find the following sums: (*a*) $A + B$   (*b*) $C + D$   (*c*) $E + F$   (*d*) $G + H$.

**5.45.** Find the following differences: (*a*) $A - B$   (*b*) $C - D$   (*c*) $E - F$   (*d*) $G - H$.

## MATRIX MULTIPLICATION

**5.46.** Using the original matrices above, find the following product matrices: (*a*) $AB$, (*b*) $CE$, (*c*) $FD$, (*d*) $GH$, (*e*) $EA$, (*f*) $GF$, (*g*) $CH$, and (*h*) $BD$.

## GAUSSIAN ELIMINATION METHOD

**5.47.** Use the Gaussian elimination method to solve each of the following systems of linear equations:

(*a*)   $4x + 7y = 131$
         $8x - 3y = 41$

(*b*)   $6x + 5y = 51$
         $-x + 8y = 124$

(*c*)   $11x - 4y = 256$
         $3x + 6y = 6$

(*d*)   $-13x + 9y = 15$
         $7x - 2y = -28$

**5.48.** Solve the following equations using the Gaussian elimination method:

(*a*)   $3x_1 + 6x_2 - 5x_3 = -5$
         $4x_1 - 7x_2 + 2x_3 = -6$
         $-x_1 + 8x_2 + 9x_3 = 93$

(*b*)   $5x_1 + 8x_2 + 2x_3 = 26$
         $7x_2 - 4x_3 = 6$
         $-x_1 \quad + 9x_3 = 89$

(*c*)   $12x_2 + 9x_3 = 84$
         $-5x_1 + 6x_2 - 2x_3 = -122$
         $8x_1 \quad - 3x_3 = 48$

(*d*)   $11x_1 - 8x_2 - 6x_3 = 133$
         $4x_1 + \quad 7x_3 = -41$
         $2x_1 + 5x_2 - 3x_3 = 71$

# Answers to Supplementary Problems

**5.40.**   (*a*)   $A = 2 \times 2$      (*b*)   $C = 2 \times 3$      (*c*)   $E = 3 \times 2$      (*d*)   $G = 3 \times 3$

**5.41.** (a) $a_{21} = 3$     (b) $b_{12} = 6$     (c) $c_{23} = -5$   (d) $d_{12} = 15$   (e) $e_{31} = 4$

(f) $f_{12} = -9$   (g) $g_{13} = -15$   (h) $g_{32} = -6$   (i) $h_{23} = 5$     (j) $h_{31} = 13$

**5.42.** (a) $B' = \begin{bmatrix} 7 & -2 \\ 6 & 1 \end{bmatrix}$

(b) $D' = \begin{bmatrix} 9 & 2 \\ 15 & -8 \\ 4 & 7 \end{bmatrix}$

(c) $F' = \begin{bmatrix} 4 & 8 & 3 \\ -9 & -2 & 16 \end{bmatrix}$

(d) $H' = \begin{bmatrix} 6 & -2 & 13 \\ 8 & 14 & -9 \\ -1 & 5 & 3 \end{bmatrix}$

**5.43.** (a) $B' = 2 \times 2$     (b) $D' = 3 \times 2$     (c) $F' = 2 \times 3$     (d) $H' = 3 \times 3$

**5.44.** (a) $A + B = \begin{bmatrix} 15 & 1 \\ 1 & 5 \end{bmatrix}$

(b) $C + D = \begin{bmatrix} 12 & 7 & 17 \\ 8 & 3 & 2 \end{bmatrix}$

(c) $E + F = \begin{bmatrix} 9 & -16 \\ 17 & 20 \\ 7 & 11 \end{bmatrix}$

(d) $G + H = \begin{bmatrix} 2 & 17 & 14 \\ 3 & 26 & -4 \\ 31 & -15 & 23 \end{bmatrix}$

**5.45.** (a) $A - B = \begin{bmatrix} 1 & -11 \\ 5 & 3 \end{bmatrix}$

(b) $C - D = \begin{bmatrix} -6 & -23 & 9 \\ 4 & 19 & -12 \end{bmatrix}$

(c) $E - F = \begin{bmatrix} 1 & 2 \\ 1 & 24 \\ 1 & -21 \end{bmatrix}$

(d) $G - H = \begin{bmatrix} -10 & 1 & 16 \\ 7 & -2 & -14 \\ 5 & 3 & 17 \end{bmatrix}$

**5.46.** (a) $AB = \begin{bmatrix} 66 & 43 \\ 13 & 22 \end{bmatrix}$

(b) $CE = \begin{bmatrix} -5 & -262 \\ 109 & 225 \end{bmatrix}$

(c) $FD = \begin{bmatrix} 18 & 132 & -47 \\ 68 & 136 & 18 \\ 59 & -83 & 124 \end{bmatrix}$

(d) $GH = \begin{bmatrix} 153 & -41 & 94 \\ -111 & 289 & 28 \\ 380 & -120 & 12 \end{bmatrix}$

(e) $EA = \begin{bmatrix} 19 & -53 \\ 138 & 43 \\ 17 & -40 \end{bmatrix}$

(f) $GF = \begin{bmatrix} 101 & 258 \\ 89 & -213 \\ 84 & 170 \end{bmatrix}$

(g) $CH = \begin{bmatrix} 203 & -205 & -4 \\ -51 & 247 & 34 \end{bmatrix}$

(h) $BD = \begin{bmatrix} 75 & 57 & 70 \\ -16 & -38 & -1 \end{bmatrix}$

**5.47.** (a) $x = 10, y = 13$

(b) $x = -4, y = 15$

(c) $x = 20, y = -9$

(d) $x = -6, y = -7$

**5.48.** (a) $x_1 = 2, x_2 = 4, x_3 = 7$

(b) $x_1 = -8, x_2 = 6, x_3 = 9$

(c) $x_1 = 12, x_2 = -5, x_3 = 16$

(d) $x_1 = 9, x_2 = 4, x_3 = -11$

# Chapter 6

# Solving Linear Equations with Matrix Algebra

## 6.1 DETERMINANTS AND LINEAR INDEPENDENCE

For a unique solution to exist, a system of equations must have as many independent and consistent equations as variables, as was explained in Chapter 4. To test a system of linear equations for linear dependence or independence, the determinant of the coefficient matrix $A$ is used. The determinant $|A|$ of a $2 \times 2$ matrix, called a *second-order determinant*, is derived by taking the product of the two elements on the principal diagonal and subtracting from it the product of the two elements off the principal diagonal. Given a general $2 \times 2$ matrix

$$A = \begin{bmatrix} a_{11} & a_{12} \\ a_{21} & a_{22} \end{bmatrix}$$

the determinant is

$$|A| = \begin{vmatrix} a_{11} & a_{12} \\ a_{21} & a_{22} \end{vmatrix} \begin{matrix} (-) \\ (+) \end{matrix} = a_{11}a_{22} - a_{12}a_{21}$$

The determinant $|A|$, which can be found only for square matrices, is a single number or scalar. If $|A| = 0$, the determinant is said to *vanish* and the matrix is called *singular*. A *singular matrix* is one in which there is linear dependence between at least two of the rows or columns. If $|A| \neq 0$, the matrix is *nonsingular*, which means that all its rows and columns are linearly independent. The *rank* $\rho$ of a matrix is determined by the maximum number of linearly independent rows or columns in the matrix. See Example 1 and Problem 6.1.

**EXAMPLE 1.** Determinants are calculated as follows, given

$$A = \begin{bmatrix} 7 & 4 \\ 6 & 5 \end{bmatrix} \qquad B = \begin{bmatrix} 8 & 12 \\ 2 & 3 \end{bmatrix}$$

Multiplying the elements on the principal diagonal from left to right $(7 \cdot 5)$, and subtracting from that product the product of the elements off the principal diagonal $(4 \cdot 6)$, we have

$$|A| = 7(5) - 4(6) = 11$$

Since $|A| \neq 0$, the matrix is nonsingular, which means that there is no linear dependence between its rows or columns. With both rows and columns linearly independent, the rank of $A$ is 2, written $\rho(A) = 2$. By way of contrast,

$$|B| = 8(3) - 12(2) = 0$$

Since $|B| = 0$, $B$ is singular and linear dependence exists between its rows and columns. Closer inspection reveals that row 1 is equal to 4 times row 2 and column 2 equals 1.5 times column 1. With only one row and column independent, $\rho(B) = 1$.

## 6.2 THIRD-ORDER DETERMINANTS

The determinant of a $3 \times 3$ matrix, called a *third-order determinant*, can be found in several different ways. The easist method is presented here with references to more sophisticated means listed below. Given

$$A = \begin{bmatrix} a_{11} & a_{12} & a_{13} \\ a_{21} & a_{22} & a_{23} \\ a_{31} & a_{32} & a_{33} \end{bmatrix}$$

1. Write down the matrix as given, and to the right of it repeat the first two columns, as in Fig. 6-1.
2. Multiply each of the three elements in the first row of the original matrix by the two elements to which they are connected by a downward-sloping line and *add* their products.
3. Multiply each of the three elements in the last row of the original matrix by the two elements to which they are connected by an upward-sloping line and *subtract* the sum of their products from the previous total. Thus, aided by Fig. 6-1,

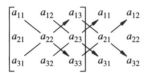

**Fig. 6-1**

$$|A| = a_{11}a_{22}a_{33} + a_{12}a_{23}a_{31} + a_{13}a_{21}a_{32} - [a_{31}a_{22}a_{13} + a_{32}a_{23}a_{11} + a_{33}a_{21}a_{12}]$$

$$|A| = a_{11}a_{22}a_{33} + a_{12}a_{23}a_{31} + a_{13}a_{21}a_{32} - a_{31}a_{22}a_{13} - a_{32}a_{23}a_{11} - a_{33}a_{21}a_{12}$$

For the more universal methods of minor and cofactor expansion, which are applicable to all square matrices of any dimensions, see Dowling, *Schaum's Outline of Introduction to Mathematical Economics*, Sections 11.3 and 11.4, Problems 11.23 to 11.30.

**EXAMPLE 2.**   Given

$$A = \begin{bmatrix} 9 & 1 & 8 \\ 4 & 6 & 5 \\ 3 & 7 & 2 \end{bmatrix}$$

Using Fig. 6-2,

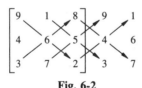

**Fig. 6-2**

$$|A| = 9 \cdot 6 \cdot 2 + 1 \cdot 5 \cdot 3 + 8 \cdot 4 \cdot 7 - [3 \cdot 6 \cdot 8 + 7 \cdot 5 \cdot 9 + 2 \cdot 4 \cdot 1]$$

$$|A| = 108 + 15 + 224 - [144 + 315 + 8] = 347 - 467 = -120$$

With $|A| \neq 0$, $A$ is nonsingular and $\rho(A) = 3$. See also Problem 6.2. For important properties of determinants and their applications, see Dowling, *Schaum's Outline of Introduction to Mathematical Economics*, Section 11.5 and Problems 11.4 to 11.20.

## 6.3  CRAMER'S RULE FOR SOLVING LINEAR EQUATIONS

Besides testing for linear dependence, determinants can also be helpful in solving a system of linear equations. One such method is Cramer's rule. *Cramer's rule* states

$$\bar{x}_i = \frac{|A_i|}{|A|}$$

where $x_i$ is the $i$th unknown variable in a system of equations, $|A|$ is the determinant of the coefficient matrix, and $|A_i|$ is the determinant of a special matrix formed from the original coefficient matrix by replacing the column of coefficients of $x_i$ with the column vector of constants. See Examples 3 and 4, and Problems 6.3 and 6.4.

**EXAMPLE 3.**   Cramer's rule is used below to solve the $2 \times 2$ system of linear equations

$$3x_1 + 7x_2 = 41$$

$$8x_1 + 9x_2 = 61$$

1. Express the equations in matrix form.

$$A \cdot X = B$$

$$\begin{bmatrix} 3 & 7 \\ 8 & 9 \end{bmatrix} \begin{bmatrix} x_1 \\ x_2 \end{bmatrix} = \begin{bmatrix} 41 \\ 61 \end{bmatrix}$$

2. Find the determinant of $A$.

$$|A| = 3(9) - 7(8) = -29$$

With $|A| \neq 0$, $A$ is nonsingular and the two equations are linearly independent. The possibility of a unique solution exists.

3. To solve for $x_1$, replace column 1 of $A$, which contains the coefficients of $x_1$, with the column vector of constants $B$, forming a new matrix $A_1$.

$$A_1 = \begin{bmatrix} 41 & 7 \\ 61 & 9 \end{bmatrix}$$

Find the determinant of $A_1$,

$$|A_1| = 41(9) - 7(61) = -58$$

and use the formula from Cramer's rule.

$$\bar{x}_1 = \frac{|A_1|}{|A|} = \frac{-58}{-29} = 2$$

4. To solve for $x_2$, replace column 2 of $A$, which contains the coefficients of $x_2$, with the column vector of constants $B$, forming a new matrix $A_2$.

$$A_2 = \begin{bmatrix} 3 & 41 \\ 8 & 61 \end{bmatrix}$$

Taking the determinant,

$$|A_2| = 3(61) - 41(8) = -145$$

and using Cramer's formula,

$$\bar{x}_2 = \frac{|A_2|}{|A|} = \frac{-145}{-29} = 5$$

**EXAMPLE 4.**   The $3 \times 3$ system of linear equations below is solved with Cramer's rule.

$$4x_1 + 2x_2 + 7x_3 = 35$$

$$3x_1 + x_2 + 8x_3 = 25$$

$$5x_1 + 3x_2 + x_3 = 40$$

1. Converting to matrix form,

$$A \cdot X = B$$

$$\begin{bmatrix} 4 & 2 & 7 \\ 3 & 1 & 8 \\ 5 & 3 & 1 \end{bmatrix} \begin{bmatrix} x_1 \\ x_2 \\ x_3 \end{bmatrix} = \begin{bmatrix} 35 \\ 25 \\ 40 \end{bmatrix}$$

2. Finding the determinant of $A$, using the method from Example 2,

$$|A| = 4 \cdot 1 \cdot 1 + 2 \cdot 8 \cdot 5 + 7 \cdot 3 \cdot 3 - [5 \cdot 1 \cdot 7 + 3 \cdot 8 \cdot 4 + 1 \cdot 3 \cdot 2]$$

$$= 4 + 80 + 63 - [35 + 96 + 6] = 147 - 137 = 10 \neq 0$$

3. Forming a new matrix $A_1$ by replacing column 1 of $A$, the coefficients of $x_1$, with the column of constants $B$ in preparation to find $x_1$,

$$A_1 = \begin{bmatrix} 35 & 2 & 7 \\ 25 & 1 & 8 \\ 40 & 3 & 1 \end{bmatrix}$$

Finding the determinant of $A_1$,

$$|A_1| = 35 \cdot 1 \cdot 1 + 2 \cdot 8 \cdot 40 + 7 \cdot 25 \cdot 3 - [40 \cdot 1 \cdot 7 + 3 \cdot 8 \cdot 35 + 1 \cdot 25 \cdot 2]$$

$$= 35 + 640 + 525 - [280 + 840 + 50] = 1200 - 1170 = 30$$

and using the formula from Cramer's rule,

$$\bar{x}_1 = \frac{|A_1|}{|A|} = \frac{30}{10} = 3$$

4. Replacing column 2 of the *original* matrix $A$ with the column of constants $B$ to find $x_2$,

$$A_2 = \begin{bmatrix} 4 & 35 & 7 \\ 3 & 25 & 8 \\ 5 & 40 & 1 \end{bmatrix}$$

$$|A_2| = 4 \cdot 25 \cdot 1 + 35 \cdot 8 \cdot 5 + 7 \cdot 3 \cdot 40 - [5 \cdot 25 \cdot 7 + 40 \cdot 8 \cdot 4 + 1 \cdot 3 \cdot 35]$$

$$= 100 + 1400 + 840 - [875 + 1280 + 105] = 2340 - 2260 = 80$$

and $$\bar{x}_2 = \frac{|A_2|}{|A|} = \frac{80}{10} = 8$$

5. Forming a new matrix $A_3$ by replacing column 3 of $A$ with the column of constants in preparation to find $x_3$,

$$A_3 = \begin{bmatrix} 4 & 2 & 35 \\ 3 & 1 & 25 \\ 5 & 3 & 40 \end{bmatrix}$$

$$|A_3| = 4 \cdot 1 \cdot 40 + 2 \cdot 25 \cdot 5 + 35 \cdot 3 \cdot 3 - [5 \cdot 1 \cdot 35 + 3 \cdot 25 \cdot 4 + 40 \cdot 3 \cdot 2]$$

$$= 160 + 250 + 315 - [175 + 300 + 240] = 725 - 715 = 10$$

and $$\bar{x}_3 = \frac{|A_3|}{|A|} = \frac{10}{10} = 1$$

A benefit of Cramer's rule is that it allows for solution of one variable without solving the whole system. See Problems 6.8 to 6.14.

## 6.4  INVERSE MATRICES

An *inverse matrix* $A^{-1}$, which exists only for a square, nonsingular matrix $A$, is a unique matrix satisfying the relationship

$$AA^{-1} = I = A^{-1}A$$

Multiplying a matrix by its inverse reduces it to an identity matrix. In this way, the inverse matrix in linear algebra performs much the same function as the reciprocal in ordinary algebra. Inverting a matrix with the Gaussian method is explained in Section 6.5. Deriving an inverse matrix from the adjoint matrix is demonstrated in Dowling, *Schaum's Outline of Introduction to Mathematical Economics*, Section 11.7 and Problem 11.31.

## 6.5  GAUSSIAN METHOD OF FINDING AN INVERSE MATRIX

The Gaussian method is a convenient way to invert a matrix. Simply set up an augmented matrix with the identity matrix to the right. Then apply row operations until the coefficient matrix on the left is converted to an identity matrix. At that point, the matrix on the right will be the inverse matrix.

The rationale for this method can be seen in a few mathematical steps. Starting with the augmented matrix $A|I$ and multiplying both sides by the inverse matrix $A^{-1}$,

$$(A^{-1})A|I(A^{-1})$$

From Section 6.4 and Problem 5.31, this reduces to

$$I|A^{-1}$$

where the identity matrix is now on the left and the inverse matrix is on the right. See Example 5 and Problems 6.5 to 6.7.

**EXAMPLE 5.**    To use the Gaussian elimination method to find the inverse for

$$A = \begin{bmatrix} 4 & 2 & 7 \\ 3 & 1 & 8 \\ 5 & 3 & 1 \end{bmatrix}$$

set up an augmented matrix with the identity matrix on the right:

$$A|I = \begin{bmatrix} 4 & 2 & 7 & 1 & 0 & 0 \\ 3 & 1 & 8 & 0 & 1 & 0 \\ 5 & 3 & 1 & 0 & 0 & 1 \end{bmatrix}$$

Then reduce the coefficient matrix on the left to an identity matrix by applying the row operations outlined in Sections 5.9 and 5.10, as follows:

1a.   Multiply row 1 by $\frac{1}{4}$ to obtain 1 in the $a_{11}$ position.

$$\begin{bmatrix} 1 & \frac{1}{2} & \frac{7}{4} & \frac{1}{4} & 0 & 0 \\ 3 & 1 & 8 & 0 & 1 & 0 \\ 5 & 3 & 1 & 0 & 0 & 1 \end{bmatrix}$$

1b.   Subtract 3 times row 1 from row 2 and 5 times row 1 from row 3 to clear the first column.

$$\begin{bmatrix} 1 & \frac{1}{2} & \frac{7}{4} & \frac{1}{4} & 0 & 0 \\ 0 & -\frac{1}{2} & \frac{11}{4} & -\frac{3}{4} & 1 & 0 \\ 0 & \frac{1}{2} & -\frac{31}{4} & -\frac{5}{4} & 0 & 1 \end{bmatrix}$$

2a.   Multiply row 2 by $-2$ to obtain 1 in the $a_{22}$ position.

$$\begin{bmatrix} 1 & \frac{1}{2} & \frac{7}{4} & \frac{1}{4} & 0 & 0 \\ 0 & 1 & -\frac{11}{2} & \frac{3}{2} & -2 & 0 \\ 0 & \frac{1}{2} & -\frac{31}{4} & -\frac{5}{4} & 0 & 1 \end{bmatrix}$$

2b.   Subtract $\frac{1}{2}$ times row 2 from row 1 and from row 3 to clear the second column.

$$\begin{bmatrix} 1 & 0 & \frac{9}{2} & -\frac{1}{2} & 1 & 0 \\ 0 & 1 & -\frac{11}{2} & \frac{3}{2} & -2 & 0 \\ 0 & 0 & -5 & -2 & 1 & 1 \end{bmatrix}$$

3a.  Multiply row 3 by $-\frac{1}{5}$ to obtain 1 in the $a_{33}$ position.

$$
\begin{bmatrix}
1 & 0 & \frac{9}{2} & -\frac{1}{2} & 1 & 0 \\
0 & 1 & -\frac{11}{2} & \frac{3}{2} & -2 & 0 \\
0 & 0 & 1 & \frac{2}{5} & -\frac{1}{5} & -\frac{1}{5}
\end{bmatrix}
$$

3b.  Subtract $\frac{9}{2}$ times row 3 from row 1 and add $\frac{11}{2}$ times row 3 to row 2 to clear the third column.

$$
\begin{bmatrix}
1 & 0 & 0 & -\frac{23}{10} & \frac{19}{10} & \frac{9}{10} \\
0 & 1 & 0 & \frac{37}{10} & -\frac{31}{10} & -\frac{11}{10} \\
0 & 0 & 1 & \frac{2}{5} & -\frac{1}{5} & -\frac{1}{5}
\end{bmatrix}
$$

With the identity matrix now on the left, we can identify the inverse matrix on the right, which, when simplified, reads

$$
A^{-1} = \begin{bmatrix}
-2.3 & 1.9 & 0.9 \\
3.7 & -3.1 & -1.1 \\
0.4 & -0.2 & -0.2
\end{bmatrix}
$$

**EXAMPLE 6.**  Since by definition $AA^{-1} = I$, the accuracy of an inverse can always be checked by multiplying the original matrix by the inverse to be sure the product is an identity matrix. Using $A$ and $A^{-1}$ from Example 5,

$$
AA^{-1} = \begin{bmatrix}
4 & 2 & 7 \\
3 & 1 & 8 \\
5 & 3 & 1
\end{bmatrix}
\begin{bmatrix}
-2.3 & 1.9 & 0.9 \\
3.7 & -3.1 & -1.1 \\
0.4 & -0.2 & -0.2
\end{bmatrix}
$$

$$
AA^{-1} = \begin{bmatrix}
4(-2.3) + 2(3.7) + 7(0.4) & 4(1.9) + 2(-3.1) + 7(-0.2) & 4(0.9) + 2(-1.1) + 7(-0.2) \\
3(-2.3) + 1(3.7) + 8(0.4) & 3(1.9) + 1(-3.1) + 8(-0.2) & 3(0.9) + 1(-1.1) + 8(-0.2) \\
5(-2.3) + 3(3.7) + 1(0.4) & 5(1.9) + 3(-3.1) + 1(-0.2) & 5(0.9) + 3(-1.1) + 1(-0.2)
\end{bmatrix}
$$

$$
AA^{-1} = \begin{bmatrix}
1 & 0 & 0 \\
0 & 1 & 0 \\
0 & 0 & 1
\end{bmatrix}
$$

Note that in checking an inverse matrix by testing the product of $A^{-1}A$, every element on the principal diagonal of the product matrix must equal 1 and every element off the principal diagonal must equal 0. This is convenient because, if a mistake has been made, only those elements corresponding to the elements which do not equal 0 or 1 in the proper places need be checked.

## 6.6  SOLVING LINEAR EQUATIONS WITH AN INVERSE MATRIX

An inverse matrix can be used to solve matrix equations. If

$$
A_{n \times n} X_{n \times 1} = B_{n \times 1}
$$

and the inverse $A^{-1}$ exists, multiplication of both sides of the equation by $A^{-1}$, following the laws of conformability, gives

$$
A_{n \times n}^{-1} A_{n \times n} X_{n \times 1} = A_{n \times n}^{-1} B_{n \times 1}
$$

Note above that $A^{-1}$ is conformable with both $A_{n \times n} X_{n \times 1}$ and $B_{n \times 1}$ only from the left. Now from Section 6.4, $A^{-1}A = I$, so

$$
I_{n \times n} X_{n \times 1} = A_{n \times n}^{-1} B_{n \times 1}
$$

and from Problem 5.31, $IX = X$, therefore,

$$
X_{n \times 1} = (A^{-1}B)_{n \times 1}
$$

The solution of the system of equations is given by the product of the inverse of the coefficient matrix $A^{-1}$ and the column vector of constants $B$, which itself will be a column vector. See Example 7 and Problems 6.5 to 6.7 and 6.15.

**EXAMPLE 7.**    Matrix inversion is used below to solve the system of equations from Example 4 for $x_1$, $x_2$, and $x_3$, where in matrix form

$$A \qquad \cdot \quad X \ = \ B$$

$$\begin{bmatrix} 4 & 2 & 7 \\ 3 & 1 & 8 \\ 5 & 3 & 1 \end{bmatrix} \begin{bmatrix} x_1 \\ x_2 \\ x_3 \end{bmatrix} = \begin{bmatrix} 35 \\ 25 \\ 40 \end{bmatrix}$$

From Section 6.6,

$$X = A^{-1}B$$

Substituting the inverse $A^{-1}$, which was found in Example 5, and multiplying,

$$X = \begin{bmatrix} -2.3 & 1.9 & 0.9 \\ 3.7 & -3.1 & -1.1 \\ 0.4 & -0.2 & -0.2 \end{bmatrix} \begin{bmatrix} 35 \\ 25 \\ 40 \end{bmatrix}$$

$$X = \begin{bmatrix} -2.3(35) + 1.9(25) + 0.9(40) \\ 3.7(35) - 3.1(25) - 1.1(40) \\ 0.4(35) - 0.2(25) - 0.2(40) \end{bmatrix} = \begin{bmatrix} 3 \\ 8 \\ 1 \end{bmatrix}$$

Thus,

$$X = \begin{bmatrix} \bar{x}_1 \\ \bar{x}_2 \\ \bar{x}_3 \end{bmatrix} = \begin{bmatrix} 3 \\ 8 \\ 1 \end{bmatrix}$$

Compare this answer and the work involved with that of Example 4, where the same problem was solved with Cramer's rule.

## 6.7   BUSINESS AND ECONOMIC APPLICATIONS

Cramer's rule and matrix inversion are used frequently in business and economics to solve problems involving a system of simultaneous equations such as IS-LM analysis and supply-and-demand analysis. Leaving matrix inversion and supply-and-demand analysis to later problems, we start by applying Cramer's rule to IS-LM analysis. Given the IS and LM equations:

$$0.3Y + 100i = 252$$

$$0.25Y - 200i = 176$$

Set them in matrix form.

$$A \qquad \cdot \ X \ = \ B$$

$$\begin{bmatrix} 0.3 & 100 \\ 0.25 & -200 \end{bmatrix} \begin{bmatrix} Y \\ i \end{bmatrix} = \begin{bmatrix} 252 \\ 176 \end{bmatrix}$$

where

$$|A| = 0.3(-200) - 100(0.25) = -60 - 25 = -85$$

Replacing the coefficients of $Y$ in column 1 of $A$ with the column of constants $B$ to prepare to solve for $Y$,

$$A_1 = \begin{bmatrix} 252 & 100 \\ 176 & -200 \end{bmatrix}$$

$$|A_1| = 252(-200) - 100(176) = -50,400 - 17,600 = -68,000$$

Thus,
$$Y_e = \frac{|A_1|}{|A|} = \frac{-68,000}{-85} = 800$$

Then replacing the coefficients of $i$ in column 2 of $A$ with the column of constants $B$ to solve for $i$,

$$|A_2| = \begin{bmatrix} 0.3 & 252 \\ 0.25 & 176 \end{bmatrix}$$

$$|A_2| = 0.3(176) - 252(0.25) = 52.8 - 63 = -10.2$$

and
$$i_e = \frac{|A_2|}{|A|} = \frac{-10.2}{-85} = 0.12$$

See Problem 4.21, where the same problem was solved using the elimination method. See also Problems 6.8 to 6.15.

## 6.8 SPECIAL DETERMINANTS

For special determinants which have important uses for economics, business, and mathematics in general, such as the *Jacobian, Hessian, bordered Hessian, and discriminant*, see Dowling, *Introduction to Mathematical Economics*, Chapter 12.

# Solved Problems

### DETERMINANTS

**6.1.** For each of following $2 \times 2$ matrices, find the determinant to determine whether the matrix is singular or nonsingular, and indicate the rank of the matrix.

(a)
$$A = \begin{bmatrix} 5 & 2 \\ 9 & 4 \end{bmatrix}$$

$$|A| = 5(4) - 2(9) = 2$$

With $|A| \neq 0$, $A$ is nonsingular. Both rows and columns, therefore, are linearly independent. With two rows and columns linearly independent, the rank of $A$ is 2: $\rho(A) = 2$.

(b)
$$B = \begin{bmatrix} 12 & 20 \\ 15 & 11 \end{bmatrix}$$

$$|B| = 12(11) - 20(15) = -168$$

$B$ is nonsingular; $\rho(B) = 2$.

(c)
$$C = \begin{bmatrix} 24 & 30 \\ 16 & 20 \end{bmatrix}$$

$$|C| = 24(20) - 30(16) = 0$$

With $|C| = 0$, $C$ is singular. There is linear dependence between its rows and columns. Since there is, in effect, only one independent row and one independent column, $\rho(C) = 1$. Closer inspection reveals that row 1 is 1.5 times row 2, and column 2 is 1.25 times column 1.

(d)
$$D = \begin{bmatrix} 2 & 8 & 3 \\ 9 & 5 & 4 \end{bmatrix}$$

The determinant does not exist because only square matrices have determinants. With only two rows, the highest possible rank of the matrix is 2: $\rho(D) = 2$.

**6.2.**    For each of the following $3 \times 3$ matrices, find the determinant and indicate the rank of the matrix.

(a)
$$A = \begin{bmatrix} 7 & 1 & 4 \\ 5 & 2 & 9 \\ 8 & 6 & 3 \end{bmatrix}$$

Setting down the original matrix and to the right of it repeating the first two columns,

$$\begin{bmatrix} 7 & 1 & 4 \\ 5 & 2 & 9 \\ 8 & 6 & 3 \end{bmatrix} \begin{matrix} 7 & 1 \\ 5 & 2 \\ 8 & 6 \end{matrix}$$

Then proceeding as in Example 2, with Fig. 6-3,

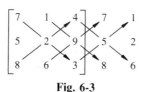

**Fig. 6-3**

$$|A| = 7 \cdot 2 \cdot 3 + 1 \cdot 9 \cdot 8 + 4 \cdot 5 \cdot 6 - [8 \cdot 2 \cdot 4 + 6 \cdot 9 \cdot 7 + 3 \cdot 5 \cdot 1]$$

$$|A| = 42 + 72 + 120 - [64 + 378 + 15] = 234 - 457 = -223$$

With $|A| \neq 0$, $A$ is nonsingular. All three rows and columns are linearly independent and $\rho(A) = 3$.

(b)
$$B = \begin{bmatrix} 4 & 1 & 9 \\ 2 & 7 & 3 \\ 5 & 8 & 6 \end{bmatrix}$$

Setting up the extended matrix in Fig. 6-4 and calculating,

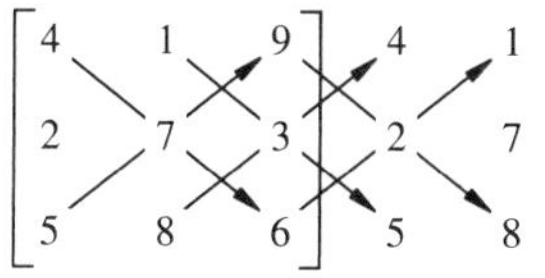

$$\begin{bmatrix} 4 & 1 & 9 \\ 2 & 7 & 3 \\ 5 & 8 & 6 \end{bmatrix} \begin{matrix} 4 & 1 \\ 2 & 7 \\ 5 & 8 \end{matrix}$$

**Fig. 6-4**

$$|B| = 4 \cdot 7 \cdot 6 + 1 \cdot 3 \cdot 5 + 9 \cdot 2 \cdot 8 - [5 \cdot 7 \cdot 9 + 8 \cdot 3 \cdot 4 + 6 \cdot 2 \cdot 1]$$

$$|B| = 168 + 15 + 144 - [315 + 96 + 12] = 327 - 423 = -96$$

$B$ is nonsingular; $\rho(B) = 3$.

(c)
$$C = \begin{bmatrix} 2 & 5 & 9 \\ 16 & 4 & 20 \\ 28 & 7 & 35 \end{bmatrix}$$

Setting up the extended matrix in Fig. 6-5 and calculating,

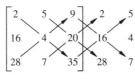

**Fig. 6-5**

$$|C| = 2 \cdot 4 \cdot 35 + 5 \cdot 20 \cdot 28 + 9 \cdot 16 \cdot 7 - [28 \cdot 4 \cdot 9 + 7 \cdot 20 \cdot 2 + 35 \cdot 16 \cdot 5]$$

$$|C| = 280 + 2800 + 1008 - [1008 + 280 + 2800] = 4088 - 4088 = 0$$

With $|C| = 0$, $C$ is singular and all three rows and columns are not linearly independent. Hence $\rho(C) \neq 3$. Though the determinant test points to the existence of linear dependence, it does not specify the nature of the dependency. Here row 3 is 1.75 times row 2. To test if any two rows or columns in $C$ are independent, apply the determinant test to the various submatrices. Starting with the $2 \times 2$ submatrix in the upper left-hand corner,

$$|C_1| = \begin{vmatrix} 2 & 5 \\ 16 & 4 \end{vmatrix} = 2(4) - 5(16) = -72 \neq 0$$

With $|C_1| \neq 0$, there are two linearly independent rows and columns in $C$ and $\rho(C) = 2$.

## CRAMER'S RULE

**6.3.**    Use Cramer's rule to solve for the unknown variables in each of the following $2 \times 2$ systems of simultaneous equations.

(a)    $8x_1 + 3x_2 = 14$

$9x_1 + 2x_2 = 24$

Expressing the equations in matrix form,

$$A \quad \cdot \quad X \quad = \quad B$$
$$\begin{bmatrix} 8 & 3 \\ 9 & 2 \end{bmatrix} \begin{bmatrix} x_1 \\ x_2 \end{bmatrix} = \begin{bmatrix} 14 \\ 24 \end{bmatrix}$$

where
$$|A| = 8(2) - 3(9) = -11$$

Replacing the first column of $A$, which contains the coefficients of the first variable $x_1$, with the column of constants $B$,

$$A_1 = \begin{bmatrix} 14 & 3 \\ 24 & 2 \end{bmatrix}$$

and taking the determinant

$$|A_1| = 14(2) - 3(24) = 28 - 72 = -44$$

Then using Cramer's rule,

$$\bar{x}_1 = \frac{|A_1|}{|A|} = \frac{-44}{-11} = 4$$

Next, replacing the second column of the original $A$ matrix, which contains the coefficients of the second variable $x_2$, with the column of constants $B$,

$$A_2 = \begin{bmatrix} 8 & 14 \\ 9 & 24 \end{bmatrix}$$

where

$$|A_2| = 8(24) - 14(9) = 192 - 126 = 66$$

$$\bar{x}_2 = \frac{|A_2|}{|A|} = \frac{66}{-11} = -6$$

(b)    $10x + 12y = -27$

      $18x - 4y = -55$

Here

$$\begin{array}{ccccc} A & \cdot & X & = & B \end{array}$$

$$\begin{bmatrix} 10 & 12 \\ 18 & -4 \end{bmatrix} \begin{bmatrix} x \\ y \end{bmatrix} = \begin{bmatrix} -27 \\ -55 \end{bmatrix}$$

and

$$|A| = 10(-4) - 12(18) = -256$$

Solving for $x$, the first variable, as in part (a),

$$A_1 = \begin{bmatrix} -27 & 12 \\ -55 & -4 \end{bmatrix}$$

$$|A_1| = -27(-4) - 12(-55) = 108 + 660 = 768$$

and

$$\bar{x} = \frac{|A_1|}{|A|} = \frac{768}{-256} = -3$$

Now solving for $y$, the second variable, also as in part (a),

$$A_2 = \begin{bmatrix} 10 & -27 \\ 18 & -55 \end{bmatrix}$$

$$|A_2| = 10(-55) - (-27)(18) = -550 + 486 = -64$$

and

$$\bar{y} = \frac{|A_2|}{|A|} = \frac{-64}{-256} = \frac{1}{4}$$

(c)    $15p_1 - 4p_2 = 39$

      $-7p_1 + 22p_2 = 163$

Here

$$A = \begin{bmatrix} 15 & -4 \\ -7 & 22 \end{bmatrix}$$

and

$$|A| = 15(22) - (-4)(-7) = 330 - 28 = 302$$

Solving for $p_1$,

$$A_1 = \begin{bmatrix} 39 & -4 \\ 163 & 22 \end{bmatrix}$$

$$|A_1| = 39(22) - (-4)(163) = 858 + 652 = 1510$$

and
$$\overline{p}_1 = \frac{|A_1|}{|A|} = \frac{1510}{302} = 5$$

Solving for $p_2$,

$$A_2 = \begin{bmatrix} 15 & 39 \\ -7 & 163 \end{bmatrix}$$

$$|A_2| = 15(163) - 39(-7) = 2445 + 273 = 2718$$

and
$$\overline{p}_2 = \frac{|A_2|}{|A|} = \frac{2718}{302} = 9$$

(d)    $20Q_1 + 45Q_2 = 950$

$\phantom{(d)}\quad 36Q_1 + 50Q_2 = 1400$

Here
$$A = \begin{bmatrix} 20 & 45 \\ 36 & 50 \end{bmatrix}$$

and
$$|A| = 20(50) - 45(36) = 1000 - 1620 = -620$$

Solving for $Q_1$,

$$A_1 = \begin{bmatrix} 950 & 45 \\ 1400 & 50 \end{bmatrix}$$

$$|A_1| = 950(50) - 45(1400) = 47,500 - 63,000 = -15,500$$

and
$$\overline{Q}_1 = \frac{|A_1|}{|A|} = \frac{-15,500}{-620} = 25$$

Solving for $Q_2$,

$$A_2 = \begin{bmatrix} 20 & 950 \\ 36 & 1400 \end{bmatrix}$$

$$|A_2| = 20(1400) - 950(36) = 28,000 - 34,200 = -6200$$

and
$$\overline{Q}_2 = \frac{|A_2|}{|A|} = \frac{-6200}{-620} = 10$$

**6.4.** Use Cramer's rule to solve for the unknown variables in each of the following $3 \times 3$ systems of simultaneous equations.

(a)    $3x_1 + 8x_2 + 2x_3 = 67$

$\phantom{(a)}\quad 4x_1 + 6x_2 + 9x_3 = 36$

$\phantom{(a)}\quad 7x_1 + \phantom{6}x_2 + 5x_3 = 49$

$$A = \begin{bmatrix} 3 & 8 & 2 \\ 4 & 6 & 9 \\ 7 & 1 & 5 \end{bmatrix}$$

Using the techniques of Section 6.2 and Problem 6.2,

$$|A| = 3 \cdot 6 \cdot 5 + 8 \cdot 9 \cdot 7 + 2 \cdot 4 \cdot 1 - [7 \cdot 6 \cdot 2 + 1 \cdot 9 \cdot 3 + 5 \cdot 4 \cdot 8]$$

$$|A| = 90 + 504 + 8 - [84 + 27 + 160] = 602 - 271 = 331$$

Solving for $x_1$,

$$A_1 = \begin{bmatrix} 67 & 8 & 2 \\ 36 & 6 & 9 \\ 49 & 1 & 5 \end{bmatrix}$$

$$|A_1| = 67 \cdot 6 \cdot 5 + 8 \cdot 9 \cdot 49 + 2 \cdot 36 \cdot 1 - [49 \cdot 6 \cdot 2 + 1 \cdot 9 \cdot 67 + 5 \cdot 36 \cdot 8]$$

$$|A_1| = 2010 + 3528 + 72 - [588 + 603 + 1440] = 5610 - 2631 = 2979$$

and
$$\bar{x}_1 = \frac{|A_1|}{|A|} = \frac{2979}{331} = 9$$

Solving for $x_2$,

$$A_2 = \begin{bmatrix} 3 & 67 & 2 \\ 4 & 36 & 9 \\ 7 & 49 & 5 \end{bmatrix}$$

$$|A_2| = 3 \cdot 36 \cdot 5 + 67 \cdot 9 \cdot 7 + 2 \cdot 4 \cdot 49 - [7 \cdot 36 \cdot 2 + 49 \cdot 9 \cdot 3 + 5 \cdot 4 \cdot 67]$$

$$|A_2| = 540 + 4221 + 392 - [504 + 1323 + 1340] = 5153 - 3167 = 1986$$

and
$$\bar{x}_2 = \frac{|A_2|}{|A|} = \frac{1986}{331} = 6$$

Solving for $x_3$,

$$A_3 = \begin{bmatrix} 3 & 8 & 67 \\ 4 & 6 & 36 \\ 7 & 1 & 49 \end{bmatrix}$$

$$|A_3| = 3 \cdot 6 \cdot 49 + 8 \cdot 36 \cdot 7 + 67 \cdot 4 \cdot 1 - [7 \cdot 6 \cdot 67 + 1 \cdot 36 \cdot 3 + 49 \cdot 4 \cdot 8]$$

$$|A_3| = 882 + 2016 + 268 - [2814 + 108 + 1568] = 3166 - 4490 = -1324$$

and
$$\bar{x}_3 = \frac{|A_3|}{|A|} = \frac{-1324}{331} = -4$$

(b)
$$-5p_1 + p_2 + 2p_3 = -6$$
$$2p_1 - 9p_2 + p_3 = -14$$
$$3p_1 + p_2 - 8p_3 = -58$$

$$A = \begin{bmatrix} -5 & 1 & 2 \\ 2 & -9 & 1 \\ 3 & 1 & -8 \end{bmatrix}$$

$$|A| = (-5)(-9)(-8) + (1)(1)(3) + (2)(2)(1)$$
$$- [(3)(-9)(2) + (1)(1)(-5) + (-8)(2)(1)]$$
$$|A| = -360 + 3 + 4 - [-54 - 5 - 16] = -353 + 75 = -278$$

Solving for $p_1$,

$$A_1 = \begin{bmatrix} -6 & 1 & 2 \\ -14 & -9 & 1 \\ -58 & 1 & -8 \end{bmatrix}$$

$$|A_1| = (-6)(-9)(-8) + (1)(1)(-58) + (2)(-14)(1)$$
$$- [(-58)(-9)(2) + (1)(1)(-6) + (-8)(-14)(1)]$$
$$|A_1| = -432 - 58 - 28 - [1044 - 6 + 112] = -518 - 1150 = -1668$$

and
$$\bar{p}_1 = \frac{|A_1|}{|A|} = \frac{-1668}{-278} = 6$$

Solving for $p_2$,

$$A_2 = \begin{bmatrix} -5 & -6 & 2 \\ 2 & -14 & 1 \\ 3 & -58 & -8 \end{bmatrix}$$

$$|A_2| = (-5)(-14)(-8) + (-6)(1)(3) + (2)(2)(-58)$$
$$- [(3)(-14)(2) + (-58)(1)(-5) + (-8)(2)(-6)]$$
$$|A_2| = -560 - 18 - 232 - [-84 + 290 + 96] = -810 - 302 = -1112$$

and

$$\overline{p}_2 = \frac{|A_2|}{|A|} = \frac{-1112}{-278} = 4$$

Solving for $p_3$,

$$A_3 = \begin{bmatrix} -5 & 1 & -6 \\ 2 & -9 & -14 \\ 3 & 1 & -58 \end{bmatrix}$$

$$|A_3| = (-5)(-9)(-58) + (1)(-14)(3) + (-6)(2)(1)$$
$$- [(3)(-9)(-6) + (1)(-14)(-5) + (-58)(2)(1)]$$
$$|A_3| = -2610 - 42 - 12 - [162 + 70 - 116] = -2664 - 116 = -2780$$

and

$$\overline{p}_3 = \frac{|A_3|}{|A|} = \frac{-2780}{-278} = 10$$

(c) 
$$16x - 3y + 10z = 8$$
$$-4x + 18y + 14z = 12$$
$$28x + 42y - 6z = 18$$

$$A = \begin{bmatrix} 16 & -3 & 10 \\ -4 & 18 & 14 \\ 28 & 42 & -6 \end{bmatrix}$$

$$|A| = (16)(18)(-6) + (-3)(14)(28) + (10)(-4)(42)$$
$$- [(28)(18)(10) + (42)(14)(16) + (-6)(-4)(-3)]$$
$$|A| = -1728 - 1176 - 1680 - [5040 + 9408 - 72] = -18,960$$

Solving for $x$,

$$A_1 = \begin{bmatrix} 8 & -3 & 10 \\ 12 & 18 & 14 \\ 18 & 42 & -6 \end{bmatrix}$$

$$|A_1| = (8)(18)(-6) + (-3)(14)(18) + (10)(12)(42)$$
$$- [(18)(18)(10) + (42)(14)(8) + (-6)(12)(-3)]$$
$$|A_1| = -864 - 756 + 5040 - [3240 + 4704 + 216] = -4740$$

and

$$\overline{x} = \frac{|A_1|}{|A|} = \frac{-4740}{-18,960} = \frac{1}{4}$$

Solving for $y$,

$$A_2 = \begin{bmatrix} 16 & 8 & 10 \\ -4 & 12 & 14 \\ 28 & 18 & -6 \end{bmatrix}$$

$$|A_2| = (16)(12)(-6) + (8)(14)(28) + (10)(-4)(18)$$

$$- [(28)(12)(10) + (18)(14)(16) + (-6)(-4)(8)]$$

$$|A_2| = -1152 + 3136 - 720 - [3360 + 4032 + 192] = -6320$$

and

$$\bar{y} = \frac{|A_2|}{|A|} = \frac{-6320}{-18,960} = \frac{1}{3}$$

Solving for $z$,

$$A_3 = \begin{bmatrix} 16 & -3 & 8 \\ -4 & 18 & 12 \\ 28 & 42 & 18 \end{bmatrix}$$

$$|A_3| = (16)(18)(18) + (-3)(12)(28) + (8)(-4)(42)$$

$$- [(28)(18)(8) + (42)(12)(16) + (18)(-4)(-3)]$$

$$|A_3| = 5184 - 1008 - 1344 - [4032 + 8064 + 216] = -9480$$

and

$$\bar{z} = \frac{|A_3|}{|A|} = \frac{-9480}{-18,960} = \frac{1}{2}$$

## INVERSE MATRICES

**6.5.**  (a) Find the inverse matrix $A^{-1}$ and (b) solve for the unknown variables using the inverse matrix, given

$$3x_1 + 7x_2 = 41$$

$$8x_1 + 9x_2 = 61$$

(a)  Using the Gaussian elimination method from Section 6.5 and Example 5, and setting up the augmented matrix,

$$\left[ \begin{array}{cc|cc} 3 & 7 & 1 & 0 \\ 8 & 9 & 0 & 1 \end{array} \right]$$

1a.  Multiplying row 1 by $\frac{1}{3}$ to obtain 1 in the $a_{11}$ position,

$$\left[ \begin{array}{cc|cc} 1 & \frac{7}{3} & \frac{1}{3} & 0 \\ 8 & 9 & 0 & 1 \end{array} \right]$$

1b.  Subtracting 8 times row 1 from row 2 to clear column 1,

$$\left[ \begin{array}{cc|cc} 1 & \frac{7}{3} & \frac{1}{3} & 0 \\ 0 & -\frac{29}{3} & -\frac{8}{3} & 1 \end{array} \right]$$

2a.  Multiplying row 2 by $-\frac{3}{29}$ to obtain 1 in the $a_{22}$ position,

$$\left[ \begin{array}{cc|cc} 1 & \frac{7}{3} & \frac{1}{3} & 0 \\ 0 & 1 & \frac{8}{29} & -\frac{3}{29} \end{array} \right]$$

2b.    Subtracting $\frac{7}{3}$ times row 2 from row 1 to clear column 2,

$$\begin{bmatrix} 1 & 0 & \Big| & -\frac{9}{29} & \frac{7}{29} \\ 0 & 1 & \Big| & \frac{8}{29} & -\frac{3}{29} \end{bmatrix}$$

With the identity matrix now on the left-hand side of the bar,

$$A^{-1} = \begin{bmatrix} -\frac{9}{29} & \frac{7}{29} \\ \frac{8}{29} & -\frac{3}{29} \end{bmatrix}$$

(b)    Following the procedure from Section 6.6 and Example 7,

$$X = A^{-1}B$$

$$X = \begin{bmatrix} -\frac{9}{29} & \frac{7}{29} \\ \frac{8}{29} & -\frac{3}{29} \end{bmatrix}\begin{bmatrix} 41 \\ 61 \end{bmatrix}$$

Then simply multiplying the two matrices using ordinary row-column operations,

$$X = \begin{bmatrix} -\frac{9}{29}(41) + \frac{7}{29}(61) \\ \frac{8}{29}(41) - \frac{3}{29}(61) \end{bmatrix} = \begin{bmatrix} \frac{58}{29} \\ \frac{145}{29} \end{bmatrix} = \begin{bmatrix} 2 \\ 5 \end{bmatrix}$$

Compare the answers and the work involved with Example 3 where the same problem was solved using Cramer's rule.

**6.6.**    Redo Problem 6.5, given

$$8x_1 + 3x_2 = 14$$

$$9x_1 + 2x_2 = 24$$

(a)    Setting up the augmented matrix,

$$\begin{bmatrix} 8 & 3 & \Big| & 1 & 0 \\ 9 & 2 & \Big| & 0 & 1 \end{bmatrix}$$

1a.    Multiplying row 1 by $\frac{1}{8}$ to obtain 1 in the $a_{11}$ position,

$$\begin{bmatrix} 1 & \frac{3}{8} & \Big| & \frac{1}{8} & 0 \\ 9 & 2 & \Big| & 0 & 1 \end{bmatrix}$$

1b.    Subtracting 9 times row 1 from row 2 to clear column 1,

$$\begin{bmatrix} 1 & \frac{3}{8} & \Big| & \frac{1}{8} & 0 \\ 0 & -\frac{11}{8} & \Big| & -\frac{9}{8} & 1 \end{bmatrix}$$

2a.    Multiplying row 2 by $-\frac{8}{11}$ to obtain 1 in the $a_{22}$ position,

$$\begin{bmatrix} 1 & \frac{3}{8} & \Big| & \frac{1}{8} & 0 \\ 0 & 1 & \Big| & \frac{9}{11} & -\frac{8}{11} \end{bmatrix}$$

2b.    Subtracting $\frac{3}{8}$ times row 2 from row 1 to clear column 2,

$$\begin{bmatrix} 1 & 0 & \Big| & -\frac{2}{11} & \frac{3}{11} \\ 0 & 1 & \Big| & \frac{9}{11} & -\frac{8}{11} \end{bmatrix}$$

With the identity matrix now on the left-hand side of the bar,

$$A^{-1} = \begin{bmatrix} -\frac{2}{11} & \frac{3}{11} \\ \frac{9}{11} & -\frac{8}{11} \end{bmatrix}$$

(b)

$$X = \qquad A^{-1} \qquad B$$

$$X = \begin{bmatrix} -\frac{2}{11} & \frac{3}{11} \\ \frac{9}{11} & -\frac{8}{11} \end{bmatrix} \begin{bmatrix} 14 \\ 24 \end{bmatrix}$$

Multiplying the two matrices using simple row-column operations,

$$X = \begin{bmatrix} -\frac{2}{11}(14) + \frac{3}{11}(24) \\ \frac{9}{11}(14) - \frac{8}{11}(24) \end{bmatrix} = \begin{bmatrix} \frac{44}{11} \\ -\frac{66}{11} \end{bmatrix} = \begin{bmatrix} 4 \\ -6 \end{bmatrix}$$

Compare the answers here with those of Problem 6.3 (a).

**6.7.** Redo Problem 6.5, given

$$3x_1 + 8x_2 + 2x_3 = 67$$

$$4x_1 + 6x_2 + 9x_3 = 36$$

$$7x_1 + \phantom{0}x_2 + 5x_3 = 49$$

(a)

$$\begin{bmatrix} 3 & 8 & 2 & | & 1 & 0 & 0 \\ 4 & 6 & 9 & | & 0 & 1 & 0 \\ 7 & 1 & 5 & | & 0 & 0 & 1 \end{bmatrix}$$

1a.   Multiply row 1 by $\frac{1}{3}$ to obtain 1 in the $a_{11}$ position.

$$\begin{bmatrix} 1 & \frac{8}{3} & \frac{2}{3} & | & \frac{1}{3} & 0 & 0 \\ 4 & 6 & 9 & | & 0 & 1 & 0 \\ 7 & 1 & 5 & | & 0 & 0 & 1 \end{bmatrix}$$

1b.   Subtract 4 times row 1 from row 2 and 7 times row 1 from row 3 to clear the first column.

$$\begin{bmatrix} 1 & \frac{8}{3} & \frac{2}{3} & | & \frac{1}{3} & 0 & 0 \\ 0 & -\frac{14}{3} & \frac{19}{3} & | & -\frac{4}{3} & 1 & 0 \\ 0 & -\frac{53}{3} & \frac{1}{3} & | & -\frac{7}{3} & 0 & 1 \end{bmatrix}$$

2a.   Multiply row 2 by $-\frac{3}{14}$ to obtain 1 in the $a_{22}$ position.

$$\begin{bmatrix} 1 & \frac{8}{3} & \frac{2}{3} & | & \frac{1}{3} & 0 & 0 \\ 0 & 1 & -\frac{19}{14} & | & \frac{2}{7} & -\frac{3}{14} & 0 \\ 0 & -\frac{53}{3} & \frac{1}{3} & | & -\frac{7}{3} & 0 & 1 \end{bmatrix}$$

2b.   Subtract $\frac{8}{3}$ times row 2 from row 1 and add $\frac{53}{3}$ times row 2 to row 3 to clear the second column.

$$\begin{bmatrix} 1 & 0 & \frac{30}{7} & | & -\frac{3}{7} & \frac{4}{7} & 0 \\ 0 & 1 & -\frac{19}{14} & | & \frac{2}{7} & -\frac{3}{14} & 0 \\ 0 & 0 & -\frac{331}{14} & | & \frac{19}{7} & -\frac{53}{14} & 1 \end{bmatrix}$$

3a.   Multiply row 3 by $-\frac{14}{331}$ to obtain 1 in the $a_{33}$ position.

$$\begin{bmatrix} 1 & 0 & \frac{30}{7} & | & -\frac{3}{7} & \frac{4}{7} & 0 \\ 0 & 1 & -\frac{19}{14} & | & \frac{2}{7} & -\frac{3}{14} & 0 \\ 0 & 0 & 1 & | & -\frac{38}{331} & \frac{53}{331} & -\frac{14}{331} \end{bmatrix}$$

3b.   Subtract $\frac{30}{7}$ times row 3 from row 1 and add $\frac{19}{14}$ times row 3 to row 2 to clear the third column.

$$\begin{bmatrix} 1 & 0 & 0 & | & \frac{21}{331} & -\frac{38}{331} & \frac{60}{331} \\ 0 & 1 & 0 & | & \frac{43}{331} & \frac{1}{331} & -\frac{19}{331} \\ 0 & 0 & 1 & | & -\frac{38}{331} & \frac{53}{331} & -\frac{14}{331} \end{bmatrix}$$

With the identity matrix now on the left,

$$A^{-1} = \begin{bmatrix} \frac{21}{331} & -\frac{38}{331} & \frac{60}{331} \\ \frac{43}{331} & \frac{1}{331} & -\frac{19}{331} \\ -\frac{38}{331} & \frac{53}{331} & -\frac{14}{331} \end{bmatrix} = \frac{1}{331} \begin{bmatrix} 21 & -38 & 60 \\ 43 & 1 & -19 \\ -38 & 53 & -14 \end{bmatrix}$$

(b)          $X = \quad\quad A^{-1} \quad\quad \cdot \quad B$

$$X = \frac{1}{331} \begin{bmatrix} 21 & -38 & 60 \\ 43 & 1 & -19 \\ -38 & 53 & -14 \end{bmatrix} \begin{bmatrix} 67 \\ 36 \\ 49 \end{bmatrix}$$

$$X = \frac{1}{331} \begin{bmatrix} 21(67) - 38(36) + 60(49) \\ 43(67) + 1(36) - 19(49) \\ -38(67) + 53(36) - 14(49) \end{bmatrix} = \frac{1}{331} \begin{bmatrix} 2979 \\ 1986 \\ -1324 \end{bmatrix} = \begin{bmatrix} 9 \\ 6 \\ -4 \end{bmatrix}$$

Compare these answers with those of Problem 6.4 (a).

## ECONOMIC AND BUSINESS APPLICATIONS

**6.8.** Use Cramer's rule to solve for the equilibrium level of price $\overline{P}$ and quantity $\overline{Q}$, given

$$\text{Supply:} \ -7P + 14Q = -42$$

$$\text{Demand:} \ \ 3P + 12Q = 90$$

Converting to matrix format,

$$A \quad\quad \cdot \ X \ = \quad B$$

$$\begin{bmatrix} -7 & 14 \\ 3 & 12 \end{bmatrix} \begin{bmatrix} P \\ Q \end{bmatrix} = \begin{bmatrix} -42 \\ 90 \end{bmatrix}$$

where
$$|A| = -7(12) - 14(3) = -126$$

Solving for $P$, the first variable,

$$|A_1| = \begin{vmatrix} -42 & 14 \\ 90 & 12 \end{vmatrix} = -42(12) - 14(90) = -1764$$

and
$$\overline{P} = \frac{|A_1|}{|A|} = \frac{-1764}{-126} = 14$$

Solving for $Q$, the second variable,

$$|A_2| = \begin{vmatrix} -7 & -42 \\ 3 & 90 \end{vmatrix} = -7(90) - (-42)(3) = -504$$

and
$$\overline{Q} = \frac{|A_2|}{|A|} = \frac{-504}{-126} = 4$$

Compare with Problem 4.25.

**6.9.** Use Cramer's rule to solve for the equilibrium level of income $\overline{Y}$ and the interest rate $\overline{i}$, given

$$\text{IS:} \ \ \ 0.3Y + 100i - 252 = 0$$

$$\text{LM:} \ 0.25Y - 200i - 193 = 0$$

$$A \quad\quad \cdot \quad X \quad = \quad B$$

$$\begin{bmatrix} 0.3 & 100 \\ 0.25 & -200 \end{bmatrix} \begin{bmatrix} Y \\ i \end{bmatrix} = \begin{bmatrix} 252 \\ 193 \end{bmatrix}$$

where

$$|A| = 0.3(-200) - 100(0.25) = -85$$

Solving for $Y$,

$$|A_1| = \begin{vmatrix} 252 & 100 \\ 193 & -200 \end{vmatrix} = 252(-200) - 100(193) = -69,700$$

and

$$\overline{Y} = \frac{|A_1|}{|A|} = \frac{-69,700}{-85} = 820$$

Solving for $i$,

$$|A_2| = \begin{vmatrix} 0.3 & 252 \\ 0.25 & 193 \end{vmatrix} = 0.3(193) - 252(0.25) = -5.1$$

and

$$\overline{i} = \frac{|A_2|}{|A|} = \frac{-5.1}{-85} = 0.06$$

Compare with Problem 4.22.

**6.10.** Redo Problem 6.9, given

$$\text{IS:} \quad 0.4Y + 150i - 209 = 0$$

$$\text{LM:} \ 0.1Y - 250i - \ \ 35 = 0$$

$$A \quad\quad \cdot \quad X \quad = \quad B$$

$$\begin{bmatrix} 0.4 & 150 \\ 0.1 & -250 \end{bmatrix} \begin{bmatrix} Y \\ i \end{bmatrix} = \begin{bmatrix} 209 \\ 35 \end{bmatrix}$$

where

$$|A| = 0.4(-250) - 150(0.1) = -115$$

Solving for $Y$,

$$|A_1| = \begin{vmatrix} 209 & 150 \\ 35 & -250 \end{vmatrix} = 209(-250) - 150(35) = -57,500$$

and

$$\overline{Y} = \frac{|A_1|}{|A|} = \frac{-57,500}{-115} = 500$$

Solving for $i$,

$$|A_2| = \begin{vmatrix} 0.4 & 209 \\ 0.1 & 35 \end{vmatrix} = 0.4(35) - 209(0.1) = -6.9$$

and

$$\overline{i} = \frac{|A_2|}{|A|} = \frac{-6.9}{-115} = 0.06$$

Compare with Problem 4.23.

**6.11.** Redo Problem 6.10 for a drop in autonomous investment, when

$$\text{IS:} \quad 0.4Y + 150i - 186 = 0$$

$$\text{LM:} \ 0.1Y - 250i - \ \ 35 = 0$$

$$\begin{array}{ccc} A & \cdot\ X & =\ B \end{array}$$

$$\begin{bmatrix} 0.4 & 150 \\ 0.1 & -250 \end{bmatrix} \begin{bmatrix} Y \\ i \end{bmatrix} = \begin{bmatrix} 186 \\ 35 \end{bmatrix}$$

where

$$|A| = 0.4(-250) - 150(0.1) = -115$$

Solving for $Y$,

$$|A_1| = \begin{vmatrix} 186 & 150 \\ 35 & -250 \end{vmatrix} = 186(-250) - 150(35) = -51{,}750$$

and

$$\overline{Y} = \frac{|A_1|}{|A|} = \frac{-51{,}750}{-115} = 450$$

Solving for $i$,

$$|A_2| = \begin{vmatrix} 0.4 & 186 \\ 0.1 & 35 \end{vmatrix} = 0.4(35) - 186(0.1) = -4.6$$

and

$$\overline{i} = \frac{|A_2|}{|A|} = \frac{-4.6}{-115} = 0.04$$

A fall in autonomous investment leads to a drop in income and a decline in the interest rate. See Problem 4.24.

**6.12.** Use Cramer's rule to solve for $\overline{P}$ and $\overline{Q}$ in each of the following three interconnected markets:

$$\begin{array}{ll} Q_{s1} = 6P_1 - 8 & Q_{d1} = -5P_1 + P_2 + P_3 + 23 \\ Q_{s2} = 3P_2 - 11 & Q_{d2} = P_1 - 3P_2 + 2P_3 + 15 \\ Q_{s3} = 3P_3 - 5 & Q_{d3} = P_1 + 2P_2 - 4P_3 + 19 \end{array}$$

From Problem 4.14, the markets are simultaneously in equilibrium when

$$11P_1 - P_2 - P_3 = 31$$

$$-P_1 + 6P_2 - 2P_3 = 26$$

$$-P_1 - 2P_2 + 7P_3 = 24$$

where

$$A = \begin{bmatrix} 11 & -1 & -1 \\ -1 & 6 & -2 \\ -1 & -2 & 7 \end{bmatrix}$$

$$|A| = (11)(6)(7) + (-1)(-2)(-1) + (-1)(-1)(-2)$$

$$- [(-1)(6)(-1) + (-2)(-2)(11) + (7)(-1)(-1)]$$

$$|A| = 462 - 2 - 2 - [6 + 44 + 7] = 458 - 57 = 401$$

Solving for $P_1$,

$$A_1 = \begin{bmatrix} 31 & -1 & -1 \\ 26 & 6 & -2 \\ 24 & -2 & 7 \end{bmatrix}$$

$$|A_1| = (31)(6)(7) + (-1)(-2)(24) + (-1)(26)(-2)$$

$$- [(24)(6)(-1) + (-2)(-2)(31) + (7)(26)(-1)]$$

$$|A_1| = 1302 + 48 + 52 - [-144 + 124 - 182] = 1402 + 202 = 1604$$

and
$$\overline{P}_1 = \frac{|A_1|}{|A|} = \frac{1604}{401} = 4$$

Solving for $P_2$,

$$A_2 = \begin{bmatrix} 11 & 31 & -1 \\ -1 & 26 & -2 \\ -1 & 24 & 7 \end{bmatrix}$$

$$|A_2| = (11)(26)(7) + (31)(-2)(-1) + (-1)(-1)(24)$$

$$- [(-1)(26)(-1) + (24)(-2)(11) + (7)(-1)(31)]$$

$$|A_2| = 2002 + 62 + 24 - [26 - 528 - 217] = 2088 + 719 = 2807$$

and
$$\overline{P}_2 = \frac{|A_2|}{|A|} = \frac{2807}{401} = 7$$

Solving for $P_3$,

$$A_3 = \begin{bmatrix} 11 & -1 & 31 \\ -1 & 6 & 26 \\ -1 & -2 & 24 \end{bmatrix}$$

$$|A_3| = (11)(6)(24) + (-1)(26)(-1) + (31)(-1)(-2)$$

$$- [(-1)(6)(31) + (-2)(26)(11) + (24)(-1)(-1)]$$

$$|A_3| = 1584 + 26 + 62 - [-186 - 572 + 24] = 1672 + 734 = 2406$$

and
$$\overline{P}_3 = \frac{|A_3|}{|A|} = \frac{2406}{401} = 6$$

Compare with Problem 4.14.

**6.13.** Use Cramer's rule to find the equilibrium values for $x$, $y$, and $\lambda$, given the following first-order conditions for constrained optimization in Problem 13.21:

$$10x - 2y - \lambda = 0$$

$$-2x + 16y - \lambda = 0$$

$$60 - x - y = 0$$

Rearranging and setting in matrix form,

$$\begin{bmatrix} 10 & -2 & -1 \\ -2 & 16 & -1 \\ -1 & -1 & 0 \end{bmatrix} \begin{bmatrix} x \\ y \\ \lambda \end{bmatrix} = \begin{bmatrix} 0 \\ 0 \\ -60 \end{bmatrix}$$

where
$$|A| = (10)(16)(0) + (-2)(-1)(-1) + (-1)(-2)(-1)$$

$$- [(-1)(16)(-1) + (-1)(-1)(10) + (0)(-2)(-2)]$$

$$|A| = 0 + (-2) + (-2) - [16 + 10 + 0] = -4 - 26 = -30$$

Solving for $x$, the first variable,

$$A_1 = \begin{bmatrix} 0 & -2 & -1 \\ 0 & 16 & -1 \\ -60 & -1 & 0 \end{bmatrix}$$

$$|A_1| = (0)(16)(0) + (-2)(-1)(-60) + (-1)(0)(-1)$$

$$- [(-60)(16)(-1) + (-1)(-1)(0) + (0)(0)(-2)]$$

$$|A_1| = 0 - 120 + 0 - [960 + 0 - 0] = -120 - 960 = -1080$$

Thus

$$\bar{x} = \frac{|A_1|}{|A|} = \frac{-1080}{-30} = 36$$

Next solving for $y$, the second variable,

$$A_2 = \begin{bmatrix} 10 & 0 & -1 \\ -2 & 0 & -1 \\ -1 & -60 & 0 \end{bmatrix}$$

$$|A_2| = (10)(0)(0) + (0)(-1)(-1) + (-1)(-2)(-60)$$

$$- [(-1)(0)(-1) + (-60)(-1)(10) + (0)(-2)(0)]$$

$$|A_2| = 0 + 0 - 120 - [0 + 600 - 0] = -120 - 600 = -720$$

Hence

$$\bar{y} = \frac{|A_2|}{|A|} = \frac{-720}{-30} = 24$$

Finally, solving for the third variable $\lambda$,

$$A_3 = \begin{bmatrix} 10 & -2 & 0 \\ -2 & 16 & 0 \\ -1 & -1 & -60 \end{bmatrix}$$

$$|A_3| = (10)(16)(-60) + (-2)(0)(-1) + (0)(-2)(-1)$$

$$- [(-1)(16)(0) + (-1)(0)(10) + (-60)(-2)(-2)]$$

$$|A_3| = -9600 + 0 + 0 - [0 - 0 - 240] = -9600 + 240 = -9360$$

and

$$\bar{\lambda} = \frac{|A_3|}{|A|} = \frac{-9360}{-30} = 312$$

**6.14.** Use Cramer's rule to find the equilibrium values for $x$, $y$, and $\lambda$, given the following first-order conditions for constrained optimization in Problem 13.33:

$$14x - 2y - \lambda = 0$$

$$-2x + 10y - \lambda = 0$$

$$77 - x - y = 0$$

In matrix form,

$$\begin{bmatrix} 14 & -2 & -1 \\ -2 & 10 & -1 \\ -1 & -1 & 0 \end{bmatrix} \begin{bmatrix} x \\ y \\ \lambda \end{bmatrix} = \begin{bmatrix} 0 \\ 0 \\ -77 \end{bmatrix}$$

Here

$$|A| = (14)(10)(0) + (-2)(-1)(-1) + (-1)(-2)(-1)$$

$$- [(-1)(10)(-1) + (-1)(-1)(14) + (0)(-2)(-2)]$$

$$|A| = 0 + (-2) + (-2) - [10 + 14 + 0] = -4 - 24 = -28$$

Solving for $x$,

$$A_1 = \begin{bmatrix} 0 & -2 & -1 \\ 0 & 10 & -1 \\ -77 & -1 & 0 \end{bmatrix}$$

$$|A_1| = (0)(10)(0) + (-2)(-1)(-77) + (-1)(0)(-1)$$

$$- [(-77)(10)(-1) + (-1)(-1)(0) + (0)(0)(-2)]$$

$$|A_1| = 0 - 154 + 0 - [770 + 0 - 0] = -154 - 770 = -924$$

and

$$\bar{x} = \frac{|A_1|}{|A|} = \frac{-924}{-28} = 33$$

Solving for $y$,

$$A_2 = \begin{bmatrix} 14 & 0 & -1 \\ -2 & 0 & -1 \\ -1 & -77 & 0 \end{bmatrix}$$

$$|A_2| = (14)(0)(0) + (0)(-1)(-1) + (-1)(-2)(-77)$$

$$- [(-1)(0)(-1) + (-77)(-1)(14) + (0)(-2)(0)]$$

$$|A_2| = 0 + 0 - 154 - [0 + 1078 - 0] = -154 - 1078 = -1232$$

and

$$\bar{y} = \frac{|A_2|}{|A|} = \frac{-1232}{-28} = 44$$

Solving for $\lambda$,

$$A_3 = \begin{bmatrix} 14 & -2 & 0 \\ -2 & 10 & 0 \\ -1 & -1 & -77 \end{bmatrix}$$

$$|A_3| = (14)(10)(-77) + (-2)(0)(-1) + (0)(-2)(-1)$$

$$- [(-1)(10)(0) + (-1)(0)(14) + (-77)(-2)(-2)]$$

$$|A_3| = -10,780 + 0 + 0 - [0 - 0 - 308] = -10,780 + 308 = -10,472$$

and

$$\bar{\lambda} = \frac{|A_3|}{|A|} = \frac{-10,472}{-28} = 374$$

**6.15.** Use matrix inversion to solve for $\bar{Y}$ and $\bar{i}$, given

$$\text{IS:} \quad 0.4Y + 150i = 209$$

$$\text{LM:} \quad 0.1Y - 250i = 35$$

Setting up the augmented matrix,

$$\begin{bmatrix} 0.4 & 150 & | & 1 & 0 \\ 0.1 & -250 & | & 0 & 1 \end{bmatrix}$$

$1a.$ Multiplying row 1 by 2.5 to obtain 1 in the $a_{11}$ position,

$$\begin{bmatrix} 1 & 375 & | & 2.5 & 0 \\ 0.1 & -250 & | & 0 & 1 \end{bmatrix}$$

$1b.$ Subtracting 0.1 times row 1 from row 2 to clear column 1,

$$\begin{bmatrix} 1 & 375 & | & 2.5 & 0 \\ 0 & -287.5 & | & -0.25 & 1 \end{bmatrix}$$

2a.   Multiplying row 2 by $-0.00348$ to obtain 1 in the $a_{22}$ position and allowing for small rounding errors,

$$\left[\begin{array}{cc|cc} 1 & 375 & 2.5 & 0 \\ 0 & 1 & 0.00087 & -0.00348 \end{array}\right]$$

2b.   Subtracting 375 times row 2 from row 1 to clear column 2,

$$\left[\begin{array}{cc|cc} 1 & 0 & 2.17375 & 1.30500 \\ 0 & 1 & 0.00087 & -0.00348 \end{array}\right]$$

With the identity matrix now on the left,

$$A^{-1} = \left[\begin{array}{cc} 2.17375 & 1.30500 \\ 0.00087 & -0.00348 \end{array}\right]$$

and

$$X = \qquad A^{-1} \qquad \cdot \quad B$$

$$X = \left[\begin{array}{cc} 2.17375 & 1.30500 \\ 0.00087 & -0.00348 \end{array}\right]\left[\begin{array}{c} 209 \\ 35 \end{array}\right]$$

$$X = \left[\begin{array}{c} 2.17375(209) + 1.30500(35) \\ 0.00087(209) - 0.00348(35) \end{array}\right] = \left[\begin{array}{c} 500 \\ 0.06 \end{array}\right] = \left[\begin{array}{c} \overline{Y} \\ \overline{i} \end{array}\right]$$

Compare with Problems 4.23 and 6.10. While matrix inversion has important uses in economic theory and the Gaussian elimination method is crucial for the simplex algorithm method of linear programming, practical problems are generally more easily solved by Cramer's rule than matrix inversion. If you wish more practice with matrix inversion, apply the method to Problems 6.8 to 6.11 on your own.

# Supplementary Problems

## DETERMINANTS

**6.16.**   Find the determinants of each of the following $2 \times 2$ matrices:

(a)   $A = \left[\begin{array}{cc} 6 & 13 \\ 4 & -5 \end{array}\right]$ 　　　　　　　　(b)   $B = \left[\begin{array}{cc} -9 & 16 \\ 11 & -2 \end{array}\right]$

(c)   $C = \left[\begin{array}{cc} 15 & 22 \\ -8 & 3 \end{array}\right]$ 　　　　　　　　(d)   $D = \left[\begin{array}{cc} 18 & 8 \\ -9 & -4 \end{array}\right]$

**6.17.**   Find the determinants of each of the following $3 \times 3$ matrices:

(a)   $A = \left[\begin{array}{ccc} 2 & 7 & 8 \\ 6 & 3 & 5 \\ 1 & 4 & 9 \end{array}\right]$ 　　　　　　(b)   $B = \left[\begin{array}{ccc} -4 & 9 & 2 \\ 5 & 6 & -1 \\ 3 & 2 & 7 \end{array}\right]$

(c)   $C = \left[\begin{array}{ccc} 10 & -2 & 3 \\ -4 & 5 & 8 \\ 0 & 9 & 6 \end{array}\right]$ 　　　　　　(d)   $D = \left[\begin{array}{ccc} 8 & 0 & 1 \\ 6 & -4 & 7 \\ 3 & 9 & 2 \end{array}\right]$

## CRAMER'S RULE

**6.18.**    Use Cramer's rule to solve each of the following equations:

    (a)  $4x + 5y = 92$                        (b)  $13x - 4y = 29$

          $7x + 6y = 128$                       $-8x + 9y = 41$

    (c)  $2x - 7y = -26$                     (d)  $-6x + 7y = -244$

          $3x + 5y = -8$                       $15x + 8y = 202$

**6.19.**    Solve each of the following equations using Cramer's rule:

    (a)  $3x_1 + 7x_2 - 4x_3 = 11$           (b)  $5x_1 - 8x_2 + 2x_3 = -59$

          $2x_1 - 8x_2 + 5x_3 = 18$               $6x_1 + 4x_2 + 7x_3 = 52$

          $9x_1 + 6x_2 - 2x_3 = 53$              $-x_1 + 9x_2 + 4x_3 = 90$

    (c)  $4x_1 - 9x_2 - 2x_3 = 28$           (d)      $-4x_2 + 3x_3 = -57$

          $2x_1 \qquad - 7x_3 = 53$                 $8x_1 \qquad - 5x_3 = 7$

             $5x_2 - 8x_3 = 50$                 $6x_1 + 7x_2 \qquad = 78$

## INVERSE MATRICES

**6.20.**    Find the inverse matrix for each of the following $2 \times 2$ matrices:

    (a)  $A = \begin{bmatrix} 4 & 5 \\ 7 & 6 \end{bmatrix}$                 (b)  $B = \begin{bmatrix} 13 & -4 \\ -8 & 9 \end{bmatrix}$

    (c)  $C = \begin{bmatrix} 2 & -7 \\ 3 & 5 \end{bmatrix}$                (d)  $D = \begin{bmatrix} -6 & 7 \\ 15 & 8 \end{bmatrix}$

**6.21.**    Find the inverse matrix for each of the following $3 \times 3$ matrices:

    (a)  $A = \begin{bmatrix} 3 & 7 & -4 \\ 2 & -8 & 5 \\ 9 & 6 & -2 \end{bmatrix}$        (b)  $B = \begin{bmatrix} 5 & -8 & 2 \\ 6 & 4 & 7 \\ -1 & 9 & 4 \end{bmatrix}$

    (c)  $C = \begin{bmatrix} 4 & -9 & -2 \\ 2 & 0 & -7 \\ 0 & 5 & -8 \end{bmatrix}$        (d)  $D = \begin{bmatrix} 0 & -4 & 3 \\ 8 & 0 & -5 \\ 6 & 7 & 0 \end{bmatrix}$

## BUSINESS AND ECONOMIC APPLICATIONS

**6.22.**    Use Cramer's rule or matrix inversion to find the equilibrium price $P_e$ and quantity $Q_e$ in each of the following markets:

    (a)  Supply:  $-8P + 16Q = 400$         (b)  Supply:  $-7P + 21Q = 546$

          Demand:  $5P + 20Q = 710$            Demand:  $22P + 11Q = 440$

    (c)  Supply:  $-30P + 6Q = -492$      (d)  Supply:  $-6P + 24Q = 2130$

          Demand:  $16P + 2Q = 304$           Demand:  $15P + 5Q = 525$

**6.23.**    Use Cramer's rule or matrix inversion to find the equilibrium level of income $Y_e$ and interest rate $i_e$, given

    (a)  IS:  $0.35Y + 150i - 1409 = 0$       (b)  IS:  $0.25Y + 175i - 1757 = 0$

          LM:  $0.2Y - 100i - 794 = 0$          LM:  $0.12Y - 75i - 837 = 0$

(c)   IS:   $0.12Y + 120i - 789 = 0$          (d)   IS:   $0.18Y + 125i - 955 = 0$

LM: $0.08Y - 80i - 514 = 0$                     LM: $0.14Y - 50i - 731 = 0$

# Answers to Supplementary Problems

**6.16.**   (a)   $|A| = -82$       (b)   $|B| = -158$       (c)   $|C| = 221$       (d)   $|D| = 0$

**6.17.**   (a)   $|A| = -161$       (b)   $|B| = -534$       (c)   $|C| = -576$       (d)   $|D| = -502$

**6.18.**   (a)   $x = 8, y = 12$       (b)   $x = 5, y = 9$       (c)   $x = -6, y = 2$
            (d)   $x = 22, y = -16$

**6.19.**   (a)   $x_1 = 5, x_2 = 4, x_3 = 8$                (b)   $x_1 = -3, x_2 = 7, x_3 = 6$
            (c)   $x_1 = 9, x_2 = 2, x_3 = -5$               (d)   $x_1 = -1, x_2 = 12, x_3 = -3$

**6.20.**   (a)   $A^{-1} = -\frac{1}{11}\begin{bmatrix} 6 & -5 \\ -7 & 4 \end{bmatrix}$          (b)   $B^{-1} = \frac{1}{85}\begin{bmatrix} 9 & 4 \\ 8 & 13 \end{bmatrix}$

            (c)   $C^{-1} = \frac{1}{31}\begin{bmatrix} 5 & 7 \\ -3 & 2 \end{bmatrix}$          (d)   $D^{-1} = -\frac{1}{153}\begin{bmatrix} 8 & -7 \\ -15 & -6 \end{bmatrix}$

**6.21**   (a)   $A^{-1} = -\frac{1}{35}\begin{bmatrix} -14 & -10 & 3 \\ 49 & 30 & -23 \\ 84 & 45 & -38 \end{bmatrix}$          (b)   $B^{-1} = \frac{1}{129}\begin{bmatrix} -47 & 50 & -64 \\ -31 & 22 & -23 \\ 58 & -37 & 68 \end{bmatrix}$

            (c)   $C^{-1} = -\frac{1}{24}\begin{bmatrix} 35 & -82 & 63 \\ 16 & -32 & 24 \\ 10 & -20 & 18 \end{bmatrix}$          (d)   $D^{-1} = \frac{1}{288}\begin{bmatrix} 35 & 21 & 20 \\ -30 & -18 & 24 \\ 56 & -24 & 32 \end{bmatrix}$

**6.22.**   (a)   $P_e = 14, Q_e = 32$          (b)   $P_e = 6, Q_e = 28$
            (c)   $P_e = 18, Q_e = 8$           (d)   $P_e = 5, Q_e = 90$

**6.23.**   (a)   $Y_e = 4000, i_e = 0.06$          (b)   $Y_e = 7000, i_e = 0.04$
            (c)   $Y_e = 6500, i_e = 0.075$         (d)   $Y_e = 5250, i_e = 0.08$

# Chapter 7

# Linear Programming: Using Graphs

## 7.1 USE OF GRAPHS

*Linear programming* has as its purpose the optimal allocation of scarce resources among competing products or activities. It is singularly helpful in business and economics where it is often necessary to optimize a profit or cost function subject to several inequality constraints. If the constraints are limited to two variables, the easiest solution is through the use of graphs. The graphic approach for maximization and minimization is illustrated in Sections 7.2 and 7.4, respectively.

## 7.2 MAXIMIZATION USING GRAPHS

The graphic approach for maximization is outlined below in four simple steps. The presentation is set in the context of a concrete example to facilitate the explanation. Assume a firm that makes sleeping bags $x_1$ and tents $x_2$. Each sleeping bag requires 2 hours for cutting $A$, 5 hours for sewing $B$, and 1 hour for waterproofing $C$. Each tent requires 1 hour for cutting, 5 hours for sewing, and 3 hours for waterproofing. Given current resources, the company has at most 14 hours for cutting, 40 hours for sewing, and 18 hours for waterproofing a day. Its profit margin is $50 per sleeping bag and $30 per tent. With this information the optimal allocation of resources to maximize profits is found in the following way:

1. Express the data as equations or inequalities. The function to be optimized, called the *objective function*, becomes

$$\pi = 50x_1 + 30x_2 \tag{7.1}$$

subject to the constraints

Constraint $A$: $\qquad\qquad 2x_1 + \ x_2 \leq 14$

Constraint $B$: $\qquad\qquad 5x_1 + 5x_2 \leq 40$

Constraint $C$: $\qquad\qquad x_1 + 3x_2 \leq 18$

Nonnegativity constraint: $\qquad x_1, x_2 \geq 0$

Here the variables $x_1$ and $x_2$ are called *decision variables*. The first three constraints are *technical constraints* determined by the state of technology and the availability of inputs. The fourth constraint is a *nonnegativity constraint* imposed on every problem to preclude negative (hence unacceptable) values from the solution.

2. Treat the inequality constraints as equations. Solve each one for $x_2$ in terms of $x_1$ and graph. Thus,

From $A$, $\qquad\qquad\qquad x_2 = 14 - 2x_1$

From $B$, $\qquad\qquad\qquad x_2 = 8 - x_1$

From $C$, $\qquad\qquad\qquad x_2 = 6 - \frac{1}{3}x_1$

The graph of the original "less than or equal to" inequality will include all the points *on the line and to the left of it.* See Fig. 7-1(a). The nonnegativity constraints $x_1, x_2 \geq 0$ are represented by the vertical and horizontal axes, respectively. The shaded area is called the *feasible region*. It contains all the points that satisfy all three constraints plus the nonnegativity constraints.

3. To find the optimal solution within the feasible region, if it exists, graph the objective function as a series of isoprofit lines. From (7.1),

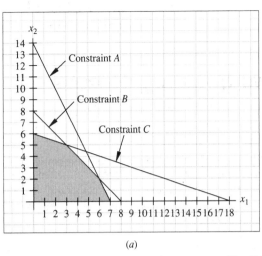

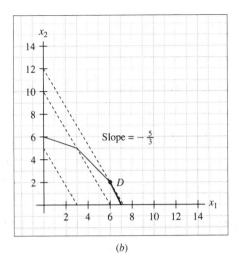

(a)                                                        (b)

**Fig. 7-1**

$$x_2 = -\frac{5}{3}x_1 + \frac{\pi}{30}$$

The isoprofit line has a slope of $-\frac{5}{3}$. Drawing a series of (dashed) lines allowing for larger and larger profits in Fig. 7-1(b), we see the isoprofit line representing the largest possible profit is tangent to the feasible region at $D$, where $x_1 = 6$ and $x_2 = 2$.

4.  Substitute the optimal values $x_1 = 6$ and $x_2 = 2$ in (7.1) to find the maximum possible profit,

$$\pi = 50(6) + 30(2) = 360$$

See also Problems 7.1 to 7.5 and 7.9 to 7.13.

## 7.3   THE EXTREME-POINT THEOREM

In Fig. 7-1(b), profit is maximized at the intersection of two constraints, called an *extreme point*. The *extreme-point theorem* states that if an optimal feasible value of the objective function exists, it will always be found at one of the extreme (or corner) points of the boundary. In Fig. 7-1(a) we can find 10 such extreme points: (0, 14), (0, 8), (0, 6), (7, 0), (8, 0), (18, 0), (3, 5), (6, 2), (4.8, 4.4), and (0, 0), the last being the intersection of the nonnegativity constraints. All 10 are *basic solutions*, but only the 5 which violate none of the constraints are basic *feasible* solutions. They are (0, 6), (3, 5), (6, 2), (7, 0), and (0, 0). Usually only one of the basic feasible solutions will prove optimal. At (3, 5), for instance, $\pi = 50(3) + 30(5) = 300$, which is lower than $\pi = 360$ above.

## 7.4   MINIMIZATION USING GRAPHS

The graphic approach to minimization is demonstrated below with a few simple adaptations to the four steps presented in Section 7.2 for maximization. The explanation is once again couched in terms of a concrete example to elucidate the procedure. Assume a botanist wishing to mix fertilizer that will provide a minimum of 45 units of phosphates $A$, 48 units of potash $B$, and 84 units of nitrates $C$. One brand of fertilizer $y_1$ provides 3 units of phosphates, 4 units of potash, and 14 units of nitrates. A second brand $y_2$ provides 9 units of phosphates, 6 units of potash, and 7 units of nitrates. The cost of $y_1$ is \$12; the cost of $y_2$ is \$20. The least-cost combination of $y_1$ and $y_2$ that will fulfill all minimum requirements is found as follows.

1.  The objective function to be minimized is

$$c = 12y_1 + 20y_2 \qquad\qquad (7.2)$$

subject to the constraints

Constraint $A$:                     $3y_1 + 9y_2 \geq 45$

Constraint $B$:                     $4y_1 + 6y_2 \geq 48$

Constraint $C$:                     $14y_1 + 7y_2 \geq 84$

Nonnegativity constraint:              $y_1, \ y_2 \geq 0$

where the technical constraints read $\geq$ since the minimum requirements must be fulfilled but may be exceeded.

2.  Treat the inequality constraints as equations. Solve each one for $y_2$ in terms of $y_1$ and graph. Thus

From $A$,                     $y_2 = 5 - \frac{1}{3}y_1$

From $B$,                     $y_2 = 8 - \frac{2}{3}y_1$

From $C$,                     $y_2 = 12 - 2y_1$

The graph of the original "greater than or equal to" inequality will include all the points *on the line and to the right of it*. See Fig. 7-2(*a*). The shaded area is the feasible region, containing all the points that satisfy all three constraints plus the nonnegativity constraints.

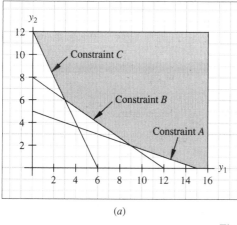 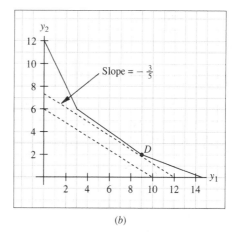

(*a*)                                                (*b*)

**Fig. 7-2**

3.  To find the optimal solution within the feasible region, graph the objective function as a series of (dashed) isocost lines. From (*7.2*),

$$y_2 = -\frac{3}{5}y_1 + \frac{c}{20}$$

In Fig. 7-2(b), we see that the lowest isocost line is tangent to the feasible region at $D$, where $y_1 = 9$ and $y_2 = 2$.

4.   Substitute the optimal values $y_1 = 9$ and $y_2 = 2$ in (7.2) to find the least cost,

$$c = 12(9) + 20(2) = 148$$

Note that for minimization problems $(0, 0)$ is not in the feasible region and no other feasible combination offers lower cost. At $(3, 6)$, for instance, $c = 12(3) + 20(6) = 156$. See also Problems 7.6 to 7.8 and 7.14 to 7.19.

## 7.5   SLACK AND SURPLUS VARIABLES

Problems involving more than two variables are beyond the scope of the two-dimensional graphic approach presented in the preceding sections. Because equations are needed, the system of linear inequalities must be converted to a system of linear equations. This is done by incorporating a separate slack or surplus variable into each inequality in the system.

A "less than or equal to" inequality such as $9x_1 + 2x_2 \leq 86$ can be converted to an equation by *adding a slack variable* $s \geq 0$, such that $9x_1 + 2x_2 + s = 86$. If $9x_1 + 2x_2 = 86$, the slack variable $s = 0$. If $9x_1 + 2x_2 < 86$, $s$ is a positive value equal to the difference between 86 and $9x_1 + 2x_2$.

A "greater than or equal to" inequality such as $3y_1 + 8y_2 \geq 55$ can be converted to an equation by *subtracting a surplus variable* $s \geq 0$, such that $3y_1 + 8y_2 - s = 55$. If $3y_1 + 8y_2 = 55$, the surplus variable $s = 0$. If $3y_1 + 8y_2 > 55$, $s$ is a positive value equal to the difference between $3y_1 + 8y_2$ and 55.

**EXAMPLE 1.**   Since the technical constraints in Section 7.2 all involve "less than or equal to" inequalities, a separate slack variable is added to each of the constraints as follows:

$$2x_1 + x_2 + s_1 = 14 \qquad 5x_1 + 5x_2 + s_2 = 40 \qquad x_1 + 3x_2 + s_3 = 18$$

Expressed in matrix form,

$$\begin{bmatrix} 2 & 1 & 1 & 0 & 0 \\ 5 & 5 & 0 & 1 & 0 \\ 1 & 3 & 0 & 0 & 1 \end{bmatrix} \begin{bmatrix} x_1 \\ x_2 \\ s_1 \\ s_2 \\ s_3 \end{bmatrix} = \begin{bmatrix} 14 \\ 40 \\ 18 \end{bmatrix}$$

In contrast, the constraints in Section 7.4 are all "greater than or equal to." Hence separate surplus variables are subtracted from each inequality constraint.

$$3y_1 + 9y_2 - s_1 = 45 \qquad 4y_1 + 6y_2 - s_2 = 48 \qquad 14y_1 + 7y_2 - s_3 = 84$$

In matrix form,

$$\begin{bmatrix} 3 & 9 & -1 & 0 & 0 \\ 4 & 6 & 0 & -1 & 0 \\ 14 & 7 & 0 & 0 & -1 \end{bmatrix} \begin{bmatrix} y_1 \\ y_2 \\ s_1 \\ s_2 \\ s_3 \end{bmatrix} = \begin{bmatrix} 45 \\ 48 \\ 84 \end{bmatrix}$$

## 7.6   THE BASIS THEOREM

As explained in Section 4.1, given a system of $n$ consistent equations and $v$ variables, where $v > n$, there will be an infinite number of solutions. Fortunately, however, the number of extreme points is finite. The *basis theorem* tells us that for a system of $n$ equations and $v$ variables, where $v > n$, a solution in which at least $v - n$ variables equal zero is an extreme point. Thus by setting $v - n$ variables equal to zero and solving the $n$ equations for the remaining $n$ variables, an extreme point, or basic solution, can be found. The number $N$ of basic solutions is given by the formula

$$N = \frac{v!}{n!(v-n)!}$$

where $v!$ reads $v$ *factorial* and is explained in Example 3.

**EXAMPLE 2.**    Reducing the inequalities to equations in Example 1 left two sets of three equations with five variables each $(x_1, x_2, s_1, s_2, s_3)$ and $(y_1, y_2, s_1, s_2, s_3)$. The procedure to determine the number of variables that must be set equal to zero to find a basic solution is demonstrated below.

Since there are three equations and five variables in each set of equations, and the basis theorem says that $v - n$ variables must equal zero for a basic solution, $5 - 3$ or 2 variables must equal zero in each set of equations in order to have a basic solution or extreme point. An initial basic solution can always be read directly from the original matrix. For example, by setting $x_1 = 0$ and $x_2 = 0$ in the first set of equations in Example 1 and multiplying, we find the initial basic solution: $s_1 = 14$, $s_2 = 40$, and $s_3 = 18$. Similarly, by setting $y_1 = 0$ and $y_2 = 0$ in the second set of equations and multiplying, we find that initial basic solution: $s_1 = -45$, $s_2 = -48$, and $s_3 = -84$. Note, however, that in minimization problems the initial basic solution is not feasible because it violates the nonnegativity constraint.

**EXAMPLE 3.**    The calculations necessary to determine the total number of basic solutions $N$ that exist are illustrated below.

Using the formula for the number of basic solutions, $v!/[n!(v - n)!]$, and substituting the given parameters from Example 1 where, in both sets of equations, $v = 5$ and $n = 3$,

$$N = \frac{5!}{3!(2)!}$$

where $5! = 5(4)(3)(2)(1)$. Thus,

$$N = \frac{5(4)(3)(2)(1)}{3(2)(1)(2)(1)} = 10 \text{ basic solutions}$$

This confirms the result obtained in Section 7.3 by the more primitive method of counting the extreme points in Fig. 7-1($a$). Since the number of parameters in Section 7.4 are the same, we know that there must also be 10 extreme points in that problem which we can check by identifying them in Fig. 7-2($a$): (0, 12), (0, 8), (0, 5), (6, 0), (12, 0), (15, 0), (3, 6), (9, 2), (4.2, 3.6), and (0, 0). Note that since (0, 0) is not in the feasible region, there are only four basic feasible solutions: (0, 12), (3, 6), (9, 2), and (15, 0).

# Solved Problems

## MATHEMATICAL INTERPRETATION OF ECONOMIC PROBLEMS

**7.1.**    A manufacturer makes two products $x_1$ and $x_2$. The first requires 5 hours for processing, 3 hours for assembling, and 4 hours for packaging. The second requires 2 hours for processing, 12 hours for assembling, and 8 hours for packaging. The plant has 40 hours available for processing, 60 for assembling, and 48 for packaging. The profit margin for $x_1$ is \$7; for $x_2$ it is \$21. Express the data in equations and inequalities necessary to determine the output mix that will maximize profits.

Maximize                         $\pi = 7x_1 + 21x_2$

subject to                       $5x_1 + \phantom{0}2x_2 \leq 40$     (processing constraint)

                                 $3x_1 + 12x_2 \leq 60$     (assembling constraint)

                                 $4x_1 + \phantom{0}8x_2 \leq 48$     (packaging constraint)

                                 $x_1, x_2 \geq \phantom{0}0$

For a graphic solution, see Problem 7.9.

**7.2.** An aluminum plant turns out two types of aluminum $x_1$ and $x_2$. Type 1 takes 6 hours for melting, 3 hours for rolling, and 1 hour for cutting. Type 2 takes 2 hours for melting, 5 hours for rolling, and 4 hours for cutting. The plant has 36 hours of melting time available, 30 hours of rolling time, and 20 hours of cutting time. The profit margin is \$10 for $x_1$ and \$8 for $x_2$. Reduce the data to equations and inequalities suitable for finding the profit-maximizing output mix.

Maximize $\qquad\qquad\qquad\qquad \pi = 10x_1 + 8x_2$

subject to
$$6x_1 + 2x_2 \leq 36 \qquad \text{(melting constraint)}$$
$$3x_1 + 5x_2 \leq 30 \qquad \text{(rolling constraint)}$$
$$x_1 + 4x_2 \leq 20 \qquad \text{(cutting constraint)}$$
$$x_1, x_2 \geq 0$$

For a graphic solution, see Problem 7.10.

**7.3.** A costume jeweler makes necklaces $x_1$ and bracelets $x_2$. Necklaces have a profit margin of \$32; bracelets \$24. Necklaces take 2 hours for stonecutting, 7 hours for setting, and 6 hours for polishing. Bracelets take 5 hours for stonecutting, 7 hours for setting, and 3 hours for polishing. The jeweler has 40 hours for stonecutting, 70 hours for setting, and 48 hours for polishing. Convert the data to equations and inequalities needed to find the profit-maximizing output mix.

Maximize $\qquad\qquad\qquad\qquad \pi = 32x_1 + 24x_2$

subject to
$$2x_1 + 5x_2 \leq 40 \qquad \text{(stonecutting constraint)}$$
$$7x_1 + 7x_2 \leq 70 \qquad \text{(setting constraint)}$$
$$6x_1 + 3x_2 \leq 48 \qquad \text{(polishing constraint)}$$
$$x_1, x_2 \geq 0$$

**7.4.** A potter makes pitchers $x_1$, bowls $x_2$, and platters $x_3$, with profit margins of \$18, \$10, and \$12, respectively. Pitchers require 5 hours of spinning and 3 hours of glazing; bowls, 2 hours of spinning and 1 of glazing; platters, 3 hours of spinning and 2 of glazing. The potter has 55 hours of spinning time and 36 hours of glazing time. Reduce the information to equations and inequalities needed to find the optimal output mix.

Maximize $\qquad\qquad\qquad\qquad \pi = 18x_1 + 10x_2 + 12x_3$

subject to
$$5x_1 + 2x_2 + 3x_3 \leq 55 \qquad \text{(spinning constraint)}$$
$$3x_1 + x_2 + 2x_3 \leq 36 \qquad \text{(glazing constraint)}$$
$$x_1, x_2 \geq 0$$

**7.5.** A carpenter makes three types of cabinets: provincial $x_1$, colonial $x_2$, and modern $x_3$. The provincial model requires 8 hours for fabricating, 5 hours for sanding, and 6 hours for staining. The colonial model requires 6 hours for fabricating, 4 for sanding, and 2 for staining. The modern model requires 5 hours for fabricating, 2 for sanding, and 4 for staining. The carpenter has 96 hours for fabricating, 44 for sanding, and 58 for staining. Profit margins are \$38, \$26, and \$22 on $x_1$, $x_2$, and $x_3$, respectively. Express the data in mathematical form suitable for finding the optimal output mix.

Maximize $\qquad\qquad\qquad\qquad \pi = 38x_1 + 26x_2 + 22x_3$

subject to
$$8x_1 + 6x_2 + 5x_3 \leq 96 \qquad \text{(fabricating constraint)}$$
$$5x_1 + 4x_2 + 2x_3 \leq 44 \qquad \text{(sanding constraint)}$$
$$6x_1 + 2x_2 + 4x_3 \leq 58 \qquad \text{(staining constraint)}$$
$$x_1, x_2 \geq 0$$

**7.6.** A game warden wants his animals to get a minimum of 36 milligrams (mg) of iodine, 84 mg of iron, and 16 mg of zinc each day. One feed $y_1$ provides 3 mg of iodine, 6 mg of iron, and 1 mg of zinc; a second feed $y_2$ provides 2 mg of iodine, 6 mg of iron, and 4 mg of zinc. The first type of feed costs $20; the second, $15. In terms of equations and graphs, what is the least-cost combination of feeds guaranteeing daily requirements?

Minimize $\qquad\qquad\qquad\quad c = 20y_1 + 15y_2$

subject to $\qquad\qquad\qquad\qquad\quad 3y_1 + 2y_2 \geq 36 \qquad$ (iodine requirement)

$\qquad\qquad\qquad\qquad\qquad\qquad\quad 6y_1 + 6y_2 \geq 84 \qquad$ (iron requirement)

$\qquad\qquad\qquad\qquad\qquad\qquad\qquad y_1 + 4y_2 \geq 16 \qquad$ (zinc requirement)

$\qquad\qquad\qquad\qquad\qquad\qquad\qquad\qquad\quad y_1, y_2 \geq 0$

For a solution using graphs, see Problem 7.14.

**7.7.** A nutritionist wishes her clients to have a daily minimum of 30 units of vitamin A, 20 units of vitamin D, and 24 units of vitamin E. One dietary supplement $y_1$ costs $80 per kilogram ($80/kg) and provides 2 units of vitamin A, 5 units of vitamin D, and 2 units of vitamin E. A second $y_2$ costs $160/kg and provides 6 units of vitamin A, 1 unit of vitamin D, and 3 units of vitamin E. Describe the least-cost combination of supplements meeting daily requirements in terms of equations and inequalities.

Minimize $\qquad\qquad\qquad\qquad c = 80y_1 + 160y_2$

subject to $\qquad\qquad\qquad\qquad\qquad\quad 2y_1 + 6y_2 \geq 30 \quad$ (vitamin A requirement)

$\qquad\qquad\qquad\qquad\qquad\qquad\quad 5y_1 + \ y_2 \geq 20 \qquad$ (vitamin D requirement)

$\qquad\qquad\qquad\qquad\qquad\qquad\quad 2y_1 + 3y_2 \geq 24 \qquad$ (vitamin E requirement)

$\qquad\qquad\qquad\qquad\qquad\qquad\qquad\quad y_1, y_2 \geq 0$

See Problem 7.15 for a graphic solution.

**7.8.** A food processor wishes to make a least-cost package mixture consisting of three ingredients $y_1$, $y_2$, and $y_3$. The first provides 4 units of carbohydrates and 3 units of protein and costs 25 cents an ounce. The second provides 6 units of carbohydrates and 2 units of protein and costs 32 cents an ounce. The third provides 9 units of carbohydrates and 5 units of protein and costs 55 cents an ounce. The package mix must have at least 60 units of carbohydrates and 45 units of protein. Express the information mathematically.

Minimize $\qquad\quad c = 0.25y_1 + 0.32y_2 + 0.55y_3$

subject to $\qquad\qquad\qquad\qquad\quad 4y_1 + 6y_2 + 9y_3 \geq 60 \qquad$ (carbohydrate requirement)

$\qquad\qquad\qquad\qquad\qquad\quad 3y_1 + 2y_2 + 5y_3 \geq 45 \qquad$ (protein requirement)

$\qquad\qquad\qquad\qquad\qquad\qquad\quad y_1, y_2, y_3 \geq 0$

## GRAPHING MAXIMIZATION PROBLEMS

**7.9.** Using the data below derived from Problem 7.1,

(1) Graph the inequality constraints by first solving each for $x_2$ in terms of $x_1$.

(2) Regraph and darken in the feasible region.

(3) Compute the slope of the objective function. Set a ruler with this slope, move it to the point where it is tangent to the feasible region, and construct a dashed line.

(4) Read the critical values for $x_1$ and $x_2$ at the point of tangency, and evaluate the objective function at these values.

Maximize $\qquad\qquad\qquad \pi = 7x_1 + 21x_2$

subject to $\qquad\qquad\qquad 5x_1 + 2x_2 \le 40 \qquad$ (constraint $A$)

$\qquad\qquad\qquad\qquad\quad\; 3x_1 + 12x_2 \le 60 \qquad$ (constraint $B$)

$\qquad\qquad\qquad\qquad\quad\; 4x_1 + 8x_2 \le 48 \qquad$ (constraint $C$)

$\qquad\qquad\qquad\qquad\qquad\qquad\; x_1, x_2 \ge 0$

The inequality constraints should be graphed as shown in Fig. 7-3($a$). From constraint $A$, $x_2 = -2.5x_1 + 20$; from $B$, $x_2 = -\frac{1}{4}x_1 + 5$; from $C$, $x_2 = -\frac{1}{2}x_1 + 6$. The nonnegativity constraints merely limit the analysis to the first quadrant.

The feasible region is graphed as shown in Fig. 7-3($b$). From the objective function, $x_2 = -\frac{1}{3}x_1 + \pi/21$ and the slope is $-\frac{1}{3}$. At the point of tangency with the feasible region, $x_1 = 4$ and $x_2 = 4$. Thus, $\pi = 7(4) + 21(4) = 112$.

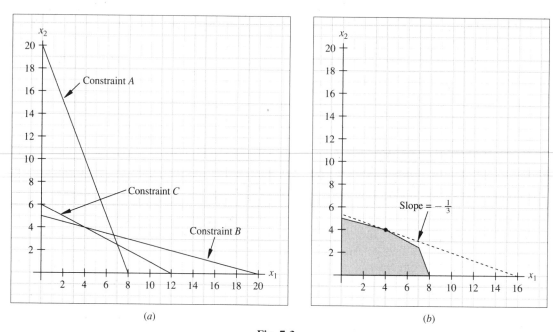

(a)                                          (b)

**Fig. 7-3**

**7.10.** Redo Problem 7.9, using the following data derived from Problem 7.2:

Maximize $\qquad\qquad\qquad \pi = 10x_1 + 8x_2$

subject to $\qquad\qquad\qquad 6x_1 + 2x_2 \le 36 \qquad$ (constraint $A$)

$\qquad\qquad\qquad\qquad\quad\; 3x_1 + 5x_2 \le 30 \qquad$ (constraint $B$)

$\qquad\qquad\qquad\qquad\quad\;\; x_1 + 4x_2 \le 20 \qquad$ (constraint $C$)

$\qquad\qquad\qquad\qquad\qquad\qquad\; x_1, x_2 \ge 0$

See Fig. 7-4($a$) for the graphed constraints and Fig. 7-4($b$) for the feasible region.

From the objective function, $x_2 = -\frac{5}{4}x_1 + \pi/8$ and the slope is $-\frac{5}{4}$. In Fig. 7-4($b$) the point of tangency occurs at (5, 3). Hence $x_1 = 5$ and $x_2 = 3$, and $\pi = 10(5) + 8(3) = 74$.

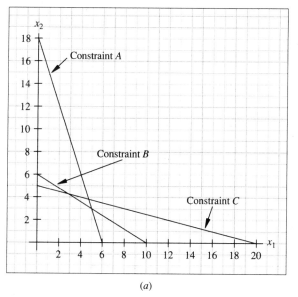

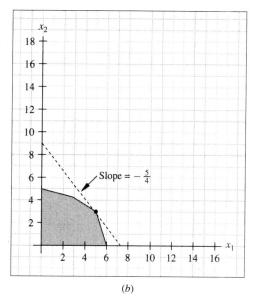

(a)                                              (b)

**Fig. 7-4**

**7.11.** Redo Problem 7.9 for the following data:

Maximize                     $\pi = 8x_1 + 6x_2$

subject to                   $2x_1 + 5x_2 \le 40$      (constraint $A$)

                             $3x_1 + 3x_2 \le 30$      (constraint $B$)

                             $8x_1 + 4x_2 \le 64$      (constraint $C$)

                             $x_1, x_2 \ge 0$

The inequalities are graphed in Fig. 7-5(a) and the feasible region in Fig. 7-5(b).

From the objective function, $x_2 = -\frac{4}{3}x_1 + \pi/6$; the slope is $-\frac{4}{3}$. In Fig. 7-5(b), the point of tangency occurs at $x_1 = 6$ and $x_2 = 4$. Thus, $\pi = 8(6) + 6(4) = 72$.

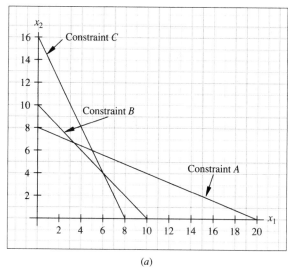

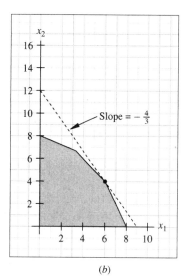

(a)                                              (b)

**Fig. 7-5**

**7.12.** Redo Problem 7.9 for the following data:

Maximize $\qquad\qquad\qquad\qquad \pi = 15x_1 + 20x_2$

subject to $\qquad\qquad\qquad\quad\; 4x_1 + 10x_2 \le 60 \qquad$ (constraint $A$)

$\qquad\qquad\qquad\qquad\qquad\; 6x_1 + \; 3x_2 \le 42 \qquad$ (constraint $B$)

$\qquad\qquad\qquad\qquad\qquad\qquad\qquad\; x_1 \le 6 \qquad$ (constraint $C$)

$\qquad\qquad\qquad\qquad\qquad\qquad\; x_1, x_2 \ge 0$

See Fig. 7-6($a$) for the graphed constraints and Fig. 7-6($b$) for the feasible region.
In Fig. 7-6($b$) the point of tangency occurs at $x_1 = 5$ and $x_2 = 4$. $\pi = 15(5) + 20(4) = 155$.

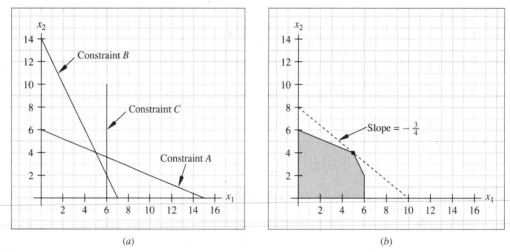

**Fig. 7-6**

**7.13.** Redo Problem 7.9 for the following data:

Maximize $\qquad\qquad\qquad\qquad \pi = 25x_1 + 50x_2$

subject to $\qquad\qquad\qquad\qquad 9x_1 + 12x_2 \le 144 \qquad$ (constraint $A$)

$\qquad\qquad\qquad\qquad\quad\; 10x_1 + \; 6x_2 \le 120 \qquad$ (constraint $B$)

$\qquad\qquad\qquad\qquad\qquad\qquad\qquad x_2 \le 9 \qquad$ (constraint $C$)

$\qquad\qquad\qquad\qquad\qquad\quad\; x_1, x_2 \ge 0$

See Fig. 7-7. From the critical values, $x_1 = 4$ and $x_2 = 9$; $\pi = 25(4) + 50(9) = 550$.

## MINIMIZATION USING GRAPHS

**7.14.** Using the following data from Problem 7.6, graph the inequality constraints by first solving each for $y_2$ in terms of $y_1$. Regraph and darken in the feasible region. Compute the slope of the objective function and construct a dashed line as in Problem 7.9. Read the critical values for $y_1$ and $y_2$ at the point of tangency, and evaluate the objective function at these values.

Minimize $\qquad\qquad\qquad\qquad c = 20y_1 + 15y_2$

subject to $\qquad\qquad\qquad\qquad 3y_1 + \; 2y_2 \ge 36 \qquad$ (constraint $A$)

$\qquad\qquad\qquad\qquad\qquad 6y_1 + \; 6y_2 \ge 84 \qquad$ (constraint $B$)

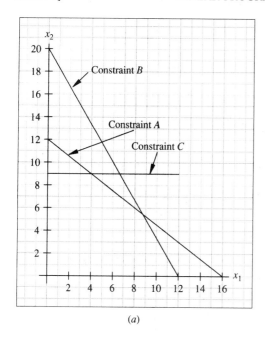

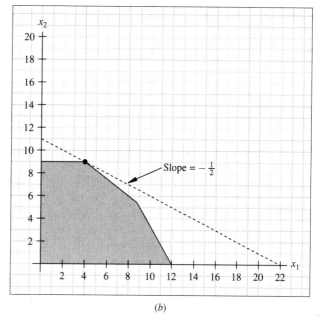

(a)                                    (b)

**Fig. 7-7**

$$y_1 + 4y_2 \geq 16 \qquad \text{(constraint } C)$$

$$y_1, y_2 \geq 0$$

See Fig. 7-8. From the objective function, $y_2 = -\frac{4}{3}y_1 + c/15$. The slope $= -\frac{4}{3}$. From Fig. 7-8(b), $y_1 = 8$ and $y_2 = 6$. Hence $c = 20(8) + 15(6) = 250$.

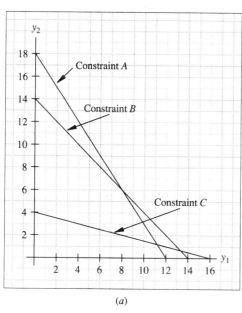

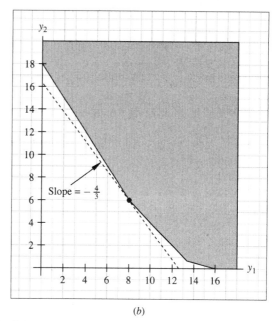

(a)                                    (b)

**Fig. 7-8**

**7.15.** Redo Problem 7.14, using the data derived from Problem 7.7.

Minimize
$$c = 80y_1 + 160y_2$$

subject to

$$2y_1 + 6y_2 \geq 30 \quad \text{(constraint } A)$$

$$5y_1 + y_2 \geq 20 \quad \text{(constraint } B)$$

$$2y_1 + 3y_2 \geq 24 \quad \text{(constraint } C)$$

$$y_1, y_2 \geq 0$$

The constraints are graphed in Fig. 7-9(a) and the feasible region in Fig. 7-9(b). In Fig. 7-9(b) the slope of the isocost line is $-\frac{1}{2}$. Hence $y_1 = 9$, $y_2 = 2$, $c = 80(9) + 160(2) = 1040$.

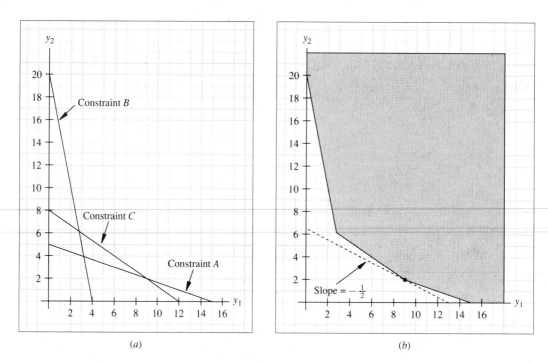

(a)                              (b)

**Fig. 7-9**

**7.16.** Redo Problem 7.14, using the following data:

Minimize
$$c = 6y_1 + 3y_2$$

subject to

$$y_1 + 2y_2 \geq 14 \quad \text{(constraint } A)$$

$$y_1 + y_2 \geq 12 \quad \text{(constraint } B)$$

$$3y_1 + y_2 \geq 18 \quad \text{(constraint } C)$$

$$y_1, y_2 \geq 0$$

See Fig. 7-10. From Fig. 7-10(b), $y_1 = 3$ and $y_2 = 9$. Hence $c = 6(3) + 3(9) = 45$.

**7.17.** Redo Problem 7.14, using the following data:

Minimize
$$c = 15y_1 + 12y_2$$

subject to

$$4y_1 + 8y_2 \geq 56 \quad \text{(constraint } A)$$

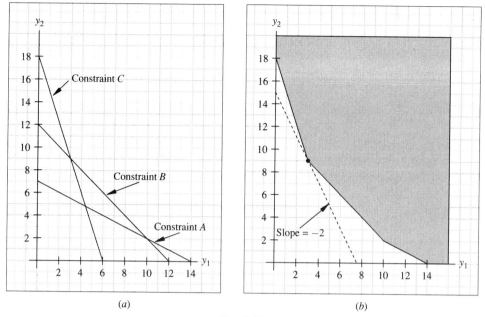

Fig. 7-10

$$3y_1 + 2y_2 \geq 30 \qquad \text{(constraint } B\text{)}$$

$$y_1 \geq 4 \qquad \text{(constraint } C\text{)}$$

$$y_1, y_2 \geq 0$$

See Fig. 7-11. From Fig. 7-11(b), $y_1 = 8$ and $y_2 = 3$. Hence $c = 15(8) + 12(3) = 156$.

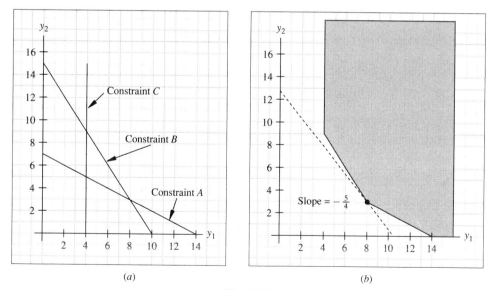

Fig. 7-11

**7.18.** Redo Problem 7.14, using the following data:

Minimize $\qquad\qquad c = 7y_1 + 28y_2$

subject to $\qquad\qquad\qquad 3y_1 + 3y_2 \geq 24 \qquad$ (constraint $A$)

$\qquad\qquad\qquad\qquad 5y_1 + \ \ y_2 \geq 20 \qquad$ (constraint $B$)

$\qquad\qquad\qquad\qquad\qquad\ \ y_2 \geq 2 \qquad$ (constraint $C$)

$\qquad\qquad\qquad\qquad y_1, y_2 \geq 0$

See Fig. 7-12. From Fig. 7-12($b$), $y_1 = 6$ and $y_2 = 2$. Hence $c = 7(6) + 28(2) = 98$.

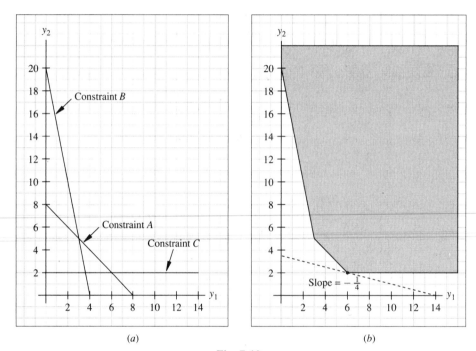

$\qquad\qquad\qquad (a) \qquad\qquad\qquad\qquad\qquad\qquad\qquad (b)$

**Fig. 7-12**

## MULTIPLE SOLUTIONS

**7.19.** Redo Problem 7.14, using the following data:

Minimize $\qquad\qquad c = 16y_1 + 20y_2$

subject to $\qquad\qquad\qquad 2y_1 + 2.5y_2 \geq 30 \qquad$ (constraint $A$)

$\qquad\qquad\qquad\qquad 3y_1 + 7.5y_2 \geq 60 \qquad$ (constraint $B$)

$\qquad\qquad\qquad\qquad 8y_1 + \ \ 5y_2 \geq 80 \qquad$ (constraint $C$)

$\qquad\qquad\qquad\qquad y_1, y_2 \geq 0$

In Fig. 7-13, with the isocost line tangent to constraint $A$, there is no *unique* optimal feasible solution. Any point on the line between (5, 8) and (10, 4) will minimize the objective function subject to the constraints. Multiple optimal solutions occur whenever there is linear dependence between the objective function and one of the constraints. In this case, the objective function and constraint $A$ both have a slope of $-\frac{4}{5}$ and are linearly dependent. Multiple optimal solutions, however, in no way contradict the extreme point theorem since the extreme points (5, 8) and (10, 4) are also included in the optimal solutions: $c = 16(5) + 20(8) = 240$ and $c = 16(10) + 20(4) = 240$.

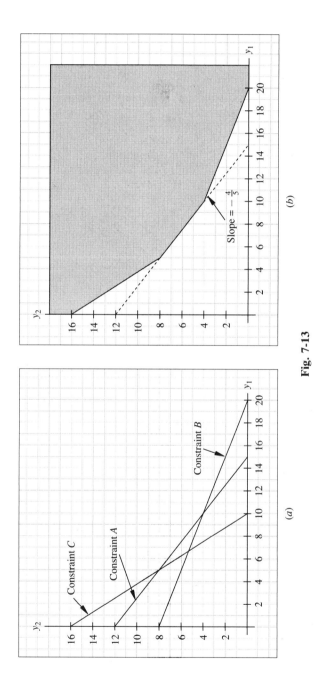

**Fig. 7-13**

**SLACK AND SURPLUS VARIABLES**

**7.20.** (*a*) Convert the inequality constraints in the following data to equations by adding slack variables or subtracting surplus variables and express the equations in matrix form. (*b*) Determine the number of variables that must be set equal to zero to find a basic solution and read the first basic solution from the matrix.

Maximize $\qquad\qquad\qquad\qquad\qquad \pi = 7x_1 + 21x_2$

subject to $\qquad\qquad\qquad\qquad\quad 5x_1 + \phantom{0}2x_2 \leq 40$

$\qquad\qquad\qquad\qquad\qquad\qquad 3x_1 + 12x_2 \leq 60$

$\qquad\qquad\qquad\qquad\qquad\qquad 4x_1 + \phantom{0}8x_2 \leq 48$

$\qquad\qquad\qquad\qquad\qquad\qquad\qquad x_1, x_2 \geq 0$

(*a*)   For "less than or equal to" inequalities, add slack variables.

$$5x_1 + 2x_2 + s_1 = 40 \qquad 3x_1 + 12x_2 + s_2 = 60 \qquad 4x_1 + 8x_2 + s_3 = 48$$

Expressed in matrix form,

$$\begin{bmatrix} 5 & 2 & 1 & 0 & 0 \\ 3 & 12 & 0 & 1 & 0 \\ 4 & 8 & 0 & 0 & 1 \end{bmatrix} \begin{bmatrix} x_1 \\ x_2 \\ s_1 \\ s_2 \\ s_3 \end{bmatrix} = \begin{bmatrix} 40 \\ 60 \\ 48 \end{bmatrix}$$

(*b*)   Since there are five variables and three equations, $v - n = 5 - 3 = 2$ variables must be set equal to zero to have a basic solution. Setting $x_1 = x_2 = 0$ and multiplying, the initial basic solution is $s_1 = 40$, $s_2 = 60$, and $s_3 = 48$.

**7.21.** Redo Problem 7.20, given

Minimize $\qquad\qquad\qquad\qquad\qquad c = 20y_1 + 15y_2$

subject to $\qquad\qquad\qquad\qquad\qquad 3y_1 + 2y_2 \geq 36$

$\qquad\qquad\qquad\qquad\qquad\qquad 6y_1 + 6y_2 \geq 84$

$\qquad\qquad\qquad\qquad\qquad\qquad\phantom{6}y_1 + 4y_2 \geq 16$

$\qquad\qquad\qquad\qquad\qquad\qquad\qquad y_1, y_2 \geq 0$

(*a*)   For "greater than or equal to" inequalities, subtract surplus variables.

$$3y_1 + 2y_2 - s_1 = 36 \qquad 6y_1 + 6y_2 - s_2 = 84 \qquad y_1 + 4y_2 - s_3 = 16$$

In matrix form,

$$\begin{bmatrix} 3 & 2 & -1 & 0 & 0 \\ 6 & 6 & 0 & -1 & 0 \\ 1 & 4 & 0 & 0 & -1 \end{bmatrix} \begin{bmatrix} y_1 \\ y_2 \\ s_1 \\ s_2 \\ s_3 \end{bmatrix} = \begin{bmatrix} 36 \\ 84 \\ 16 \end{bmatrix}$$

(*b*)   With $v = 5$ and $n = 3$, $5 - 3 = 2$ variables must be set equal to zero. Setting $y_1 = y_2 = 0$ and multiplying, the initial basic solution is $s_1 = -36$, $s_2 = -84$, and $s_3 = -16$. Since this initial basic solution contains negative values, it is not feasible.

**7.22.** Redo Problem 7.20, given

Maximize $\qquad\qquad\qquad\qquad\qquad \pi = 14x_1 + 12x_2 + 18x_3$

subject to $\qquad\qquad\qquad\qquad\qquad 2x_1 + x_2 + x_3 \leq 2$

$$x_1 + x_2 + 3x_3 \leq 4$$

$$x_1, x_2, x_3 \geq 0$$

(a)       $2x_1 + x_2 + x_3 + s_1 = 2$     $x_1 + x_2 + 3x_3 + s_2 = 4$

In matrix form,

$$\begin{bmatrix} 2 & 1 & 1 & 1 & 0 \\ 1 & 1 & 3 & 0 & 1 \end{bmatrix} \begin{bmatrix} x_1 \\ x_2 \\ x_3 \\ s_1 \\ s_2 \end{bmatrix} = \begin{bmatrix} 2 \\ 4 \end{bmatrix}$$

(b)   With $v = 5$ and $n = 2$, $5 - 2 = 3$ variables must be set equal to zero for a basic solution. Setting $x_1 = x_2 = x_3 = 0$, the initial basic solution is $s_1 = 2$ and $s_2 = 4$.

**7.23.** Find the total number of basic solutions that exist in Problem 7.22.

Substituting $v = 5$ and $n = 2$ in the formula,

$$N = \frac{v!}{n!(v-n)!} = \frac{5!}{2!(3)!} = \frac{5(4)(3)(2)(1)}{2(1)(3)(2)(1)} = 10$$

# Supplementary Problems

## MATHEMATICAL EXPRESSION OF BUSINESS AND ECONOMIC PROBLEMS

**7.24.** A bakery makes $4 profit on its wedding cakes $x_1$ and $3 on its birthday cakes $x_2$. Wedding cakes take 4 minutes for mixing, 90 minutes for baking, and 8 minutes for icing. Birthday cakes take 6 minutes for mixing, 15 minutes for baking, and 4 minutes for icing. The bakery has 120 minutes of mixing time, 900 minutes of baking time, and 96 minutes of icing time. Express the data in terms of equations and inequalities necessary to determine the combination of wedding cakes and birthday cakes that will maximize profit subject to the constraints.

**7.25.** A maker of fine preserves earns $15 profit on its premium brand $x_1$ and $6 profit on its standard brand $x_2$. The premium-brand preserves take 7.5 minutes for peeling, 20 minutes for stewing, and 8 minutes for canning. The standard-brand preserves take 5 minutes for peeling, 30 minutes for stewing, and 2 minutes for canning. The manufacturer has 150 minutes for peeling, 540 minutes for stewing, and 120 minutes for canning. Reduce the data to equations and inequalities suitable for finding the combination of brands that will maximize profit.

**7.26.** A cereal manufacturer wants to make a new brand of cereal combining two natural grains $x_1$ and $x_2$. The new cereal must have a minimum of 128 units of carbohydrates, 168 units of protein, and 120 units of fructose. Grain 1 has 24 units of carbohydrates, 14 units of protein, and 8 units of fructose. Grain 2 has 4 units of carbohydrates, 7 units of protein, and 32 units of fructose. Grain 1 costs $7 a bushel, grain 2 costs $2. Express the data in equations and inequalities amenable to finding the least-cost combination of grains that will fulfill all the nutritional requirements.

**7.27.** A landscaper wants to mix her own fertilizer containing a minimum of 50 units of phosphates, 240 units of nitrates, and 210 units of calcium. Brand 1 contains 1 unit of phosphates, 6 units of nitrates, and 15 units of calcium. Brand 2 contains 5 units of phosphates, 8 units of nitrates, and 6 units of calcium. Brand 1 costs $2.50 a pound; brand 2 costs $5. Express the data in equations and inequalities suitable to determine the least-cost combination of fertilizers that will meet her requirements.

## MAXIMIZATION USING GRAPHS

Use graphs to solve the following linear programming problems.

**7.28.**    Maximize                                    $\pi = 2x_1 + 3x_2$

subject to                                 $2x_1 + 2x_2 \leq \ 32$

$3x_1 + 9x_2 \leq 108$

$6x_1 + 4x_2 \leq \ 84$

$x_1, x_2 \geq \ \ 0$

**7.29.**    Maximize $\pi = 5x_1 + 4x_2$ subject to the same constraints in Problem 7.28

**7.30.**    Maximize                                    $\pi = 4x_1 + 3x_2$

subject to                                 $8x_1 + 4x_2 \leq \ 96$

$4x_1 + 6x_2 \leq 120$

$18x_1 + 3x_2 \leq 180$

$x_1, x_2 \geq \ \ 0$

**7.31.**    Maximize $\pi = 4x_1 + x_2$ subject to the same constraints in Problem 7.30.

**7.32.**    Maximize                                    $\pi = 5x_1 + 10x_2$

subject to                                 $2x_1 + \ 3x_2 \leq \ 48$

$4x_1 + 12x_2 \leq 168$

$8x_1 + \ 6x_2 \leq 144$

$x_1, x_2 \geq \ \ 0$

**7.33.**    Maximize $\pi = 11x_1 + 10x_2$ subject to the same constraints in Problem 7.32.

**7.34.**    Maximize                                    $\pi = 5x_1 + 4x_2$

subject to                                 $7.5x_1 + 5x_2 \leq 150$

$3x_1 + 4x_2 \leq 108$

$8x_1 + 2x_2 \leq 120$

$x_1, x_2 \geq \ \ 0$

**7.35.**    Maximize $\pi = 15x_1 + 6x_2$ subject to the same constraints in Problem 7.34.

## MINIMIZATION USING GRAPHS

Use graphs to solve the following linear programming problems.

**7.36.**    Minimize                                    $c = 7y_1 + 4y_2$

subject to                                 $3y_1 + 2y_2 \geq \ 48$

$9y_1 + 4y_2 \geq 108$

$2y_1 + 5y_2 \geq \ 65$

$y_1, y_2 \geq \ \ 0$

**7.37.** Minimize $c = 8y_1 + 10y_2$ subject to the same constraints in Problem 7.36.

**7.38.** Minimize
$$c = 12y_1 + 20y_2$$
subject to
$$4y_1 + 5y_2 \geq 100$$
$$24y_1 + 15y_2 \geq 360$$
$$2y_1 + 10y_2 \geq 80$$
$$y_1, y_2 \geq 0$$

**7.39.** Minimize $c = 30y_1 + 25y_2$ subject to the same constraints in Problem 7.38.

**7.40.** Minimize
$$c = 10y_1 + 5y_2$$
subject to
$$4y_1 + 3y_2 \geq 84$$
$$16y_1 + 6y_2 \geq 192$$
$$6y_1 + 9y_2 \geq 180$$
$$y_1, y_2 \geq 0$$

**7.41.** Minimize $c = 3y_1 + 2.5y_2$ subject to the same constraints in Problem 7.40.

**7.42.** Minimize
$$c = 5y_1 + 4y_2$$
subject to
$$7y_1 + 8y_2 \geq 168$$
$$14y_1 + 8y_2 \geq 224$$
$$2y_1 + 4y_2 \geq 60$$
$$y_1, y_2 \geq 0$$

**7.43.** Minimize $c = 15y_1 + 20y_2$ subject to the same constraints in Problem 7.42.

# Answers to Supplementary Problems

**7.24.** Maximize
$$\pi = 4x_1 + 3x_2$$
subject to
$$4x_1 + 6x_2 \leq 120$$
$$90x_1 + 15x_2 \leq 900$$
$$8x_1 + 4x_2 \leq 96$$
$$x_1, x_2 \geq 0$$

**7.25.** Maximize
$$\pi = 15x_1 + 6x_2$$
subject to
$$7.5x_1 + 5x_2 \leq 150$$
$$20x_1 + 30x_2 \leq 540$$
$$8x_1 + 2x_2 \leq 120$$
$$x_1, x_2 \geq 0$$

**7.26.**    Minimize                                              $c = 7y_1 + 2y_2$

subject to                                          $24y_1 + \phantom{0}4y_2 \geq 128$

$14y_1 + \phantom{0}7y_2 \geq 168$

$8y_1 + 32y_2 \geq 120$

$y_1, y_2 \geq \phantom{00}0$

**7.27.**    Minimize                                              $c = 2.5y_1 + 5y_2$

subject to                                          $y_1 + 5y_2 \geq \phantom{0}50$

$6y_1 + 8y_2 \geq 240$

$15y_1 + 6y_2 \geq 210$

$y_1, y_2 \geq \phantom{00}0$

**7.28.**    $x_1 = 6, x_2 = 10, \pi = 42$

**7.29.**    $x_1 = 10, x_2 = 6, \pi = 74$

**7.30.**    $x_1 = 3, x_2 = 18, \pi = 66$

**7.31.**    $x_1 = 9, x_2 = 6, \pi = 42$

**7.32.**    $x_1 = 6, x_2 = 12, \pi = 150$

**7.33.**    $x_1 = 12, x_2 = 8, \pi = 212$

**7.34.**    $x_1 = 4, x_2 = 24, \pi = 116$

**7.35.**    $x_1 = 12, x_2 = 12, \pi = 252$

**7.36.**    $y_1 = 4, y_2 = 18, c = 100$

**7.37.**    $y_1 = 10, y_2 = 9, c = 170$

**7.38.**    $y_1 = 20, y_2 = 4, c = 320$

**7.39.**    $y_1 = 5, y_2 = 16, c = 550$

**7.40.**    $y_1 = 3, y_2 = 24, c = 150$

**7.41.**    $y_1 = 12, y_2 = 12, c = 66$

**7.42.**    $y_1 = 8, y_2 = 14, c = 96$

**7.43.**    $y_1 = 16, y_2 = 7, c = 380$

# Chapter 8

# Linear Programming: The Simplex Algorithm and the Dual

## 8.1 THE SIMPLEX ALGORITHM

An *algorithm* is a systematic procedure or set of rules for finding a solution to a problem. The *simplex algorithm* is a computational method that (1) seeks out basic feasible solutions for a system of linear equations and (2) tests the solutions for optimality. Since a minimum of $v - n$ variables must equal zero for a basic solution, $v - n$ variables are set equal to zero in each step of the procedure and a basic solution is found by solving the $n$ equations for the remaining $n$ variables. The algorithm moves from one basic feasible solution to another, always improving on the previous solution until the optimal solution is found. The variables set equal to zero in a particular step are called *not in the basis*, or *not in the solution*. Those not set equal to zero are called *in the basis*, *in the solution*, or, more simply, *basic solutions*. The simplex method for maximization is demonstrated in Section 8.2. Minimization is discussed in Section 8.4.

## 8.2 MAXIMIZATION

The simplex algorithm method for maximization is explained below in four easy steps, using the following concrete example:

Maximize
$$\pi = 8x_1 + 6x_2$$

subject to
$$2x_1 + 5x_2 \le 40 \qquad 8x_1 + 4x_2 \le 64$$
$$3x_1 + 3x_2 \le 30 \qquad x_1, x_2 \ge 0$$

I.   *The Initial Simplex Tableau (or Table)*

1.   Convert the inequalities to equations by adding slack variables.

$$2x_1 + 5x_2 + s_1 = 40$$
$$3x_1 + 3x_2 + s_2 = 30 \qquad\qquad (8.1)$$
$$8x_1 + 4x_2 + s_3 = 64$$

2.   Express the constraint equations in matrix form.

$$\begin{bmatrix} 2 & 5 & 1 & 0 & 0 \\ 3 & 3 & 0 & 1 & 0 \\ 8 & 4 & 0 & 0 & 1 \end{bmatrix} \begin{bmatrix} x_1 \\ x_2 \\ s_1 \\ s_2 \\ s_3 \end{bmatrix} = \begin{bmatrix} 40 \\ 30 \\ 64 \end{bmatrix}$$

3.   Set up an initial simplex tableau which will be the framework for the algorithm. The initial tableau represents the first basic feasible solution when $x_1$ and $x_2$ equal zero. It is composed of the coefficient matrix of the constraint equations and the column vector of constants set above a row of *indicators* which are the negatives of the coefficients of the decision variables in the objective function and a zero coefficient for each slack variable. The constant column entry in the last row is also zero,

197

corresponding to the value of the objective function when $x_1$ and $x_2$ equal zero. The initial simplex tableau is as follows:

| $x_1$ | $x_2$ | $s_1$ | $s_2$ | $s_3$ | Constant |
|---|---|---|---|---|---|
| 2 | 5 | 1 | 0 | 0 | 40 |
| 3 | 3 | 0 | 1 | 0 | 30 |
| ⑧ | 4 | 0 | 0 | 1 | 64 |
| −8 | −6 | 0 | 0 | 0 | 0 |

↑   Indicators

4.  By setting $x_1 = x_2 = 0$, the first basic feasible solution can be read directly from the initial tableau: $s_1 = 40$, $s_2 = 30$, and $s_3 = 64$. Since $x_1$ and $x_2$ are initially set equal to zero, the objective function has a value of zero.

II.  *The Pivot Element and a Change of Basis*

To increase the value of the objective function, a new basic solution is examined. To move to a new basic feasible solution, a new variable must be introduced into the basis and one of the variables formerly in the basis must be excluded. The process of selecting the variable to be included and the variable to be excluded is called *change of basis*.

1.  The negative indicator with the largest absolute value determines the variable to enter the basis. Since $-8$ in the first (or $x_1$) column is the negative indicator with the largest absolute value, $x_1$ is brought into the basis. The $x_1$ column becomes the *pivot column* and is denoted by an arrow.

2.  The variable to be eliminated is determined by the smallest *displacement ratio*. Displacement ratios are found by dividing the elements of the constant column by the elements of the pivot column. The row with the smallest displacement ratio, ignoring ratios less than or equal to zero, becomes the pivot row and determines the variable to leave the basis. Since $\frac{64}{8}$ provides the smallest ratio ($\frac{64}{8} < \frac{30}{3} < \frac{40}{2}$), row 3 is the *pivot row*. Since the *unit column vector* with 1 in the third row appears under the $s_3$ column, $s_3$ leaves the basis. The *pivot element* is ⑧, the element at the intersection of the column of the variable entering the basis and the row associated with the variable leaving the basis (i.e., the element at the intersection of the pivot row and the pivot column).

III.  *Pivoting*

*Pivoting* is the process of solving the $n$ equations for the $n$ variables presently in the basis. Since only one new variable enters the basis at each step of the process and the previous step always involves an identity matrix (although the columns are often out of normal order), pivoting simply involves converting the pivot element to 1 and all the other elements in the pivot column to zero, as in the Gaussian elimination method of finding an inverse matrix (see Section 6.5), as follows:

1.  Multiply the pivot row by the reciprocal of the pivot element. In this case, multiply row 3 of the initial tableau by $\frac{1}{8}$:

| $x_1$ | $x_2$ | $s_1$ | $s_2$ | $s_3$ | Constant |
|---|---|---|---|---|---|
| 2 | 5 | 1 | 0 | 0 | 40 |
| 3 | 3 | 0 | 1 | 0 | 30 |
| 1 | $\frac{1}{2}$ | 0 | 0 | $\frac{1}{8}$ | 8 |
| −8 | −6 | 0 | 0 | 0 | 0 |

2.  Having reduced the pivot element to 1, clear the pivot column. Here subtract 2 times row 3 from row 1, 3 times row 3 from row 2, and add 8 times row 3 to row 4. This gives the second tableau:

| $x_1$ | $x_2$ | $s_1$ | $s_2$ | $s_3$ | Constant |
|---|---|---|---|---|---|
| 0 | 4 | 1 | 0 | $-\frac{1}{4}$ | 24 |
| 0 | $\frac{3}{2}$ | 0 | 1 | $-\frac{3}{8}$ | 6 |
| 1 | $\frac{1}{2}$ | 0 | 0 | $\frac{1}{8}$ | 8 |
| 0 | $-2$ | 0 | 0 | 1 | 64 |

The second basic feasible solution can be read directly from the second tableau. Setting equal to zero all the variables heading columns which are not composed of unit vectors (in this case $x_2$ and $s_3$), and mentally rearranging the unit column vectors to form an identity matrix, we see that $s_1 = 24$, $s_2 = 6$, and $x_1 = 8$. With $x_1 = 8$, $\pi = 64$, as is indicated by the last element of the last row.

### IV.  *Optimization*

The objective function is maximized when there are no negative indicators in the last row. Changing the basis and pivoting continue according to the rules above until this is achieved. Since $-2$ in the second column is the only negative indicator, $x_2$ is brought into the basis and column 2 becomes the pivot column. Dividing the constant column by the pivot column shows that the smallest ratio is in the second row. Thus, $\left(\frac{3}{2}\right)$ becomes the new pivot element. Since the unit column vector with 1 in the second row is under $s_2$, $s_2$ will leave the basis. To pivot, perform the following steps:

1.  Multiply row 2 by $\frac{2}{3}$.

| $x_1$ | $x_2$ | $s_1$ | $s_2$ | $s_3$ | Constant |
|---|---|---|---|---|---|
| 0 | 4 | 1 | 0 | $-\frac{1}{4}$ | 24 |
| 0 | 1 | 0 | $\frac{2}{3}$ | $-\frac{1}{4}$ | 4 |
| 1 | $\frac{1}{2}$ | 0 | 0 | $\frac{1}{8}$ | 8 |
| 0 | $-2$ | 0 | 0 | 1 | 64 |

2.  Then subtract 4 times row 2 from row 1, $\frac{1}{2}$ times row 2 from row 3, and add 2 times row 2 to row 4, deriving the third tableau:

| $x_1$ | $x_2$ | $s_1$ | $s_2$ | $s_3$ | Constant |
|---|---|---|---|---|---|
| 0 | 0 | 1 | $-\frac{8}{3}$ | $\frac{3}{4}$ | 8 |
| 0 | 1 | 0 | $\frac{2}{3}$ | $-\frac{1}{4}$ | 4 |
| 1 | 0 | 0 | $-\frac{1}{3}$ | $\frac{1}{4}$ | 6 |
| 0 | 0 | 0 | $\frac{4}{3}$ | $\frac{1}{2}$ | 72 |

Setting all the variables heading non-unit vector columns equal to zero (i.e., $s_2 = s_3 = 0$), and mentally rearranging the unit column vectors to form an identity matrix, we see that $s_1 = 8$, $x_2 = 4$, and $x_1 = 6$. Since there are no negative indicators left in the last row, this is the optimal solution. The last element in the last row indicates that at $\bar{x}_1 = 6$, $\bar{x}_2 = 4$, $\bar{s}_1 = 8$, $\bar{s}_2 = 0$, and $\bar{s}_3 = 0$, the objective function reaches a maximum at $\bar{\pi} = 72$. With $\bar{s}_2 = 0$ and $\bar{s}_3 = 0$, we also know from (8.1) that there is no slack in the last two constraints and the last two inputs are all used up. With $\bar{s}_1 = 8$, however, eight units of the first input remain unused. For a graphic representation, see Problem 7.11. For similar problems, see Problems 8.1 to 8.3.

## 8.3  MARGINAL VALUE OR SHADOW PRICING

The value of the indicator under each slack variable in the final tableau expresses the marginal value or *shadow price* of the input associated with the variable, that is, how much the value of the objective function would change as a result of a 1-unit change in the availability of the input. Thus, in Section 8.2 profits would increase by $\frac{4}{3}$ units or approximately \$1.33 for a 1-unit increment in the constant of constraint 2; by $\frac{1}{2}$ unit or 50¢ for a 1-unit increase in the constant of constraint 3; and by 0 for a 1-unit increase in the constant of constraint 1. Since constraint 1 has a positive slack variable, it is not fully utilized in the optimal solution and its marginal value therefore is zero (i.e., the addition of still another unit would add nothing to the profit function).

An important point to note is that the optimal value of the objective function will always equal the sum of the marginal value of each input times the total amount available of each input.

**EXAMPLE 1.**   The optimal solution in any linear programming problem can easily be checked by (1) substituting the critical values in the objective function and the constraint equations and (2) computing the sum of the marginal values of all the resources times their respective availabilities. Using the data from Section 8.2 by way of illustration and substituting the critical values $\bar{x}_1 = 6$, $\bar{x}_2 = 4$, $\bar{s}_1 = 8$, $\bar{s}_2 = 0$, and $\bar{s}_3 = 0$ in step 1 we have

$$(1) \qquad \begin{aligned} \pi &= 8x_1 + 6x_2 & 3x_1 + 3x_2 + s_2 &= 30 \\ \pi &= 8(6) + 6(4) = 72 & 3(6) + 3(4) + 0 &= 30 \end{aligned}$$

$$\begin{aligned} 2x_1 + 5x_2 + s_1 &= 40 & 8x_1 + 4x_2 + s_3 &= 64 \\ 2(6) + 5(4) + 8 &= 40 & 8(6) + 4(4) + 0 &= 64 \end{aligned}$$

Letting $A$, $B$, $C$ respectively symbolize the constants of the constraints 1, 2, 3 and substituting the pertinent data in step 2, we have

$$(2) \qquad \pi = MV_A(A) + MV_B(B) + MV_C(C)$$

$$\pi = 0(4) + \tfrac{4}{3}(30) + \tfrac{1}{2}(64) = 72$$

## 8.4  MINIMIZATION

If the simplex algorithm is used for a minimization problem, the negative values generated by the surplus variables present a special problem. The first basic solution will consist totally of negative numbers and so will not be feasible. Consequently, still other variables, called *artificial variables*, must be introduced to generate an initial basic feasible solution. It is frequently easier, therefore, to solve minimization problems by using the dual, which will be explained in the next section. For a thorough treatment of the simplex algorithm approach to minimization, see Dowling, *Introduction to Mathematical Economics*, Section 14.3 and Problems 14.4 to 14.7.

## 8.5  THE DUAL

Every minimization problem in linear programming has a corresponding maximization problem, and every maximization problem has a corresponding minimization problem. The original problem is called the *primal*, the corresponding problem is called the *dual*. The relationship between the two can most easily be seen in terms of the parameters they share in common. Given an original primal problem,

Minimize $\qquad\qquad c = g_1 y_1 + g_2 y_2 + g_3 y_3$

subject to $\qquad\qquad a_{11} y_1 + a_{12} y_2 + a_{13} y_3 \geq h_1$

$$a_{21}y_1 + a_{22}y_2 + a_{23}y_3 \geq h_2$$

$$a_{31}y_1 + a_{32}y_2 + a_{33}y_3 \geq h_3$$

$$y_1, y_2, y_3 \geq 0$$

the related dual problem is

Maximize $\qquad\qquad\qquad\quad \pi = h_1x_1 + h_2x_2 + h_3x_3$

subject to $\qquad\qquad\qquad\quad a_{11}x_1 + a_{21}x_2 + a_{31}x_3 \leq g_1$

$$a_{12}x_1 + a_{22}x_2 + a_{32}x_3 \leq g_2$$

$$a_{13}x_1 + a_{23}x_2 + a_{33}x_3 \leq g_3$$

$$x_1, x_2, x_3 \geq 0$$

## 8.6   RULES OF TRANSFORMATION TO OBTAIN THE DUAL

In the formulation of a dual from a primal problem,

1.   The direction of optimization is reversed. Minimization becomes maximization in the dual and vice versa.
2.   The inequality signs of the technical constraints are reversed, but the nonnegativity constraints on decision variables always remain in effect.
3.   The rows of the coefficient matrix of the constraints in the primal are transposed to columns for the coefficient matrix of constraints in the dual.
4.   The row vector of coefficients in the objective function in the primal is transposed to a column vector of constants for the dual constraints.
5.   The column vector of constants from the primal constraints is transposed to a row vector of coefficients for the objective function in the dual.
6.   Primal decision variables $x_i$ or $y_i$ are replaced by the corresponding dual decision variables $y_i$ or $x_i$.

Application of these steps is illustrated in Example 2.

**EXAMPLE 2.**   The dual of the linear programming problem

Minimize $\qquad\qquad\qquad\qquad\qquad c = 24y_1 + 15y_2 + 32y_3$

subject to $\qquad\qquad\qquad\qquad\qquad 4y_1 + \ \ y_2 + 8y_3 \geq 56$

$$6y_1 + 3y_2 + 2y_3 \geq 49$$

$$y_1, y_2, y_3 \geq 0$$

is

Maximize $\qquad\qquad\qquad\qquad\qquad \pi = 56x_1 + 49x_2$

subject to $\qquad\qquad\qquad\qquad\qquad 4x_1 + 6x_2 \leq 24$

$$x_1 + 3x_2 \leq 15$$

$$8x_1 + 2x_2 \leq 32$$

$$x_1, x_2 \geq 0$$

**EXAMPLE 3.**    The dual of the linear programming problem

Maximize

$$\pi = 120x_1 + 360x_2$$

subject to

$$7x_1 + 2x_2 \leq 28$$

$$3x_1 + 9x_2 \leq 36$$

$$6x_1 + 4x_2 \leq 48$$

$$x_1, x_2 \geq 0$$

is

Minimize

$$c = 28y_1 + 36y_2 + 48y_3$$

subject to

$$7y_1 + 3y_2 + 6y_3 \geq 120$$

$$2y_1 + 9y_2 + 4y_3 \geq 360$$

$$y_1, y_2, y_3 \geq 0$$

Note that if the dual of the dual were taken here or in the examples above, the corresponding primal would be obtained.

## 8.7   THE DUAL THEOREMS

The following two dual theorems are crucial in linear programming:

1.   The optimal value of the primal objective function always equals the optimal value of the corresponding dual objective function, provided an optimal feasible solution exists.
2.   If in the optimal feasible solution (*a*) a decision variable in the primal program has a nonzero value, the corresponding slack (or surplus) variable in the dual program must have an optimal value of zero, or (*b*) a slack (or surplus) variable in the primal has a nonzero value, the corresponding decision variable in the dual program must have an optimal value of zero.

Application of these theorems is demonstrated in Example 4 and Problems 8.4 to 8.7 and 8.9 to 8.11.

**EXAMPLE 4.**    Given the following linear programming problem

Minimize

$$c = 14y_1 + 40y_2 + 18y_3$$

subject to

$$2y_1 + 5y_2 + y_3 \geq 50$$

$$y_1 + 5y_2 + 3y_3 \geq 30$$

$$y_1, y_2, y_3 \geq 0$$

The dual is

Maximize

$$\pi = 50x_1 + 30x_2$$

subject to

$$2x_1 + x_2 \leq 14$$

$$5x_1 + 5x_2 \leq 40$$

$$x_1 + 3x_2 \leq 18$$

$$x_1, x_2 \geq 0$$

where the optimal values of the dual were found graphically in Fig. 7-1 of Section 7.2 to be $\bar{x}_1 = 6, \bar{x}_2 = 2$, and $\bar{\pi} = 360$.

The dual theorems are used as follows to find the optimal value of (1) the primal objective function and (2) the primal decision variables.

1. Since the optimal value of the objective function in the dual is $\bar{\pi} = 360$, it should be clear from the first dual theorem that the optimal value of the primal objective function must be $\bar{c} = 360$.

2. To find the optimal values of the primal decision variables, convert the inequality constraints to equations by subtracting surplus variables $s$ from the primal (I) and adding slack variables $t$ to the dual (II).

$$\text{I.} \quad \begin{aligned} 2y_1 + 5y_2 + y_3 - s_1 &= 50 \\ y_1 + 5y_2 + 3y_3 - s_2 &= 30 \end{aligned} \tag{8.2}$$

$$\text{II.} \quad \begin{aligned} 2x_1 + x_2 + t_1 &= 14 \\ 5x_1 + 5x_2 + t_2 &= 40 \\ x_1 + 3x_2 + t_3 &= 18 \end{aligned} \tag{8.3}$$

Substituting the optimal values $\bar{x}_1 = 6$ and $\bar{x}_2 = 2$ in (8.3) to find $t_1$, $t_2$, $t_3$,

$$2(6) + (2) + t_1 = 14 \qquad (\bar{t}_1 = 0)$$

$$5(6) + 5(2) + t_2 = 40 \qquad (\bar{t}_2 = 0)$$

$$(6) + 3(2) + t_3 = 18 \qquad (\bar{t}_3 = 6)$$

With $\bar{t}_1 = \bar{t}_2 = 0$, according to the second dual theorem, the corresponding primal decision variables $y_1$ and $y_2$ must have nonzero values. With $\bar{t}_3 \neq 0$, its corresponding decision variable $y_3$ must equal zero. Therefore, $\bar{y}_3 = 0$ in (8.2).

The second dual theorem also states that if the optimal dual decision variables, here $\bar{x}_1$ and $\bar{x}_2$, do not equal zero in the dual, their corresponding primal surplus/slack variables, here $\bar{s}_1$ and $\bar{s}_2$, in the primal must equal zero. Substituting the relevant values $\bar{s}_1 = 0$, $\bar{s}_2 = 0$, and $\bar{y}_3 = 0$ in (8.2),

$$2y_1 + 5y_2 + 0 - 0 = 50$$

$$y_1 + 5y_2 + 0 - 0 = 30$$

and solving simultaneously,

$$\bar{y}_1 = 20 \qquad \bar{y}_2 = 2$$

Thus, the optimal decision variables are $\bar{y}_1 = 20$, $\bar{y}_2 = 2$, $\bar{y}_3 = 0$, which can easily be checked by substituting into the objective function: $\bar{c} = 14(20) + 40(2) + 18(0) = 360$. See also Problems 8.4 to 8.7.

## 8.8  SHADOW PRICES IN THE DUAL

When using the dual to solve the primal, the marginal value or shadow price of the $i$th resource in the primal equals the optimal value of the corresponding decision variable in the objective function of the dual. Thus, in terms of Example 4 where the optimal value of the dual program was found to be $\bar{x}_1 = 6$ and $\bar{x}_2 = 2$, the marginal value of the first resource in the primal is 6 and the marginal value of the second resource is 2. Letting $A$ and $B$ represent the constants in the primal constraint equations, we see that the optimal value of the primal objective function also equals the sum of the available resources times their marginal values:

$$c = MV_A(A) + MV_B(B) = 6(50) + 2(30) = 360$$

See also Problem 8.8.

## 8.9  INTEGER PROGRAMMING

The discussion of linear programming thus far has been based on the premise that decision variables can assume either integer or noninteger values. Some linear programming problems, however, such as those involved with allocating discrete units of output or assigning individuals to different tasks must by their very nature restrict decision variables to integer values. *Integer programming* is a subfield of linear programming in which some or all the variables are constrained to be integers. A *pure-integer programming model* restricts all variables to integer values. A *mixed-integer programming model* merely requires that some of the variables be integers.

Two integer programming theorems, to be illustrated in Example 5, are of interest:

1.  The set of feasible solutions in an integer programming model is necessarily a subset of the feasible solutions in the corresponding ordinary linear programming model. This means that the feasible region in an integer programming model will never be larger than the feasible region of the comparable noninteger linear programming model.

2.  The optimal value of the objective function in an integer programming model cannot be better than the optimal value of the parallel regular linear programming model.

If the optimal value of the objective function in an integer model differs from the optimal value in a noninteger model, the difference is referred to as the *cost of indivisibility*.

**EXAMPLE 5.**    The following model is solved graphically in Fig. 8-1 in order to demonstrate some of the salient features of integer programming and illustrate the two theorems cited in Section 8.9.

Maximize $\qquad\qquad\qquad\qquad\qquad \pi = 15x_1 + 60x_2$

subject to $\qquad\qquad\qquad\qquad\quad 2.5x_1 + 5x_2 \le 27.5$

$\qquad\qquad\qquad\qquad\qquad\quad 12x_1 + 4x_2 \le 36$

$\qquad\qquad\qquad\qquad\quad x_1, x_2 = \text{ nonnegative integers}$

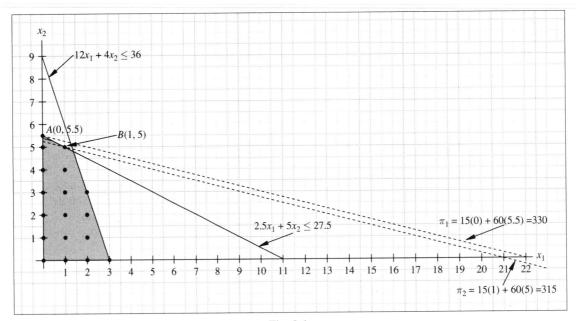

**Fig. 8-1**

For noninteger programming in which decision variables can assume continuous values, the entire shaded area in Fig. 8-1 is the feasible region. For integer programming, the feasible region consists solely of the 17 points within the shaded area which are delineated by large dots. This illustrates graphically that the feasible region in integer programming cannot exceed that of ordinary linear programming (theorem 1).

The point of tangency between the dashed isoprofit line $\pi_1$ and the noninteger feasible region at $A$ indicates that the problem in regular linear programming would be maximized at $(0, 5.5)$. Shifting the isoprofit line in parallel toward the origin, the first point of tangency between the new isoprofit line $\pi_2$ and the integer feasible region occurs at $B$, indicating the integer programming problem is maximized at $(1, 5)$. With $\pi_1 = 15(0)+60(5.5) = 330$ and $\pi_2 = 15(1)+60(5) = 315$, this illustrates the fact that profit cannot be higher in integer programming than in noninteger programming (theorem 2). In this example the cost of indivisibility, that is, the loss of profit from being unable to divide the decision variables into sections, is $\pi_1 - \pi_2 = 330 - 315 = 15$.

Integer programming involving more than two constraints is considerably more complex than the above analysis and not easily consigned to a graph. It is generally best left to the computer.*

## 8.10  ZERO-ONE PROGRAMMING

*Zero-one programming* is a subsector of integer programming in which decision variables are restricted to integer values of 0 or 1. Zero-one programming is helpful in problems involving constraints which are mutually or partially exclusive and constraints which are mutually or serially dependent. Some of the more common uses are demonstrated in Examples 6 to 9.

**EXAMPLE 6.**  *Mutual Exclusivity between Two Choices.* Assume a trucking firm wants to optimize a function subject to a constraint involving the choice of which of two models of trucks to use for a particular shipment. Truck 1 has a capacity of 2000 cubic meters $(m^3)$; truck 2, a capacity of 2500 $m^3$. In ordinary linear programming the constraints are

$$15x_1 + 20x_2 + 25x_3 + 35x_4 \leq 2000 \quad \text{if truck 1 is used}$$

$$15x_1 + 20x_2 + 25x_3 + 35x_4 \leq 2500 \quad \text{if truck 2 is used}$$

Since only one truck can be used the constraints can be rewritten

$$15x_1 + 20x_2 + 25x_3 + 35x_4 \leq 2000 + My$$

$$15x_1 + 20x_2 + 25x_3 + 35x_4 \leq 2500 + M(1 - y)$$

$$y = 0 \text{ or } 1$$

where $M$ is constrained to be a very large positive number. If $y = 1$, the right-hand side of the first constraint becomes very large, making the constraint redundant and directing the choice to the second truck. If $y = 0$, the second constraint becomes redundant and the first truck is selected.

**EXAMPLE 7.**  *Mutual Exclusivity among Several Choices.* The model in Example 6 can easily be adjusted to accommodate a situation involving several options of which only one can be implemented. Assume two more truck models with capacities of 3000 and 4000 $m^3$, respectively, are added to the fleet in Example 6. The constraints can then be written

$$15x_1 + 20x_2 + 25x_3 + 35x_4 \leq 2000 + My_1$$

$$15x_1 + 20x_2 + 25x_3 + 35x_4 \leq 2500 + My_2$$

$$15x_1 + 20x_2 + 25x_3 + 35x_4 \leq 3000 + My_3$$

$$15x_1 + 20x_2 + 25x_3 + 35x_4 \leq 4000 + My_4$$

$$y_1 + y_2 + y_3 + y_4 = 3$$

$$y_1, y_2, y_3, y_4 = 0 \text{ or } 1$$

The addition of the two last constraints ensures that three of the four 0/1 variables will equal 1, thus excluding their corresponding constraints.

---

*An excellent linear programming package, reasonably priced for the student pc edition, is Lindo/PC available from The Scientific Press, 661 Gateway Blvd., Suite 1100, South San Francisco, CA 94080-7014, phone (415) 583-8840.

**EXAMPLE 8.**   *Partial Exclusivity among Several Choices.* The model in Example 7 can also be adjusted to accommodate a situation in which several of a number of possible choices must be selected. Assume a fifth truck with a 4500-m$^3$ capacity is added to the fleet and the company can only use three of the five on any given project. The problem can be written

$$15x_1 + 20x_2 + 25x_3 + 35x_4 \leq 2000 + My_1$$

$$15x_1 + 20x_2 + 25x_3 + 35x_4 \leq 2500 + My_2$$

$$15x_1 + 20x_2 + 25x_3 + 35x_4 \leq 3000 + My_3$$

$$15x_1 + 20x_2 + 25x_3 + 35x_4 \leq 4000 + My_4$$

$$15x_1 + 20x_2 + 25x_3 + 35x_4 \leq 4500 + My_5$$

$$y_1 + y_2 + y_3 + y_4 + y_5 = 2$$

$$y_1, y_2, y_3, y_4, y_5 = 0 \text{ or } 1$$

The last two constraints will see to it that two of the five 0/1 variables will equal 1 and three will equal zero, thus excluding two constraints and including the desired three.

**EXAMPLE 9.**   *Interdependence between Variables.* Models in which decision variables are interconnected in such a way that they must be selected or rejected together are also easily accommodated. Let $x_i = 1$ if the $i$th variable is selected and $x_i = 0$ if the ith variable is rejected. Then if $x_1$ and $x_2$ must be used together or not at all, the constraint can simply be written

$$x_1 = x_2$$

$$x_1, x_2 = 0 \text{ or } 1$$

The constraints force both variables to equal one or zero at the same time, meaning they must be selected or rejected together.

If one decision variable precludes the other so that they cannot both be selected together, the constraint becomes

$$x_1 + x_2 \leq 1$$

$$x_1, x_2 = 0 \text{ or } 1$$

And if $x_1$ cannot be selected unless $x_2$ is selected, the constraint is simply

$$x_1 \leq x_2$$

$$x_1, x_2 = 0 \text{ or } 1$$

# Solved Problems

## MAXIMIZATION USING THE SIMPLEX ALGORITHM

**8.1.**   Maximize                                    $\pi = 50x_1 + 30x_2$

subject to                        $2x_1 + x_2 \leq 14$      $x_1 + 3x_2 \leq 18$

$$5x_1 + 5x_2 \leq 40 \qquad x_1, x_2 \geq 0$$

1.   Construct the initial simplex tableau.
(a)   Add slack variables to the inequalities to make them equations.

$$2x_1 + x_2 + s_1 = 14 \qquad 5x_1 + 5x_2 + s_2 = 40 \qquad x_1 + 3x_2 + s_3 = 18$$

(b)   Express the constraint equations in matrix form.

$$\begin{bmatrix} 2 & 1 & 1 & 0 & 0 \\ 5 & 5 & 0 & 1 & 0 \\ 1 & 3 & 0 & 0 & 1 \end{bmatrix} \begin{bmatrix} x_1 \\ x_2 \\ s_1 \\ s_2 \\ s_3 \end{bmatrix} = \begin{bmatrix} 14 \\ 40 \\ 18 \end{bmatrix}$$

(c)  Form the initial simplex tableau composed of the coefficient matrix of the constraint equations and the column vector of constants set above a row of indicators which are the negatives of the coefficients of the objective function and zero coefficients for the slack variables. The initial tableau is

| $x_1$ | $x_2$ | $s_1$ | $s_2$ | $s_3$ | Constant |
|---|---|---|---|---|---|
| ②  | 1 | 1 | 0 | 0 | 14 |
| 5 | 5 | 0 | 1 | 0 | 40 |
| 1 | 3 | 0 | 0 | 1 | 18 |
| −50 ↑ | −30 | 0 | 0 | 0 | 0 |

Setting $x_1 = x_2 = 0$, the first basic feasible solution is $s_1 = 14$, $s_2 = 40$, and $s_3 = 18$. At the first basic feasible solution, $\pi = 0$.

2.  Change the basis. The negative indicator with the largest absolute value (arrow) determines the pivot column. The smallest displacement ratio arising from the division of the elements of the constant column by the elements of the pivot column decides the pivot row. Thus ② becomes the pivot element, the element at the intersection of the pivot row and the pivot column.

3.  Pivot.
   (a)  Convert the pivot row to 1 by multiplying row 1 by $\frac{1}{2}$.

| $x_1$ | $x_2$ | $s_1$ | $s_2$ | $s_3$ | Constant |
|---|---|---|---|---|---|
| 1 | $\frac{1}{2}$ | $\frac{1}{2}$ | 0 | 0 | 7 |
| 5 | 5 | 0 | 1 | 0 | 40 |
| 1 | 3 | 0 | 0 | 1 | 18 |
| −50 | −30 | 0 | 0 | 0 | 0 |

   (b)  Clear the pivot column by subtracting 5 times row 1 from row 2, row 1 from row 3, and adding 50 times row 1 to row 4. The second tableau is

| $x_1$ | $x_2$ | $s_1$ | $s_2$ | $s_3$ | Constant |
|---|---|---|---|---|---|
| 1 | $\frac{1}{2}$ | $\frac{1}{2}$ | 0 | 0 | 7 |
| 0 | $\frac{5}{2}$ | $-\frac{5}{2}$ | 1 | 0 | 5 |
| 0 | $\frac{5}{2}$ | $-\frac{1}{2}$ | 0 | 1 | 11 |
| 0 | −5 ↑ | 25 | 0 | 0 | 350 |

4.  Change the basis and pivot again. Column 2 is the pivot column, row 2 is the pivot row, and $\left(\frac{5}{2}\right)$ is the pivot element.
   (a)  Multiply row 2 by $\frac{2}{5}$.

| $x_1$ | $x_2$ | $s_1$ | $s_2$ | $s_3$ | Constant |
|---|---|---|---|---|---|
| 1 | $\frac{1}{2}$ | $\frac{1}{2}$ | 0 | 0 | 7 |
| 0 | 1 | −1 | $\frac{2}{5}$ | 0 | 2 |
| 0 | $\frac{5}{2}$ | $-\frac{1}{2}$ | 0 | 1 | 11 |
| 0 | −5 | 25 | 0 | 0 | 350 |

(b) Clear the pivot column by subtracting $\frac{1}{2}$ times row 2 from row 1, $\frac{5}{2}$ times row 2 from row 3, and adding 5 times row 2 to row 4. The third tableau is

| $x_1$ | $x_2$ | $s_1$ | $s_2$ | $s_3$ | Constant |
|---|---|---|---|---|---|
| 1 | 0 | 1 | $-\frac{1}{5}$ | 0 | 6 |
| 0 | 1 | $-1$ | $\frac{2}{5}$ | 0 | 2 |
| 0 | 0 | 2 | $-1$ | 1 | 6 |
| 0 | 0 | 20 | 2 | 0 | 360 |

With no negative indicators left, this is the final tableau. Setting the variables ($s_1$ and $s_2$) above non-unit column vectors equal to zero, and rearranging mentally to form the identity matrix, we see that $\bar{x}_1 = 6$, $\bar{x}_2 = 2$, $\bar{s}_1 = 0$, $\bar{s}_2 = 0$, $\bar{s}_3 = 6$, and $\bar{\pi} = 360$. The shadow prices of the inputs are 20, 2, and 0, respectively. See Fig. 7-1 for a graphic interpretation.

**8.2.** Redo Problem 8.1 for the equation and inequalities specified below:

Maximize
$$\pi = 56x_1 + 24x_2 + 18x_3$$

subject to
$$4x_1 + 2x_2 + 3x_3 \le 240 \qquad 8x_1 + 2x_2 + x_3 \le 120$$
$$x_1, x_2, x_3 \ge 0$$

Add slack variables and express the constraint equations in matrix form.

$$4x_1 + 2x_2 + 3x_3 + s_1 = 240 \qquad 8x_1 + 2x_2 + x_3 + s_2 = 120$$

$$\begin{bmatrix} 4 & 2 & 3 & 1 & 0 \\ 8 & 2 & 1 & 0 & 1 \end{bmatrix} \begin{bmatrix} x_1 \\ x_2 \\ x_3 \\ s_1 \\ s_2 \end{bmatrix} = \begin{bmatrix} 240 \\ 120 \end{bmatrix}$$

Then set up the initial tableau:

| $x_1$ | $x_2$ | $x_3$ | $s_1$ | $s_2$ | Constant |
|---|---|---|---|---|---|
| 4 | 2 | 3 | 1 | 0 | 240 |
| ⑧ | 2 | 1 | 0 | 1 | 120 |
| $-56$ | $-24$ | $-18$ | 0 | 0 | 0 |

Then change the basis and pivot, as follows. (1) Multiply row 2 by $\frac{1}{8}$.

| $x_1$ | $x_2$ | $x_3$ | $s_1$ | $s_2$ | Constant |
|---|---|---|---|---|---|
| 4 | 2 | 3 | 1 | 0 | 240 |
| 1 | $\frac{1}{4}$ | $\frac{1}{8}$ | 0 | $\frac{1}{8}$ | 15 |
| $-56$ | $-24$ | $-18$ | 0 | 0 | 0 |

(2) Clear the pivot column by subtracting 4 times row 2 from row 1 and adding 56 times row 2 to row 3. Set up the second tableau:

| $x_1$ | $x_2$ | $x_3$ | $s_1$ | $s_2$ | Constant |
|---|---|---|---|---|---|
| 0 | 1 | $\frac{5}{2}$ | 1 | $-\frac{1}{2}$ | 180 |
| 1 | $\frac{1}{4}$ | $\frac{1}{8}$ | 0 | $\frac{1}{8}$ | 15 |
| 0 | $-10$ | $-11$ | 0 | 7 | 840 |

Change the basis and pivot again. (1) Multiply row 1 by $\frac{2}{5}$.

| $x_1$ | $x_2$ | $x_3$ | $s_1$ | $s_2$ | Constant |
|---|---|---|---|---|---|
| 0 | $\frac{2}{5}$ | 1 | $\frac{2}{5}$ | $-\frac{1}{5}$ | 72 |
| 1 | $\frac{1}{4}$ | $\frac{1}{8}$ | 0 | $\frac{1}{8}$ | 15 |
| 0 | $-10$ | $-11$ | 0 | 7 | 840 |

(2) Clear the pivot column by subtracting $\frac{1}{8}$ times row 1 from row 2 and adding 11 times row 1 to row 3. Set up the third tableau:

| $x_1$ | $x_2$ | $x_3$ | $s_1$ | $s_2$ | Constant |
|---|---|---|---|---|---|
| 0 | $\frac{2}{5}$ | 1 | $\frac{2}{5}$ | $-\frac{1}{5}$ | 72 |
| 1 | $\left(\frac{1}{5}\right)$ | 0 | $-\frac{1}{20}$ | $\frac{3}{20}$ | 6 |
| 0 | $-\frac{28}{5}$ | 0 | $\frac{22}{5}$ | $\frac{24}{5}$ | 1632 |

Since there is still a negative indicator, pivot again. Multiply row 2 by 5.

| $x_1$ | $x_2$ | $x_3$ | $s_1$ | $s_2$ | Constant |
|---|---|---|---|---|---|
| 0 | $\frac{2}{5}$ | 1 | $\frac{2}{5}$ | $-\frac{1}{5}$ | 72 |
| 5 | 1 | 0 | $-\frac{1}{4}$ | $\frac{3}{4}$ | 30 |
| 0 | $-\frac{28}{5}$ | 0 | $\frac{22}{5}$ | $\frac{24}{5}$ | 1632 |

Subtract $\frac{2}{5}$ times row 2 from row 1 and add $\frac{28}{5}$ times row 2 to row 3. Set up the final tableau:

| $x_1$ | $x_2$ | $x_3$ | $s_1$ | $s_2$ | Constant |
|---|---|---|---|---|---|
| $-2$ | 0 | 1 | $\frac{1}{2}$ | $-\frac{1}{2}$ | 60 |
| 5 | 1 | 0 | $-\frac{1}{4}$ | $\frac{3}{4}$ | 30 |
| 28 | 0 | 0 | 3 | 9 | 1800 |

Here $\bar{x}_1 = 0$, $\bar{x}_2 = 30$, $\bar{x}_3 = 60$, $\bar{s}_1 = 0$, $\bar{s}_2 = 0$, and $\bar{\pi} = 1800$. The shadow price of the first input is 3; of the second, 9. Notice that $x_1$, which was brought into the basis by the first pivot, leaves the basis on the third pivot. A variable which has entered the basis may leave the basis, as the algorithm seeks further improvement in the value of the objective function.

**8.3.**  Redo Problem 8.1, given the data below:

Maximize
$$\pi = 250x_1 + 200x_2$$

subject to
$$6x_1 + 2x_2 \leq 36 \qquad x_1 + 4x_2 \leq 20$$
$$3x_1 + 5x_2 \leq 30 \qquad x_1, x_2 \geq 0$$

1.   Set up the initial tableau:

| $x_1$ | $x_2$ | $s_1$ | $s_2$ | $s_3$ | Constant |
|---|---|---|---|---|---|
| $\left(6\right)$ | 2 | 1 | 0 | 0 | 36 |
| 3 | 5 | 0 | 1 | 0 | 30 |
| 1 | 4 | 0 | 0 | 1 | 20 |
| $-250$ | $-200$ | 0 | 0 | 0 | 0 |

2.  Change the basis and pivot.
    (a)   Multiply row 1 by $\frac{1}{6}$.

| $x_1$ | $x_2$ | $s_1$ | $s_2$ | $s_3$ | Constant |
|---|---|---|---|---|---|
| 1 | $\frac{1}{3}$ | $\frac{1}{6}$ | 0 | 0 | 6 |
| 3 | 5 | 0 | 1 | 0 | 30 |
| 1 | 4 | 0 | 0 | 1 | 20 |
| $-250$ | $-200$ | 0 | 0 | 0 | 0 |

(b)   Subtract 3 times row 1 from row 2, subtract row 1 from row 3, and add 250 times row 1 to row 4. Set up the second tableau:

| $x_1$ | $x_2$ | $s_1$ | $s_2$ | $s_3$ | Constant |
|---|---|---|---|---|---|
| 1 | $\frac{1}{3}$ | $\frac{1}{6}$ | 0 | 0 | 6 |
| 0 | ④ | $-\frac{1}{2}$ | 1 | 0 | 12 |
| 0 | $\frac{11}{3}$ | $-\frac{1}{6}$ | 0 | 1 | 14 |
| 0 | $-\frac{350}{3}$ | $\frac{125}{3}$ | 0 | 0 | 1500 |

3.  Change the basis and pivot again.
    (a)   Multiply row 2 by $\frac{1}{4}$.

| $x_1$ | $x_2$ | $s_1$ | $s_2$ | $s_3$ | Constant |
|---|---|---|---|---|---|
| 1 | $\frac{1}{3}$ | $\frac{1}{6}$ | 0 | 0 | 6 |
| 0 | 1 | $-\frac{1}{8}$ | $\frac{1}{4}$ | 0 | 3 |
| 0 | $\frac{11}{3}$ | $-\frac{1}{6}$ | 0 | 1 | 14 |
| 0 | $-\frac{350}{3}$ | $\frac{125}{3}$ | 0 | 0 | 1500 |

(b)   Subtract $\frac{1}{3}$ times row 2 from row 1, $\frac{11}{3}$ times row 2 from row 3, and add $\frac{350}{3}$ times row 2 to row 4. Set up the final tableau:

| $x_1$ | $x_2$ | $s_1$ | $s_2$ | $s_3$ | Constant |
|---|---|---|---|---|---|
| 1 | 0 | $\frac{5}{24}$ | $-\frac{1}{12}$ | 0 | 5 |
| 0 | 1 | $-\frac{1}{8}$ | $\frac{1}{4}$ | 0 | 3 |
| 0 | 0 | $\frac{7}{24}$ | $-\frac{11}{12}$ | 1 | 3 |
| 0 | 0 | $\frac{325}{12}$ | $\frac{175}{6}$ | 0 | 1850 |

Thus $\bar{x}_1 = 5$, $\bar{x}_2 = 3$, $\bar{s}_1 = 0$, $\bar{s}_2 = 0$, $\bar{s}_3 = 3$, and $\bar{\pi} = 1850$. The shadow prices of the inputs are $\frac{325}{12} \approx 27.08$ and $\frac{175}{6} \approx 29.17$, respectively.

## SOLVING THE PRIMAL THROUGH THE DUAL

**8.4.**   For the following primal problem, (a) formulate the dual, (b) solve the dual graphically. Then use the dual solution to find the optimal values of (c) the primal objective function and (d) the primal decision variables.

Minimize $\qquad\qquad\qquad\qquad\qquad c = 36y_1 + 30y_2 + 20y_3$

subject to $\qquad\qquad\qquad\qquad 6y_1 + 3y_2 + \ y_3 \geq 10$

$$2y_1 + 5y_2 + 4y_3 \geq \ 8$$

$$y_1, y_2, y_3 \geq \ 0$$

(a)  The dual is

Maximize $\qquad\qquad\qquad\qquad \pi = 10x_1 + 8x_2$

subject to $\qquad\qquad\qquad\qquad 6x_1 + 2x_2 \leq 36$

$$3x_1 + 5x_2 \leq 30$$

$$x_1 + 4x_2 \leq 20$$

$$x_1, x_2 \geq 0$$

(b)  The dual was solved graphically as a primal in Problem 7.10, where $\bar{x}_1 = 5$, $\bar{x}_2 = 3$, and $\bar{\pi} = 74$.

(c)  With $\bar{\pi} = 74$, $\bar{c} = 74$.

(d)  To find the primal decision variables $y_1$, $y_2$, $y_3$ from the dual decision variables $x_1$, $x_2$, first convert the primal (I) and dual (II) inequalities to equations.

I.  $6y_1 + 3y_2 + \ y_3 - s_1 = 10$ $\qquad\qquad$ II.  $6x_1 + 2x_2 + t_1 = 36$

$\qquad 2y_1 + 5y_2 + 4y_3 - s_2 = 8$ $\qquad\qquad\qquad\qquad 3x_1 + 5x_2 + t_2 = 30$

$\qquad\qquad\qquad\qquad\qquad\qquad\qquad\qquad\qquad\qquad\quad\ x_1 + 4x_2 + t_3 = 20$

Then substitute $\bar{x}_1 = 5$, $\bar{x}_2 = 3$ into the dual constraint equations to solve for the slack variables $t_1$, $t_2$, $t_3$.

$$6(5) + 2(3) + t_1 = 36 \qquad (\bar{t}_1 = 0)$$

$$3(5) + 5(3) + t_2 = 30 \qquad (\bar{t}_2 = 0)$$

$$(5) + 4(3) + t_3 = 20 \qquad (\bar{t}_3 = 3)$$

With $\bar{t}_3 \neq 0$, the corresponding decision variable $\bar{y}_3$ must equal zero. With $\bar{t}_1 = \bar{t}_2 = 0$, $\bar{y}_1$, $\bar{y}_2 \neq 0$.

Since the optimal values of the dual decision variables $\bar{x}_1$, $\bar{x}_2$ do not equal zero, the corresponding primal surplus variables $\bar{s}_1$, $\bar{s}_2$ must equal zero. Incorporating the knowledge that $\bar{y}_3 = \bar{s}_1 = \bar{s}_2 = 0$ into the primal constraint equations,

$$6y_1 + 3y_2 + \ (0) - (0) = 10$$

$$2y_1 + 5y_2 + 4(0) - (0) = 8$$

Solved simultaneously, $\bar{y}_1 = \frac{13}{12} \approx 1.08$ and $\bar{y}_2 = \frac{7}{6} \approx 1.17$. Thus, the primal decision variables, allowing for rounding, are $\bar{y}_1 = 1.08$, $\bar{y}_2 = 1.17$, and $\bar{y}_3 = 0$.

**8.5.**  Redo Problem 8.4 for the following primal problem:

Minimize $\qquad\qquad\qquad\qquad\qquad c = 40y_1 + 60y_2 + 48y_3$

subject to $\qquad\qquad 5y_1 + 3y_2 + 4y_3 \geq 7 \qquad 2y_1 + 12y_2 + 8y_3 \geq 21$

$$y_1, y_2, y_3 \geq 0$$

(a)  The dual is

Maximize $\qquad\qquad\qquad\qquad\qquad \pi = 7x_1 + 21x_2$

subject to $\qquad\qquad\qquad 5x_1 + \ 2x_2 \leq 40 \qquad 4x_1 + 8x_2 \leq 48$

$$3x_1 + 12x_2 \leq 60 \qquad\qquad x_1, x_2 \geq \ 0$$

(b) The dual was solved graphically as a primal in Problem 7.9, where $\bar{x}_1 = 4$, $\bar{x}_2 = 4$, and $\bar{\pi} = 112$.

(c) With $\bar{\pi} = 112$, $\bar{c} = 112$.

(d)  I.  $5y_1 + 3y_2 + 4y_3 - s_1 = 7$     II.  $5x_1 + 2x_2 + t_1 = 40$

$2y_1 + 12y_2 + 8y_3 - s_2 = 21$     $3x_1 + 12x_2 + t_2 = 60$

$4x_1 + 8x_2 + t_3 = 48$

Substituting $\bar{x}_1 = \bar{x}_2 = 4$ in II,

$$5(4) + 2(4) + t_1 = 40 \qquad (\bar{t}_1 = 12)$$

$$3(4) + 12(4) + t_2 = 60 \qquad (\bar{t}_2 = 0)$$

$$4(4) + 8(4) + t_3 = 48 \qquad (\bar{t}_3 = 0)$$

With $\bar{t}_1 \neq 0$, $\bar{y}_1 = 0$. Since $\bar{t}_2 = \bar{t}_3 = 0$, $\bar{y}_2$, $\bar{y}_3 \neq 0$. With $\bar{x}_1$, $\bar{x}_2 \neq 0$, $\bar{s}_1 = \bar{s}_2 = 0$. Substituting in I,

$$5(0) + 3y_2 + 4y_3 - (0) = 7$$

$$2(0) + 12y_2 + 8y_3 - (0) = 21$$

Solved simultaneously, $\bar{y}_2 = \frac{7}{6} \approx 1.167$ and $\bar{y}_3 = \frac{7}{8} = 0.875$. Thus, the primal decision variables are $\bar{y}_1 = 0$, $\bar{y}_2 = 1.167$, and $\bar{y}_3 = 0.875$.

**8.6.**   Redo Problem 8.4 for the following primal problem:

Maximize                               $\pi = 30x_1 + 20x_2 + 24x_3$

subject to                  $2x_1 + 5x_2 + 2x_3 \leq 80 \qquad 6x_1 + x_2 + 3x_3 \leq 160$

$$x_1, x_2, x_3 \geq 0$$

(a)   The dual is

Minimize                               $c = 80y_1 + 160y_2$

subject to                  $2y_1 + 6y_2 \geq 30 \qquad 2y_1 + 3y_2 \geq 24$

$5y_1 + y_2 \geq 20 \qquad y_1, y_2 \geq 0$

(b)   The dual was solved graphically as a primal in Problem 7.15, where $\bar{y}_1 = 9$, $\bar{y}_2 = 2$, and $\bar{c} = 1040$.

(c)   With $\bar{c} = 1040$, $\bar{\pi} = 1040$.

(d)  I.  $2x_1 + 5x_2 + 2x_3 + s_1 = 80$     II.  $2y_1 + 6y_2 - t_1 = 30$

$6x_1 + x_2 + 3x_3 + s_2 = 160$     $5y_1 + y_2 - t_2 = 20$

$2y_1 + 3y_2 - t_3 = 24$

Substituting $\bar{y}_1 = 9$ and $\bar{y}_2 = 2$ in II,

$$2(9) + 6(2) - t_1 = 30 \qquad (\bar{t}_1 = 0)$$

$$5(9) + (2) - t_2 = 20 \qquad (\bar{t}_2 = 27)$$

$$2(9) + 3(2) - t_3 = 24 \qquad (\bar{t}_3 = 0)$$

With $\bar{t}_1 = \bar{t}_3 = 0$, $\bar{x}_1$, $\bar{x}_3 \neq 0$. Since $\bar{t}_2 \neq 0$, $\bar{x}_2 = 0$. With $\bar{y}_1$, $\bar{y}_2 \neq 0$, $\bar{s}_1 = \bar{s}_2 = 0$. Substituting in I,

$$2x_1 + 5(0) + 2x_3 + (0) = 80$$

$$6x_1 + (0) + 3x_3 + (0) = 160$$

Solved simultaneously, $\bar{x}_1 = \frac{40}{3} \approx 13.33$ and $\bar{x}_3 = \frac{80}{3} \approx 26.67$. Thus, $\bar{x}_1 = 13.33$, $\bar{x}_2 = 0$, and $\bar{x}_3 = 26.67$.

**8.7.**  Redo Problem 8.4 for the following primal problem:

Maximize $\qquad\qquad\qquad\qquad \pi = 36x_1 + 84x_2 + 16x_3$

subject to $\qquad\qquad\qquad 3x_1 + 6x_2 + x_3 \leq 20 \qquad 2x_1 + 6x_2 + 4x_3 \leq 15$

$$x_1, x_2, x_3 \geq 0$$

(a)  The dual is

Minimize $\qquad\qquad\qquad\qquad\qquad c = 20y_1 + 15y_2$

subject to $\qquad\qquad\qquad\qquad 3y_1 + 2y_2 \geq 36 \qquad y_1 + 4y_2 \geq 16$

$$6y_1 + 6y_2 \geq 84 \qquad y_1, y_2 \geq 0$$

(b)  The dual was solved graphically as a primal in Problem 7.14, where $\bar{y}_1 = 8$, $\bar{y}_2 = 6$, and $\bar{c} = 250$.

(c)  With $\bar{c} = 250$, $\bar{\pi} = 250$.

(d)  I.   $3x_1 + 6x_2 + x_3 + s_1 = 20$ $\qquad\qquad\qquad$ II.   $3y_1 + 2y_2 - t_1 = 36$

$\qquad\quad 2x_1 + 6x_2 + 4x_3 + s_2 = 15$ $\qquad\qquad\qquad\qquad\quad 6y_1 + 6y_2 - t_2 = 84$

$$y_1 + 4y_2 - t_3 = 16$$

Substituting $\bar{y}_1 = 8$ and $\bar{y}_2 = 6$ in II,

$$3(8) + 2(6) - t_1 = 36 \qquad (\bar{t}_1 = 0)$$

$$6(8) + 6(6) - t_2 = 84 \qquad (\bar{t}_2 = 0)$$

$$(8) + 4(6) - t_3 = 16 \qquad (\bar{t}_3 = 16)$$

With $\bar{t}_1 = \bar{t}_2 = 0$, $\bar{x}_1, \bar{x}_2 \neq 0$. Since $\bar{t}_3 \neq 0$, $\bar{x}_3 = 0$. With $\bar{y}_1, \bar{y}_2 \neq 0$, $\bar{s}_1 = \bar{s}_2 = 0$. Substituting in I,

$$3x_1 + 6x_2 + (0) + (0) = 20$$

$$2x_1 + 6x_2 + 4(0) + (0) = 15$$

Solved simultaneously, $\bar{x}_1 = 5$ and $\bar{x}_2 = \frac{5}{6}$. Thus, $\bar{x}_1 = 5$, $\bar{x}_2 = \frac{5}{6}$, and $\bar{x}_3 = 0$.

**8.8.**  (a) Use the duals in Problems 8.4 to 8.7 to determine the shadow prices or marginal values (MVs) of the resources in the primal constraints. (b) Using $A$ and $B$ for the resources in the constraints, show that the sum of the available resources times their respective marginal values equals the optimal value of the primal objective function.

From Problem 8.4, $\qquad$ (a) $\text{MV}_A = \bar{x}_1 = 5$ and $\text{MV}_B = \bar{x}_2 = 3$

$\qquad\qquad\qquad\qquad\quad$ (b) $\bar{c} = \text{MV}_A(A) + \text{MV}_B(B) = 5(10) + 3(8) = 74$

From Problem 8.5, $\qquad$ (a) $\text{MV}_A = \bar{x}_1 = 4$ and $\text{MV}_B = \bar{x}_2 = 4$

$\qquad\qquad\qquad\qquad\quad$ (b) $\bar{c} = \text{MV}_A(A) + \text{MV}_B(B) = 4(7) + 4(21) = 112$

From Problem 8.6, $\qquad$ (a) $\text{MV}_A = \bar{y}_1 = 9$ and $\text{MV}_B = \bar{y}_2 = 2$

$\qquad\qquad\qquad\qquad\quad$ (b) $\bar{\pi} = \text{MV}_A(A) + \text{MV}_B(B) = 9(80) + 2(160) = 1040$

From Problem 8.7, $\qquad$ (a) $\text{MV}_A = \bar{y}_1 = 8$ and $\text{MV}_B = \bar{y}_2 = 6$

$\qquad\qquad\qquad\qquad\quad$ (b) $\bar{\pi} = \text{MV}_A(A) + \text{MV}_B(B) = 8(20) + 6(15) = 250$

## THE SIMPLEX ALGORITHM AND THE DUAL

**8.9.** For the following problem (*a*) formulate the dual. (*b*) Solve the dual, using the simplex algorithm. (*c*) Use the final dual tableau to determine the optimal values of the primal objective function and decision variables.

Minimize

$$c = 14y_1 + 40y_2 + 18y_3$$

subject to

$$2y_1 + 5y_2 + y_3 \geq 50$$

$$y_1 + 5y_2 + 3y_3 \geq 30$$

$$y_1, y_2, y_3 \geq 0$$

(*a*) Maximize

$$\pi = 50x_1 + 30x_2$$

subject to

$$2x_1 + x_2 \leq 14$$

$$5x_1 + 5x_2 \leq 40$$

$$x_1 + 3x_2 \leq 18$$

$$x_1, x_2 \geq 0$$

(*b*)  The dual was solved as a primal in Problem 8.1, where the final tableau was

| $x_1$ | $x_2$ | $s_1$ | $s_2$ | $s_3$ | Constant |
|---|---|---|---|---|---|
| 1 | 0 | 1 | $-\frac{1}{5}$ | 0 | 6 |
| 0 | 1 | $-1$ | $\frac{2}{5}$ | 0 | 2 |
| 0 | 0 | 2 | $-1$ | 1 | 6 |
| 0 | 0 | 20 | 2 | 0 | 360 |

(*c*)  The primal decision variables are given by the shadow prices or marginal values of the corresponding resources in the dual which are boxed above for easy identification. Thus, $\bar{y}_1 = 20$, $\bar{y}_2 = 2$, $\bar{y}_3 = 0$, and $\bar{c} = 14(20) + 40(2) + 18(0) = 360$.

**8.10.** Redo Problem 8.9 for the following linear programming problem:

Minimize

$$c = 240y_1 + 120y_2$$

subject to

$$4y_1 + 8y_2 \geq 56$$

$$2y_1 + 2y_2 \geq 24$$

$$3y_1 + y_2 \geq 18$$

$$y_1, y_2 \geq 0$$

(*a*)  Maximize

$$\pi = 56x_1 + 24x_2 + 18x_3$$

subject to

$$4x_1 + 2x_2 + 3x_3 \leq 240$$

$$8x_1 + 2x_2 + x_3 \leq 120$$

$$x_1, x_2, x_3 \geq 0$$

(b)    The dual was solved as a primal in Problem 8.2. The final tableau is

| $x_1$ | $x_2$ | $x_3$ | $s_1$ | $s_2$ | Constant |
|---|---|---|---|---|---|
| $-2$ | $0$ | $1$ | $\frac{1}{2}$ | $-\frac{1}{2}$ | $60$ |
| $5$ | $1$ | $0$ | $-\frac{1}{4}$ | $\frac{3}{4}$ | $30$ |
| $28$ | $0$ | $0$ | $\boxed{3}$ | $\boxed{9}$ | $1800$ |

(c)    Thus, $\bar{y}_1 = 3$, $\bar{y}_2 = 9$, and $\bar{c} = 240(3) + 120(9) = 1800$.

**8.11.**    Redo Problem 8.9, given

Minimize
$$c = 36y_1 + 30y_2 + 20y_3$$

subject to
$$6y_1 + 3y_2 + y_3 \geq 250$$
$$2y_1 + 5y_2 + 4y_3 \geq 200$$
$$y_1, y_2, y_3 \geq 0$$

(a)    Maximize
$$\pi = 250x_1 + 200x_2$$

subject to
$$6x_1 + 2x_2 \leq 36$$
$$3x_1 + 5x_2 \leq 30$$
$$x_1 + 4x_2 \leq 20$$
$$x_1, x_2 \geq 0$$

(b)    The dual was solved as a primal in Problem 8.3. The final tableau is

| $x_1$ | $x_2$ | $s_1$ | $s_2$ | $s_3$ | Constant |
|---|---|---|---|---|---|
| $1$ | $0$ | $\frac{5}{24}$ | $-\frac{1}{12}$ | $0$ | $5$ |
| $0$ | $1$ | $-\frac{1}{8}$ | $\frac{1}{4}$ | $0$ | $3$ |
| $0$ | $0$ | $\frac{7}{24}$ | $-\frac{11}{12}$ | $1$ | $3$ |
| $0$ | $0$ | $\boxed{\frac{325}{12}}$ | $\boxed{\frac{175}{6}}$ | $0$ | $1850$ |

(c)    Thus, $\bar{y}_1 = \frac{325}{12} \approx 27.08$, $\bar{y}_2 = \frac{175}{6} \approx 29.17$, $\bar{y}_3 = 0$, and $\bar{c} = 36\left(\frac{325}{12}\right) + 30\left(\frac{175}{6}\right) + 20(0) = 1850$.

# Supplementary Problems

## MAXIMIZATION WITH THE SIMPLEX ALGORITHM

Use the simplex algorithm to solve each of the following linear programming problems.

**8.12.**    Maximize
$$\pi = 20x_1 + 8x_2$$

subject to
$$4x_1 + 5x_2 \leq 200$$
$$6x_1 + 3x_2 \leq 180$$
$$8x_1 + 2x_2 \leq 160$$
$$x_1, x_2 \geq 0$$

**8.13.**     Maximize

$$\pi = 4x_1 + 3x_2$$

subject to

$$3x_1 + 9x_2 \leq 207$$
$$6x_1 + 4x_2 \leq 120$$
$$15x_1 + 5x_2 \leq 225$$
$$x_1, x_2 \geq 0$$

**8.14.**     Maximize

$$\pi = 8x_1 + 2x_2$$

subject to

$$5x_1 + 4x_2 \leq 216$$
$$6x_1 + 3x_2 \leq 180$$
$$12x_1 + 2x_2 \leq 312$$
$$x_1, x_2 \geq 0$$

**8.15.**     Maximize

$$\pi = 9x_1 + 5x_2$$

subject to

$$2x_1 + 4x_2 \leq 280$$
$$6x_1 + 5x_2 \leq 450$$
$$15x_1 + 6x_2 \leq 720$$
$$x_1, x_2 \geq 0$$

**8.16.**     Maximize

$$\pi = 39x_1 + 70x_2 + 16x_3$$

subject to

$$x_1 + 2x_2 + 5x_3 \leq 180$$
$$5x_1 + 2.5x_2 + x_3 \leq 300$$
$$x_1, x_2, x_3 \geq 0$$

**8.17.**     Maximize

$$\pi = 168x_1 + 222x_2 + 60x_3$$

subject to

$$7x_1 + 14x_2 + 2x_3 \leq 90$$
$$8x_1 + 8x_2 + 4x_3 \leq 120$$
$$x_1, x_2, x_3 \geq 0$$

**THE DUAL**

**8.18.**     Find the dual of Problem 8.12.

**8.19.**     Find the dual of Problem 8.16.

**MINIMIZATION WITH THE DUAL**

Solve each of the following problems by first finding the dual and then using the simplex algorithm.

**8.20.**    Minimize

$$c = 225y_1 + 180y_2$$

subject to

$$8y_1 + \ y_2 \geq \ 32$$
$$7y_1 + 4y_2 \geq 112$$
$$y_1 + 6y_2 \geq \ 54$$
$$y_1, y_2, \geq \ 0$$

**8.21.**    Minimize

$$c = 540y_1 + 900y_2$$

subject to

$$3y_1 + 15y_2 \geq 195$$
$$4y_1 + \ 5y_2 \geq 140$$
$$10y_1 + \ 2y_2 \geq \ 80$$
$$y_1, y_2 \geq \ 0$$

**8.22.**    Minimize

$$c = 24y_1 + 61y_2 + 60y_3$$

subject to

$$2y_1 + 2y_2 + 6y_3 \geq 60$$
$$y_1 + 3y_2 + \ y_3 \geq 15$$
$$y_1, y_2, y_3 \geq \ 0$$

**8.23.**    Minimize

$$c = 48y_1 + 168y_2 + 145y_3$$

subject to

$$2y_1 + 4y_2 + 8y_3 \geq 48$$
$$3y_1 + 12y_2 + 6y_3 \geq 96$$
$$y_1, y_2, y_3 \geq \ 0$$

# Answers to Supplementary Problems

**8.12.**    $\bar{x}_1 = 12.5, \bar{x}_2 = 30, \bar{\pi} = 490$

**8.13.**    $\bar{x}_1 = 6, \bar{x}_2 = 21, \bar{\pi} = 87$

**8.14.**    $\bar{x}_1 = 24, \bar{x}_2 = 12, \bar{\pi} = 216$

**8.15.**    $\bar{x}_1 = 25, \bar{x}_2 = 57.5, \bar{\pi} = 512.5$

**8.16.**    $\bar{x}_1 = 20, \bar{x}_2 = 80, \bar{x}_3 = 0, \bar{\pi} = 6380$

**8.17.**    $\bar{x}_1 = 10, \bar{x}_2 = 0, \bar{x}_3 = 10, \bar{\pi} = 2280$

**8.18.** Minimize $c = 200y_1 + 180y_2 + 160y_3$

subject to

$$4y_1 + 6y_2 + 8y_3 \geq 20$$

$$5y_1 + 3y_2 + 2y_3 \geq 8$$

$$y_1, y_2, y_3 \geq 0$$

**8.19.** Minimize $c = 180y_1 + 300y_2$

subject to

$$y_1 + 5y_2 \geq 39$$

$$2y_1 + 2.5y_2 \geq 70$$

$$5y_1 + y_2 \geq 16$$

$$y_1, y_2 \geq 0$$

**8.20.** $\bar{y}_1 = 12, \bar{y}_2 = 7, \bar{c} = 3960$

**8.21.** $\bar{y}_1 = 25, \bar{y}_2 = 8, \bar{c} = 20, 700$

**8.22.** $\bar{y}_1 = 7.5, \bar{y}_2 = 0, \bar{y}_3 = 7.5, \bar{c} = 630$

**8.23.** $\bar{y}_1 = 16, \bar{y}_2 = 4, \bar{y}_3 = 0, \bar{c} = 1440$

# Chapter 9

# Differential Calculus: The Derivative and the Rules of Differentiation

## 9.1 LIMITS

If the functional values $f(x)$ of a function $f$ draw closer to one and only one finite real number $L$ for all values of $x$ as $x$ draws closer to $a$ from both sides, but does not equal $a$, $L$ is defined as the *limit* of $f(x)$ as $x$ approaches $a$, and is written

$$\lim_{x \to a} f(x) = L$$

Assuming that $\lim_{x \to a} f(x)$ and $\lim_{x \to a} g(x)$ both exist, the *rules of limits* are given below, explained in Example 2, and treated in Problems 9.1 to 9.5.

1.  $\lim_{x \to a} k = k$  ($k$ = a constant)
2.  $\lim_{x \to a} x^n = a^n$  ($n$ = a positive integer)
3.  $\lim_{x \to a} kf(x) = k \lim_{x \to a} f(x)$  ($k$ = a constant)
4.  $\lim_{x \to a} [f(x) \pm g(x)] = \lim_{x \to a} f(x) \pm \lim_{x \to a} g(x)$
5.  $\lim_{x \to a} [f(x) \cdot g(x)] = \lim_{x \to a} f(x) \cdot \lim_{x \to a} g(x)$
6.  $\lim_{x \to a} [f(x) \div g(x)] = \lim_{x \to a} f(x) \div \lim_{x \to a} g(x)$  $[\lim_{x \to a} g(x) \neq 0]$
7.  $\lim_{x \to a} [f(x)]^n = [\lim_{x \to a} f(x)]^n$  ($n > 0$)

**EXAMPLE 1.** (a) To find $\lim_{x \to 2} f(x)$, if it exists, given

$$f(x) = \frac{x^2 - 4}{x - 2} \qquad (x \neq 2)$$

we construct a schedule and draw a graph, as in Fig. 9-1. Note that the parenthetical expression ($x \neq 2$) and the open circle in Fig. 9-1 both signify that $f(x)$ is not defined at $x = 2$ since division by zero is mathematically impermissible.

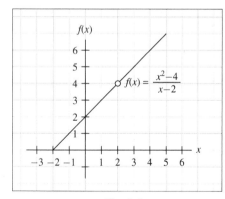

| $x$ | $f(x)$ |
|------|--------|
| $-2$ | 0 |
| $-1$ | 1 |
| 0 | 2 |
| 1 | 3 |
| 1.5 | 3.5 |
| 1.75 | 3.75 |
| 2.25 | 4.25 |
| 2.5 | 4.5 |
| 3 | 5 |
| 4 | 6 |

**Fig. 9-1**

In Fig. 9-1, note that even though $f(x)$ is not defined at $x = 2$, as $x$ approaches 2 from the left (from values $< 2$), written $x \to 2^-$, $f(x)$ approaches 4; and as $x$ approaches 2 from the right (from values $> 2$), written $x \to 2^+$, $f(x)$ also approaches 4. The limit exists, therefore, since the limit of a function as $x$ approaches a number depends only on the values of *x* *close to* that number. It is written

$$\lim_{x \to 2} \frac{x^2 - 4}{x - 2} = 4$$

(*b*) To find $\lim_{x \to 6} g(x)$, if it exists, given

$$g(x) = \begin{cases} \frac{1}{2}x & \text{when} \quad x < 6 \\ \frac{1}{2}x + 2 & \text{when} \quad x \geq 6 \end{cases}$$

we construct schedules and draw a graph, as in Fig. 9-2.

$g(x) = \frac{1}{2}x$

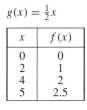

| x | f(x) |
|---|------|
| 0 | 0 |
| 2 | 1 |
| 4 | 2 |
| 5 | 2.5 |

$g(x) = \frac{1}{2}x + 2$

| x | f(x) |
|----|------|
| 6 | 5 |
| 8 | 6 |
| 10 | 7 |

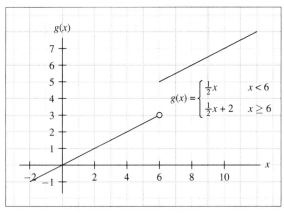

**Fig. 9-2**

In Fig. 9-2, as *x* approaches 6 from the left ($x \to 6^-$), $g(x)$ approaches 3, called a *one-sided limit*; however, as *x* approaches 6 from the right ($x \to 6^+$), $g(x)$ approaches 5, another one-sided limit. The limit does not exist, therefore, since $g(x)$ does not approach a *single* number as *x* approaches 6 from *both* sides.

**EXAMPLE 2.**    In the absence of graphs, limits can be found by using the rules of limits enumerated above.

(*a*)  $\lim_{x \to 5} 24 = 24$                                                                      (Rule 1)

(*b*)  $\lim_{x \to 7} x^2 = (7)^2 = 49$                                                          (Rule 2)

(*c*)  $\lim_{x \to 2} 4x^3 = 4 \lim_{x \to 2} x^3 = 4(2)^3 = 32$                        (Rules 2 and 3)

(*e*)  $\lim_{x \to 3} (x^4 + 5x) = \lim_{x \to 3} x^4 + 5 \lim_{x \to 3} x$          (Rule 4)

$\qquad\qquad = (3)^4 + 5(3) = 96$

(*e*)  $\lim_{x \to 4} [(x + 8)(x - 5)] = \lim_{x \to 4} (x + 8) \cdot \lim_{x \to 4} (x - 5)$          (Rule 5)

$\qquad\qquad = (4 + 8) \cdot (4 - 5) = -12$

## 9.2   CONTINUITY

A *continuous* function is one which has no breaks in its curve. It can be drawn without lifting the pencil from the paper. A function *f* is continuous at *x* = *a* if

1.  $f(x)$ is defined, that is, exists, at $x = a$
2.  $\lim\limits_{x \to a} f(x)$ exists
3.  $\lim\limits_{x \to a} f(x) = f(a)$

As a point of helpful information, all polynomial functions are continuous, as are all rational functions, except where undefined, that is, where their denominators equal zero. See Problem 9.6.

**EXAMPLE 3.**    Recalling that the graph of a continuous function can be sketched without ever removing the pencil from the paper and that an open circle means the function is not defined at that point, it is clear that in Fig. 9-3(a) $f(x)$ is discontinuous at $x = 4$ and in Fig. 9-3(b) $g(x)$ is discontinuous at $x = 6$.

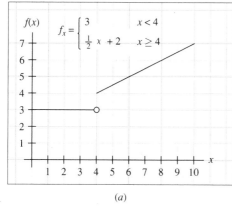
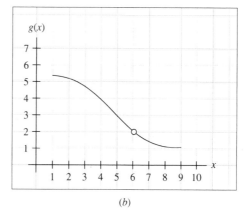

(a)                                                (b)

**Fig. 9-3**

Note, however, that at the point of discontinuity in Fig. 9-3(a) $\lim_{x \to 4} f(x)$ does not exist, but at the point of discontinuity in Fig. 9-3(b) $\lim_{x \to 6} g(x)$ does exist. Limits and continuity are not synonymous, therefore. A limit can exist at a point without the function being continuous at that point; a function cannot be continuous at a point, however, unless a limit exists at that point. In brief, a limit is a necessary but not a sufficient condition for continuity.

## 9.3    THE SLOPE OF A CURVILINEAR FUNCTION

The slope of a curvilinear function is not constant. It differs at different points on the curve. In geometry, the slope of a curvilinear function at a given point is measured by the slope of a line drawn tangent to the function at that point. A *tangent line* to a curvilinear function is a straight line which touches the curve at only one point in the area immediately surrounding the point. Measuring the slope of a curvilinear function at different points requires separate tangent lines, as seen in Fig. 9-4, where the slope of the curve gets flatter (grows smaller) from $A$ to $B$ to $C$.

The slope of a tangent line is derived from the slopes of a family of secant lines. A *secant line S* is a straight line that intersects a curve at two points, as in Fig. 9-5, where

$$\text{Slope } S = \frac{y_2 - y_1}{x_2 - x_1}$$

By letting $x_2 = x_1 + \Delta x$ and $y_2 = f(x_1 + \Delta x)$, the slope of the secant line can also be expressed by a *difference equation*:

$$\text{Slope } S = \frac{f(x_1 + \Delta x) - f(x_1)}{(x_1 + \Delta x) - x_1}$$

$$\text{Slope } S = \frac{f(x_1 + \Delta x) - f(x_1)}{\Delta x}$$

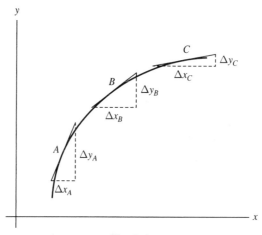

**Fig. 9-4**

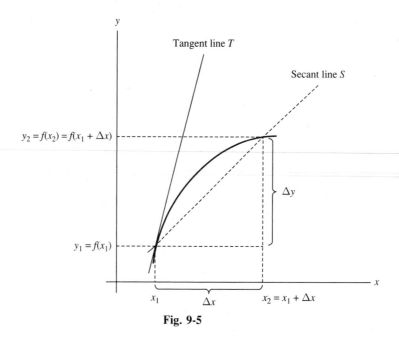

**Fig. 9-5**

If the distance between $x_2$ and $x_1$ is made smaller and smaller, that is, if $\Delta x \to 0$, the secant line pivots back to the left and draws progressively closer to the tangent line. If the slope of the secant line approaches a limit as $\Delta x \to 0$, the limit is the slope of the tangent line $T$, which is also the slope of the function at the point. It is written

$$\text{Slope } T = \lim_{\Delta x \to 0} \frac{f(x_1 + \Delta x) - f(x_1)}{\Delta x} \qquad (9.1)$$

*Note:* In many texts $h$ is used in place of $\Delta x$, giving

$$\text{Slope } T = \lim_{h \to 0} \frac{f(x_1 + h) - f(x_1)}{h} \qquad (9.1a)$$

**EXAMPLE 4.**    The slope of a curvilinear function, such as $f(x) = 3x^2$, is found as follows: (1) employ the specific function in the algebraic formula $(9.1)$ or $(9.1a)$ and substitute the arguments $x_1 + \Delta x$ (or $x_1 + h$) and $x_1$, respectively; (2) simplify the function; and (3) evaluate the limit of the function in its simplified form. Thus, from $(9.1)$,

$$\text{Slope } T = \lim_{\Delta x \to 0} \frac{f(x_1 + \Delta x) - f(x_1)}{\Delta x}$$

1)  Employing the specific function $f(x) = 3x^2$ and substituting the arguments,

$$\text{Slope } T = \lim_{\Delta x \to 0} \frac{3(x + \Delta x)^2 - 3x^2}{\Delta x}$$

2)  Simplifying the results,

$$\text{Slope } T = \lim_{\Delta x \to 0} \frac{3[(x^2 + 2x(\Delta x) + (\Delta x)^2] - 3x^2}{\Delta x}$$

$$\text{Slope } T = \lim_{\Delta x \to 0} \frac{6x(\Delta x) + 3(\Delta x)^2}{\Delta x}$$

Dividing through by $\Delta x$,

$$\text{Slope } T = \lim_{\Delta x \to 0} (6x + 3\Delta x)$$

3)  Taking the limit of the specified expression,

$$\text{Slope } T = 6x$$

*Note:* The value of the slope depends on the value of $x$ chosen. At $x = 1$, slope $T = 6(1) = 6$; at $x = 2$, slope $T = 6(2) = 12$.

## 9.4  THE DERIVATIVE

Given a function $y = f(x)$, the *derivative* of the function at $x$, written $f'(x)$, $y'$, $df/dx$, or $dy/dx$, is defined as

$$f'(x) = \lim_{\Delta x \to 0} \frac{f(x + \Delta x) - f(x)}{\Delta x} \tag{9.2}$$

provided the limit exists. Or using (*9.1a*),

$$f'(x) = \lim_{h \to 0} \frac{f(x + h) - f(x)}{h} \tag{9.2a}$$

where $f'(x)$ is read "the derivative of $f$ with respect to $x$" or "$f$ prime of $x$."

The derivative of a function, $f'(x)$ or simply $f'$ is itself a function which measures both the slope and the instantaneous rate of change of the original function at a given point.

## 9.5  DIFFERENTIABILITY AND CONTINUITY

A function is *differentiable* at a point if the derivative exists (may be taken) at that point. To be differentiable at a point, a function must (1) be continuous at that point and (2) have a unique tangent at that point. In Fig. 9-6, $f(x)$ is not differentiable at $a$ and $c$ because gaps exist in the function at those points and the derivative cannot be taken at any point where the function is discontinuous.

Continuity alone, however, does not ensure (is not a sufficient condition for) differentiability. In Fig. 9-6, $f(x)$ is continuous at $b$, but it is not differentiable at $b$ because at a sharp point or *cusp* an infinite number of tangent lines (and no one unique tangent line) can be drawn.

## 9.6  DERIVATIVE NOTATION

The derivative of a function can be written in many different ways. If $y = f(x)$, the derivative can be expressed as

$$f'(x) \qquad f' \qquad y' \qquad \frac{df}{dx} \qquad \frac{dy}{dx} \qquad \frac{d}{dx}[f(x)] \qquad \text{or} \qquad D_x[f(x)]$$

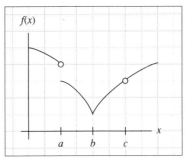

**Fig. 9-6**

If $y = \phi(t)$, the derivative can be written

$$\phi'(t) \qquad \phi' \qquad y' \qquad \frac{dy}{dt} \qquad \frac{d\phi}{dt} \qquad \frac{d}{dt}[\phi(t)] \qquad \text{or} \qquad D_t[\phi(t)]$$

If the derivative of $y = f(x)$ is evaluated at $x = a$, proper notation includes $f'(a)$ and $dy/dx|_a$. See Problems 9.7 and 9.8.

**EXAMPLE 5.**   If $y = 5x^2 + 7x + 12$, the derivative can be written

$$y' \qquad \frac{dy}{dx} \qquad \frac{d}{dx}[5x^2 + 7x + 12] \qquad \text{or} \qquad D_x[5x^2 + 7x + 12]$$

If $z = \sqrt{8t - 3}$, the derivative can be expressed as

$$z' \qquad \frac{dz}{dt} \qquad \frac{d}{dt}[\sqrt{8t - 3}] \qquad \text{or} \qquad D_t[\sqrt{8t - 3}]$$

## 9.7   RULES OF DIFFERENTIATION

*Differentiation* is the process of finding the derivative of a function. It simply involves applying a few basic rules or formulas to a given function. In explaining the rules of differentiation for a function such as $y = f(x)$, other functions such as $g(x)$ and $h(x)$ are commonly used, where $g$ and $h$ are both unspecified functions of $x$ and assumed to be differentiable. The rules of differentiation are listed below and treated in Problems 9.7 to 9.22. Select proofs are found in Problems 9.23 to 9.25.

### 9.7.1   The Constant Function Rule

The derivative of a constant function, $f(x) = k$ where $k$ is a constant, is zero.

$$\text{Given } f(x) = k, \qquad f'(x) = 0$$

**EXAMPLE 6.**

$$\text{Given } f(x) = 5, \qquad f'(x) = 0$$
$$\text{Given } f(x) = -9, \qquad f'(x) = 0$$

### 9.7.2   The Linear Function Rule

The derivative of a linear function $f(x) = mx + b$ is equal to $m$, the coefficient of $x$. The derivative of a variable raised to the first power is always equal to the coefficient of the variable, while the derivative of a constant is simply zero.

$$\text{Given } f(x) = mx + b, \qquad f'(x) = m$$

**EXAMPLE 7.**

$$\text{Given } f(x) = 6x + 7, \qquad f'(x) = 6$$

$$\text{Given } f(x) = 9 - \tfrac{1}{4}x, \qquad f'(x) = -\tfrac{1}{4}$$

$$\text{Given } f(x) = 18x, \qquad f'(x) = 18$$

### 9.7.3 The Power Function Rule

The derivative of a power function, $f(x) = kx^n$, where $k$ is a constant and $n$ is any real number, is equal to the coefficient $k$ times the exponent $n$, multiplied by the variable $x$ raised to the $(n-1)$ power.

$$\text{Given } f(x) = kx^n, \qquad f'(x) = k \cdot n \cdot x^{n-1}$$

**EXAMPLE 8.**

$$\text{Given } f(x) = 6x^3, \qquad f'(x) = 6 \cdot 3 \cdot x^{3-1} = 18x^2$$

$$\text{Given } f(x) = 7x^2, \qquad f'(x) = 7 \cdot 2 \cdot x^{2-1} = 14x$$

$$\text{Given } f(x) = x^5, \qquad f'(x) = (1) \cdot 5 \cdot x^{5-1} = 5x^4$$

See also Problem 9.8.

### 9.7.4 The Rule for Sums and Differences

The derivative of a sum of two functions, is equal to the sum of the derivatives of the individual functions. Similarly, the derivative of the difference of two functions is equal to the difference of the derivatives of the two functions. Given $f(x) = g(x) \pm h(x)$,

$$f'(x) = g'(x) \pm h'(x)$$

**EXAMPLE 9.**

$$\text{Given } f(x) = 16x^4 - 5x^3, \qquad f'(x) = 64x^3 - 15x^2$$

$$\text{Given } f(x) = 6x^2 + 4x - 9, \qquad f'(x) = 12x + 4$$

See Problem 9.9. For derivation of the rule, see Problem 9.23.

### 9.7.5 The Product Rule

The derivative of a product of two functions is equal to the first function times the derivative of the second function plus the second function times the derivative of the first function. Given $f(x) = g(x) \cdot h(x)$,

$$f'(x) = g(x) \cdot h'(x) + h(x) \cdot g'(x) \tag{9.3}$$

**EXAMPLE 10.**   Given $f(x) = 4x^5(3x - 2)$, let $g(x) = 4x^5$ and $h(x) = (3x - 2)$. Taking the individual derivatives, $g'(x) = 20x^4$ and $h'(x) = 3$. Then substituting the appropriate values in the specified places in the product rule formula (9.3),

$$f'(x) = 4x^5(3) + (3x - 2)(20x^4)$$

and simplifying algebraically,

$$f'(x) = 12x^5 + 60x^5 - 40x^4 = 72x^5 - 40x^4$$

See Problems 9.10 to 9.12; for derivation of the rule, Problem 9.24.

### 9.7.6  The Quotient Rule

The derivative of a quotient of two functions is equal to the denominator times the derivative of the numerator, minus the numerator times the derivative of the denominator, all divided by the denominator squared. Given $f(x) = g(x)/h(x)$, where $h(x) \neq 0$,

$$f'(x) = \frac{h(x) \cdot g'(x) - g(x) \cdot h'(x)}{[h(x)]^2} \tag{9.4}$$

**EXAMPLE 11.**

$$\text{Given } f(x) = \frac{6x^3}{2x + 5} \qquad \left(x \neq -\frac{5}{2}\right)$$

where $g(x) = 6x^3$ and $h(x) = 2x + 5$; $g'(x) = 18x^2$ and $h'(x) = 2$. Substituting these values in the quotient rule formula (9.4),

$$f'(x) = \frac{(2x + 5)(18x^2) - 6x^3(2)}{(2x + 5)^2}$$

Simplifying algebraically,

$$f'(x) = \frac{36x^3 + 90x^2 - 12x^3}{(2x + 5)^2} = \frac{24x^3 + 90x^2}{(2x + 5)^2} = \frac{6x^2(4x + 15)}{(2x + 5)^2}$$

See Problems 9.13 and 9.14; for derivation of the rule, Problem 9.25.

### 9.7.7  The Generalized Power Function Rule

The derivative of a function raised to a power $f(x) = [g(x)]^n$, where $g(x)$ is a differentiable function and $n$ is any real number, is equal to the exponent $n$ times the function $g(x)$ raised to the $(n - 1)$ power, multiplied in turn by the derivative of the function itself $g'(x)$. Given $f(x) = [g(x)]^n$,

$$f'(x) = n[g(x)]^{n-1} \cdot g'(x) \tag{9.5}$$

**EXAMPLE 12.**   Given $f(x) = (x^2 + 8)^3$, let $g(x) = x^2 + 8$, then $g'(x) = 2x$. Substituting these values in the generalized power function formula (9.5),

$$f'(x) = 3(x^2 + 8)^{3-1} \cdot 2x$$

Simplifying algebraically,

$$f'(x) = 6x(x^2 + 8)^2$$

*Note:* The generalized power function rule is derived from the *chain rule*, which follows below, but it is generally presented first because it is easier to understand. See Problems 9.15 and 9.16.

### 9.7.8  The Chain Rule

Given a *composite function*, also called a *function of a function*, in which $y$ is a function of $u$ and $u$ in turn is a function of $x$, that is, $y = f(u)$ and $u = g(x)$, then $y = f[g(x)]$ and the derivative of $y$ with respect to $x$ is equal to the derivative of the first function with respect to $u$ times the derivative of the second function with respect to $x$:

$$\frac{dy}{dx} = \frac{dy}{du} \cdot \frac{du}{dx} \qquad (9.6)$$

See Problems 9.17 and 9.18.

**EXAMPLE 13.**    Consider the function $y = (4x^3 + 7)^5$. To use the chain rule, let $y = u^5$ and $u = 4x^3 + 7$. Then $dy/du = 5u^4$ and $du/dx = 12x^2$. Substituting these values in $(9.6)$,

$$\frac{dy}{dx} = 5u^4 \cdot 12x^2 = 60x^2 u^4$$

Then to express the derivative in terms of a single variable, simply substitute $(4x^3 + 7)$ for $u$.

$$\frac{dy}{dx} = 60x^2 (4x^3 + 7)^4$$

For more complicated functions, different combinations of the basic rules must be used. See Problems 9.19 and 9.20.

## 9.8  HIGHER-ORDER DERIVATIVES

The second-order derivative, written $f''(x)$, measures the slope and the rate of change of the first derivative, just as the first derivative measures the slope and the rate of change of the original or *primitive function*. The third-order derivative $[f'''(x)]$ measures the slope and rate of change of the second-order derivative, and so forth. Higher-order derivatives are found by applying the rules of differentiation to lower-order derivatives, as illustrated in Example 14 and treated in Problems 9.21 and 9.22.

**EXAMPLE 14.**    Given $y = f(x)$, common notation for the second-order derivative includes $f''(x)$, $y''$, $d^2y/dx^2$, $D^2y$; for the third-order derivative, $f'''(x)$, $y'''$, $d^3y/dx^3$, $D^3y$; for the fourth-order derivative, $f^{(4)}(x)$, $y^{(4)}$, $d^4y/dx^4$, $D^4y$; and so on.

Higher-order derivatives are found by successively applying the rules of differentiation to derivatives of the previous order. Thus, if $f(x) = 5x^4 + 8x^3 + 7x^2$,

$$f'(x) = 20x^3 + 24x^2 + 14x$$

$$f''(x) = 60x^2 + 48x + 14$$

$$f'''(x) = 120x + 48$$

$$f^{(4)}(x) = 120 \qquad f^{(5)}(x) = 0$$

See Problems 9.21 and 9.22.

## 9.9  IMPLICIT FUNCTIONS

The rules of differentiation were presented above in terms of explicit functions. An *explicit function* is one in which the dependent variable is to the left of the equality sign and the independent variable and parameters are to the right. Occasionally, however, one encounters an *implicit function* in which both variables are on the same side of the equality sign. See Example 15; for differentiation of implicit functions, see Section 13.11.

**EXAMPLE 15.**    Samples of explicit and implicit functions include:

*Explicit:*    $y = 9x$,        $y = x^2 + 5x - 8$,        $y = \dfrac{x^4 - 7x^3}{x^2 - 48}$        $\left(x \neq \pm\sqrt{48}\right)$

*Implicit:*    $3x + 8y = 54$,        $4x^2 - 7xy - 6y = 82$,        $78x^5 y^9 = 429$

# Solved Problems

## LIMITS AND CONTINUITY

**9.1.**    Use the rules of limits to find the limits for the following functions:

(a)  $\lim\limits_{x \to 3}[x^4(x+5)]$

$$\lim\limits_{x \to 3}[x^4(x+5)] = \lim\limits_{x \to 3} x^4 \cdot \lim\limits_{x \to 3}(x+5) \qquad \text{(Rule 5)}$$

$$= (3)^4 \cdot (3+5) = 81 \cdot 8 = 648$$

(b)  $\lim\limits_{x \to 5} \dfrac{7x^2 - 9x}{x + 8}$

$$\lim\limits_{x \to 5} \frac{7x^2 - 9x}{x + 8} = \frac{\lim_{x \to 5}(7x^2 - 9x)}{\lim_{x \to 5}(x + 8)} \qquad \text{(Rule 6)}$$

$$= \frac{7(5)^2 - 9(5)}{5 + 8} = \frac{175 - 45}{13} = 10$$

(c)  $\lim\limits_{x \to 4} \sqrt{2x^3 - 7}$

$$\lim\limits_{x \to 4} \sqrt{2x^3 - 7} = \lim\limits_{x \to 4}(2x^3 - 7)^{1/2} \qquad \text{(Rule 7)}$$

$$= \left[\lim\limits_{x \to 4}(2x^3 - 7)\right]^{1/2}$$

$$= [2(4)^3 - 7]^{1/2} = (121)^{1/2} = 11$$

**9.2.**    Find the limits for the following polynomial and rational functions.

(a)  $\lim\limits_{x \to 4}(3x^2 - 5x + 9)$

From the properties of limits it can be shown that for all polynomial functions and all rational functions, where defined, $\lim_{x \to a} f(x) = f(a)$. The limits can be taken, therefore, by simply evaluating the functions at the given level of $a$.

$$\lim\limits_{x \to 4}(3x^2 - 5x + 9) = 3(4)^2 - 5(4) + 9 = 37$$

(b)  $\lim\limits_{x \to -3}(6x^2 + 8x - 13)$

$$\lim\limits_{x \to -3}(6x^2 + 8x - 13) = 6(-3)^2 + 8(-3) - 13 = 17$$

(c)  $\lim\limits_{x \to 8} \dfrac{2x^2 - 3x - 9}{4x^2 + 29}$

$$\lim\limits_{x \to 8} \frac{2x^2 - 3x - 9}{4x^2 + 29} = \frac{2(8)^2 - 3(8) - 9}{4(8)^2 + 29} = \frac{95}{285} = \frac{1}{3}$$

**9.3.**    Find the limits of the following rational functions, aware that if the limit of the denominator equals zero, neither Rule 6 nor the generalized rule for rational functions used above apply.

(a)  $\lim\limits_{x \to 9} \dfrac{x - 9}{x^2 - 81}$

The limit of the denominator is zero, so Rule 6 cannot be used. Since we are only interested in the function as $x$ draws *near* to 9, however, the limit can be found if by factoring and canceling, the problem of zero in the denominator is resolved.

$$\lim_{x \to 9} \frac{x-9}{x^2-81} = \lim_{x \to 9} \frac{x-9}{(x+9)(x-9)}$$

$$= \lim_{x \to 9} \frac{1}{x+9} = \frac{1}{18}$$

(*b*)  $\lim_{x \to -9} \dfrac{x-9}{x^2-81}$

$$\lim_{x \to -9} \frac{x-9}{x^2-81} = \lim_{x \to -9} \frac{x-9}{(x+9)(x-9)}$$

$$= \lim_{x \to -9} \frac{1}{x+9}$$

With the limit of the denominator still equal to zero and no possibility of further factoring, the limit does not exist.

(*c*)  $\lim_{x \to 4} \dfrac{x^2-3x-4}{x-4}$

With the limit of the denominator equal to zero, factor.

$$\lim_{x \to 4} \frac{x^2-3x-4}{x-4} = \lim_{x \to 4} \frac{(x+1)(x-4)}{x-4} = \lim_{x \to 4} (x+1) = 5$$

**9.4.**  Graph the following functions and explain the significance of the graphs in terms of Problem 9.3 (*c*):

$$f(x) = \frac{x^2-3x-4}{x-4} \qquad g(x) = x+1$$

From the graphs in Fig. 9-7(*a*) and (*b*) it is clear that $f(x)$ and $g(x)$ are the same everywhere that $f(x)$ is defined. This justifies the factoring in Problem 9.3 (*c*) and enables us to say

$$\lim_{x \to 4} \frac{x^2-3x-4}{x-4} = \lim_{x \to 4} (x+1) = 5$$

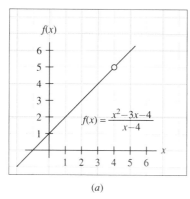

(*a*)

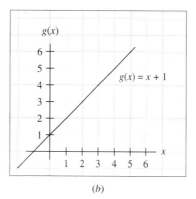

(*b*)

**Fig. 9-7**

**9.5.** Find the limits of the following functions, noting the role that infinity plays.

(a)  $\lim\limits_{x \to 0} \dfrac{2}{x}$    $(x \neq 0)$

> As seen in Fig. 3-2, as $x$ approaches zero from the right $[x \to 0^+]$, $f(x)$ approaches positive infinity; as $x$ approaches zero from the left $[x \to 0^-]$, $f(x)$ approaches negative infinity. If a limit approaches either positive or negative infinity, the limit does *not* exist and is written
>
> $$\lim\limits_{x \to 0^+} \frac{2}{x} = \infty \qquad \lim\limits_{x \to 0^-} \frac{2}{x} = -\infty \qquad \text{The limit does not exist.}$$

(b)  $\lim\limits_{x \to \infty} \dfrac{2}{x} \qquad \lim\limits_{x \to -\infty} \dfrac{2}{x}$

> As also seen in Fig. 3-2, as $x$ approaches $\infty$, $f(x)$ approaches 0; as $x$ approaches $-\infty$, $f(x)$ also approaches 0. The limit exists in both cases since zero is a legitimate limit, and is written
>
> $$\lim\limits_{x \to \infty} \frac{2}{x} = 0 \qquad \lim\limits_{x \to -\infty} \frac{2}{x} = 0$$

(c)  $\lim\limits_{x \to \infty} \dfrac{4x^2 - 9x}{5x^2 - 34}$

> As $x \to \infty$, both numerator and denominator become infinite, leaving matters unclear. A trick in such circumstances is to divide all terms by the highest power of $x$ which appears in the function. Here dividing all terms by $x^2$ leaves
>
> $$\lim\limits_{x \to \infty} \frac{4x^2 - 9x}{5x^2 - 34} = \lim\limits_{x \to \infty} \frac{4 - (9/x)}{5 - (34/x^2)} = \frac{4 - (0)}{5 - (0)} = \frac{4}{5}$$

**9.6.** Indicate whether the following functions are continuous at the specified points by determining whether at the given point all of the following conditions from Section 9.2 hold: (1) $f(a)$ is defined, (2) $\lim_{x \to a} f(x)$ exists, and (3) $\lim_{x \to a} f(x) = f(a)$.

(a)  $f(x) = 3x^2 - 7x + 4$    at $x = 5$

> (1)   $f(5) = 3(5)^2 - 7(5) + 4 = 44$
>
> (2)   $\lim_{x \to 5} 3x^2 - 7x + 4 = 3(5)^2 - 7(5) + 4 = 44$
>
> (3)   $\lim_{x \to 5} f(x) = 44 = f(5)$    $f(x)$ is continuous.

(b)  $f(x) = \dfrac{x^2 + 5x + 26}{x - 4}$    at    $x = 3$

> (1)   $f(3) = \dfrac{(3)^2 + 5(3) + 26}{(3) - 4} = \dfrac{50}{-1} = -50$
>
> (2)   $\lim\limits_{x \to 3} \dfrac{x^2 + 5x + 26}{x - 4} = -50$
>
> (3)   $\lim\limits_{x \to 3} f(x) = -50 = f(3)$    $f(x)$ is continuous.

(c)  $f(x) = \dfrac{x - 2}{x^2 - 4}$    at    $x = 2$

> (1) $f(2) = \dfrac{2 - 2}{(2)^2 - 4}$
>
> With the denominator equal to zero, $f(x)$ is not defined at $x = 2$ and so cannot be continuous at $x = 2$ even though the limit exists at $x = 2$. $\lim_{x \to 2} f(x) = \frac{1}{4}$.

## DERIVATIVE NOTATION AND SIMPLE DERIVATIVES

**9.7.**  Differentiate each of the following functions and practice the use of the different notations for a derivative.

(a)  $f(x) = 29$

$$f'(x) = 0 \quad \text{(constant rule)}$$

(b)  $y = -18$

$$\frac{dy}{dx} = 0$$

(c)  $y = 6x + 13$

$$y' = 6 \quad \text{(linear function rule)}$$

(d)  $f(x) = -7x + 2$

$$f' = -7$$

**9.8.**  Differentiate each of the following functions, using the power function rule. Continue to use the different notations.

(a)  $y = 9x^4$

$$\frac{d}{dx}(9x^4) = 36x^3$$

(b)  $f(x) = -5x^3$

$$f' = -15x^2$$

(c)  $f(x) = 7x^{-2}$

$$f'(x) = 7(-2) \cdot x^{[-2-(1)]} = -14x^{-3} = -\frac{14}{x^3}$$

(d)  $y = -8x^{-3}$

$$\frac{dy}{dx} = -8(-3) \cdot x^{[-3-(1)]} = 24x^{-4} = \frac{24}{x^4}$$

(e)  $y = \dfrac{6}{x} = 6x^{-1}$

$$D_x(6x^{-1}) = 6(-1)x^{-2} = -6x^{-2} = -\frac{6}{x^2}$$

(f)  $f(x) = 42\sqrt{x} = 42x^{1/2}$

$$\frac{df}{dx} = 42\left(\frac{1}{2}\right) \cdot x^{[1/2-1]} = 21x^{-1/2} = \frac{21}{\sqrt{x}}$$

**9.9.**  Use the rule for sums and differences to differentiate the following functions. Simply treat the dependent variable on the left as $y$ and the independent variable on the right as $x$.

(a)  $R = 9t^2 + 7t - 4$

$$\frac{dR}{dt} = 18t + 7$$

(b)  $C = 5t^3 - 8t^2 + 36t + 79$

$$C' = 15t^2 - 16t + 36$$

(c)  $p = 6q^4 - 2q^3$

$$\frac{dp}{dq} = 24q^3 - 6q^2$$

(d)  $q = 5p^3 + 16p^{-2}$

$$D_p(5p^3 + 16p^{-2}) = 15p^2 - 32p^{-3}$$

## THE PRODUCT RULE

**9.10.**  Given $y = f(x) = 6x^3(4x - 9)$, (a) use the product rule to find the derivative. (b) Simplify the original function first and then find the derivative. (c) Compare the two derivatives.

(a)  Recalling the formula for the product rule from (9.3),

$$f'(x) = g(x) \cdot h'(x) + h(x) \cdot g'(x)$$

let $g(x) = 6x^3$ and $h(x) = 4x - 9$. Then $g'(x) = 18x^2$; $h'(x) = 4$. Substituting these values in the product rule formula,

$$y' = f'(x) = 6x^3(4) + (4x - 9)(18x^2)$$

Simplifying algebraically,

$$y' = 24x^3 + 72x^3 - 162x^2 = 96x^3 - 162x^2$$

(b)    Simplifying the original function by multiplication,

$$y = 6x^3(4x - 9) = 24x^4 - 54x^3$$

Taking the derivative,

$$y' = 96x^3 - 162x^2$$

(c)    The derivatives found in parts (a) and (b) are identical. The derivative of a product can be found by either method, but as the functions grow more complicated, the product rule becomes more useful. Knowledge of another method helps to check answers.

**9.11.**   Redo Problem 9.10, given $y = f(x) = (x^6 + 4)(x^3 + 15)$.

(a)    Let $g(x) = x^6 + 4$ and $h(x) = x^3 + 15$. Then $g'(x) = 6x^5$ and $h'(x) = 3x^2$. Substituting these values in (9.3),

$$y' = f'(x) = (x^6 + 4)(3x^2) + (x^3 + 15)(6x^5)$$

$$y' = 3x^8 + 12x^2 + 6x^8 + 90x^5 = 9x^8 + 90x^5 + 12x^2$$

(b)    Simplifying first through multiplication,

$$y = (x^6 + 4)(x^3 + 15) = x^9 + 15x^6 + 4x^3 + 60$$

Then,

$$y' = 9x^8 + 90x^5 + 12x^2$$

(c)    The derivatives are, of course, identical.

**9.12.**   Differentiate each of the following functions, using the product rule.

*Note:* The choice of problems is purposely kept simple in this and other sections of the book to enable students to see how various rules work. While proper and often easier to simplify a function algebraically before taking the derivative, applying the rules to the problems as given will in the long run help the student to master the rules more efficiently.

(a)    $y = 7x^4(4x^2 - 10)$

$$\frac{dy}{dx} = 7x^4(8x) + (4x^2 - 10)(28x^3)$$

$$\frac{dy}{dx} = 56x^5 + 112x^5 - 280x^3 = 168x^5 - 280x^3$$

(b)    $f(x) = (4x^3 + 1)(6x^5)$

$$\frac{df}{dx} = (4x^3 + 1)(30x^4) + 6x^5(12x^2)$$

$$\frac{df}{dx} = 120x^7 + 30x^4 + 72x^7 = 192x^7 + 30x^4$$

(c)    $y = (5x^4 + 6)(2x^5 - 8)$

$$y' = (5x^4 + 6)(10x^4) + (2x^5 - 8)(20x^3)$$

$$y' = 50x^8 + 60x^4 + 40x^8 - 160x^3 = 90x^8 + 60x^4 - 160x^3$$

(d)    $f(x) = (4 - 6x^5)(3 + 2x^8)$

$$f' = (4 - 6x^5)(16x^7) + (3 + 2x^8)(-30x^4)$$

$$f' = 64x^7 - 96x^{12} - 90x^4 - 60x^{12} = -156x^{12} + 64x^7 - 90x^4$$

**QUOTIENT RULE**

**9.13.** Given

$$f(x) = \frac{18x^5 - 9x^4}{3x} \qquad (x \neq 0)$$

(a) Find the derivative directly, using the quotient rule. (b) Simplify the function by division and then take its derivative. (c) Compare the two derivatives.

(a) From (9.4), the formula for the quotient rule is

$$f'(x) = \frac{h(x) \cdot g'(x) - g(x) \cdot h'(x)}{[h(x)]^2}$$

where $g(x) =$ the numerator $= 18x^5 - 9x^4$ and $h(x) =$ the denominator $= 3x$. Taking the individual derivatives,

$$g'(x) = 90x^4 - 36x^3 \qquad h'(x) = 3$$

Substituting in the formula,

$$f'(x) = \frac{3x(90x^4 - 36x^3) - (18x^5 - 9x^4)(3)}{(3x)^2}$$

$$f'(x) = \frac{270x^5 - 108x^4 - 54x^5 + 27x^4}{9x^2} = \frac{216x^5 - 81x^4}{9x^2}$$

$$f'(x) = 24x^3 - 9x^2 = 3x^2(8x - 3)$$

(b) Simplifying the original function first by division,

$$f(x) = \frac{18x^5 - 9x^4}{3x} = 6x^4 - 3x^3$$

$$f'(x) = 24x^3 - 9x^2$$

(c) The derivatives will always be the same if done correctly, but as functions grow in complexity, the quotient rule becomes more important. A second method is also a way to check answers.

**9.14.** Differentiate each of the following functions by means of the quotient rule. Continue to apply the rules to the functions as given. Later, when all the rules have been mastered, the functions can be simplified first and the easiest rule applied.

(a)  $y = \dfrac{9x^7 - 8x^6}{5x^4} \qquad (x \neq 0)$

Here $g(x) = 9x^7 - 8x^6$ and $h(x) = 5x^4$. Thus, $g'(x) = 63x^6 - 48x^5$ and $h'(x) = 20x^3$. Substituting in the quotient formula,

$$y' = \frac{5x^4(63x^6 - 48x^5) - (9x^7 - 8x^6)(20x^3)}{(5x^4)^2}$$

$$y' = \frac{315x^{10} - 240x^9 - 180x^{10} + 160x^9}{25x^8} = \frac{135x^{10} - 80x^9}{25x^8} = 5.4x^2 - 3.2x$$

(b)  $y = \dfrac{6x^8}{1 - 4x} \qquad \left( x \neq \dfrac{1}{4} \right)$

$$\frac{dy}{dx} = \frac{(1 - 4x)(48x^7) - 6x^8(-4)}{(1 - 4x)^2}$$

$$\frac{dy}{dx} = \frac{48x^7 - 192x^8 + 24x^8}{(1 - 4x)^2} = \frac{48x^7 - 168x^8}{(1 - 4x)^2}$$

(c)  $f(x) = \dfrac{10x^3}{4x^2 + 9x - 2}$    $\left(x \neq \dfrac{-9 \pm \sqrt{113}}{8}\right)$

$$\frac{df}{dx} = \frac{(4x^2 + 9x - 2)(30x^2) - 10x^3(8x + 9)}{(4x^2 + 9x - 2)^2}$$

$$\frac{df}{dx} = \frac{120x^4 + 270x^3 - 60x^2 - 80x^4 - 90x^3}{(4x^2 + 9x - 2)^2} = \frac{40x^4 + 180x^3 - 60x^2}{(4x^2 + 9x - 2)^2}$$

(d)  $f = \dfrac{5x - 4}{9x + 2}$    $\left(x \neq -\dfrac{2}{9}\right)$

$$f' = \frac{(9x + 2)(5) - (5x - 4)(9)}{(9x + 2)^2}$$

$$f' = \frac{45x + 10 - 45x + 36}{(9x + 2)^2} = \frac{46}{(9x + 2)^2}$$

(e)  $y = \dfrac{4x^2 + 7x - 9}{3x^2 - 5}$    $\left(x \neq \pm\dfrac{\sqrt{15}}{3}\right)$

$$\frac{dy}{dx} = \frac{(3x^2 - 5)(8x + 7) - (4x^2 + 7x - 9)(6x)}{(3x^2 - 5)^2}$$

$$\frac{dy}{dx} = \frac{24x^3 + 21x^2 - 40x - 35 - 24x^3 - 42x^2 + 54x}{(3x^2 - 5)^2} = \frac{-21x^2 + 14x - 35}{(3x^2 - 5)^2}$$

## THE GENERALIZED POWER FUNCTION RULE

**9.15.**  Given $f(x) = (6x + 7)^2$, (a) use the generalized power function rule to find the derivative. (b) Simplify the function first by squaring it and then taking the derivative. (c) Compare answers.

(a)  From the generalized power function rule in (9.5), if $f(x) = [g(x)]^n$, we obtain

$$f'(x) = n[g(x)]^{n-1} \cdot g'(x)$$

Here $g(x) = 6x + 7$, $g'(x) = 6$, and $n = 2$. Substituting these values in the generalized power function rule,

$$f'(x) = 2(6x + 7)^{2-1} \cdot 6 = 12(6x + 7) = 72x + 84$$

(b)  Squaring the function first and then taking the derivative,

$$f(x) = (6x + 7)(6x + 7) = 36x^2 + 84x + 49$$

$$f'(x) = 72x + 84$$

(c)  The derivatives are identical. But for higher, negative, and fractional values of $n$, the generalized power function rule is faster and more practical.

**9.16.**  Find the derivative for each of the following functions with the help of the generalized power function rule.

(a)  $y = (3x^3 + 8)^5$

Here $g(x) = 3x^3 + 8$, $g'(x) = 9x^2$, and $n = 5$. Substituting in the generalized power function rule,

$$y' = 5(3x^3 + 8)^{5-1} \cdot 9x^2$$

$$y' = 5(3x^3 + 8)^4 \cdot 9x^2 = 45x^2(3x^3 + 8)^4$$

(b)  $y = (3x^2 - 5x + 9)^4$

$$y' = 4(3x^2 - 5x + 9)^3 \cdot (6x - 5)$$

$$y' = (24x - 20)(3x^2 - 5x + 9)^3$$

(c)   $y = \dfrac{1}{9x^3 + 11x + 4}$

First converting the function to an easier equivalent form,

$$y = (9x^3 + 11x + 4)^{-1}$$

then using the generalized power function rule,

$$y' = -1(9x^3 + 11x + 4)^{-2} \cdot (27x^2 + 11)$$

$$y' = -(27x^2 + 11)(9x^3 + 11x + 4)^{-2}$$

$$y' = \frac{-(27x^2 + 11)}{(9x^3 + 11x + 4)^2}$$

(d)   $y = \sqrt{21 - 4x^3}$

Converting the radical to a power function, then differentiating,

$$y = (21 - 4x^3)^{1/2}$$

$$y' = \frac{1}{2}(21 - 4x^3)^{-1/2} \cdot (-12x^2)$$

$$y' = -6x^2(21 - 4x^3)^{-1/2} = \frac{-6x^2}{\sqrt{21 - 4x^3}}$$

(e)   $y = \dfrac{1}{\sqrt{7x^2 + 66}}$

Converting to an equivalent form, then taking the derivative,

$$y = (7x^2 + 66)^{-1/2}$$

$$y' = -\frac{1}{2}(7x^2 + 66)^{-3/2} \cdot (14x) = -7x(7x^2 + 66)^{-3/2} = \frac{-7x}{(7x^2 + 66)^{3/2}} = \frac{-7x}{\sqrt{(7x^2 + 66)^3}}$$

## CHAIN RULE

**9.17.**   Use the chain rule to find the derivative $dy/dx$ for each of the following functions of a function. Check each answer on your own with the generalized power function rule, noting that the generalized power function rule is simply a specialized use of the chain rule.

(a)   $y = (2x^5 + 9)^7$

Let $y = u^7$ and $u = 2x^5 + 9$. Then $dy/du = 7u^6$ and $du/dx = 10x^4$. From the chain rule in (9.6),

$$\frac{dy}{dx} = \frac{dy}{du}\frac{du}{dx}$$

Substituting,

$$\frac{dy}{dx} = 7u^6 \cdot 10x^4 = 70x^4 u^6$$

But $u = 2x^5 + 9$. Substituting again,

$$\frac{dy}{dx} = 70x^4(2x^5 + 9)^6$$

(b)   $y = (6x + 1)^2$

Let $y = u^2$ and $u = 6x + 1$, then $dy/du = 2u$ and $du/dx = 6$. Substituting these values in the chain rule,

$$\frac{dy}{dx} = 2u \cdot 6 = 12u$$

Then substituting $(6x + 1)$ for $u$,

$$\frac{dy}{dx} = 12(6x + 1) = 72x + 12$$

(c)   $y = (7x^3 - 4)^5$

Let $y = u^5$ and $u = 7x^3 - 4$; then $dy/du = 5u^4$, $du/dx = 21x^2$, and

$$\frac{dy}{dx} = 5u^4 \cdot 21x^2 = 105x^2 u^4$$

Substituting $u = (7x^3 - 4)$,

$$\frac{dy}{dx} = 105x^2 (7x^3 - 4)^4$$

**9.18.**   Redo Problem 9.17, given

(a)   $y = (x^2 + 5x - 8)^6$

Let $y = u^6$, $u = x^2 + 5x - 8$, then $dy/du = 6u^5$ and $du/dx = 2x + 5$. Substituting in (9.6),

$$\frac{dy}{dx} = 6u^5 (2x + 5) = (12x + 30)u^5$$

But $u = x^2 + 5x - 8$. Substituting, therefore,

$$\frac{dy}{dx} = (12x + 30)(x^2 + 5x - 8)^5$$

(b)   $y = -4(x^2 - 3x + 9)^5$

Let $y = -4u^5$ and $u = x^2 - 3x + 9$. Then, $dy/du = -20u^4$, $du/dx = 2x - 3$, and

$$\frac{dy}{dx} = -20u^4 (2x - 3) = (-40x + 60)u^4$$

$$\frac{dy}{dx} = (-40x + 60)(x^2 - 3x + 9)^4$$

## COMBINATION OF RULES

**9.19.**   Use whatever combination of rules is necessary to find the derivatives of the following functions. Do not simplify the original functions first. They are deliberately kept simple to facilitate the practice of the rules.

(a)   $y = \dfrac{4x(3x - 1)}{2x - 5}$     $\left( x \neq \dfrac{5}{2} \right)$

The function involves a quotient with a product in the numerator. Hence both the quotient rule and the product rule are required. Starting with the quotient rule from (9.4),

$$y' = \frac{h(x) \cdot g'(x) - g(x) \cdot h'(x)}{[h(x)]^2}$$

where $g(x) = 4x(3x - 1)$, $h(x) = 2x - 5$, and $h'(x) = 2$. Then using the product rule from (9.3) for $g'(x)$,

$$g'(x) = 4x \cdot 3 + (3x - 1) \cdot 4 = 24x - 4$$

Substituting the appropriate values in the quotient rule,

$$y' = \frac{(2x - 5)(24x - 4) - [4x(3x - 1)] \cdot 2}{(2x - 5)^2}$$

Simplifying algebraically,

$$y' = \frac{48x^2 - 8x - 120x + 20 - 24x^2 + 8x}{(2x - 5)^2} = \frac{24x^2 - 120x + 20}{(2x - 5)^2}$$

*Note:* To check this answer one could let, among other things,

$$y = 4x \cdot \frac{3x - 1}{2x - 5} \qquad \text{or} \qquad y = \frac{4x}{2x - 5} \cdot (3x - 1)$$

and use the product rule involving a quotient.

(*b*)  $y = 5x(3x - 4)^2$

The function involves a product in which one function is raised to a power. Both the product rule and the generalized power function rule are needed. Starting with the product rule,

$$y' = g(x) \cdot h'(x) + h(x) \cdot g'(x)$$

where $g(x) = 5x$, $h(x) = (3x - 4)^2$, and $g'(x) = 5$. Using the generalized power function rule for $h'(x)$,

$$h'(x) = 2(3x - 4) \cdot 3 = 6(3x - 4) = 18x - 24$$

Substituting the appropriate values in the product rule,

$$y' = 5x \cdot (18x - 24) + (3x - 4)^2 \cdot (5)$$

and simplifying algebraically,

$$y' = 90x^2 - 120x + 5(9x^2 - 24x + 16) = 135x^2 - 240x + 80$$

(*c*)  $y = (2x - 7) \cdot \dfrac{4x + 1}{3x + 5} \qquad \left( x \neq -\dfrac{5}{3} \right)$

Here we have a product involving a quotient. Both product and quotient rules are needed. Starting with the product rule,

$$y' = g(x) \cdot h'(x) + h(x) \cdot g'(x)$$

where $g(x) = 2x - 7$, $h(x) = (4x + 1)/(3x + 5)$, $g'(x) = 2$, and using the quotient rule for $h'(x)$,

$$h'(x) = \frac{(3x + 5)(4) - (4x + 1)(3)}{(3x + 5)^2} = \frac{17}{(3x + 5)^2}$$

Substituting the appropriate values in the product rule,

$$y' = (2x - 7) \cdot \frac{17}{(3x + 5)^2} + \frac{4x + 1}{3x + 5} \cdot 2 = \frac{34x - 119}{(3x + 5)^2} + \frac{8x + 2}{3x + 5}$$

$$y' = \frac{34x - 119 + (8x + 2)(3x + 5)}{(3x + 5)^2} = \frac{24x^2 + 80x - 109}{(3x + 5)^2}$$

By letting $y = [(2x - 7)(4x + 1)]/(3x + 5)$ and using the quotient rule involving a product, one can easily check this answer.

(*d*)  $y = \dfrac{(7x - 3)^4}{(5x + 9)} \qquad \left( x \neq -\dfrac{9}{5} \right)$

Starting with the quotient rule, where $g(x) = (7x - 3)^4$, $h(x) = 5x + 9$, $h'(x) = 5$, and using the generalized power function rule for $g'(x)$,

$$g'(x) = 4(7x - 3)^3 \cdot 7 = 28(7x - 3)^3$$

Substituting these values in the quotient rule,

$$y' = \frac{(5x + 9) \cdot 28(7x - 3)^3 - (7x - 3)^4 \cdot 5}{(5x + 9)^2}$$

$$y' = \frac{(140x + 252)(7x - 3)^3 - 5(7x - 3)^4}{(5x + 9)^2}$$

To check this answer, one could let $y = (7x - 3)^4 \cdot (5x + 9)^{-1}$, and use the product rule involving the generalized power function rule twice.

(e)   $y = \left(\dfrac{5x + 4}{3x + 2}\right)^2$    $\left(x \neq -\dfrac{2}{3}\right)$

Starting with the generalized power function rule,

$$y' = 2\left(\frac{5x + 4}{3x + 2}\right) \cdot \frac{d}{dx}\left(\frac{5x + 4}{3x + 2}\right)$$    (9.7)

Then using the quotient rule,

$$\frac{d}{dx}\left(\frac{5x + 4}{3x + 2}\right) = \frac{(3x + 2)(5) - (5x + 4)(3)}{(3x + 2)^2} = \frac{-2}{(3x + 2)^2}$$

and substituting this value in (9.7),

$$y' = 2\left(\frac{5x + 4}{3x + 2}\right) \cdot \frac{-2}{(3x + 2)^2} = \frac{-4(5x + 4)}{(3x + 2)^3} = \frac{-20x - 16}{(3x + 2)^3}$$

To check this answer, let $y = (5x + 4)^2 \cdot (3x + 2)^{-2}$, and use the product rule involving the generalized power function rule twice.

**9.20.**   Differentiate each of the following, using whatever rules are necessary:

(a)   $y = \dfrac{(4x^2 - 7)(6x + 5)}{3x}$    $(x \neq 0)$

Using the quotient rule and the product rule,

$$y' = \frac{3x[(4x^2 - 7)(6) + (6x + 5)(8x)] - (4x^2 - 7)(6x + 5)(3)}{(3x)^2}$$

Simplifying algebraically,

$$y' = \frac{3x[24x^2 - 42 + 48x^2 + 40x] - 3[24x^3 + 20x^2 - 42x - 35]}{9x^2}$$

$$y' = \frac{144x^3 + 60x^2 + 105}{9x^2}$$

(b)   $f(x) = (9x + 4)(2x - 7)^4$

Using the product rule and the generalized power function rule,

$$f' = (9x + 4)[4(2x - 7)^3(2)] + (2x - 7)^4(9)$$

Simplifying algebraically,

$$f' = (9x + 4)(8)(2x - 7)^3 + 9(2x - 7)^4$$

$$f' = (72x + 32)(2x - 7)^3 + 9(2x - 7)^4$$

(c)   $y = \dfrac{12x + 7}{(5x + 2)^2}$    $\left(x \neq -\dfrac{5}{2}\right)$

Using the quotient and generalized power function rules,

$$\frac{dy}{dx} = \frac{(5x + 2)^2(12) - (12x + 7)[2(5x + 2)(5)]}{(5x + 2)^4}$$

Simplifying algebraically,

$$\frac{dy}{dx} = \frac{12(5x + 2)^2 - (12x + 7)(50x + 20)}{(5x + 2)^4} = \frac{-300x^2 - 350x - 92}{(5x + 2)^4}$$

(d)   $f(x) = \left(\dfrac{4x - 5}{3x + 1}\right)^2$    $\left(x \neq -\dfrac{1}{3}\right)$

Using the generalized power function rule and the quotient rule,

$$\frac{df}{dx} = 2\left(\frac{4x-5}{3x+1}\right)\frac{(3x+1)(4)-(4x-5)(3)}{(3x+1)^2}$$

Simplifying algebraically,

$$\frac{df}{dx} = 2\left(\frac{4x-5}{3x+1}\right)\frac{19}{(3x+1)^2} = \frac{152x-190}{(3x+1)^3}$$

(e)   $y = (5x+6)\dfrac{3x}{8x-1}$      $\left(x \neq -\dfrac{1}{8}\right)$

Using the product rule and quotient rule,

$$D_y = (5x+6)\frac{(8x-1)(3)-3x(8)}{(8x-1)^2} + \frac{3x}{8x-1}(5)$$

Simplifying algebraically,

$$D_y = (5x+6)\frac{(24x-3-24x)}{(8x-1)^2} + \frac{15x}{8x-1}$$

Adding the two terms, using the common denominator $(8x-1)^2$,

$$D_y = \frac{120x^2-30x-18}{(8x-1)^2}$$

## HIGHER-ORDER DERIVATIVES

**9.21.** For each of the following functions, (1) find the second-order derivative and (2) evaluate it at $x = 3$. Practice the use of the different second-order notations.

(a)   $y = 10x^3 + 8x^2 + 19$

(1)   $\dfrac{dy}{dx} = 30x^2 + 16x$          (2)   At $x = 3$,

$\dfrac{d^2y}{dx^2} = 60x + 16$          $\dfrac{d^2y}{dx^2} = 60(3) + 16 = 196$

(b)   $f(x) = 3x^4 + 5x^3 + 6x$

(1)   $f'(x) = 12x^3 + 15x^2 + 6$          (2)   At $x = 3$,

$f''(x) = 36x^2 + 30x$          $f''(3) = 36(3)^2 + 30(3) = 414$

(c)   $f(x) = (4x-1)(3x^2+2)$

(1)   $f' = (4x-1)(6x) + (3x^2+2)(4)$          (2)   At $x = 3$,

$f' = 24x^2 - 6x + 12x^2 + 8$          $f''(3) = 72(3) - 6$

$f' = 36x^2 - 6x + 8$          $f''(3) = 210$

$f'' = 72x - 6$

(d)  $y = \dfrac{3x}{2x - 1}$  $(x \neq \frac{1}{2})$

(1)  $y' = \dfrac{(2x - 1)(3) - 3x(2)}{(2x - 1)^2}$

$y' = \dfrac{-3}{(2x - 1)^2}$

$y'' = \dfrac{(2x - 1)^2(0) - (-3)[2(2x - 1)(2)]}{(2x - 1)^4}$

$y'' = \dfrac{24x - 12}{(2x - 1)^4}$

(2)  $y''(3) = \dfrac{24(3) - 12}{[2(3) - 1]^4}$

$y''(3) = \dfrac{60}{625} = \dfrac{12}{125}$

(e)  $f(x) = (6x - 5)^3$

(1)  $f' = 3(6x - 5)^2(6)$

$f' = 18(6x - 5)^2$

$f'' = 18[2(6x - 5)(6)] = 216(6x - 5)$

(2)  $f''(3) = 216[6(3) - 5]$

$f''(3) = 216(13) = 2808$

**9.22.**  For each of the following functions, (1) investigate the successive derivatives and (2) evaluate them at $x = 2$.

(a)  $f(x) = 2x^3 + 7x^2 + 9x - 2$

(1)  $f'(x) = 6x^2 + 14x + 9$

$f''(x) = 12x + 14$

$f'''(x) = 12$

$f^{(4)}(x) = 0$

(2)  $f'(2) = 6(2)^2 + 14(2) + 9 = 61$

$f''(2) = 12(2) + 14 = 38$

$f'''(2) = 12$

$f^{(4)}(2) = 0$

(b)  $y = (6x + 7)(3x - 8)$

(1)  $y' = (6x + 7)(3) + (3x - 8)(6)$

$y' = 36x - 27$

$y'' = 36$

$y''' = 0$

(2)  $y'(2) = 36(2) - 27$

$y'(2) = 45$

$y''(2) = 36$

$y'''(2) = 0$

(c)  $f(x) = (8 - x)^4$

(1)  $D_x = 4(8 - x)^3(-1) = -4(8 - x)^3$

$D_x^2 = -12(8 - x)^2(-1) = 12(8 - x)^2$

$D_x^3 = 24(8 - x)(-1) = -24(8 - x)$

$D_x^4 = -24(-1) = 24$

$D_x^5 = 0$

(2)  $D_x(2) = -4(6)^3 = -864$

$D_x^2(2) = 12(6)^2 = 432$

$D_x^3(2) = -24(6) = -144$

$D_x^4(2) = 24$

$D_x^5(2) = 0$

## DERIVATION OF THE RULES OF DIFFERENTIATION

**9.23.**  Given $f(x) = g(x) + h(x)$, where $g(x)$ and $h(x)$ are both differentiable functions, prove the rule of sums by demonstrating that $f'(x) = g'(x) + h'(x)$.

From (9.2) the derivative of $f(x)$ is

$$f'(x) = \lim_{\Delta x \to 0} \frac{f(x + \Delta x) - f(x)}{\Delta x}$$

Substituting $f(x) = g(x) + h(x)$,

$$f'(x) = \lim_{\Delta x \to 0} \frac{[g(x + \Delta x) + h(x + \Delta x)] - [g(x) + h(x)]}{\Delta x}$$

Rearranging terms,

$$f'(x) = \lim_{\Delta x \to 0} \frac{g(x + \Delta x) - g(x) + h(x + \Delta x) - h(x)}{\Delta x}$$

Separating terms, and taking the limits,

$$f'(x) = \lim_{\Delta x \to 0} \left[ \frac{g(x + \Delta x) - g(x)}{\Delta x} + \frac{h(x + \Delta x) - h(x)}{\Delta x} \right]$$

$$f'(x) = \lim_{\Delta x \to 0} \frac{g(x + \Delta x) - g(x)}{\Delta x} + \lim_{\Delta x \to 0} \frac{h(x + \Delta x) - h(x)}{\Delta x}$$

$$f'(x) = g'(x) + h'(x)$$

**9.24.**  Given $f(x) = g(x) \cdot h(x)$, where $g'(x)$ and $h'(x)$ both exist, prove the product rule by demonstrating that $f'(x) = g(x) \cdot h'(x) + h(x) \cdot g'(x)$.

From (9.2), the derivative of $f(x)$ is:

$$f'(x) = \lim_{\Delta x \to 0} \frac{f(x + \Delta x) - f(x)}{\Delta x}$$

Substituting $f(x) = g(x) \cdot h(x)$,

$$f'(x) = \lim_{\Delta x \to 0} \frac{g(x + \Delta x) \cdot h(x + \Delta x) - g(x) \cdot h(x)}{\Delta x}$$

Subtracting and adding $g(x + \Delta x) \cdot h(x)$,

$$f'(x) = \lim_{\Delta x \to 0} \frac{g(x + \Delta x)h(x + \Delta x) - g(x + \Delta x)h(x) + g(x + \Delta x)h(x) - g(x)h(x)}{\Delta x}$$

Partially factoring out $g(x + \Delta x)$ and $h(x)$,

$$f'(x) = \lim_{\Delta x \to 0} \frac{g(x + \Delta x)[h(x + \Delta x) - h(x)] + h(x)[g(x + \Delta x) - g(x)]}{\Delta x}$$

$$f'(x) = \lim_{\Delta x \to 0} \frac{g(x + \Delta x)[h(x + \Delta x) - h(x)]}{\Delta x} + \lim_{\Delta x \to 0} \frac{h(x)[g(x + \Delta x) - g(x)]}{\Delta x}$$

$$f'(x) = \lim_{\Delta x \to 0} g(x + \Delta x) \cdot \lim_{\Delta x \to 0} \frac{h(x + \Delta x) - h(x)}{\Delta x} + \lim_{\Delta x \to 0} h(x) \cdot \lim_{\Delta x \to 0} \frac{g(x + \Delta x) - g(x)}{\Delta x}$$

$$f'(x) = g(x) \cdot h'(x) + h(x) \cdot g'(x)$$

**9.25.**  Given $f(x) = g(x)/h(x)$, where $g'(x)$ and $h'(x)$ both exist and $h(x) \neq 0$, prove the quotient rule by demonstrating

$$f'(x) = \frac{h(x) \cdot g'(x) - g(x) \cdot h'(x)}{[h(x)]^2}$$

Starting with $f(x) = g(x)/h(x)$ and solving for $g(x)$,

$$g(x) = f(x) \cdot h(x)$$

Then taking the derivative of $g(x)$, using the product rule,

$$g'(x) = f(x) \cdot h'(x) + h(x) \cdot f'(x)$$

and solving algebraically for $f'(x)$.

$$h(x) \cdot f'(x) = g'(x) - f(x) \cdot h'(x)$$

$$f'(x) = \frac{g'(x) - f(x) \cdot h'(x)}{h(x)}$$

Substituting $g(x)/h(x)$ for $f(x)$,

$$f'(x) = \frac{g'(x) - [g(x) \cdot h'(x)/h(x)]}{h(x)}$$

Now multiplying both numerator and denominator by $h(x)$,

$$f'(x) = \frac{h(x) \cdot g'(x) - g(x) \cdot h'(x)}{[h(x)]^2}$$

# Supplementary Problems

## LIMITS

**9.26.**   Find the limits of the following functions:

(a)  $\lim\limits_{x \to 6} (5x^2 - 2x - 18)$

(b)  $\lim\limits_{x \to -3} (4x^2 + 9x - 5)$

(c)  $\lim\limits_{x \to 2} \dfrac{3x^2 - 4x + 6}{x^2 + 8x - 15}$

(d)  $\lim\limits_{x \to -4} \dfrac{7x^2 - 12}{8x^2 + 3x + 4}$

(e)  $\lim\limits_{x \to 5} \sqrt{x^3 + 9x - 1}$

(f)  $\lim\limits_{x \to -9} \sqrt[3]{2x^2 + 5x + 8}$

**9.27.**   Find the limits of each of the following functions in which the limit of the denominator approaches zero:

(a)  $\lim\limits_{x \to -12} \dfrac{x + 12}{x^2 - 144}$

(b)  $\lim\limits_{x \to 7} \dfrac{x - 7}{2x^2 - 98}$

(c)  $\lim\limits_{x \to 5} \dfrac{x^2 - 11x + 30}{x - 5}$

(d)  $\lim\limits_{x \to 8} \dfrac{x^2 + 7x - 120}{x - 8}$

**9.28.**   Find the limits of each of the following functions:

(a)  $\lim\limits_{x \to \infty} \dfrac{7x - 22}{21x + 8}$

(b)  $\lim\limits_{x \to \infty} \dfrac{8x^2 - 49}{3x^3 + 16}$

(c)  $\lim\limits_{x \to \infty} \dfrac{9x^2 - 4x + 2}{4x^2 + 6x - 7}$

(d)  $\lim\limits_{x \to \infty} \dfrac{x^4 - 9}{x^3 + 8}$

## DERIVATIVES

**9.29.**   Find the first derivative for each of the following functions:

(a)  $y = 6x^5$

(b)  $f(x) = 2x + 9$

(c)  $f(x) = 17$

(d)  $y = 8x^3 + 4x^2 + 9x + 3$

(e)  $y = 6x^{-4}$

(f)  $f(x) = -7x^{-2}$

(g)  $y = 18x^{1/3}$

(h)  $f(x) = 5x^{-1/2}$

To facilitate mastery of the basic rules of differentiation, apply the appropriate rule or rules to the following functions in the remainder of this chapter exactly as they are given and do not attempt to simplify them algebraically first.

## PRODUCT RULE

**9.30.** Use the product rule to differentiate each of the following functions:

(a)  $f(x) = (8x - 9)(4x^5)$

(b)  $f(x) = 12x^3(7x + 3)$

(c)  $y = (6x^2 + 11)(9x^3 - 4)$

(d)  $y = (15 - 8x^4)(6x^6 - 5)$

## QUOTIENT RULE

**9.31.** Differentiate each of the following functions, using the quotient rule:

(a)  $y = \dfrac{22x^4 - 15}{8x - 1}$

(b)  $y = \dfrac{8x^6}{3x + 5}$

(c)  $y = \dfrac{5x^3}{6x^2 - 7x + 2}$

(d)  $y = \dfrac{8x^2 + 3x - 9}{7x^2 - 4}$

## GENERALIZED POWER FUNCTION RULE

**9.32.** Use the generalized power function rule to differentiate each of the following functions:

(a)  $f(x) = (9x - 4)^5$

(b)  $f(x) = (7x^3 + 6)^4$

(c)  $f(x) = \dfrac{1}{(3x^2 - 11)^2} = (3x^2 - 11)^{-2}$

(d)  $f(x) = \dfrac{-50}{6x^2 - 4x - 9} = -50(6x^2 - 4x - 9)^{-1}$

(e)  $f(x) = \sqrt{14x^4 - 45} = (14x^4 - 45)^{1/2}$

(f)  $f(x) = \dfrac{1}{\sqrt{22 - 9x^6}} = (22 - 9x^6)^{-1/2}$

## CHAIN RULE

**9.33.** Using the chain rule, find the first derivative of each of the following functions:

(a)  $y = (6x^4 - 35)^8$

(b)  $y = (27 - 8x^3)^5$

(c)  $f(x) = (18x^2 + 23)^{1/3}$

(d)  $f(x) = (122x^3 - 49)^{-4}$

## COMBINATION OF RULES

**9.34.** Differentiate each of the following functions using whatever combination of rules is necessary.

(a)  $y = 5x^2(4x - 9)^3$

(b)  $y = \dfrac{6x(x^2 + 7)}{4x + 1}$

(c)  $f(x) = \dfrac{(8x - 5)^3}{3x + 2}$

(d)  $f(x) = (7x - 4)(3x + 8)^4$

## HIGHER-ORDER DERIVATIVES

**9.35.** Find the successive derivatives of each of the following functions:

(a)  $y = 3x^4 - 5x^3 + 8x^2 - 7x - 13$

(b)  $y = (8x + 9)(10x - 3)$

(c)  $f(x) = (6x - 7)^4$

(d)  $f(x) = (5 - 2x)^4$

# Answers to Supplementary Problems

**9.26.**  (*a*)  150    (*b*)  4    (*c*)  2    (*d*)  $\frac{5}{6}$    (*e*)  13    (*f*)  5

**9.27.**  (*a*)  $-\frac{1}{24}$    (*b*)  $\frac{1}{28}$    (*c*)  $-1$    (*d*)  23

**9.28.**  (*a*)  $\frac{1}{3}$    (*b*)  0    (*c*)  2.25    (*d*)  The limit does not exist.

**9.29.**  (*a*)  $y' = 30x^4$    (*b*)  $f' = 2$    (*c*)  $f' = 0$    (*d*)  $y' = 24x^2 + 8x + 9$
(*e*)  $y' = -24x^{-5}$    (*f*)  $f' = 14x^{-3}$    (*g*)  $y' = 6x^{-2/3}$    (*h*)  $f' = -2.5x^{-3/2}$

**9.30.**  (*a*)  $f' = 192x^5 - 180x^4$    (*b*)  $f' = 336x^3 + 108x^2$
(*c*)  $y' = 270x^4 + 297x^2 - 48x$    (*d*)  $y' = -480x^9 + 540x^5 + 160x^3$

**9.31.**  (*a*)  $y' = \dfrac{528x^4 - 88x^3 + 120}{(8x - 1)^2}$    (*b*)  $y' = \dfrac{120x^6 + 240x^5}{(3x + 5)^2}$

(*c*)  $y' = \dfrac{30x^4 - 70x^3 + 30x^2}{(6x^2 - 7x + 2)^2}$    (*d*)  $y' = \dfrac{-21x^2 + 62x - 12}{(7x^2 - 4)^2}$

**9.32.**  (*a*)  $f' = 45(9x - 4)^4$    (*b*)  $f' = 84x^2(7x^3 + 6)^3$

(*c*)  $f' = -12x(3x^2 - 11)^{-3} = \dfrac{-12x}{(3x^2 - 11)^3}$

(*d*)  $f' = (600x - 200)(6x^2 - 4x - 9)^{-2} = \dfrac{600x - 200}{(6x^2 - 4x - 9)^2}$

(*e*)  $f' = 28x^3(14x^4 - 45)^{-1/2} = \dfrac{28x^3}{\sqrt{14x^4 - 45}}$

(*f*)  $f' = 27x^5(22 - 9x^6)^{-3/2} = \dfrac{27x^5}{\sqrt{(22 - 9x^6)^3}}$    or    $\dfrac{27x^5}{\left(\sqrt{22 - 9x^6}\right)^3}$

**9.33.**  (*a*)  $y' = 192x^3(6x^4 - 35)^7$    (*b*)  $y' = -120x^2(27 - 8x^3)^4$

(*c*)  $f'(x) = 12x(18x^2 + 23)^{-2/3} = \dfrac{12x}{\sqrt[3]{(18x^2 + 23)^2}}$    or    $\dfrac{12x}{\left(\sqrt[3]{18x^2 + 23}\right)^2}$

(*d*)  $f'(x) = -1464x^2(122x^3 - 49)^{-5} = \dfrac{-1464x^2}{(122x^3 - 49)^5}$

**9.34.**  (*a*)  $y' = 60x^2(4x - 9)^2 + 10x(4x - 9)^3$

(*b*)  $y' = \dfrac{48x^3 + 18x^2 + 42}{(4x + 1)^2}$

(*c*)  $f'(x) = \dfrac{(72x + 48)(8x - 5)^2 - 3(8x - 5)^3}{(3x + 2)^2}$

(*d*)  $f'(x) = (84x - 48)(3x + 8)^3 + 7(3x + 8)^4$

**9.35.**  (*a*)  $y' = 12x^3 - 15x^2 + 16x - 7$    (*b*)  $y' = 160x + 66$

$y'' = 36x^2 - 30x + 16$    $y'' = 160$

$y''' = 72x - 30$    $y''' = 0$

$y^{(4)} = 72$

$y^{(5)} = 0$

(c)   $f'(x) = 24(6x - 7)^3$

  $f''(x) = 432(6x - 7)^2$

  $f'''(x) = 5184(6x - 7)$

  $f^{(4)}(x) = 31, 104$

  $f^{(5)}(x) = 0$

(d)   $f'(x) = -8(5 - 2x)^3$

  $f''(x) = 48(5 - 2x)^2$

  $f'''(x) = -192(5 - 2x)$

  $f^{(4)}(x) = 384$

  $f^{(5)}(x) = 0$

# Chapter 10

# Differential Calculus: Uses of the Derivative

## 10.1 INCREASING AND DECREASING FUNCTIONS

A function $f(x)$ is said to be *increasing (decreasing)* at $x = a$ if in the immediate vicinity of the point $[a, f(a)]$ the graph of the function rises (falls) as it moves from left to right. Since the first derivative measures both the rate of change and slope of a function, a positive first derivative at $x = a$ indicates that the function is increasing at $a$; a negative first derivative indicates it is decreasing. In short, as seen in Fig. 10-1,

$$f'(a) > 0: \quad \text{increasing function at} \quad x = a$$

$$f'(a) < 0: \quad \text{decreasing function at} \quad x = a$$

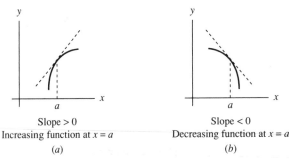

| Slope > 0 | Slope < 0 |
|-----------|-----------|
| Increasing function at $x = a$ | Decreasing function at $x = a$ |
| (a) | (b) |

**Fig. 10-1**

A function that increases (or decreases) over its entire domain is called a *monotonic function*. It is said to increase (decrease) *monotonically*. See Problems 10.1 to 10.3.

## 10.2 CONCAVITY AND CONVEXITY

A function $f(x)$ is *concave* at $x = a$ if in some small region close to the point $[a, f(a)]$, the graph of the function lies completely below its tangent line. A function is *convex* at $x = a$ if in the area very close to $[a, f(a)]$, the graph of the function lies completely above its tangent line. A positive second derivative at $x = a$ denotes the function is convex at $x = a$; a negative second derivative at $x = a$ denotes the function is concave at $a$. The sign of the first derivative is irrelevant for concavity. In brief, as seen in Fig. 10-2 and Problems 10.1 to 10.4,

$$f''(a) < 0: \quad f(x) \text{ is concave at} \quad x = a$$

$$f''(a) > 0: \quad f(x) \text{ is convex at} \quad x = a$$

If $f''(x) < 0$ for all $x$ in the domain, $f(x)$ is *strictly concave*. If $f''(x) > 0$ for all $x$ in the domain, $f(x)$ is *strictly convex*.

## 10.3 RELATIVE EXTREMA

A *relative* or *local extremum* is a point where a function reaches a relative maximum or minimum. To be at a relative maximum or minimum at a point $a$, the function must be at a relative *plateau*, that is, neither

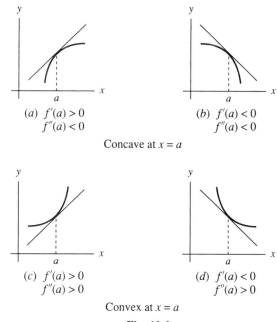

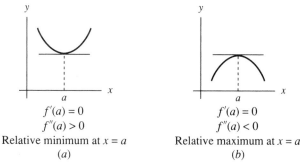

Fig. 10-2

increasing nor decreasing at $a$. If the function is neither increasing nor decreasing at $a$, the first derivative of the function at $a$ must equal zero or be undefined. A point in the domain of a function where the derivative equals zero or is undefined is called a *critical point* or *value*.

To distinguish mathematically between a relative maximum and minimum, the *second-derivative* test is used. Assuming $f'(a) = 0$,

1.  If $f''(a) > 0$, indicating that the function is convex and the graph of the function lies completely above its (horizontal) tangent line at $x = a$, the function must be at a relative minimum at $x = a$, as is seen in Fig. 10-3($a$).
2.  If $f''(a) < 0$, denoting that the function is concave and the graph of the function lies completely below its (horizontal) tangent line at $x = a$, the function must be at a relative maximum at $x = a$, as seen in Fig. 10-3($b$).
3.  If $f''(a) = 0$, the test is inconclusive.

Fig. 10-3

For functions which are differentiable at all values of $x$, called *differentiable or smooth functions*, which one typically encounters in business and economics, one need only consider cases where $f'(x) = 0$ in looking for critical points. To sum up,

$$f'(a) = 0, \quad f''(a) > 0 \qquad \text{relative minimum at } x = a$$

$$f'(a) = 0, \quad f''(a) < 0 \qquad \text{relative maximum at } x = a$$

See Example 1 and Problems 10.5 and 10.6.

**EXAMPLE 1.**   Keeping in mind that (*a*) a positive second derivative (++) at a critical point means that the function is moving up from a plateau (⌣) and (*b*) a negative second derivative (−−) at a critical point means that the function is moving down from a plateau (⌢), a simple mnemonic to remember the rule is presented in Fig. 10-4.

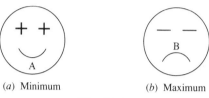

(*a*)  Minimum          (*b*)  Maximum

**Fig. 10-4**

## 10.4   INFLECTION POINTS

An *inflection point* is a point on the graph where the function crosses its tangent line and changes from concave to convex or vice versa. Inflection points are possible only where the *second* derivative equals zero or is undefined. The sign of the first derivative is immaterial. In brief, for an inflection point at *a*, as seen in Fig. 10-5,

1.   $f''(a) = 0$ or is undefined.
2.   Concavity changes at $x = a$.
3.   Graph crosses its tangent line at $x = a$.

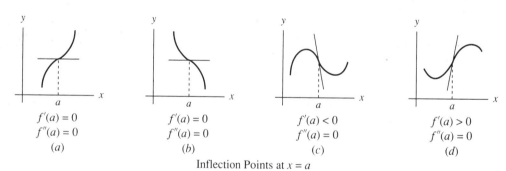

Inflection Points at $x = a$

**Fig. 10-5**

See also Example 2 and Problems 10.7, 10.9, and 10.14.

## 10.5   CURVE SKETCHING

The first and second derivatives provide useful information about the general shape of a curve and so facilitate graphing. A relatively complicated function can be roughly sketched in a few easy steps. Given $f(x)$, a differentiable function,

1.  Seek out any relative extremum by looking for points where the first derivative equals zero:
    $f'(x) = 0$.
2.  Determine concavity at the critical point(s) by testing the sign of the second derivative to distinguish
    between a relative maximum $[f''(x) < 0]$ and a relative minimum $[f''(x) > 0]$.
3.  Check for inflection points where $f''(x) = 0$ or $f''(x)$ is undefined, and where concavity changes.

**EXAMPLE 2.**    The function $y = f(x) = -x^3 + 3x^2 + 45x$ is sketched below using the procedure outlined above in
Section 10.5.

(a)  Take the first derivative,

$$f'(x) = -3x^2 + 6x + 45$$

set it equal to zero and solve for $x$,

$$f'(x) = -3x^2 + 6x + 45 = 0$$

Factoring,

$$-3(x^2 - 2x - 15) = 0$$

$$(x + 3)(x - 5) = 0$$

$$x = -3 \qquad x = 5 \qquad \text{critical points}$$

(b)  Take the second derivative, evaluate it at the critical points, and test for concavity by checking the signs.

$$f''(x) = -6x + 6$$

$$f''(-3) = -6(-3) + 6 = 24 > 0 \qquad \text{convex, relative minimum}$$

$$f''(5) = -6(5) + 6 = -24 < 0 \qquad \text{concave, relative maximum}$$

(c)  Look for inflection points where $f''(x) = 0$ and concavity changes.

$$f''(x) = -6x + 6 = 0$$

$$x = 1$$

With $f''(x) = 0$ at $x = 1$ and $f(x)$ changing from convex at $x = -3$ to concave at $x = 5$, an inflection point
exists at $x = 1$.

(d)  Evaluate $f(x)$ at $x = -3$, 1, and 5 and graph as in Fig. 10-6.

$$f(-3) = -(-3)^3 + 3(-3)^2 + 45(-3) = -81$$

$$f(1) = -(1)^3 + 3(1)^2 + 45(1) = 47$$

$$f(5) = -(5)^3 + 3(5)^2 + 45(5) = 175$$

See Problems 10.6 to 10.14.

## 10.6  OPTIMIZATION OF FUNCTIONS

*Optimization* is the process of finding the relative maximum or minimum of a function. Without the aid
of a graph, this is done with the techniques developed in Sections 10.3 to 10.5 and outlined below. Given
the usual differentiable function,

1.  Take the first derivative, set it equal to zero, and solve for the critical point(s). This step is known
    as the *first-order condition* and the *necessary condition*.
2.  Take the second derivative, evaluate it at the critical point(s), and check the sign(s). If at a critical
    point $a$,

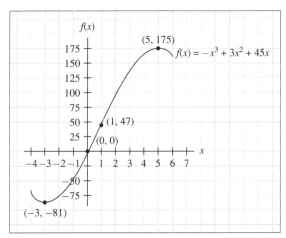

**Fig. 10-6**

$$f''(a) > 0: \quad \text{convex, relative minimum}$$

$$f''(a) < 0: \quad \text{concave, relative maximum}$$

$$f''(a) = 0: \quad \text{test inconclusive}$$

This step, the second-derivative test, is also called the *second-order condition* and the *sufficient condition*. In sum,

| Relative Maximum | Relative Minimum |
|---|---|
| $f'(a) = 0$ | $f'(a) = 0$ |
| $f''(a) < 0$ | $f''(a) > 0$ |

Recall Example 1. See also Example 3 and Problems 10.15 and 10.16.

**EXAMPLE 3.**    To optimize $f(x) = 3x^3 - 54x^2 + 288x - 22$,

(a)  Find the critical points by taking the first derivative, setting it equal to zero, and solving for $x$.

$$f'(x) = 9x^2 - 108x + 288 = 0$$

$$9(x - 4)(x - 8) = 0$$

$$x = 4 \quad x = 8 \quad \text{critical points}$$

(b)  Test for concavity by taking the second derivative, evaluating it at the critical points, and checking the signs to distinguish between a relative maximum and a relative minimum.

$$f''(x) = 18x - 108$$

$$f''(4) = 18(4) - 108 = -36 < 0 \quad \text{concave, relative maximum}$$

$$f''(8) = 18(8) - 108 = 36 \quad > 0 \quad \text{convex, relative minimum}$$

The function is maximized at $x = 4$ and minimized at $x = 8$.

## 10.7   THE SUCCESSIVE-DERIVATIVE TEST

The second-derivative test would be inconclusive for functions such as those illustrated in Fig. 10-5(*a*) and (*b*), where inflection points occur at the critical values. In the event that $f''(a) = 0$ and without a graph for guidance, apply the *successive-derivative* test:

1.  If, when evaluated at a critical point, the first nonzero higher-order derivative is an odd-numbered derivative (third, fifth, etc.), the function is at an inflection point. See Problems 10.7, 10.9, and 10.14.
2.  If, when evaluated at a critical value, the first nonzero higher-order derivative is an even-numbered derivative, the function is at a relative extremum. A positive value of this derivative indicates a relative minimum; a negative value, a relative maximum. See Problems 10.6, 10.8, 10.13, and 10.16.

## 10.8   MARGINAL CONCEPTS IN ECONOMICS

*Marginal revenue* in economics is defined as the change in total revenue brought about by the sale of an additional good. *Marginal cost* is defined as the change in total cost incurred from the production of an extra unit. Since total revenue TR and total cost TC are both functions of their levels of output $Q$, marginal revenue MR and marginal cost MC can each be expressed mathematically as derivatives of their respective total functions:

$$\text{Given TR} = \text{TR}(Q), \qquad \text{MR} = \frac{d\text{TR}}{dQ}$$

$$\text{Given TC} = \text{TC}(Q), \qquad \text{MC} = \frac{d\text{TC}}{dQ}$$

In brief, the marginal expression of any economic function can be written as the derivative of the total function.

**EXAMPLE 4.**   If $TC = Q^2 + 5Q + 72$, then

$$MC = \frac{d\text{TC}}{dQ} = 2Q + 5$$

If $\text{TR} = -3Q^2 + 95Q$,

$$MR = -6Q + 95$$

**EXAMPLE 5.**   Given the demand function $P = 80 - 3Q$, the marginal revenue can be found by first finding the total revenue function and then taking the derivative of that function with respect to $Q$. Thus

$$\text{TR} = PQ = (80 - 3Q)Q = 80Q - 3Q^2$$

and

$$MR = \frac{d\text{TR}}{dQ} = 80 - 6Q$$

If $Q = 5$, MR $= 80 - 6(5) = 50$; if $Q = 7$, MR $= 80 - 6(7) = 38$. See Problems 10.17 to 10.19.

## 10.9   OPTIMIZING ECONOMIC FUNCTIONS FOR BUSINESS

People in business and economists are frequently asked to help firms solve such problems as maximizing profits, levels of physical output, or productivity and minimizing costs, levels of pollution or noise, and the use of scarce natural resources. The task is facilitated by the techniques developed earlier in the chapter and illustrated in Example 6 and Problems 10.20 to 10.25.

**EXAMPLE 6.**   To maximize profits $\pi$ for a firm, given total revenue $R = 3300Q - 26Q^2$ and total cost $C = Q^3 - 2Q^2 + 420Q + 750$, assuming $Q > 0$,

(*a*)   Set up the profit function: $\pi = R - C$.

$$\pi = 3300Q - 26Q^2 - (Q^3 - 2Q^2 + 420Q + 750)$$

$$\pi = -Q^3 - 24Q^2 + 2880Q - 750$$

(*b*)   Take the first derivative, set it equal to zero, and solve for $Q$ to find the critical points.

$$\pi' = -3Q^2 - 48Q + 2880 = 0$$

$$= -3(Q^2 + 16Q - 960) = 0$$

$$= -3(Q - 24)(Q + 40) = 0$$

$$Q = 24 \qquad Q = -40 \qquad \text{critical points}$$

(*c*)   Take the second derivative and evaluate it at the positive critical value. Then check the sign for concavity to be sure it represents a relative maximum. Ignore the negative critical value. It has no economic significance and will prove mathematically to be a relative minimum.

$$\pi'' = -6Q - 48$$

$$\pi''(24) = -6(24) - 48 = -192 < 0 \qquad \text{concave, relative maximum}$$

Profit is maximized at $Q = 24$ where

$$\pi(24) = -(24)^3 - 24(24)^2 + 2880(24) - 750 = 40,722$$

## 10.10   RELATIONSHIPS AMONG TOTAL, MARGINAL, AND AVERAGE FUNCTIONS

People in business and economists frequently deal with functions involving costs, revenues, and profits. The critical relationships among total, average, and marginal functions are perhaps most easily understood in terms of graphs. Using the curve sketching techniques developed earlier in the chapter, Example 7 illustrates by way of example the connections that exist among all interrelated total cost, average cost, and marginal cost functions.

**EXAMPLE 7.**   Given the total cost function, $TC = Q^3 - 24Q^2 + 600Q$, the relationships among total, average, and marginal cost functions are demonstrated as follows:

(*a*)   Take the first and second derivatives of the total cost function,

$$TC' = 3Q^2 - 48Q + 600$$

$$TC'' = 6Q - 48$$

and check for (1) concavity and (2) inflection points, using the second derivative,

(1)                              For $Q < 8, TC'' < 0$      concave

                                 For $Q > 8, TC'' > 0$      convex

(2)                                     $6Q - 48 = 0$

                                        $Q = 8$

$$TC(8) = (8)^3 - 24(8)^2 + 600(8) = 3776$$

With $TC(Q)$ changing from concave to convex at $Q = 8$,

$$(8, 3776) \qquad \text{inflection point}$$

(b)   Find the average cost function AC and the relative extrema.

$$AC = \frac{TC}{Q} = Q^2 - 24Q + 600$$

$$AC' = 2Q - 24 = 0$$

$$Q = 12 \qquad \text{critical value}$$

$$AC'' = 2 > 0 \qquad \text{convex, relative minimum}$$

(c)   Do the same thing for the marginal cost function.

$$MC = TC' = 3Q^2 - 48Q + 600$$

$$MC' = 6Q - 48 = 0$$

$$Q = 8 \qquad \text{critical value}$$

$$MC'' = 6 > 0 \qquad \text{convex, relative minimum}$$

(d)   Sketch the graph as in Fig. 10-7.

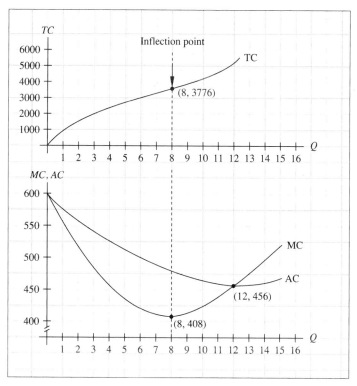

**Fig. 10-7**

*Note:*

1. MC decreases when TC is concave and increasing at a decreasing rate, increases when TC is convex and increasing at an increasing rate, and is at a minimum when TC is at an inflection point and changing concavity;

2. AC decreases over the whole region where MC < AC, is at a minimum when MC = AC, and increases when MC > AC.

3. If MC < AC, AC decreases. If MC = AC, AC is at a minimum. If MC > AC, AC increases.

See also Problem 10.26.

# Solved Problems

## INCREASING AND DECREASING FUNCTIONS, CONCAVITY AND CONVEXITY

**10.1.** From the graphs in Fig. 10-8, indicate which graphs are (1) increasing for all *x*, (2) decreasing for all *x*, (3) convex for all *x*, (4) concave for all *x*, (5) which have relative maxima or minima, and (6) which have inflection points.

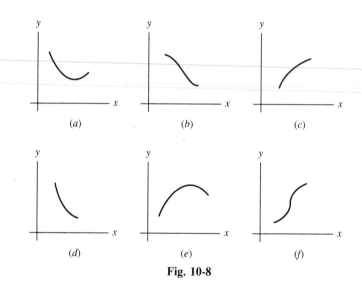

**Fig. 10-8**

(1)    *c, f*; increasing for all *x*.

(2)    *b, d*; decreasing for all *x*.

(3)    *a, d*; convex for all *x*.

(4)    *c, e*; concave for all *x*.

(5)    *a, e*; exhibit a relative maximum or minimum.

(6)    *b, f*; have an inflection point.

**10.2.** Indicate with respect to the graphs in Fig. 10-9 which functions have (1) positive first derivatives for all *x*, (2) negative first derivatives for all *x*, (3) positive second derivatives for all *x*, (4) negative

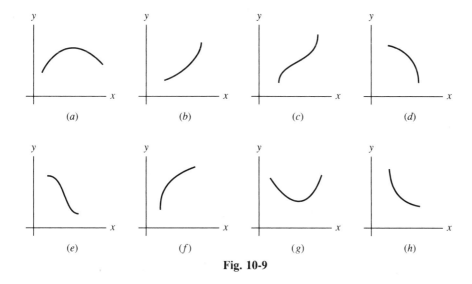

Fig. 10-9

second derivatives for all $x$, (5) first derivatives equal to zero or undefined at some point, and (6) second derivatives equal to zero or undefined at some point.

(1)    $b, c, f$; the graphs all move up from left to right.

(2)    $d, e, h$; the graphs all move down from left to right.

(3)    $b, g, h$; the graphs are all convex.

(4)    $a, d, f$; the graphs are all concave.

(5)    $a, g$; the graphs reach a plateau (at an extreme point).

(6)    $c, e$; the graphs have inflection points.

**10.3.**    Test to see whether the following functions are increasing, decreasing, or stationary at $x = 3$.

(a)    $y = 5x^2 - 12x + 8$

$$y' = 10x - 12$$

$$y'(3) = 10(3) - 12 = 18 > 0 \qquad \text{function is increasing}$$

(b)    $y = x^3 - 4x^2 - 9x + 19$

$$y' = 3x^2 - 8x - 9$$

$$y'(3) = 3(3)^2 - 8(3) - 9 = -6 < 0 \qquad \text{function is decreasing}$$

(c)    $y = x^4 - 9x^3 + 22.5x^2 - 92$

$$y' = 4x^3 - 27x^2 + 45x$$

$$y'(3) = 4(3)^3 - 27(3)^2 + 45(3) = 0 \qquad \text{function is stationary}$$

**10.4.**    Test to see if the following functions are concave or convex at $x = 2$.

(a)    $y = -4x^3 + 5x^2 + 3x - 25$

$$y' = -12x^2 + 10x + 3$$

$$y'' = -24x + 10$$

$$y''(2) = -24(2) + 10 = -38 < 0 \qquad \text{concave}$$

(b)   $y = (3x^2 - 4)^2$

$$y' = 2(3x^2 - 4)(6x) = 12x(3x^2 - 4) = 36x^3 - 48x$$

$$y'' = 108x^2 - 48$$

$$y''(2) = 108(2)^2 - 48 = 384 > 0 \qquad \text{convex}$$

## RELATIVE EXTREMA

**10.5.** Find the relative extrema for the following functions by (1) finding the critical value(s) and (2) determining whether at the critical value(s) the function is at a relative maximum or minimum.

(a)   $f(x) = -9x^2 + 126x - 45$

(1)   Take the first derivative, set it equal to zero, and solve for $x$ to find the critical value(s).

$$f'(x) = -18x + 126 = 0$$

$$x = 7 \qquad \text{critical value}$$

(2)   Take the second derivative, evaluate it at the critical value(s), and check for concavity to distinguish between a relative maximum and minimum.

$$f''(x) = -18$$

$$f''(7) = -18 < 0 \qquad \text{concave, relative maximum}$$

(b)   $f(x) = 2x^3 - 18x^2 + 48x - 29$

(1)
$$f'(x) = 6x^2 - 36x + 48 = 0$$

$$f'(x) = 6(x^2 - 6x + 8) = 0$$

$$f'(x) = 6(x - 2)(x - 4) = 0$$

$$x = 2 \qquad x = 4 \qquad \text{critical values}$$

(2)
$$f''(x) = 12x - 36$$

$$f''(2) = 12(2) - 36 = -12 < 0 \qquad \text{concave, relative maximum}$$

$$f''(4) = 12(4) - 36 = 12 > 0 \qquad \text{convex, relative minimum}$$

(c)   $f(x) = x^4 + 8x^3 - 80x^2 + 195$

(1)
$$f'(x) = 4x^3 + 24x^2 - 160x = 0$$

$$f'(x) = 4x(x^2 + 6x - 40) = 0$$

$$f'(x) = 4x(x - 4)(x + 10) = 0$$

$$x = 0 \qquad x = 4 \qquad x = -10 \qquad \text{critical values}$$

(2)   $f''(x) = 12x^2 + 48x - 160$

$$f''(-10) = 12(-10)^2 + 48(-10) - 160 = 560 > 0 \qquad \text{convex, relative minimum}$$

$$f''(0) = 12(0)^2 + 48(0) - 160 = -160 < 0 \qquad \text{concave, relative maximum}$$

$$f''(4) = 12(4)^2 + 48(4) - 160 = 224 > 0 \qquad \text{convex, relative minimum}$$

**10.6.** For the function $y = (x - 5)^4$, (1) find the critical values and (2) test to see if at the critical values the function is at a relative maximum, minimum, or possible inflection point.

(1)   Take the first derivative, set it equal to zero, and solve for $x$ to obtain the critical value(s).

$$y' = 4(x - 5)^3 = 0$$

$$x - 5 = 0$$

$$x = 5 \quad \text{critical value}$$

(2)   Take the second derivative, evaluate it at the critical value, and check the sign for concavity to distinguish between a relative maximum, minimum, or inflection point.

$$y'' = 12(x - 5)^2$$

$$y''(5) = 12(5 - 5)^2 = 0 \quad \text{test inconclusive}$$

If the second derivative test is inconclusive, continue to take successively higher derivatives and evaluate them at the critical value until you reach the first nonzero higher-order derivative:

$$y''' = 24(x - 5)$$

$$y'''(5) = 24(5 - 5) = 0 \quad \text{test inconclusive}$$

$$y^{(4)} = 24$$

$$y^{(4)}(5) = 24 > 0$$

As explained in Section 10.7, with the first nonzero higher-order derivative an even-numbered derivative, $y$ is at a relative extremum. With that derivative positive, $y$ is convex and at a relative minimum. See Fig. 10-10.

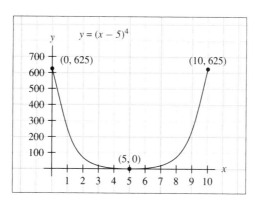

**Fig. 10-10**

**10.7.**   Redo Problem 10.6, given $y = (6 - x)^3$.

(1)   $$y' = 3(6 - x)^2(-1) = -3(6 - x)^2 = 0$$

$$x = 6 \quad \text{critical value}$$

(2)   $$y'' = -6(6 - x)(-1) = 6(6 - x)$$

$$y''(6) = 6(6 - 6) = 0 \quad \text{test inconclusive}$$

Continuing to take successively higher-order derivatives and evaluating them at the critical value in search of the first higher-order derivative that does not equal zero,

$$y''' = -6$$

$$y'''(6) = -6 < 0$$

As explained in Section 10.7, with the first nonzero higher-order derivative an odd-numbered derivative, $y$ is at an inflection point and not at an extreme point. See Fig. 10-11.

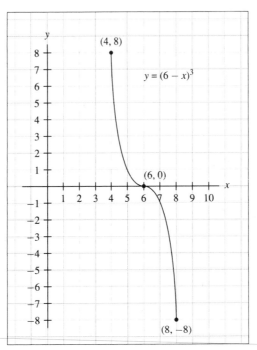

**Fig. 10-11**

**10.8.** Redo Problem 10.6, given $y = -3(x - 4)^6$

(1)
$$y' = -18(x - 4)^5 = 0$$

$$x = 4 \qquad \text{critical value}$$

(2)
$$y'' = -90(x - 4)^4$$

$$y''(4) = -90(0)^4 = 0 \qquad \text{test inconclusive}$$

Continuing on,

$$y''' = -360(x - 4)^3 \qquad y'''(4) \;= 0 \qquad \text{test inconclusive}$$

$$y^{(4)} = -1080(x - 4)^2 \qquad y^{(4)}(4) = 0 \qquad \text{test inconclusive}$$

$$y^{(5)} = -2160(x - 4) \qquad y^{(5)}(4) \;= 0 \qquad \text{test inconclusive}$$

$$y^{(6)} = -2160 \qquad y^{(6)}(4) \;\;= -2160 < 0$$

With the first nonzero higher-order derivative an even-numbered derivative, $y$ is at an extreme point; with $y^{(6)}(4) < 0$, $y$ is concave and at a relative maximum. See Fig. 10-12.

**10.9.** Redo Problem 10.6, given $y = (x - 3)^5$

(1)
$$y' = 5(x - 3)^4 = 0$$

$$x = 3 \qquad \text{critical value}$$

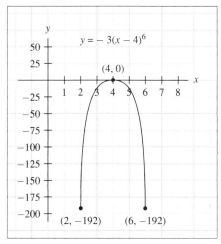

**Fig. 10-12**

(2)                                      $$y'' = 20(x - 3)^3$$

$$y''(3) = 20(0)^3 = 0 \qquad \text{test inconclusive}$$

Moving on to the third and higher-order derivatives,

$$y''' = 60(x - 3)^2 \qquad y'''(3) = 0 \qquad \text{test inconclusive}$$

$$y^{(4)} = 120(x - 3) \qquad y^{(4)}(3) = 0 \qquad \text{test inconclusive}$$

$$y^{(5)} = 120 \qquad\qquad y^{(5)}(3) = 120 > 0$$

With the first nonzero higher-order derivative an odd-numbered derivative, $y$ is at an inflection point. See Fig. 10-13.

## SKETCHING CURVES

**10.10.** From the information below, describe and then draw a rough sketch of the function around the point indicated.

(*a*)   $f(4) = 2,\ f'(4) = 3,\ f''(4) = 5$

With $f(4) = 2$, the function passes through the point $(4, 2)$. With $f'(4) = 3 > 0$, the function is increasing; with $f''(4) = 5 > 0$, the function is convex. See Fig. 10-14(*a*).

(*b*)   $f(5) = 4,\ f'(5) = -6,\ f''(5) = -9$

With $f(5) = 4$, the function passes through the point $(5, 4)$. Since $f'(5) = -6$, the function is decreasing; with $f''(5) = -9$, it is concave. See Fig. 10-14(*b*).

(*c*)   $f(4) = 5,\ f'(4) = 0,\ f''(4) = -7$

The function passes through the point $(4, 5)$. From $f'(4) = 0$, we know that the function is at a relative plateau at $x = 4$; and from $f''(4) = -7$, we know it is concave at $x = 4$. Hence it is at a relative maximum. See Fig. 10-14(*c*).

(*d*)   $f(2) = 4,\ f'(2) = -5,\ f''(2) = 6$

The function passes through the point $(2, 4)$ where it is decreasing and convex. See Fig. 10-14(*d*).

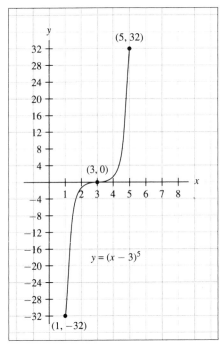

**Fig. 10-13**

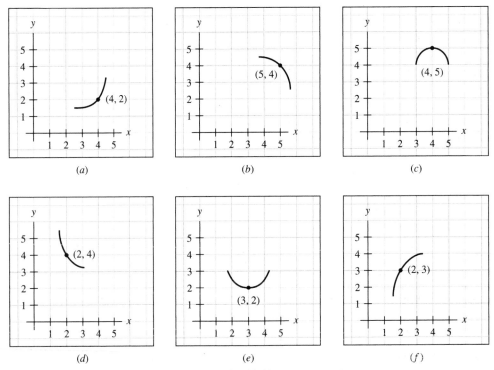

**Fig. 10-14**

(e)   $f(3) = 2, f'(3) = 0, f''(3) = 8$

The function passes through the point (3, 2), where it is at a relative plateau and convex. The function, therefore, is at a relative minimum at (3, 2). See Fig. 10-14(e).

(f)   $f(2) = 3, f'(2) = 7, f''(2) = -9$

The function passes through the point (2, 3), where it is increasing and concave. See Fig. 10-14(f).

**10.11.** (a) Find the critical values, (b) test for concavity to determine relative maxima or minima, (c) check for inflection points, (d) evaluate the function at the critical values and inflection points, and (e) graph the function, given

$$f(x) = x^3 - 18x^2 + 81x - 58$$

(a)                    $f'(x) = 3x^2 - 36x + 81 = 3(x^2 - 12x + 27) = 0$

$$f'(x) = 3(x - 3)(x - 9) = 0$$

$$x = 3 \qquad x = 9 \qquad \text{critical values}$$

(b)                    $f''(x) = 6x - 36$

$$f''(3) = 6(3) - 36 = -18 < 0 \qquad \text{concave, relative maximum}$$

$$f''(9) = 6(9) - 36 = 18 > 0 \qquad \text{convex, relative minimum}$$

(c)                    $f'' = 6x - 36 = 0$

$$x = 6 \qquad \text{possible inflection point}$$

With $f''(6) = 0$ and concavity changing between $x = 3$ and $x = 9$, as seen in part (b), there is an inflection point at $x = 6$.

(d)                $f(3) = (3)^3 - 18(3)^2 + 81(3) - 58 = 50 \qquad (3, 50) \qquad \text{relative maximum}$

$$f(6) = (6)^3 - 18(6)^2 + 81(6) - 58 = -4 \qquad (6, -4) \qquad \text{inflection point}$$

$$f(9) = (9)^3 - 18(9)^2 + 81(9) - 58 = -58 \qquad (9, -58) \qquad \text{relative minimum}$$

(e)   See Fig. 10-15.

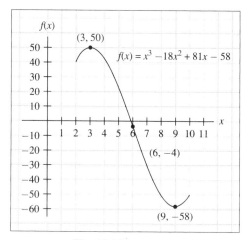

**Fig. 10-15**

**10.12.** Redo Problem 10.11, given $f(x) = -2x^3 + 12x^2 + 72x - 70$.

(a)          $f'(x) = -6x^2 + 24x + 72 = -6(x^2 - 4x - 12) = 0$

$f'(x) = -6(x + 2)(x - 6) = 0$

$x = -2 \quad x = 6 \quad$ critical values

(b)          $f''(x) = -12x + 24$

$f''(-2) = -12(-2) + 24 = 48 > 0 \quad$ convex, relative minimum

$f''(6) = -12(6) + 24 = -48 < 0 \quad$ concave, relative maximum

(c)    $f''(x) = -12x + 24 = 0$

$x = 2 \quad$ inflection point at $x = 2$

(d)          $f(-2) = -150 \quad (-2, -150) \quad$ relative minimum

$f(2) = 106 \quad (2, 106) \quad$ inflection point

$f(6) = 362 \quad (6, 362) \quad$ relative maximum

(e)    See Fig. 10-16.

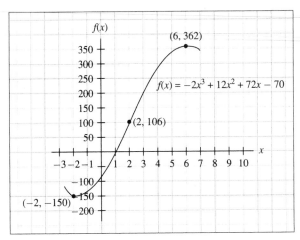

**Fig. 10-16**

**10.13.** Redo Problem 10.11, given $f(x) = (3 - x)^4$

(a)    $f'(x) = 4(3 - x)^3(-1) = -4(3 - x)^3 = 0$

$x = 3 \quad$ critical value

(b)          $f''(x) = 12(3 - x)^2$

$f''(3) = 12(3 - 3)^2 = 0 \quad$ test inconclusive

Continuing on, as explained in Section 10.7,

$$f'''(x) = -24(3 - x)$$

$$f'''(3) = -24(3 - 3) = 0 \qquad \text{test inconclusive}$$

$$f^{(4)}(x) = 24$$

$$f^{(4)}(3) = 24 > 0 \qquad \text{relative minimum}$$

With the first nonzero higher-order derivative even-numbered and greater than 0, $f(x)$ is minimized at $x = 3$.

(c)    There is no inflection point.

(d)    $$f(3) = 0 \qquad (3, 0) \qquad \text{relative minimum}$$

(e)    See Fig. 10-17.

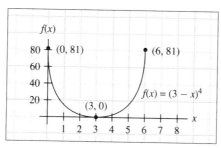

**Fig. 10-17**

**10.14.** Redo Problem 10.11, given $f(x) = (2x - 8)^3$

(a)    $$f'(x) = 3(2x - 8)^2(2) = 6(2x - 8)^2 = 0$$

$$x = 4 \qquad \text{critical value}$$

(b)    $$f''(x) = 12(2x - 8)(2) = 24(2x - 8)$$

$$f''(4) = 24(8 - 8) = 0 \qquad \text{test inconclusive}$$

Continuing on to successively higher-order derivatives,

$$f''' = 48$$

$$f'''(4) = 48 \neq 0$$

(c)    As explained in Section 10.7, with the first nonzero higher-order derivative an odd-numbered derivative, the function is at an inflection point at $x = 4$. With an inflection point at the only critical value, there is no relative maximum or minimum.

(d)    $$f(4) = 0 \qquad (4, 0) \qquad \text{inflection point}$$

Testing for concavity to the left ($x = 3$) and right ($x = 5$) of the inflection point,

$$f''(3) = 24[2(3) - 8] = -48 < 0 \qquad \text{concave}$$

$$f''(5) = 24[2(5) - 8] = 48 > 0 \qquad \text{convex}$$

(e)    See Fig. 10-18.

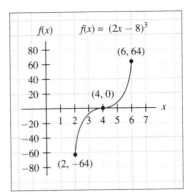

**Fig. 10-18**

## OPTIMIZATION

**10.15.** Optimize the following quadratic and cubic functions by (1) finding the critical value(s) at which the function is optimized and (2) testing the second-order condition to distinguish between a relative maximum or minimum.

(a)  $y = 9x^2 + 126x - 74$

    (1)   Take the first derivative, set it equal to zero, and solve for $x$ to find the critical value(s).

$$y' = 18x + 126 = 0$$

$$x = -7 \qquad \text{critical value}$$

    (2)   Take the second derivative, evaluate it at the critical value, and check the sign for a relative maximum or minimum.

$$y'' = 18$$

$$y''(-7) = 18 > 0 \qquad \text{convex, relative minimum}$$

(b)  $y = -5x^2 + 90x - 73$

    (1)               $y' = -10x + 90 = 0$

$$x = 9 \qquad \text{critical value}$$

    (2)               $y'' = -10$

$$y''(9) = -10 < 0 \qquad \text{concave, relative maximum}$$

(c)  $y = x^3 + 6x^2 - 96x + 23$

    (1)               $y' = 3x^2 + 12x - 96 = 0$

$$y' = 3(x^2 + 4x - 32) = 0$$

$$y' = 3(x - 4)(x + 8) = 0$$

$$x = 4 \qquad x = -8 \qquad \text{critical values}$$

    (2)               $y'' = 6x + 12$

$$y''(4) = 6(4) + 12 = 36 > 0 \qquad \text{convex, relative minimum}$$

$$y''(-8) = 6(-8) + 12 = -36 < 0 \qquad \text{concave, relative maximum}$$

(d)   $y = -3x^3 + 40.5x^2 - 162x + 39$

(1)
$$y' = -9x^2 + 81x - 162 = 0$$
$$y' = -9(x^2 - 9x + 18)$$
$$y' = -9(x - 3)(x - 6) = 0$$

$x = 3$     $x = 6$     critical values

(2)
$$y'' = -18x + 81$$

$y''(3) = -18(3) + 81 = 27 > 0$     convex, relative minimum

$y''(6) = -18(6) + 81 = -27 < 0$     concave, relative maximum

**10.16.** Optimize the following higher-order polynomial functions, using the same procedure as in Problem 10.15.

(a)   $y = 2x^4 - 8x^3 - 40x^2 + 79$

(1)
$$y' = 8x^3 - 24x^2 - 80x = 0$$
$$y' = 8x(x^2 - 3x - 10) = 0$$
$$y' = 8x(x + 2)(x - 5) = 0$$

$x = 0$     $x = -2$     $x = 5$     critical values

(2)
$$y'' = 24x^2 - 48x - 80$$

$y''(-2) = 24(-2)^2 - 48(-2) - 80 = 112 > 0$     convex, relative minimum

$y''(0) = 24(0)^2 - 48(0) - 80 = -80 < 0$     concave, relative maximum

$y''(5) = 24(5)^2 - 48(5) - 80 = 280 > 0$     convex, relative minimum

(b)   $y = -5x^4 + 20x^3 + 280x^2 - 19$

(1)
$$y' = -20x^3 + 60x^2 + 560x = 0$$
$$y' = -20x(x^2 - 3x - 28) = 0$$
$$y' = -20x(x + 4)(x - 7) = 0$$

$x = 0$     $x = -4$     $x = 7$     critical values

(2)
$$y'' = -60x^2 + 120x + 560$$

$y''(-4) = -60(-4)^2 + 120(-4) + 560 = -880 < 0$     concave, relative maximum

$y''(0) = -60(0)^2 + 120(0) + 560 = 560 > 0$     convex, relative minimum

$y''(7) = -60(7)^2 + 120(7) + 560 = -1540 < 0$     concave, relative maximum

(c)   $y = (7 - 2x)^4$

(1)
$$y' = 4(7 - 2x)^3(-2) = -8(7 - 2x)^3 = 0$$
$$7 - 2x = 0 \qquad x = 3.5 \qquad \text{critical value}$$

(2)
$$y'' = -24(7 - 2x)^2(-2) = 48(7 - 2x)^2$$

$y''(3.5) = 48[7 - 2(3.5)]^2 = 48(0)^2 = 0$     test inconclusive

Continuing on, as explained in Section 10.7 and Problem 10.6,

$$y''' = 96(7 - 2x)(-2) = -192(7 - 2x)$$

$$y'''(3.5) = -192(0) = 0 \quad \text{test inconclusive}$$

$$y^{(4)} = 384$$

$$y^{(4)}(3.5) = 384 > 0 \quad \text{convex, relative minimum}$$

(d)  $y = -2(x + 19)^4$

(1)
$$y' = -8(x + 19)^3 = 0$$

$$x + 19 = 0 \quad x = -19 \quad \text{critical value}$$

(2)
$$y'' = -24(x + 19)^2$$

$$y''(-19) = -24(-19 + 19)^2 = 0 \quad \text{test inconclusive}$$

$$y''' = -48(x + 19)$$

$$y'''(-19) = -48(0) = 0 \quad \text{test inconclusive}$$

$$y^{(4)} = -48$$

$$y^{(4)}(-19) = -48 < 0 \quad \text{concave, relative maximum}$$

## MARGINAL AND AVERAGE CONCEPTS

**10.17.** Find (1) the marginal and (2) the average functions for each of the following total functions. Evaluate them at $Q = 2$ and $Q = 4$.

(a)
$$\text{TC} = Q^2 + 9Q + 16$$

(1)  $\text{MC} = \dfrac{d\text{TC}}{dQ} = 2Q + 9$

(2)  $\text{AC} = \dfrac{\text{TC}}{Q} = Q + 9 + \dfrac{16}{Q}$

$\text{MC}(2) = 2(2) + 9 = 13$

$\text{AC}(2) = (2) + 9 + \dfrac{16}{2} = 19$

$\text{MC}(4) = 2(4) + 9 = 17$

$\text{AC}(4) = (4) + 9 + \dfrac{16}{4} = 17$

*Note:* When finding the average function, be sure to divide the constant term by $Q$.

(b)
$$\text{TR} = 24Q - Q^2$$

(1)  $\text{MR} = \dfrac{dTR}{dQ} = 24 - 2Q$

(2)  $\text{AR} = \dfrac{\text{TR}}{Q} = 24 - Q$

$\text{MR}(2) = 24 - 2(2) = 20$

$\text{AR}(2) = 24 - 2 = 22$

$\text{MR}(4) = 24 - 2(4) = 16$

$\text{AR}(4) = 24 - 4 = 20$

(c)
$$\pi = -Q^2 + 75Q - 12$$

(1)  $M\pi = \dfrac{d\pi}{dQ} = -2Q + 75$

(2)  $A\pi = \dfrac{\pi}{Q} = -Q + 75 - \dfrac{12}{Q}$

$M\pi(2) = -2(2) + 75 = 71$

$A\pi(2) = -(2) + 75 - \dfrac{12}{2} = 67$

$M\pi(4) = -2(4) + 75 = 67$

$A\pi(4) = -(4) + 75 - \dfrac{12}{4} = 68$

**10.18.** Find the MR functions associated with each of the following demand functions and evaluate them at $Q = 20$ and $Q = 40$.

(a)
$$P = -0.1Q + 25$$

To find the MR function, given simply a demand function, first find the TR function in terms of $Q$ and then take its derivative with respect to $Q$.

$$TR = PQ = (-0.1Q + 25)Q = -0.1Q^2 + 25Q$$

$$MR = dTR/dQ = -0.2Q + 25$$

and
$$MR(20) = -0.2(20) + 25 = 21$$

$$MR(40) = -0.2(40) + 25 = 17$$

(b)
$$P = -0.5Q + 48$$

$$TR = (-0.5Q + 48)Q = -0.5Q^2 + 48Q$$

$$MR = dTR/dQ = -Q + 48$$

and
$$MR(20) = -(20) + 48 = 28$$

$$MR(40) = -(40) + 48 = 8$$

(c)
$$Q = -4P + 240$$

If the demand function is given as a function of $P$, first find the inverse function and then derive the TR function. Finding the inverse,

$$P = -0.25Q + 60$$

then
$$TR = (-0.25Q + 60)Q = -0.25Q^2 + 60Q$$

$$MR = -0.5Q + 60$$

and
$$MR(20) = -0.5(20) + 60 = 50$$

$$MR(40) = -0.5(40) + 60 = 40$$

**10.19.** For each of the following consumption functions, use the derivative to find the marginal propensity to consume $MPC = dC/dY$.

(a)   $C = bY + C_0$

   $MPC = dC/dY = b$

(b)   $C = 0.85Y + 1250$

   $MPC = dC/dY = 0.85$

## OPTIMIZING BUSINESS AND ECONOMIC FUNCTIONS

**10.20.** Maximize the following total revenue TR and total profit $\pi$ functions by (1) finding the critical value(s), (2) testing the second-order conditions, and (3) calculating the maximum TR or $\pi$.

(a)   $TR = 96Q - 2Q^2$

   (1)   $TR' = 96 - 4Q = 0$

   $$Q = 24 \quad \text{critical value}$$

   (2)   $TR'' = -4 < 0$      concave, relative maximum
   (3)   $TR = 96(24) - 2(24)^2 = 1152$

(b)   $\pi = -Q^2 + 25Q - 12$

   (1)   $\pi' = -2Q + 25 = 0$

   $$Q = 12.5 \quad \text{critical value}$$

(2)   $\pi'' = -2 < 0$      concave, relative maximum
(3)   $\pi = -(12.5)^2 + 25(12.5) - 12 = 144.25$

(c)   $\pi = -\frac{1}{3}Q^3 - 7.5Q^2 + 450Q - 200$

(1)   $\pi' = -Q^2 - 15Q + 450 = 0$                                        $(10.1)$

$-1(Q^2 + 15Q - 450) = 0$                                    $(10.2)$

$(Q - 15)(Q + 30) = 0$

$Q = 15 \qquad Q = -30$      critical values

(2)   $\pi'' = -2Q - 15$

$\pi''(15) = -2(15) - 15 = -45 < 0$      concave, relative maximum

$\pi''(-30) = -2(-30) - 15 = 45 > 0$      convex, relative minimum

Negative critical values will subsequently be ignored as having no economic significance.

(3)   $\pi = -\frac{1}{3}(15)^3 - 7.5(15)^2 + 450(15) - 200 = 3737.50$

*Note:* In testing the second-order conditions, as in step 2, always take the second derivative from the original first derivative $(10.1)$ before any *negative* number has been factored out. Taking the second derivative from the first derivative after factoring out a negative, as in $(10.2)$, will reverse the second-order conditions and suggest that the function is maximized at $Q = -30$ and minimized at $Q = 15$. Test it yourself.

(d)   $\pi = -2Q^3 - 15Q^2 + 3000Q - 1200$

(1)   $\pi' = -6Q^2 - 30Q + 3000 = 0$

$= -6(Q^2 + 5Q - 500) = 0$

$= -6(Q - 20)(Q + 25) = 0$

$Q = 20 \qquad Q = -25$      critical values

(2)      $\pi'' = -12Q - 30$

$\pi''(20) = -12(20) - 30 = -270 < 0$      concave, relative maximum

(3)   $\pi = -2(20)^3 - 15(20)^2 + 3000(20) - 1200 = 36,800$

**10.21.** From each of the following total cost TC functions, (1) find the average cost AC function, (2) the critical value at which AC is minimized, and (3) the minimum average cost.

(a)   $TC = 2Q^3 - 12Q^2 + 225Q$

(1)   $AC = \dfrac{TC}{Q} = \dfrac{2Q^3 - 12Q^2 + 225Q}{Q} = 2Q^2 - 12Q + 225$

(2)   $AC' = 4Q - 12 = 0 \qquad Q = 3$

$AC'' = 4 > 0$      convex, relative minimum

(3)   $AC(3) = 2(3)^2 - 12(3) + 225 = 207$

(b)   $TC = Q^3 - 16Q^2 + 450Q$

(1)   $AC = \dfrac{Q^3 - 16Q^2 + 450Q}{Q} = Q^2 - 16Q + 450$

(2)   $AC' = 2Q - 16 = 0 \qquad\qquad Q = 8$

$AC'' = 2 > 0$      convex, relative minimum

(3)   $AC = (8)^2 - 16(8) + 450 = 386$

**10.22.** Given the following total revenue TR and total cost TC functions for different firms, maximize profit $\pi$ for the firms as follows: (1) set up the profit function $\pi = TR - TC$, (2) find the critical value(s) where $\pi$ is at a relative extremum and test the second-order condition, and (3) calculate the maximum profit.

(a)   $TR = 440Q - 3Q^2 \qquad TC = 14Q + 225$

(1)   $\pi = 440Q - 3Q^2 - (14Q + 225)$

$\pi = -3Q^2 + 426Q - 225$

(2)   $\pi' = -6Q + 426 = 0$

$Q = 71 \qquad$ critical value

$\pi'' = -6 < 0 \qquad$ concave, relative maximum

(3)   $\pi = -3(71)^2 + 426(71) - 225 = 14,898$

(b)   $TR = 800Q - 7Q^2 \qquad TC = 2Q^3 - Q^2 + 80Q + 150$

(1)   $\pi = 800Q - 7Q^2 - (2Q^3 - Q^2 + 80Q + 150)$

$= -2Q^3 - 6Q^2 + 720Q - 150$

(2)   $\pi' = -6Q^2 - 12Q + 720 = 0 \qquad\qquad\qquad\qquad (10.3)$

$= -6(Q^2 + 2Q - 120) = 0$

$= -6(Q - 10)(Q + 12) = 0$

$Q = 10 \qquad Q = -12 \qquad$ critical values

Taking the second derivative directly from (10.3), as explained in Problem 10.20 (c), and ignoring all negative critical values,

$$\pi'' = -12Q - 12$$

$$\pi''(10) = -12(10) - 12 = -132 < 0 \qquad \text{concave, relative maximum}$$

(3)   $\pi = -2(10)^3 - 6(10)^2 + 720(10) - 150 = 4450$

(c)   $TR = 3200Q - 9Q^2, \qquad TC = Q^3 - 1.5Q^2 + 50Q + 425$

(1)   $\pi = 3200Q - 9Q^2 - (Q^3 - 1.5Q^2 + 50Q + 425)$

$= -Q^3 - 7.5Q^2 + 3150Q - 425$

(2)   $\pi' = -3Q^2 - 15Q + 3150 = 0$

$= -3(Q^2 + 5Q - 1050) = 0$

$= -3(Q - 30)(Q + 35) = 0$

$Q = 30 \qquad Q = -35 \qquad$ critical values

$\pi'' = -6Q - 15$

$\pi''(30) = -6(30) - 15 = -195 < 0 \qquad$ concave, relative maximum

(3)   $\pi = -(30)^3 - 7.5(30)^2 + 3150(30) - 425 = 60,325$

(d)   $TR = 500Q - 11Q^2, \qquad TC = 3Q^3 - 2Q^2 + 68Q + 175$

(1)   $\pi = 500Q - 11Q^2 - (3Q^3 - 2Q^2 + 68Q + 175)$

$= -3Q^3 - 9Q^2 + 432Q - 175$

(2)  $\quad \pi' = -9Q^2 - 18Q + 432 = 0$

$\qquad = -9(Q^2 + 2Q - 48) = 0$

$\qquad = -9(Q - 6)(Q + 8) = 0$

$\qquad\qquad Q = 6 \quad Q = -8 \quad$ critical values

$\quad \pi'' = -18Q - 18$

$\quad \pi''(6) = -18(6) - 18 = -126 < 0 \quad$ concave, relative maximum

(3)  $\quad \pi = -3(6)^3 - 9(6)^2 + 432(6) - 175 = 1445$

**10.23.** Show that marginal revenue (MR) must equal marginal cost (MC) at the profit-maximizing level of output.

By definition, $\pi = \text{TR} - \text{TC}$. Taking the derivative and setting it equal to zero, since $d\pi/dQ$ must equal zero for $\pi$ to be at a maximum,

$$\frac{d\pi}{dQ} = \frac{d\text{TR}}{dQ} - \frac{d\text{TC}}{dQ} = 0$$

Solving algebraically, recalling that $d\text{TR}/dQ = \text{MR}$, $d\text{TC}/dQ = \text{MC}$,

$$\frac{d\text{TR}}{dQ} = \frac{d\text{TC}}{dQ}$$

$$\text{MR} = \text{MC} \qquad \text{Q.E.D.}$$

**10.24.** Use the MR = MC method to (a) maximize profit $\pi$, and (b) check the second-order conditions, given

$$\text{TR} = 800Q - 7Q^2 \qquad \text{TC} = 2Q^3 - Q^2 + 80Q + 150$$

(a) $\qquad\qquad \text{MR} = \text{TR}' = 800 - 14Q \qquad \text{MC} = \text{TC}' = 6Q^2 - 2Q + 80$

Equating MR = MC to maximize profits,

$$800 - 14Q = 6Q^2 - 2Q + 80$$

Solving for $Q$ by moving everything to the right,

$$6Q^2 + 12Q - 720 = 0$$

$$6(Q^2 + 2Q - 120) = 0$$

$$6(Q - 10)(Q + 12) = 0$$

$$Q = 10 \qquad Q = -12 \quad \text{critical values}$$

(b) $\qquad\qquad\qquad \text{TR}'' = -14 \qquad \text{TC}'' = 12Q - 2$

Since $\pi = \text{TR} - \text{TC}$ and the objective is to maximize $\pi$, be sure to *subtract* TC'' *from* TR'' or you will reverse the second-order conditions and select the wrong critical value.

$$\pi'' = \text{TR}'' - \text{TC}''$$

$$= -14 - 12Q + 2 = -12Q - 12$$

$$\pi''(10) = -12(10) - 12 = -132 < 0 \quad \text{concave, relative maximum}$$

Compare these results with Problem 10.22 (b).

**10.25.** Redo Problem 10.24, given

$$\text{TR} = 500Q - 11Q^2 \qquad \text{TC} = 3Q^3 - 2Q^2 + 68Q + 175$$

(a) $\qquad\qquad \text{MR} = \text{TR}' = 500 - 22Q \qquad \text{MC} = \text{TC}' = 9Q^2 - 4Q + 68$

Equating MR = MC,

$$500 - 22Q = 9Q^2 - 4Q + 68$$

Moving everything to the right,

$$9Q^2 + 18Q - 432 = 0$$

$$9(Q^2 + 2Q - 48) = 0$$

$$9(Q - 6)(Q + 8) = 0$$

$$Q = 6 \qquad Q = -8 \qquad \text{critical values}$$

(b)                                    $$\text{TR}'' = -22 \qquad \text{TC}'' = 18Q - 4$$

Being sure to subtract TC″ from TR″ so as not to reverse the second-order conditions, recalling that

$$\pi'' = \text{TR}'' - \text{TC}''$$

$$\pi'' = -22 - 18Q + 4 = -18Q - 18$$

$$\pi''(6) = -18(6) - 18 = -126 < 0 \qquad \text{concave, relative maximum}$$

Compare these results with Problem 10.22 (d).

## RELATIONSHIP BETWEEN FUNCTIONS AND GRAPHS

**10.26.** A *total product curve* TP of an input is a production function which allows the amounts of one input (say, labor) to vary while holding the other inputs (capital, land) constant. Given $\text{TP} = 562.5L^2 - 15L^3$, sketch a graph of the (a) total profit (TP), (b) average profit (AP), and (c) marginal product (MP) curves and (d) explain the relationships among them.

(a)   Testing the first-order conditions to find whatever critical values exist,

$$\text{TP}' = 1125L - 45L^2 = 0$$

$$45L(25 - L) = 0$$

$$L = 0 \qquad L = 25 \qquad \text{critical values}$$

Checking the second-order conditions,

$$\text{TP}'' = 1125 - 90L$$

$$\text{TP}''(0) = 1125 > 0 \qquad \text{convex, relative minimum}$$

$$\text{TP}''(25) = -1125 < 0 \qquad \text{concave, relative maximum}$$

Looking for inflection points,

$$\text{TP}'' = 1125 - 90L = 0$$

$$L = 12.5$$

For $L < 12.5$,

$$\text{TP}'' > 0 \qquad \text{convex}$$

For $L > 12.5$,

$$\text{TP}'' < 0 \qquad \text{concave}$$

See Fig. 10-19.

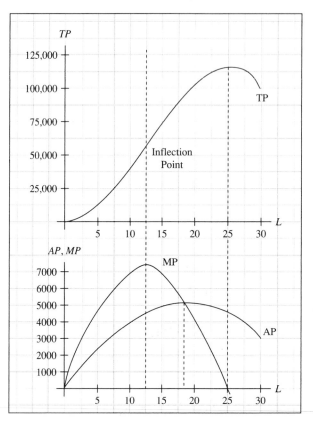

**Fig. 10-19**

(b)    $\text{AP}_L = \dfrac{\text{TP}}{L} = \dfrac{562.5L^2 - 15L^3}{L} = 562.5L - 15L^2$

Checking for extrema,

$$\text{AP}'_L = 562.5 - 30L = 0$$

$$L = 18.75$$

$$\text{AP}''_L = -30 < 0 \qquad \text{concave, relative maximum}$$

$$\text{AP}_L(18.75) \approx 5273.43$$

(c)    Recalling that $\text{MP}_L = \text{TP}' = 1125L - 45L^2$ and optimizing,

$$\text{MP}'_L = 1125 - 90L = 0$$

$$L = 12.5$$

$$\text{MP}''_L = -90 < 0 \qquad \text{concave, relative maximum}$$

$$\text{MP}_L(12.5) = 7031.25$$

(d)    We note that (1) $\text{MP}_L$ increases when TP is convex and hence increasing at an increasing rate, is at a maximum when TP is at an inflection point, and decreases when TP is concave and consequently increasing at a decreasing rate. (2) TP increases over the whole range where $\text{MP}_L$ is positive, is at a maximum where $\text{MP}_L = 0$, and decreases when $\text{MP}_L$ is negative. (3) $\text{AP}_L$ increases when $\text{MP}_L > \text{AP}_L$, decreases when $\text{MP}_L < \text{AP}_L$, and is at a maximum when $\text{MP}_L = \text{AP}_L$, which is also the point where the slope of a straight line from the origin to the TP curve is tangent to the TP curve.

**10.27.** Given $y = ax^2 + bx + c$, prove that the coordinates of the vertex of a parabola are the ordered pair $[-b/2a, (4ac - b^2)/4a]$, as was asserted in Section 3.6.

The vertex is the maximum or minimum point of a parabola. Finding the critical point $x$ at which the given function is optimized,

$$y' = 2ax + b = 0$$

$$x = -\frac{b}{2a}$$

Then substituting $x = -b/2a$ into the function to solve for $y$,

$$y = a\left(-\frac{b}{2a}\right)^2 + b\left(-\frac{b}{2a}\right) + c$$

$$y = \frac{b^2}{4a} - \frac{b^2}{2a} + c = \frac{b^2 - 2b^2 + 4ac}{4a} = \frac{4ac - b^2}{4a} \qquad \text{Q. E. D}$$

# Supplementary Problems

## INCREASING AND DECREASING FUNCTIONS

**10.28.** Indicate whether the following functions are increasing or decreasing at the indicated points:
  (a) $f(x) = 5x^2 - 4x - 89$ at $x = 3$       (b) $f(x) = 8x^3 + 6x^2 - 15x$ at $x = -4$
  (c) $f(x) = -4x^3 + 2x^2 - 7x + 9$ at $x = 5$       (d) $f(x) = 2x^3 - 17x^2 + 5x + 127$ at $x = 2$

## CONCAVITY AND CONVEXITY

**10.29.** Determine whether the following functions are concave or convex at $x = 3$.
  (a) $f(x) = 7x^2 + 19x - 24$       (b) $f(x) = 5x^3 - 81x^2 + 11x + 97$
  (c) $f(x) = (4 - 9x)^3$       (d) $f(x) = (3x^4 - 7)^2$

## RELATIVE EXTREMA AND INFLECTION POINTS

**10.30.** Find all relative extrema and inflection points for each of the following functions and sketch the graphs on your own.
  (a) $f(x) = 4x^3 - 48x^2 - 240x + 29$       (b) $f(x) = (x + 6)^3$
  (c) $f(x) = -2x^3 + 24x^2 + 288x - 35$       (d) $f(x) = (2x + 7)^4$
  (e) $f(x) = -5x^3 + 135x^2 - 675x + 107$       (f) $f(x) = -3(4x - 11)^6$
  (g) $f(x) = 3x^3 - 58.5x^2 + 360x - 49$       (h) $f(x) = 2(5x - 13)^5$

## OPTIMIZATION OF FUNCTIONS

**10.31.** Optimize the following functions and test the second-order conditions at the critical points to distinguish between a relative maximum and a relative minimum.
  (a) $y = 8x^2 - 208x + 73$       (b) $y = -12x^2 + 528x + 95$
  (c) $y = -2x^3 + 69x^2 + 1260x - 9$       (d) $y = x^3 + 27x^2 + 96x - 47$
  (e) $y = 3x^3 - 45x^2 - 675x + 13$       (f) $y = -4x^3 + 186x^2 - 1008x + 25$
  (g) $y = x^4 + 36x^3 + 280x^2 - 79$       (h) $y = 4x^4 - 48x^3 - 288x^2 + 129$

## MARGINAL AND AVERAGE FUNCTIONS

**10.32.**    Find the marginal and average functions for each of the following total functions:

    (*a*)  $TR = -Q^2 + 96Q$                         (*b*)  $TR = -3Q^2 + 198Q$

    (*c*)  $TC = Q^2 + 3Q + 55$                      (*d*)  $TC = 0.5Q^2 + 2Q + 69$

**10.33.**    Find the marginal revenue functions associated with each of the following demand functions:

    (*a*)  $P = -0.3Q + 228$                       (*b*)  $P = -8Q + 1465$

    (*c*)  $P = -2.5Q + 145$                       (*d*)  $P = -4Q + 875$

## OPTIMIZATION OF BUSINESS AND ECONOMIC FUNCTIONS

**10.34.**    Find the critical points at which each of the following total revenue TR functions and profit $\pi$ functions is maximized. Check the second-order conditions on your own.

    (*a*)  $TR = -3Q^2 + 210Q$               (*b*)  $\pi = -2.5Q^2 + 315Q - 16$

    (*c*)  $\pi = -3Q^3 - 18Q^2 + 2880Q - 125$     (*d*)  $\pi = -2Q^3 - 9Q^2 + 1080Q - 48$

**10.35.**    Find the critical points at which each of the following average cost AC functions is minimized. Check the second-order conditions on your own.

    (*a*)  $AC = 3Q^2 - 18Q + 585$           (*b*)  $AC = 2.25Q^2 - 27Q + 768$

**10.36.**    Find the critical points at which profit $\pi$ is maximized for each of the following firms given the total revenue TR and total cost TC functions. Check the second-order conditions on your own.

    (*a*)  $TR = 520Q - 2Q^2$     $TC = 28Q + 176$      (*b*)  $TR = 693Q - 4Q^2$     $TC = 33Q + 125$

    (*c*)  $TR = 985Q - 17Q^2$     $TC = 5Q^3 - 2Q^2 + 40Q + 167$

    (*d*)  $TR = 3875Q - 27Q^2$     $TC = 4Q^3 - 3Q^2 + 35Q + 223$

    (*e*)  $TR = 1526Q - 9Q^2$     $TC = Q^3 - 1.5Q^2 + 26Q + 127$

    (*f*)  $TR = 4527Q - 18Q^2$     $TC = 2Q^3 - 3Q^2 + 27Q + 324$

# Answers to Supplementary Problems

**10.28.**    (*a*)  increasing     (*b*)  increasing     (*c*)  decreasing     (*d*)  decreasing.

**10.29.**    (*a*)  convex     (*b*)  concave     (*c*)  concave     (*d*)  convex

**10.30.**    (*a*)  $x = -2$ relative maximum, $x = 4$ inflection point, $x = 10$ relative minimum

    (*b*)  $x = -6$ inflection point; no relative maximum or minimum

    (*c*)  $x = -4$ relative minimum, $x = 4$ inflection point, $x = 12$ relative maximum

    (*d*)  $x = -3.5$ relative minimum; no relative maximum or inflection point

    (*e*)  $x = 3$ relative minimum, $x = 9$ inflection point, $x = 15$ relative maximum

    (*f*)  $x = 2.75$ relative maximum; no relative minimum or inflection point

    (*g*)  $x = 5$ relative maximum, $x = 6.5$ inflection point, $x = 8$ relative minimum

    (*h*)  $x = 2.6$ inflection point; no relative maximum or minimum

**10.31.**    (*a*)  $x = 13$ relative minimum     (*b*)  $x = 22$ relative maximum

    (*c*)  $x = -7$ relative minimum, $x = 30$ relative maximum

    (*d*)  $x = -16$ relative maximum, $x = -2$ relative minimum

    (*e*)  $x = -5$ relative maximum, $x = 15$ relative minimum

    (*f*)  $x = 3$ relative minimum, $x = 28$ relative maximum

    (*g*)  $x = -20$ relative minimum, $x = -7$ relative maximum, $x = 0$ relative minimum

    (*h*)  $x = -3$ relative minimum, $x = 0$ relative maximum, $x = 12$ relative minimum

**10.32.**    (a)   MR $= -2Q + 96$, AR $= -Q + 96$     (b)   MR $= -6Q + 198$, AR $= -3Q + 198$
            (c)   MC $= 2Q + 3$, AC $= Q + 3 + 55/Q$     (d)   MC $= Q + 2$, AC $= 0.5Q + 2 + 69/Q$

**10.33.**    (a)   MR $= -0.6Q + 228$     (b)   MR $= -16Q + 1465$
            (c)   MR $= -5Q + 145$         (d)   MR $= -8Q + 875$

**10.34.**    (a)  $Q = 35$     (b)  $Q = 63$     (c)  $Q = 16$     (d)  $Q = 12$

**10.35.**    (a)  $Q = 3$     (b)  $Q = 6$

**10.36.**    (a)  $Q = 123$     (b)  $Q = 82.5$     (c)  $Q = 7$
            (d)  $Q = 16$     (e)  $Q = 20$     (f)  $Q = 25$

# Chapter 11

## Exponential and Logarithmic Functions

### 11.1 EXPONENTIAL FUNCTIONS

Previous chapters dealt mainly with *power functions*, such as $y = x^a$, in which a variable base $x$ is raised to a constant exponent $a$. In this chapter we introduce an important new function in which a constant base $a$ is raised to a variable exponent $x$. It is called an exponential function and is defined as

$$y = a^x, \qquad a > 0, \qquad a \neq 1$$

Commonly used to express rates of growth and decay, such as interest compounding and depreciation, exponential functions have the following general properties. Given $y = a^x$, $a > 0$, and $a \neq 1$,

1. The domain of the function is the set of all real numbers; the range of the function is the set of all positive real numbers, that is, for all $x$, even $x \leq 0$, $y > 0$.
2. For $a > 1$, the function is increasing and convex; for $0 < a < 1$, the function is decreasing and convex.
3. At $x = 0$, $y = 1$, independently of the base.

See Example 1 and Problems 11.1 and 11.2; for a review of exponents, see Section 1.1 and Problem 1.1.

**EXAMPLE 1.** Given (a) $y = 3^x$ and (b) $y = 3^{-x} = \left(\frac{1}{3}\right)^x$, the above properties of exponential functions can readily be seen from the tables and graphs of the functions in Fig. 11-1. More complicated exponential functions are estimated with the help of the $\boxed{y^x}$ key on pocket calculators.

(a) $y = 3^x$

| $x$ | $y$ |
|----|----|
| $-3$ | $\frac{1}{27}$ |
| $-2$ | $\frac{1}{9}$ |
| $-1$ | $\frac{1}{3}$ |
| $0$ | $1$ |
| $1$ | $3$ |
| $2$ | $9$ |
| $3$ | $27$ |

(b) $y = 3^{-x} = \left(\frac{1}{3}\right)^x$

| $x$ | $y$ |
|----|----|
| $-3$ | $27$ |
| $-2$ | $9$ |
| $-1$ | $3$ |
| $0$ | $1$ |
| $1$ | $\frac{1}{3}$ |
| $2$ | $\frac{1}{9}$ |
| $3$ | $\frac{1}{27}$ |

### 11.2 LOGARITHMIC FUNCTIONS

Interchanging the variables of an exponential function $f$ defined by $y = a^x$ gives rise to a new function $g$ defined by $x = a^y$ such that any ordered pair of numbers in $f$ will also be found in $g$ in reverse order. For example, if $f(2) = 9$, then $g(9) = 2$; if $f(3) = 27$, then $g(27) = 3$. The new function $g$, the *inverse* of the exponential function $f$, is called a *logarithmic function with base a*. Instead of $x = a^y$, the logarithmic function with base $a$ is more commonly written

$$y = \log_a x \qquad a > 0, \qquad a \neq 1$$

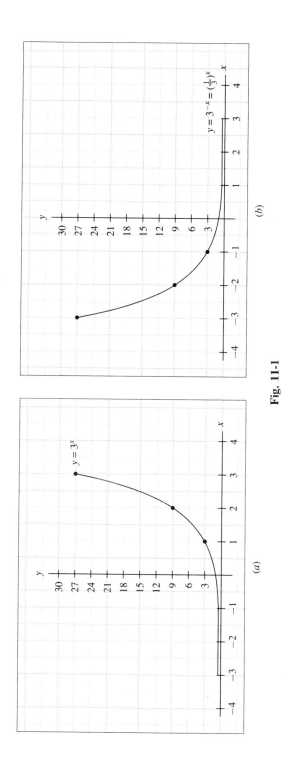

Fig. 11-1

$\text{Log}_a x$ is the exponent to which $a$ must be raised to get $x$. Any positive number except 1 may serve as the base for a logarithm. The *common logarithm of x*, written $\log_{10} x$ or simply $\log x$, is the exponent to which 10 must be raised to get $x$. Logarithms have the following properties. Given $y = \log_a x$, $a > 0$, $a \neq 1$,

1. The domain of the function is the set of all positive real numbers; the range is the set of all real numbers—the exact opposite of its inverse function, the exponential function.
2. For base $a > 1$, $g(x)$ is increasing and concave. For $0 < a < 1$, $g(x)$ is decreasing and convex.
3. At $x = 1$, $y = 0$, independently of the base.

See Examples 2 to 4 and Problems 11.5 and 11.6.

**EXAMPLE 2.**    A graph of two functions $f$ and $g$ in which $x$ and $y$ are interchanged, such as $y = (\frac{1}{2})^x$ and $x = (\frac{1}{2})^y$ in Fig. 11-2, shows that one function is a *mirror image* of the other along the 45° line $y = x$, such that if $f(x) = y$, then $g(y) = x$. Recall that $x = (\frac{1}{2})^y$ is equivalent to and generally expressed as $y = \log_{1/2} x$.

$\quad$ (a) $y = (\frac{1}{2})^x$ $\qquad$ (b) $y = \log_{1/2} x \longleftrightarrow x = (\frac{1}{2})^y$

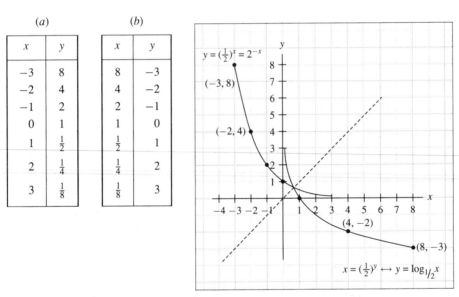

**Fig. 11-2**

**EXAMPLE 3.**    Knowing that the common logarithm of $x$ is the power to which 10 must be raised to get $x$, it follows that

$$\log 10 = 1 \text{ since } 10^{(1)} = 10 \qquad \log 1 = 0 \text{ since } 10^0 = 1$$

$$\log 100 = 2 \text{ since } 10^2 = 100 \qquad \log .1 = -1 \text{ since } 10^{-1} = .1$$

$$\log 1000 = 3 \text{ since } 10^3 = 1000 \qquad \log .01 = -2 \text{ since } 10^{-2} = .01$$

**EXAMPLE 4.**    For numbers that are exact powers of the base, logs are easily calculated without the aid of calculators.

$$\log_8 64 = 2 \text{ since } 8^2 = 64 \qquad \log_3 81 = 4 \text{ since } 3^4 = 81$$

$$\log_{25} 5 = \tfrac{1}{2} \text{ since } 25^{1/2} = 5 \qquad \log_{27} 3 = \tfrac{1}{3} \text{ since } 27^{1/3} = 3$$

$$\log_2 \tfrac{1}{16} = -4 \text{ since } 2^{-4} = \tfrac{1}{16} \qquad \log_3 \tfrac{1}{9} = -2 \text{ since } 3^{-2} = \tfrac{1}{9}$$

For numbers that are not exact powers of the base, log tables or calculators are needed. See Section 1.7.9.

## 11.3  PROPERTIES OF EXPONENTS AND LOGARITHMS

Assuming $a, b > 0$; $a, b \neq 1$, and $x$ and $y$ any real numbers,

1. $a^x \cdot a^y = a^{x+y}$          4. $(a^x)^y = a^{xy}$

2. $1/a^x = a^{-x}$                   5. $a^x \cdot b^x = (ab)^x$

3. $a^x / a^y = a^{x-y}$              6. $a^x / b^x = (a/b)^x$

For $a$, $x$ and $y$ positive real numbers, $n$ a real number, and $a \neq 1$,

1. $\log_a(xy) = \log_a x + \log_a y$          3. $\log_a x^n = n \log_a x$

2. $\log_a(x/y) = \log_a x - \log_a y$         4. $\log_a(\sqrt[n]{x}) = \frac{1}{n}(\log_a x)$

Properties of exponents were treated in Section 1.1 and Problem 1.1. Properties of logarithms are treated in Example 5 and Problems 11.12 to 11.16.

**TABLE 11.1**

| $x$ | $\log x$ | $x$ | $\log x$ | $x$ | $\log x$ | $x$ | $\log x$ |
|---|---|---|---|---|---|---|---|
| 1 | 0.0000 | 6 | 0.7782 | 11 | 1.0414 | 16 | 1.2041 |
| 2 | 0.3010 | 7 | 0.8451 | 12 | 1.0792 | 27 | 1.4314 |
| 3 | 0.4771 | 8 | 0.9031 | 13 | 1.1139 | 36 | 1.5563 |
| 4 | 0.6021 | 9 | 0.9542 | 14 | 1.1461 | 49 | 1.6902 |
| 5 | 0.6990 | 10 | 1.0000 | 15 | 1.1761 | 64 | 1.8062 |

**EXAMPLE 5.**   The problems below are kept simple and solved by means of the logarithms given in Table 11.1 in order to illustrate the properties of logarithms.

($a$)   $x = 3 \cdot 12$

$\log x = \log 3 + \log 12$

$\log x = 0.4771 + 1.0792$

$\log x = 1.5563$

$x = 36$

($b$)   $x = 64 \div 4$

$\log x = \log 64 - \log 4$

$\log x = 1.8062 - 0.6021$

$\log x = 1.2041$

$x = 16$

($c$)   $x = 7^2$

$\log x = 2 \log 7$

$\log x = 2(0.8451)$

$\log x = 1.6902$

$x = 49$

($d$)   $x = \sqrt[3]{27}$

$\log x = \frac{1}{3} \log 27$

$\log x = \frac{1}{3}(1.4314)$

$\log x = 0.4771$

$x = 3$

## 11.4  NATURAL EXPONENTIAL AND LOGARITHMIC FUNCTIONS

The most frequently used base for exponential and logarithmic functions is the irrational number $e$. Expressed mathematically,

$$e = \lim_{n \to \infty} \left(1 + \frac{1}{n}\right)^n \approx 2.71828 \qquad (11.1)$$

Exponential functions to base $e$ are called *natural exponential functions* and are written $y = e^x$; logarithmic functions to base $e$ are termed *natural logarithmic functions* and are expressed as $y = \log_e x$ or, more frequently, $\ln x$. The $\ln x$ is simply the exponent or power to which $e$ must be raised to get $x$.

As with other exponential and logarithmic functions to a common base, one function is the inverse of the other, such that the ordered pair $(a, b)$ will belong to the set of $e^x$ if and only if $(b, a)$ belongs to the set of $\ln x$. Natural exponential and logarithmic functions follow the same rules as other exponential and logarithmic functions and are estimated with the help of tables or the $\boxed{e^x}$ and $\boxed{\ln x}$ keys on pocket calculators. See Sections 1.7.10 and 1.7.12 and Problems 11.3, 11.4, and 11.6.

## 11.5  SOLVING NATURAL EXPONENTIAL AND LOGARITHMIC FUNCTIONS

Since natural exponential functions and natural logarithmic functions are inverses of each other, one is generally helpful in solving the other. Mindful that the $\ln x$ signifies the power to which $e$ must be raised to get $x$, it follows that

1.  $e$ raised to the natural log of a constant $(a > 0)$, a variable $(x > 0)$, or a function of a variable $[f(x) > 0]$, must equal that constant, variable, or function of the variable:

$$e^{\ln a} = a \qquad e^{\ln x} = x \qquad e^{\ln f(x)} = f(x) \qquad\qquad (11.2)$$

2.  Conversely, taking the natural log of $e$ raised to the power of a constant, variable, or function of a variable must also equal that constant, variable, or function of the variable:

$$\ln e^a = a \qquad \ln e^x = x \qquad \ln e^{f(x)} = f(x) \qquad\qquad (11.3)$$

See Example 6 and Problems 11.18 and 11.19.

**EXAMPLE 6.**   The principles set forth in $(11.2)$ and $(11.3)$ are used below to solve the given equations for $x$.

(a)   $7.5e^{x-3} = 150$

Solve algebraically for $e^{x-3}$,

$$7.5e^{x-3} = 150$$

$$e^{x-3} = 20$$

Take the natural log of both sides to eliminate $e$.

$$\ln e^{x-3} = \ln 20$$

From $(11.3)$,                    $$x - 3 = \ln 20$$

$$x = \ln 20 + 3$$

Enter 20 on your calculator and press the $\boxed{\ln x}$ key to find $\ln 20 = 2.99573$. Then, substitute and solve.

$$x = 2.99573 + 3 = 5.99573$$

(b)   $4 \ln x + 9 = 30.6$

Solve algebraically for $\ln x$,

$$4 \ln x = 21.6$$

$$\ln x = 5.4$$

Set both sides of the equation as exponents of $e$ to eliminate the natural log expression,

$$e^{\ln x} = e^{5.4}$$

From $(11.2)$,                    $$x = e^{5.4}$$

Enter 5.4 on your calculator and press the $\boxed{e^x}$ key to find $e^{5.4} = 221.40642$ and substitute.

$$x = 221.40642$$

*Note*: On many calculators the $\boxed{e^x}$ key is the inverse (shift, or second function) of the $\boxed{\ln x}$ key and to activate the $\boxed{e^x}$ key one must first press the $\boxed{\text{INV}}$ ($\boxed{\text{Shift}}$, or $\boxed{\text{2dF}}$) key followed by the $\boxed{\ln x}$ key.

## 11.6  LOGARITHMIC TRANSFORMATION OF NONLINEAR FUNCTIONS

Linear algebra and regression analysis using ordinary or two-stage least squares, common tools in economic analysis, assume linear functions or equations. Some important nonlinear functions, such as Cobb-Douglas production functions, are easily converted to linear functions by logarithmic transformation. Given a generalized Cobb-Douglas production function

$$q = AK^\alpha L^\beta$$

it is clear from the properties of logarithms that

$$\ln q = \ln A + \alpha \ln K + \beta \ln L \qquad (11.4)$$

which is log linear. A linear transformation of a Cobb-Douglas production function, such as in (11.4), furthermore has the nice added feature that estimates for $\alpha$ and $\beta$ provide direct measures of the *output elasticity* of $K$ and $L$, respectively.

## 11.7  DERIVATIVES OF NATURAL EXPONENTIAL AND LOGARITHMIC FUNCTIONS

The rules of differentiation for natural exponential and logarithmic functions are presented below, illustrated in Examples 7 and 8, and treated in Problems 11.20 to 11.23. For proof of the rules and more detailed treatment of the topic, see Dowling, *Schaum's Outline of Introduction to Mathematical Economics*, Chapter 9.

1) Given $f(x) = e^{g(x)}$, where $g(x)$ is a differentiable function of $x$, the derivative is

$$f'(x) = e^{g(x)} \cdot g'(x) \qquad (11.5)$$

In brief, the derivative of a natural exponential function is equal to the original natural exponential function times the derivative of the exponent.

2) Given $f(x) = \ln[g(x)]$, where $g(x)$ is positive and differentiable, the derivative is

$$f'(x) = \frac{1}{g(x)} \cdot g'(x) = \frac{g'(x)}{g(x)} \qquad (11.6)$$

**EXAMPLE 7.**    The derivatives of each of the natural exponential functions below are found as follows:

1)   $f(x) = e^x$. Let $g(x) = x$, then $g'(x) = 1$. Substituting in (11.5),

$$f'(x) = e^x \cdot 1 = e^x$$

The derivative of $e^x$ is simply $e^x$, the original function itself. The only base for which this is true is $e$. This explains why $e$ is so frequently used as the base of an exponential function.

2)   $f(x) = e^{7x-4}$. Letting $g(x) = 7x - 4$, then $g'(x) = 7$. Substituting in (11.5),

$$f'(x) = e^{7x-4} \cdot 7 = 7e^{7x-4}$$

3)   $f(x) = e^{5x^3}$. Here $g(x) = 5x^3$, so $g'(x) = 15x^2$. From (11.5),

$$f'(x) = e^{5x^3} \cdot 15x^2 = 15x^2 e^{5x^3}$$

See also Problems 11.20 and 11.21.

**EXAMPLE 8.**   Finding the derivative of a natural logarithmic function is demonstrated below:

1)   $f(x) = \ln x$. Let $g(x) = x$, then $g'(x) = 1$. Substituting in $(11.6)$,

$$f'(x) = \frac{g'(x)}{g(x)} = \frac{1}{x}$$

The derivative of $\ln x$ is simply $1/x$.

2)   $y = \ln 9x^4$. Here $g(x) = 9x^4$, so $g'(x) = 36x^3$. From $(11.6)$,

$$y' = \frac{1}{9x^4} \cdot 36x^3 = \frac{4}{x}$$

3)   $f(x) = \ln(8x^2 - 13)$. Letting $g(x) = 8x^2 - 13$, then $g'(x) = 16x$. From $(11.6)$,

$$f'(x) = \frac{g'(x)}{g(x)} = \frac{16x}{8x^2 - 13}$$

See also Problems 11.22 and 11.23.

## 11.8   INTEREST COMPOUNDING

Interest compounding is generally expressed in terms of exponential functions. A person lending out a principal $P$ at interest rate $r$ under terms of annual compounding, for example, will have a value $A$ at the end of one year equal to

$$A_1 = P + rP = P(1 + r)$$

At the end of 2 years, this person will have $A_1$ plus interest on $A_1$,

$$A_2 = A_1 + rA_1$$

$$A_2 = P(1 + r) + r[P(1 + r)]$$

Factoring out $P(1 + r)$,

$$A_2 = [P(1 + r)][1 + r] = P(1 + r)^2$$

At the end of $t$ years, following the same procedure,

$$A_t = P(1 + r)^t \tag{11.7}$$

If interest is compounded $m$ times a year with the person receiving $(r/m)$ interest $m$ times during the course of the year, at the end of $t$ years,

$$A_t = P \left(1 + \frac{r}{m}\right)^{mt} \tag{11.8}$$

And if interest is compounded continuously so that $m \to \infty$,

$$A_t = \lim_{m \to \infty} P \left(1 + \frac{r}{m}\right)^{mt}$$

Moving $P$ and multiplying the exponent by $r/r$,

$$A_t = P \lim_{m \to \infty} \left(1 + \frac{r}{m}\right)^{(m/r)rt}$$

Letting $m/r = n$,

$$A_t = P \lim_{n \to \infty} \left(1 + \frac{1}{n}\right)^{nrt}$$

But from Section 11.4, $\lim_{n \to \infty}[1 + (1/n)]^n = e$ Substituting,

$$A_t = Pe^{rt} \tag{11.9}$$

**EXAMPLE 9.**    The value $A$ of a principal $P = \$1000$ set out at an interest rate $r = 6$ percent for time $t = 5$ years when compounded (a) annually, (b) quarterly, (c) monthly, and (d) continuously is found below.

(a)   Substituting the relevant values in (11.7) and using the $\boxed{y^x}$ key of a calculator to estimate exponential functions throughout,

$$A = 1000(1 + .06)^5$$

$$A = 1000(1.33823) = \$1338.23$$

(b)   Substituting in (11.8) where $m = 4$ for quarterly compounding,

$$A = 1000\left(1 + \frac{.06}{4}\right)^{4(5)}$$

$$A = 1000(1 + .015)^{20}$$

$$A = 1000(1.34686) = \$1346.86$$

(c)   Now substituting $m = 12$ in (11.8) for monthly compounding,

$$A = 1000\left(1 + \frac{.06}{12}\right)^{12(5)}$$

$$A = 1000(1 + .005)^{60}$$

$$A = 1000(1.34885) = \$1348.85$$

(d)   Finally, using (11.9) for continuous compounding,

$$A = 1000e^{(.06)5} = 1000e^{0.3}$$

$$A = 1000(1.34986) = \$1349.86$$

See also Problems 11.24 to 11.31; for discounting problems, see Problems 11.32 to 11.34.

## 11.9   ESTIMATING GROWTH RATES FROM DATA POINTS

Given two sets of data for a function—profits, sales, or assets—growing consistently over time, annual growth rates can be measured and a natural exponential function estimated through a system of simultaneous equations. For example, if a mutual fund's assets equal $48 million in 1983 and $98.6 million in 1993, let $t = 0$ for the base year 1983, then $t = 10$ for 1993. Express each of the two sets of data points in terms of a natural exponential function $S = Pe^{rt}$, recalling that $e^{(0)} = 1$.

$$48.0 = Pe^{r(0)} = P \tag{11.10}$$

$$98.6 = Pe^{r(10)} \tag{11.11}$$

Substituting $P = 48.0$ from (11.10) in (11.11) and simplifying algebraically,

$$98.6 = 48.0e^{10r}$$

$$2.05417 = e^{10r}$$

Using (11.3) and taking the natural log of both sides,

$$\ln 2.05417 = \ln e^{10r} = 10r$$

$$0.71987 = 10r$$

$$r = 0.07199 \approx 7.2\%$$

Substituting,                                    $$S = 48.0e^{0.072t}$$

With $r = 0.072$, the rate of continuous growth per year is 7.2 percent. See Example 10 and Problems 11.35 to 11.39.

**EXAMPLE 10.**    Given the original information above, an ordinary exponential function for growth in terms of $S = P(1+r)^t$ can also be estimated directly from the data.

Setting the data in ordinary exponential form,

$$48.0 = P(1+r)^0 = P \qquad\qquad (11.12)$$

$$98.6 = P(1+r)^{10} \qquad\qquad (11.13)$$

Substituting $P = 48.0$ from ($11.12$) in ($11.13$) and simplifying,

$$98.6 = 48.0(1+r)^{10}$$

$$2.05417 = (1+r)^{10}$$

Taking the common log of both sides,

$$\log 2.05417 = 10 \log(1+r)$$

$$\tfrac{1}{10}(0.31264) = \log(1+r)$$

$$\log(1+r) = 0.03126$$

Setting both sides as exponents of 10, analogously to ($11.2$),

$$1+r = 10^{0.03126} = 1.07463$$

$$r = 1.07463 - 1 = 0.07463 \approx 7.46\%$$

Substituting,                          $$S = 48.0(1 + 0.0746)^t$$

In terms of annual compounding, assets grow by 7.46 percent a year.

# Solved Problems

## GRAPHS

**11.1.** Make a schedule for each of the following exponential functions with base $a > 1$ and then graph them on the same grid to convince yourself that (1) the functions never equal zero, (2) they all pass through (0, 1), and (3) they are all positively sloped and convex:

      (*a*)   $y = 2^x$      (*b*)   $y = 3^x$      (*c*) $y = 4^x$

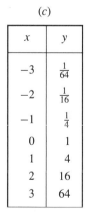

| (*a*) | | (*b*) | | (*c*) | |
|---|---|---|---|---|---|
| $x$ | $y$ | $x$ | $y$ | $x$ | $y$ |
| $-3$ | $\frac{1}{8}$ | $-3$ | $\frac{1}{27}$ | $-3$ | $\frac{1}{64}$ |
| $-2$ | $\frac{1}{4}$ | $-2$ | $\frac{1}{9}$ | $-2$ | $\frac{1}{16}$ |
| $-1$ | $\frac{1}{2}$ | $-1$ | $\frac{1}{3}$ | $-1$ | $\frac{1}{4}$ |
| $0$ | $1$ | $0$ | $1$ | $0$ | $1$ |
| $1$ | $2$ | $1$ | $3$ | $1$ | $4$ |
| $2$ | $4$ | $2$ | $9$ | $2$ | $16$ |
| $3$ | $8$ | $3$ | $27$ | $3$ | $64$ |

See Fig. 11-3.

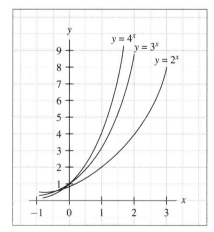

**Fig. 11-3**

**11.2.** Make a schedule for each of the following exponential functions with $0 < a < 1$ and then sketch them on the same graph to convince yourself that (1) the functions never equal zero, (2) they all pass through $(0, 1)$, and (3) they are all negatively sloped and convex:

$$(a) \quad y = \left(\tfrac{1}{2}\right)^x = 2^{-x} \qquad (b) \quad y = \left(\tfrac{1}{3}\right)^x = 3^{-x} \qquad (c) \quad y = \left(\tfrac{1}{4}\right)^x = 4^{-x}$$

|   (a)   |        |
| :-----: | :----: |
|    x    |   y    |
|   −3    |   8    |
|   −2    |   4    |
|   −1    |   2    |
|    0    |   1    |
|    1    | $\frac{1}{2}$ |
|    2    | $\frac{1}{4}$ |
|    3    | $\frac{1}{8}$ |

See Fig. 11-4.

|   (b)   |        |
| :-----: | :----: |
|    x    |   y    |
|   −3    |   27   |
|   −2    |   9    |
|   −1    |   3    |
|    0    |   1    |
|    1    | $\frac{1}{3}$ |
|    2    | $\frac{1}{9}$ |
|    3    | $\frac{1}{27}$ |

|   (c)   |        |
| :-----: | :----: |
|    x    |   y    |
|   −3    |   64   |
|   −2    |   16   |
|   −1    |   4    |
|    0    |   1    |
|    1    | $\frac{1}{4}$ |
|    2    | $\frac{1}{16}$ |
|    3    | $\frac{1}{64}$ |

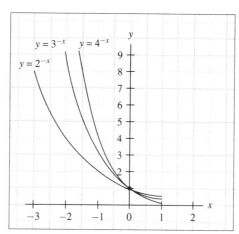

Fig. 11-4

**11.3.** Using a calculator or tables, set up a schedule for each of the following natural exponential functions $y = e^{kx}$ where $k > 0$, noting (1) the functions never equal zero, (2) they all pass through $(0, 1)$, and (3) they are all positively sloped and convex:

$$(a) \quad y = e^{0.5x}, \qquad (b) \quad y = e^x, \qquad (c) \quad y = e^{2x}$$

|   (a)   |        |
| :-----: | :----: |
|    x    |   y    |
|   −2    |  0.37  |
|   −1    |  0.61  |
|    0    |  1.00  |
|    1    |  1.65  |
|    2    |  2.72  |

See Fig. 11-5.

|   (b)   |        |
| :-----: | :----: |
|    x    |   y    |
|   −2    |  0.14  |
|   −1    |  0.37  |
|    0    |  1.00  |
|    1    |  2.72  |
|    2    |  7.39  |

|   (c)   |        |
| :-----: | :----: |
|    x    |   y    |
|   −2    |  0.02  |
|   −1    |  0.14  |
|    0    |  1.00  |
|    1    |  7.39  |
|    2    | 54.60  |

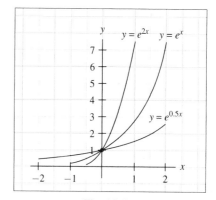

Fig. 11-5

**11.4.** Set up a schedule, rounding to two decimal places, for the following natural exponential functions $y = e^{kx}$ where $k < 0$, noting (1) the functions never equal zero, (2) they all pass through $(0, 1)$, and (3) they are all negatively sloped and convex:

$$(a) \quad y = e^{-0.5x} \qquad (b) \quad y = e^{-x} \qquad (c) \quad y = e^{-2x}$$

|     (a)     |     (b)     |     (c)     |
|:---:|:---:|:---:|

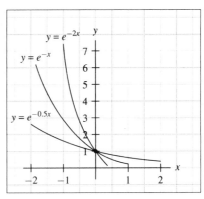

| x | y | | x | y | | x | y |
|:---:|:---:|---|:---:|:---:|---|:---:|:---:|
| −2 | 2.72 | | −2 | 7.39 | | −2 | 54.60 |
| −1 | 1.65 | | −1 | 2.72 | | −1 | 7.39 |
| 0 | 1.00 | | 0 | 1.00 | | 0 | 1.00 |
| 1 | 0.61 | | 1 | 0.37 | | 1 | 0.14 |
| 2 | 0.37 | | 2 | 0.14 | | 2 | 0.02 |

See Fig. 11-6.

**Fig. 11-6**

**11.5.** Construct a schedule and draw a graph for the following functions to show that one is the mirror image and hence the inverse of the other, noting that (1) the domain of (a) is the range of (b) and the range of (a) is the domain of (b), and (2) a logarithmic function with $a > 1$ is increasing and concave:

$$(a) \quad y = 2^x \qquad (b) \quad y = \log_2 x \longleftrightarrow x = 2^y$$

|     (a)     |     (b)     |
|:---:|:---:|

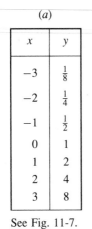  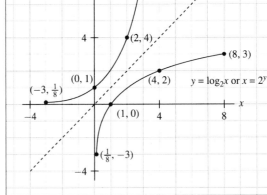

| x | y | | x | y |
|:---:|:---:|---|:---:|:---:|
| −3 | $\frac{1}{8}$ | | $\frac{1}{8}$ | −3 |
| −2 | $\frac{1}{4}$ | | $\frac{1}{4}$ | −2 |
| −1 | $\frac{1}{2}$ | | $\frac{1}{2}$ | −1 |
| 0 | 1 | | 1 | 0 |
| 1 | 2 | | 2 | 1 |
| 2 | 4 | | 4 | 2 |
| 3 | 8 | | 8 | 3 |

See Fig. 11-7.

**Fig. 11-7**

**11.6.** Given (a) $y = e^x$ and (b) $y = \ln x$, and using a calculator or tables, construct a schedule and draw a graph for each of the functions to show that one function is the mirror image or inverse of the other, noting that (1) the domain of (a) is the range of (b) while the range of (a) is the domain of (b), (2) the $\ln x$ is negative for $0 < x < 1$ and positive for $x > 1$, and (3) the $\ln x$ is an increasing function and concave downward.

(a) $y = e^x$

| $x$ | $y$ |
|-----|-----|
| $-2$ | 0.13534 |
| $-1$ | 0.36788 |
| 0 | 1.00000 |
| 1 | 2.71828 |
| 2 | 7.38906 |

See Fig. 11-8.

(b) $y = \ln x$

| $x$ | $y$ |
|-----|-----|
| 0.13534 | $-2$ |
| 0.36788 | $-1$ |
| 1.00000 | 0 |
| 2.71828 | 1 |
| 7.38906 | 2 |

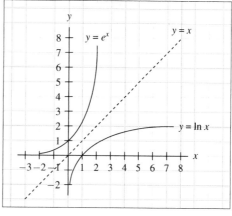

**Fig. 11-8**

## EXPONENTIAL-LOGARITHMIC CONVERSION

**11.7.** Change the following logarithms to their equivalent exponential forms:

(a)  $\log_7 49 = 2$

$49 = 7^2$

(b)  $\log_4 64 = 3$

$64 = 4^3$

(c)  $\log_9 \left(\frac{1}{9}\right) = -1$

$\frac{1}{9} = 9^{-1}$

(d)  $\log_2 \left(\frac{1}{32}\right) = -5$

$\frac{1}{32} = 2^{-5}$

(e)  $\log_{64} 8 = \frac{1}{2}$

$8 = 64^{1/2}$

(f)  $\log_{81} 3 = \frac{1}{4}$

$3 = 81^{1/4}$

(g)  $\log_a y = 5x$

$y = a^{5x}$

(h)  $\log_2 y = 4x$

$y = 2^{4x}$

**11.8.** Convert the following natural logarithms into natural exponential functions:

(a)  $\ln 24 = 3.17805$

$24 = e^{3.17805}$

(b)  $\ln 0.6 = -0.51083$

$0.6 = e^{-0.51083}$

(c)  $\ln 44 = 3.78419$

$44 = e^{3.78419}$

(d)  $\ln 4.2 = 1.43508$

$4.2 = e^{1.43508}$

(e)  $\ln y = -8x$

$y = e^{-8x}$

(f)  $\ln y = 5t - 3$

$y = e^{5t-3}$

**11.9.** Change the following exponential forms to logarithmic forms.

(a)  $36 = 6^2$

$\log_6 36 = 2$

(b)  $125 = 5^3$

$\log_5 125 = 3$

(c)  $\frac{1}{4} = 2^{-2}$

$\log_2\left(\frac{1}{4}\right) = -2$

(d)  $\frac{1}{27} = 3^{-3}$

$\log_3\left(\frac{1}{27}\right) = -3$

(e)  $3 = 81^{1/4}$

$\log_{81} 3 = \frac{1}{4}$

(f)  $12 = 144^{1/2}$

$\log_{144} 12 = \frac{1}{2}$

(g)   $64 = 16^{3/2}$

$\log_{16} 64 = \frac{3}{2}$

(h)   $27 = 81^{3/4}$

$\log_{81} 27 = \frac{3}{4}$

**11.10.** Convert the following natural exponential expressions into equivalent natural logarithmic forms:

(a)   $3.5 = e^{1.25276}$

$\ln 3.5 = 1.25276$

(b)   $26 = e^{3.25810}$

$\ln 26 = 3.25810$

(c)   $0.3 = e^{-1.20397}$

$\ln 0.3 = -1.20397$

(d)   $145 = e^{4.97673}$

$\ln 145 = 4.97673$

(e)   $y = e^{(1/4)t}$

$\ln y = \frac{1}{4}t$

(f)   $y = e^{9-t}$

$\ln y = 9 - t$

**11.11.** Solve the following for $x$, $y$, or $a$ by finding the equivalent expression:

(a)   $y = \log_{20} 400$

$400 = 20^y$

$y = 2$

(b)   $y = \log_3(\frac{1}{27})$

$\frac{1}{27} = 3^y$

$y = -3$

(c)   $\log_2 x = 4$

$x = 2^4$

$x = 16$

(d)   $\log_{256} x = \frac{3}{4}$

$x = 256^{3/4}$

$x = 64$

(e)   $\log_a 49 = 2$

$49 = a^2$

$a = 49^{1/2}$

$a = 7$

(f)   $\log_a 125 = 3$

$125 = a^3$

$a = 125^{1/3}$

$a = 5$

(g)   $\log_a 9 = \frac{2}{3}$

$9 = a^{2/3}$

$a = 9^{3/2}$

$a = 27$

(h)   $\log_a 32 = \frac{5}{3}$

$32 = a^{5/3}$

$a = 32^{3/5}$

$a = 8$

## PROPERTIES OF LOGARITHMS AND EXPONENTS

**11.12.** Use the properties of logarithms to write the following expressions as sums, differences, or products:

(a)   $\log_a(23x)$

$\log_a(23x) = \log_a 23 + \log_a x$

(b)   $\log_a(69x^3)$

$\log_a(69x^3) = \log_a 69 + 3\log_a x$

(c)   $\log_a x^4 y^5$

$\log_a x^4 y^5 = 4\log_a x + 5\log_a y$

(d)   $\log_a u^2 v^{-3}$

$\log_a u^2 v^{-3} = 2\log_a u - 3\log_a v$

(e)   $\log_a(3x/8y)$

$\log_a(3x/8y) = \log_a 3x - \log_a 8y$

$= \log_a 3 + \log_a x - (\log_a 8 + \log_a y)$

$= \log_a 3 + \log_a x - \log_a 8 - \log_a y$

(f)  $\log_a(x^9/y^4)$

$\log_a(x^9/y^4) = 9\log_a x - 4\log_a y$

(g)  $\log_a \sqrt[5]{x}$

$\log_a \sqrt[5]{x} = \frac{1}{5}\log_a x$

**11.13.** Use the properties of logarithms to write the following natural logarithmic forms as sums, differences, or products:

(a)  $\ln 49x^6$

$\ln 49x^6 = \ln 49 + 6\ln x$

(b)  $\ln x^3 y^7$

$\ln x^3 y^7 = 3\ln x + 7\ln y$

(c)  $\ln(x^9/y^2)$

$\ln(x^9/y^2) = 9\ln x - 2\ln y$

(d)  $\ln(5x/8y)$

$\ln(5x/8y) = \ln 5x - \ln 8y$

$\qquad\qquad = \ln 5 + \ln x - (\ln 8 + \ln y)$

$\qquad\qquad = \ln 5 + \ln x - \ln 8 - \ln y$

(e)  $\ln \sqrt[3]{x}$

$\ln \sqrt[3]{x} = \frac{1}{3}\ln x$

(f)  $\ln(x^4 \cdot \sqrt{y})$

$\ln(x^4 \cdot \sqrt{y}) = 4\ln x + \frac{1}{2}\ln y$

(g)  $\dfrac{8\sqrt[4]{x}}{\sqrt{y}}$

$\ln \dfrac{8\sqrt[4]{x}}{\sqrt{y}} = \ln 8 + \frac{1}{4}\ln x - \frac{1}{2}\ln y$

(h)  $\ln \sqrt{\dfrac{x^5}{y^3}}$

$\ln \sqrt{\dfrac{x^5}{y^3}} = \frac{1}{2}(5\ln x - 3\ln y)$

**11.14.** Use the properties of exponents to simplify the following exponential expressions, assuming $a, b > 0$ and $a \neq b$:

(a)  $a^x \cdot a^y$

$a^x \cdot a^y = a^{x+y}$

(b)  $a^{7x} \cdot a^{9y}$

$a^{7x} \cdot a^{9y} = a^{7x+9y}$

(c)  $\dfrac{a^{5x}}{a^{2y}}$

$\dfrac{a^{5x}}{a^{2y}} = a^{5x-2y}$

(d)  $\dfrac{a^x}{b^x}$

$\dfrac{a^x}{b^x} = \left(\dfrac{a}{b}\right)^x$

(e)  $\sqrt{a^{9x}}$

$\sqrt{a^{9x}} = (a^{9x})^{1/2} = a^{(9/2)x}$

(f)  $(a^x)^{5y}$

$(a^x)^{5y} = a^{5xy}$

**11.15.** Simplify the following natural exponential expressions.

(a)  $e^{3x} \cdot e^{8y}$

$e^{3x} \cdot e^{8y} = e^{3x+8y}$

(b)  $(e^{4x})^3$

$(e^{4x})^3 = e^{12x}$

(c)  $\dfrac{e^{7x}}{e^{4x}}$

$\dfrac{e^{7x}}{e^{4x}} = e^{7x-4x} = e^{3x}$

(d)  $\dfrac{e^{2x}}{e^{8x}}$

$\dfrac{e^{2x}}{e^{8x}} = e^{2x-8x} = e^{-6x} = \dfrac{1}{e^{6x}}$

**11.16.** Simplify the following natural logarithmic expressions:

(a)  $\ln 6 + \ln x$

$$\ln 6 + \ln x = \ln 6x$$

(b)  $\ln x^7 - \ln x^2$

$$\ln x^7 - \ln x^2 = \ln\left(\frac{x^7}{x^2}\right) = \ln x^5 = 5\ln x$$

(c)  $\ln 12 + \ln 3 - \ln 4$

$$\ln 12 + \ln 3 - \ln 4 = \ln\left(\frac{12 \cdot 3}{4}\right) = \ln 9$$

(d)  $\ln 5 - \ln 2x + \ln 8$

$$\ln 5 - \ln 2x + \ln 8 = \ln\left(\frac{5 \cdot 8}{2x}\right) = \ln\left(\frac{20}{x}\right)$$

(e)  $\frac{1}{2}\ln 36$

$$\tfrac{1}{2}\ln 36 = \ln 36^{1/2} = \ln 6$$

(f)  $4\ln\left(\frac{1}{3}\right)$

$$4\ln\tfrac{1}{3} = \ln\left(\tfrac{1}{3}\right)^4 = \ln\tfrac{1}{81}$$

(g)  $\frac{1}{2}\ln 49 + 3\ln 4$

$$\tfrac{1}{2}\ln 49 + 3\ln 4 = \ln 49^{1/2} + \ln 4^3 = \ln(7 \cdot 64) = \ln 448$$

(h)  $2\ln 9 - \frac{1}{4}\ln 81$

$$2\ln 9 - \tfrac{1}{4}\ln 81 = \ln 9^2 - \ln 81^{1/4} = \ln\left(\tfrac{81}{3}\right) = \ln 27$$

**11.17.** Simplify each of the following exponential expressions:

(a)  $e^{2\ln x}$

$$e^{2\ln x} = e^{\ln x^2}$$

From (11.2),
$$e^{\ln x^2} = x^2$$

(b)  $e^{3\ln x + 4\ln y}$

$$e^{3\ln x + 4\ln y} = e^{\ln x^3} \cdot e^{\ln y^4} = x^3 y^4$$

(c)  $e^{1/2\ln 9x}$

$$e^{1/2\ln 9x} = e^{\ln(9x)^{1/2}} = (9x)^{1/2} = \sqrt{9x} = 3\sqrt{x}$$

(d)  $e^{5\ln x - 8\ln y}$

$$e^{5\ln x - 8\ln y} = \frac{e^{\ln x^5}}{e^{\ln y^8}} = \frac{x^5}{y^8}$$

**11.18.** Use the techniques from Section 11.5 to solve the following natural exponential functions for $x$:

(a)  $7e^{3x} = 630$

Solving algebraically for $e^{3x}$,
$$e^{3x} = 90$$

Taking the natural log of both sides to eliminate $e$,

$$\ln e^{3x} = \ln 90$$

From (*11.3*),                                      $3x = \ln 90$

Entering 90 on a calculator and pressing the $\boxed{\ln x}$ key as explained in Section 1.7.10, $\ln 90 = 4.49981 \approx 4.5$. Thus

$$3x = 4.5$$

$$x = 1.5$$

(*b*)  $3e^{2x+1.2} = 4018.29$

Solving for $e^{2x+1.2}$,                   $e^{2x+1.2} = 1339.43$

Taking the natural log,                      $\ln e^{2x+1.2} = \ln 1339.43$

From (*11.3*),                               $2x + 1.2 = 7.19999 \approx 7.2$

$$2x + 1.2 = 7.2 \qquad x = 3$$

(*c*)  $\frac{1}{4}e^{x^2} = 2.37194$

Solving for $e^{x^2}$,                        $e^{x^2} = 9.48776$

Taking the natural log,                       $x^2 = 2.25$

$$x = \pm 1.5$$

**11.19.**  Using the techniques of Section 11.5, solve the following natural logarithmic functions for $x$:

(*a*)  $7 \ln x - 2.6 = 10$

Solving algebraically for $\ln x$,

$$7 \ln x = 12.6, \qquad \ln x = 1.8$$

Setting both sides of the equation as exponents of $e$ to eliminate the natural log,

$$e^{\ln x} = e^{1.8}$$

From (*11.2*),                               $x = e^{1.8}$

Using a calculator, as explained in Section 1.7.12,

$$x = 6.04965 \approx 6.05$$

(*b*)  $\ln(x - 1.51)^3 = 1.2$

Simplifying with the laws of logs, then solving for $\ln x$,

$$3 \ln(x - 1.51) = 1.2$$

$$\ln(x - 1.51) = 0.4$$

Setting both sides of the equation as exponents of $e$,

$$e^{\ln(x-1.51)} = e^{0.4}$$

From (*11.2*),                               $x - 1.51 \approx 1.49$

$$x = 3$$

(*c*)  $\ln \sqrt{x + 2.24} = 2.14$

Simplifying and solving,                     $\frac{1}{2} \ln(x + 2.24) = 2.14$

$$\ln(x + 2.24) = 4.28$$

Making both sides exponents of $e$,

$$e^{\ln(x+2.24)} = e^{4.28}$$

From (11.2),                              $x + 2.24 \approx 72.24$

$$x = 70$$

## DERIVATIVES OF NATURAL EXPONENTIAL AND LOGARITHMIC FUNCTIONS

**11.20.** Differentiate the following natural exponential functions, using the rule found in (11.5):
$d/dx[e^{g(x)}] = e^{g(x)} \cdot g'(x)$.

(a)  $f(x) = e^{6x}$                         (b)  $y = e^{-8x}$

    Let $g(x) = 6x$,  then                    Let $g(x) = -8x$,  then

    $g'(x) = 6$,  and from (11.5),            $g'(x) = -8$,  and from (11.5),

    $f'(x) = e^{6x} \cdot 6$                     $y' = e^{-8x} \cdot -8$

        $= 6e^{6x}$                            $= -8e^{-8x}$

(c)  $f(x) = e^{3x^2}$                         (d)  $y = e^{4x-9}$

    $f'(x) = e^{3x^2} \cdot 6x$                   $y' = e^{4x-9} \cdot 4$

        $= 6xe^{3x^2}$                         $= 4e^{4x-9}$

(e)  $y = 5e^{7-x^3}$                         (f)  $y = \frac{1}{3}e^{(1/2)x^4}$

    $y' = 5e^{7-x^3} \cdot -3x^2$                $y' = \frac{1}{3}e^{(1/2)x^4} \cdot \frac{1}{2}(4x^3)$

        $= -15x^2 e^{7-x^3}$                   $= \frac{2}{3}x^3 e^{(1/2)x^4}$

**11.21.** Combine rules to differentiate the following functions.

(a)  $f(x) = 7xe^{2x}$                        (b)  $y = 2x^4 e^{5x}$

    By the product rule,                         $y' = 2x^4(5e^{5x}) + e^{5x}(8x^3)$

    $f'(x) = 7x(2e^{2x}) + e^{2x}(7)$            $= 10x^4 e^{5x} + 8x^3 e^{5x}$

        $= 14xe^{2x} + 7e^{2x}$                $= 2x^3 e^{5x}(5x + 4)$

        $= 7e^{2x}(2x + 1)$

(c)  $y = (e^{-4x})^3$

    By the generalized power function rule,

$$y' = 3 \cdot (e^{-4x})^2 \cdot (-4e^{-4x})$$
$$= -12(e^{-4x})^3 = -12e^{-12x}$$

Note: By the properties of exponents, $(e^{-4x})^3 = e^{-12x}$ and the derivative of $e^{-12x}$ is, of course, simply $-12e^{-12x}$.

(d)  $f(x) = (e^{4x} + e^{-3x})^5$

$$f'(x) = 5 \cdot (e^{4x} + e^{-3x})^4 \cdot (4e^{4x} - 3e^{-3x})$$
$$= (20e^{4x} - 15e^{-3x})(e^{4x} + e^{-3x})^4$$

(e)   $f(x) = \dfrac{e^{-9x}}{1 - 9x}$

By the quotient rule,

$$f'(x) = \frac{(1 - 9x)(-9e^{-9x}) - (e^{-9x})(-9)}{(1 - 9x)^2} = \frac{81xe^{-9x}}{(1 - 9x)^2}$$

(f)   $y = \dfrac{e^{7x} - 1}{e^{7x} + 1}$

$$y' = \frac{(e^{7x} + 1)(7e^{7x}) - (e^{7x} - 1)(7e^{7x})}{(e^{7x} + 1)^2} = \frac{14e^{7x}}{(e^{7x} + 1)^2}$$

**11.22.** Differentiate the following natural logarithmic functions, using the rule found in (11.6):
$d/dx[\ln g(x)] = \dfrac{1}{g(x)} \cdot g'(x)$.

(a)   $y = \ln 7x^3$

Let $g(x) = 7x^3$,  then

$g'(x) = 21x^2$ and from (11.6),

$y' = \dfrac{1}{7x^3}(21x^2) = 3x^{-1} = \dfrac{3}{x}$

(b)   $y = \ln(4x + 9)$

Let $g(x) = 4x + 9$,  then

$g'(x) = 4$ and from (11.6),

$y' = \dfrac{1}{4x + 9}(4) = \dfrac{4}{4x + 9}$

(c)   $y = \ln(8x^2 + 3)$

$y' = \dfrac{1}{8x^2 + 3}(16x)$

$= \dfrac{16x}{8x^2 + 3}$

(d)   $y = \ln(x^2 + 7x + 15)$

$y' = \dfrac{1}{x^2 + 7x + 15}(2x + 7)$

$= \dfrac{2x + 7}{x^2 + 7x + 15}$

(e)   $y = \ln 18x$

$y' = \dfrac{1}{18x}(18) = \dfrac{1}{x}$

(f)   $y = 18 \ln x$

$y' = 18 \cdot \dfrac{1}{x}(1) = \dfrac{18}{x}$

Note how a multiplicative constant within the log expression drops out in differentiation as in (e), but a multiplicative constant outside the log expression as in (f) remains.

**11.23.** Combine rules to differentiate the following functions.

(a)   $y = \ln^2 x = (\ln x)^2$

By the generalized power function rule,

$$y' = 2(\ln x)\left(\frac{1}{x}\right) = \frac{2 \ln x}{x}$$

(b)   $y = \ln^2 4x^3 = (\ln 4x^3)^2$

By the generalized power function rule again,

$$y' = 2(\ln 4x^3)\left(\frac{1}{4x^3}\right)(12x^2) = \frac{6 \ln 4x^3}{x}$$

(c)   $y = \ln^2(21x + 8) = [\ln(21x + 8)]^2$

$$y' = 2[\ln(21x + 8)]\left(\frac{1}{21x + 8}\right)(21) = \frac{42 \ln(21x + 8)}{21x + 8}$$

(d)   $y = \ln(9x + 4)^2 \neq [\ln(9x + 4)]^2$

Let $g(x) = (9x + 4)^2$, then $g'(x) = 2(9x + 4)(9) = 18(9x + 4)$. Substituting in ($11.6$),

$$y' = \frac{1}{(9x + 4)^2}[18(9x + 4)] = \frac{18}{9x + 4}$$

(e)   $y = x^5 \ln x^3$

By the product rule,

$$y' = x^5 \left(\frac{1}{x^3}\right)(3x^2) + \ln x^3 (5x^4)$$

$$y' = 3x^4 + 5x^4 \ln x^3 = 3x^4 + 15x^4 \ln x = 3x^4(1 + 5\ln x)$$

(f)   $y = \dfrac{x}{\ln x}$

By the quotient rule,

$$y' = \frac{\ln x (1) - x (1/x)}{(\ln x)^2} = \frac{\ln x - 1}{\ln^2 x}$$

(g)   $y = e^{2x} \ln 3x$

By the product rule,

$$y' = e^{2x}\left(\frac{1}{3x} \cdot 3\right) + (\ln 3x)(e^{2x} \cdot 2)$$

$$y' = e^{2x}\left(\frac{1}{x}\right) + 2e^{2x} \ln 3x = e^{2x}\left(\frac{1}{x} + 2\ln 3x\right)$$

## INTEREST COMPOUNDING

**11.24.** Find the value $A$ of a principal $P = \$100$ set out at an interest rate $r = 12$ percent for time $t = 1$ year when compounded ($a$) annually, ($b$) semiannually, ($c$) quarterly, and ($d$) continuously; ($e$) distinguish between the nominal and the effective rate of interest. Use a calculator for exponential expressions, as explained in Section 1.7.11 and Problems 1.23 to 1.26.

(a)   From ($11.7$),          $A = 100(1 + .12)^1 = \$112.00$

(b)   From ($11.8$), with $m = 2$,     $A = 100\left(1 + \dfrac{.12}{2}\right)^{2(1)}$

$$A = 100(1 + .06)^2$$

$$A = 100(1.1236) = \$112.36$$

(c)   From ($11.8$), with $m = 4$,     $A = 100\left(1 + \dfrac{.12}{4}\right)^{4(1)}$

$$A = 100(1 + .03)^4$$

$$A = 100(1.1255) = \$112.55$$

(d)   From ($11.9$),          $A = 100e^{(0.12)(1)} = 100e^{.12}$

$$A = 100(1.1275) = \$112.75$$

(e)   In all four instances the stated or *nominal interest rate* is the same, namely, 12 percent; the actual interest earned, however, varies according to the type of compounding. The *effective interest rate* on multiple compoundings is the comparable rate the bank would have to pay if interest were paid only once a year, specifically, 12.36 percent to equal semiannual compounding, 12.55 percent to equal quarterly compounding, and 12.75 percent to equal continuous compounding.

**11.25.** Find the value $A$ of a principal $P = \$3000$ set out at an interest rate $r = 8$ percent for time $t = 6$ years when compounded (*a*) annually, (*b*) semiannually, (*c*) quarterly, and (*d*) continuously.

(*a*) From (*11.7*),

$$A = 3000(1 + .08)^6$$

$$A = 3000(1.5868743) = \$4760.62$$

(*b*) From (*11.8*), with $m = 2$,

$$A = 3000\left(1 + \frac{.08}{2}\right)^{2(6)}$$

$$A = 3000(1 + .04)^{12}$$

$$A = 3000(1.6010322) = \$4803.10$$

(*c*) From (*11.8*), with $m = 4$,

$$A = 3000\left(1 + \frac{.08}{4}\right)^{4(6)}$$

$$A = 3000(1 + .02)^{24}$$

$$A = 3000(1.6084373) = \$4825.31$$

(*d*) From (*11.9*),

$$A = 3000e^{(0.08)6} = 3000e^{0.48}$$

$$A = 3000(1.6160744) = \$4848.22$$

**11.26.** Redo Problem 11.25, given $P = \$10,000$, $r = 9$ percent, and $t = 3$.

(*a*)

$$A = 10,000(1 + .09)^3$$

$$A = 10,000(1.295029) = \$12,950.29$$

(*b*)

$$A = 10,000\left(1 + \frac{.09}{2}\right)^{2(3)}$$

$$A = 10,000(1 + .045)^6$$

$$A = 10,000(1.3022601) = \$13,022.60$$

(*c*)

$$A = 10,000\left(1 + \frac{.09}{4}\right)^{4(3)}$$

$$A = 10,000(1 + .0225)^{12}$$

$$A = 10,000(1.30605) = \$13,060.50$$

(*d*)

$$A = 10,000e^{(0.09)3} = 10,000e^{.27}$$

$$A = 10,000(1.3099645) = \$13,099.65$$

**11.27.** Find the formula for finding the effective rate of interest $r_e$ for multiple compoundings when $t > 1$.

From the explanation of the effective rate of interest in Problem 11.24 (*e*), we can write

$$P(1 + r_e)^t = P\left(1 + \frac{r}{m}\right)^{mt}$$

Dividing by $P$ and taking the $t$th root of each side,

$$1 + r_e = \left(1 + \frac{r}{m}\right)^m$$

$$r_e = \left(1 + \frac{r}{m}\right)^m - 1 \tag{11.14}$$

For continuous compounding,

$$P(1 + r_e)^t = Pe^{rt}$$

$$(1 + r_e) = e^r$$
$$r_e = e^r - 1 \qquad\qquad (11.15)$$

**11.28.** Given a nominal rate of interest $r = 8$ percent, as in Problem 11.25, find the effective rate of interest under (a) semiannual, (b) quarterly, and (c) continuous compounding. Note that in (11.14) and (11.15) time $t$ and principal $P$ do not matter.

(a)   From (11.14),              $r_e = \left(1 + \dfrac{r}{m}\right)^m - 1$

      Substituting,          $r_e = \left(1 + \dfrac{.08}{2}\right)^2 - 1 = (1 + .04)^2 - 1$

$$r_e = 1.08160 - 1 = 0.08160 = 8.16\%$$

(b)                          $r_e = \left(1 + \dfrac{.08}{4}\right)^4 - 1 = (1 + .02)^4 - 1$

$$r_e = 1.08243 - 1 = 0.08243 \approx 8.24\%$$

(c)   From (11.15),            $r_e = e^r - 1$

      Substituting,          $r_e = e^{0.08} - 1 = 1.08329 - 1 = 0.08329 \approx 8.33\%$

**11.29.** Find the effective rate of interest under (a) semiannual, (b) quarterly, and (c) continuous compounding for Problem 11.26 where $r$ was 9 percent.

(a)   From (11.14),              $r_e = \left(1 + \dfrac{r}{m}\right)^m - 1$

      Substituting,          $r_e = \left(1 + \dfrac{.09}{2}\right)^2 - 1 = (1 + .045)^2 - 1$

$$r_e = 1.09203 - 1 = 0.09203 = 9.20\%$$

(b)                          $r_e = \left(1 + \dfrac{.09}{4}\right)^4 - 1 = (1 + .0225)^4 - 1$

$$r_e = 1.09308 - 1 = 0.09308 \approx 9.31\%$$

(c)   From (11.15),            $r_e = e^r - 1$

      Substituting,          $r_e = e^{0.09} - 1$

$$r_e = 1.09417 - 1 = 0.09417 \approx 9.42\%$$

**11.30.** How many years $t$ will it take a sum of money $P$ to double at 6 percent interest compounded annually?

$$A = P(1 + .06)^t$$

For money to double, $A = 2P$. Substituting for $A$,

$$2P = P(1 + .06)^t$$

Dividing by $P$,                        $2 = (1 + .06)^t$

Taking the natural log,                $\ln 2 = t \ln 1.06$

$$0.69315 = 0.05827t$$

Dividing by 0.05827,                        $t \approx 11.9$ years

**11.31.** How long will it take money to treble at 12 percent interest compounded quarterly?

From (11.8),              $A = P\left(1 + \dfrac{.12}{4}\right)^{4(t)} = P(1 + .03)^{4t}$

If money trebles,                              $3 = (1 + .03)^{4t}$

Taking logs,                                   $\ln 3 = 4t \ln 1.03$

$$1.09861 = 4(0.02956)t = 0.11824t$$

$$t \approx 9.3 \text{ years}$$

## DISCOUNTING

**11.32.** *Discounting* is the process of determining the present value $P$ of a sum of money $A$ to be received in the future. Find the formula for discounting under (*a*) annual compounding, (*b*) multiple compounding, and (*c*) continuous compounding.

(*a*)   Under annual compounding,

$$A = P(1 + r)^t$$

Solving for $P$,                $$P = \frac{A}{(1+r)^t} = A(1+r)^{-t} \qquad\qquad (11.16)$$

(*b*)   Under multiple compounding,

$$A = P\left(1 + \frac{r}{m}\right)^{mt}$$

Solving for $P$,                $$P = A\left(1 + \frac{r}{m}\right)^{-mt} \qquad\qquad (11.17)$$

(*c*)   For continuous compounding,

$$A = Pe^{rt}$$

Hence                          $$P = Ae^{-rt} \qquad\qquad (11.18)$$

**11.33.** Find the present value of $1000 to be paid 4 years from now when the current interest rate is 6 percent if interest is compounded (*a*) annually, (*b*) quarterly, and (*c*) continuously.

(*a*)   From (*11.16*),                     $P = A(1 + r)^{-t}$

Substituting,                                $P = 1000(1 + .06)^{-4}$

Entering 1.06 on a calculator, hitting the $\boxed{y^x}$ key, then entering 4 *followed* by the $\boxed{\pm}$ key to make it negative, and finally activating the $\boxed{=}$ key, as explained in Section 1.7.11,

$$P = 1000(0.79209) = \$792.09$$

(*b*)   From (*11.17*),        $$P = A\left(1 + \frac{r}{m}\right)^{-mt}$$

Substituting,                  $$P = 1000\left(1 + \frac{.06}{4}\right)^{-4(4)} = 1000(1 + .015)^{-16}$$

$$P = 1000(0.78803) = \$788.03$$

(*c*)   From (*11.18*),        $P = Ae^{-rt}$

Substituting,                  $P = 1000e^{-(0.06)(4)} = 1000e^{-0.24}$

$$P = 1000(0.78663) = \$786.63$$

**11.34.** Redo Problem 11.33 for a present value of $1500 to be paid in 8 years when the current interest rate is 5 percent.

(*a*)                          $P = 1500(1 + .05)^{-8}$

$$P = 1500(0.67684) = \$1015.26$$

(*b*)                          $$P = 1500\left(1 + \frac{.05}{4}\right)^{-4(8)} = 1500(1 + .0125)^{-32}$$

$$P = 1500(0.67198) = \$1007.97$$

(c)                                 $$P = 1500e^{-(0.05)(8)} = 1500e^{-0.4}$$

$$P = 1500(0.67032) = \$1005.48$$

## ESTIMATING GROWTH RATES FROM DATA POINTS

**11.35.** Given two sets of points from data growing consistently over time, such as subscriptions to a magazine numbering 6.25 million in 1988 and 11.1 million in 1993, (a) express subscriptions as a natural exponential function of time $S = S_0 e^{rt}$ and (b) find the annual rate of growth $G$ in subscriptions.

(a)  Letting $t = 0$ for the base year 1988, then $t = 5$ for 1993. Expressing the two sets of data points in terms of $S = S_0 e^{rt}$, and recalling that $e^0 = 1$,

$$6.25 = S_0 e^{r(0)} = S_0 \tag{11.19}$$

$$11.1 = S_0 e^{r(5)} \tag{11.20}$$

Substituting $S_0$ from (11.19) in (11.20) and simplifying algebraically,

$$11.1 = 6.25 e^{5r}$$

$$1.776 = e^{5r}$$

Taking the natural logarithm of each side and using (11.3),

$$\ln 1.776 = \ln e^{5r} = 5r$$

$$0.57436 = 5r \qquad r = 0.11487 \approx 11.5\%$$

Setting the values of $S_0$ and $r$ in the proper form $S_0 e^{rt}$,

$$S = 6.25 e^{0.115t}$$

(b)  The growth rate $G$ of a function is given by the derivative of the natural log of the function: $d/dx[\ln f(x)] = f'(x)/f(x)$. Taking the natural log of $S$, therefore, and again using (11.3),

$$\ln S = \ln 6.25 + \ln e^{0.115t}$$

$$\ln S = \ln 6.25 + 0.115t$$

Then taking the derivative, recalling that $\ln 6.25$ is a constant,

$$G = \frac{d}{dt}(\ln S) = 0.115 = 11.5\%$$

The growth rate of a natural exponential function will always be the coefficient of the variable in the exponent.

**11.36.** A firm's profits $\pi$ have been growing consistently over time from \$3.40 million in 1985 to \$6.71 million in 1993. (a) Express profits as a natural exponential function of time $\pi = \pi_0 e^{rt}$ and (b) find the annual rate of growth.

(a)  Let 1985 be the base year with $t = 0$, then $t = 8$ for 1993, and

$$3.40 = \pi_0 e^{r(0)} = \pi_0$$

$$6.71 = \pi_0 e^{r(8)} \tag{11.21}$$

Substituting $\pi_0 = 3.40$ in (11.21) and simplifying algebraically,

$$6.71 = 3.40e^{8r}$$

$$1.97353 = e^{8r}$$

Taking logs, $\qquad 0.67982 = 8r$

$$r = 0.08498 \approx 0.085$$

and $\qquad \pi = 3.40e^{0.085t}$

(b)    The growth rate is 8.5 percent.

**11.37.**  A country's population $P$ goes from 54 million in 1987 to 63.9 million in 1993. At what rate $r$ is the population growing?

Letting $t = 0$ for 1987 and $t = 6$ for 1993, and noting from the previous problems that the value of the function at $t = 0$ is the base of the function in subsequent years,

$$63.9 = 54e^{r(6)}$$

Simplifying algebraically, $\qquad 1.18333 = e^{6r}$

Taking the natural log, $\qquad 0.16833 = 6r$

$$r = 0.02806 \approx 0.028$$

Thus, $\qquad P = 54e^{0.028t}$

and the population growth rate is 2.8 percent.

**11.38.**  A country's timberland $T$ is being cut back at a rate of 3.2 percent a year. How much will be left in 8 years?

$$T = T_0 e^{-0.032(8)} = T_0 e^{-0.256}$$

Using a calculator,

$$T = 0.77414T_0 \approx 77\% \text{ of current timberland}$$

**11.39.**  A village's arable land $L$ is eroding at a rate of 1.8 percent a year. If conditions continue, how much will be left in 15 years?

$$L = L_0 e^{-0.018(15)} = L_0 e^{-0.27}$$

$$L = 0.76338L_0 \approx 76\% \text{ of current arable land}$$

# Supplementary Problems

## EXPONENTIAL-LOGARITHMIC CONVERSIONS

**11.40.**    Convert the following logarithms into their equivalent exponential forms:
   (a)  $\log_{13} 169 = 2$        (b)  $\log_5 125 = 3$        (c)  $\log_6(\frac{1}{36}) = -2$
   (d)  $\log_{81} 9 = \frac{1}{2}$        (e)  $\log_4(\frac{1}{64}) = -3$        (f)  $\log_{16} 2 = \frac{1}{4}$

**11.41.**    Change the following natural logarithms to their equivalent natural exponential forms:
   (a)  $\ln 13 = 2.56495$        (b)  $\ln 5.8 = 1.75786$        (c)  $\ln 0.4 = -0.91629$
   (d)  $\ln 62 = 4.12713$        (e)  $\ln y = 2.35t$        (f)  $\ln y = 7 - 9t$

**11.42.**  Convert the following exponential expressions to their equivalent logarithmic forms:

(a)  $64 = 8^2$           (b)  $81 = 3^4$           (c)  $\frac{1}{8} = 2^{-3}$

(d)  $7 = 49^{1/2}$        (e)  $4 = 8^{2/3}$        (f)  $216 = 36^{3/2}$

**11.43.**  Change the following natural exponential expressions to their equivalent natural logarithmic forms:

(a)  $6.7 = e^{1.90211}$      (b)  $0.25 = e^{-1.38629}$      (c)  $122 = e^{4.80402}$

(d)  $43 = e^{3.76120}$        (e)  $y = e^{0.06t}$             (f)  $y = e^{5t+8}$

**11.44.**  Solve each of the following for the unknown variable or parameter:

(a)  $y = \log_{11} 121$      (b)  $y = \log_3 \left(\frac{1}{81}\right)$      (c)  $\log_3 x = 5$

(d)  $\log_{64} x = \frac{4}{3}$        (e)  $\log_a 9 = \frac{2}{3}$        (f)  $\log_a \left(\frac{1}{8}\right) = -\frac{1}{2}$

## PROPERTIES OF LOGARITHMS AND EXPONENTS

**11.45.**  Convert the following logarithmic expressions to sums, differences, or products:

(a)  $y = \log_a 38x^2$     (b)  $y = \log_a x^2 y^4$     (c)  $y = \log_a (7x/9y)$     (d)  $y = \log_a (x^5/y^3)$

**11.46.**  Convert the following natural logarithmic expressions to sums, differences, or products:

(a)  $y = \ln \left(\dfrac{x^4 y^2}{z^5}\right)$                     (b)  $y = \ln \sqrt[4]{x}$

(c)  $y = \ln \dfrac{\sqrt[3]{x}}{\sqrt{y}}$                      (d)  $y = \ln \sqrt{\dfrac{x^7}{y^5}}$

**11.47.**  Simplify the following exponential expressions:

(a)  $y = a^{5x} \cdot a^{8y}$     (b)  $y = \dfrac{a^{2x}}{a^{7y}}$     (c)  $y = (a^{3x})^{2y}$     (d)  $y = \sqrt[3]{a^{7x}}$

**11.48.**  Simplify the following natural exponential expressions:

(a)  $y = (e^{6x})^{-1/2}$     (b)  $y = \dfrac{e^{4x}}{e^{5y}}$     (c)  $y = \sqrt{e^{5x}}$     (d)  $y = \dfrac{e^{7x}}{e^{-4y}}$

## DERIVATIVES

**11.49.**  Differentiate each of the following natural exponential functions:

(a)  $f(x) = e^{-9x}$        (b)  $y = e^{5x^3}$             (c)  $f(x) = 7e^{8-3x^2}$

(d)  $y = 6e^{4x^3 - 17}$      (e)  $f(x) = -\frac{1}{2}e^{(1/3)x^6}$      (f)  $y = -\frac{1}{4}e^{5-x^2}$

**11.50.**  Differentiate each of the following natural logarithmic functions:

(a)  $y = \ln (2x^2 - 5)$           (b)  $f(x) = \ln 17x$          (c)  $y = \ln 9x^5$

(d)  $f(x) = \ln (x^3 + 4x^2 - 9x + 16)$      (e)  $y = 3\ln 6x^4$      (f)  $f(x) = 2\ln (5x^3 - 4)$

**11.51.**  Combine rules to differentiate the following functions:

(a)  $f(x) = (e^{-5x})^3$                  (b)  $y = 7xe^{4x}$

(c)  $f(x) = (e^{3x} + e^{2x})^3$           (d)  $y = \dfrac{e^{6x}}{2x + 1}$

(e)  $f(x) = \ln (3x - 7)^2$            (f)  $y = 8x^2 \ln x^3$

(g)  $f(x) = \ln^2 5x^4$                (h)  $y = \dfrac{e^x}{\ln x}$

## INTEREST COMPOUNDING

For each of the following questions, determine the future value of the given principals when compounded (*a*) annually, (*b*) semiannually, (*c*) quarterly, and (*d*) continuously.

**11.52.**    $1000 at 8 percent for 5 years.

**11.53.**    $2500 at 6 percent for 4 years.

**11.54.**    $500 at 12 percent for 6 years.

**11.55.**    $10,000 at 9 percent for 8 years.

**11.56.**    How long will it take money to double at 8 percent interest when compounded semiannually?

**11.57.**    How long will it take money to treble at 9 percent interest when compounded annually?

## DISCOUNTING

**11.58.**    Determine the present value of $5000 to be paid in 8 years time if current interest of 10 percent is compounded (*a*) annually, (*b*) semiannually, (*c*) quarterly, and (*d*) continuously.

**11.59.**    Repeat Problem 11.58 for $8000 to be paid in 4 years when the current interest rate is 5 percent.

**11.60.**    Given the current interest rate of 6 percent compounded annually, find the present value of $10,000 to be paid in (*a*) 1 year, (*b*) 3 years, (*c*) 5 years, and (*d*) 10 years.

**11.61.**    Given the current interest rate of 7 percent compounded semiannually, find the present value of $10,000 to be paid in (*a*) 1 year, (*b*) 3 years, (*c*) 5 years, and (*d*) 10 years.

## GROWTH RATES

**11.62.**    A company's sales have been growing consistently over time from $23.2 million in 1987 to $32.27 million in 1993. (*a*) Express sales *S* as a natural exponential function of time *t*. (*b*) Indicate the annual rate of growth *G* of sales.

**11.63.**    A firm's total revenue TR, growing consistently over time, has increased from $345 million in 1985 to $616.18 million in 1993. (*a*) Find the exponential function and (*b*) indicate the growth rate of revenue.

**11.64.**    A company with 1993 sales of $8.9 million estimates sales of $13.82 million in 1997. (*a*) Determine the exponential function used for the estimate and (*b*) indicate the projected growth rate.

**11.65.**    A country's population of 44.35 million is growing at a rate of 2.65 percent a year. Find the expected population in 5 years.

**11.66.**    What is the projected level of revenue in 3 years for a company with current revenues of $48.25 million if estimated growth under continuous compounding is 8.6 percent?

**11.67.**    Using an exponential function a consultant suggests that a newspaper's circulation will increase from 1.25 million at present to 1.73 million in 5 years. What growth rate is the consultant assuming?

**11.68.**    What will a machine worth $3.8 million today be worth in 3 years if it depreciates at a continuous rate of 15 percent a year?

# Answers to Supplementary Problems

**11.40.**    (a)   $169 = 13^2$      (b)   $125 = 5^3$      (c)   $\frac{1}{36} = 6^{-2}$

         (d)   $9 = 81^{1/2}$      (e)   $\frac{1}{64} = 4^{-3}$      (f)   $2 = 16^{1/4}$

**11.41.**    (a)   $13 = e^{2.56495}$      (b)   $5.8 = e^{1.75786}$      (c)   $0.4 = e^{-0.91629}$

         (d)   $62 = e^{4.12713}$      (e)   $y = e^{2.35t}$      (f)   $y = e^{7-9t}$

**11.42.**    (a)   $\log_8 64 = 2$      (b)   $\log_3 81 = 4$      (c)   $\log_2 \frac{1}{8} = -3$

         (d)   $\log_{49} 7 = \frac{1}{2}$      (e)   $\log_8 4 = \frac{2}{3}$      (f)   $\log_{36} 216 = \frac{3}{2}$

**11.43.**    (a)   $\ln 6.7 = 1.90211$      (b)   $\ln 0.25 = -1.38629$      (c)   $\ln 122 = 4.80402$

         (d)   $\ln 43 = 3.76120$      (e)   $\ln y = 0.06t$      (f)   $\ln y = 5t + 8$

**11.44.**    (a)   $y = 2$      (b)   $y = -4$      (c)   $x = 243$

         (d)   $x = 256$      (e)   $a = 27$      (f)   $a = 64$

**11.45.**    (a)   $y = \log_a 38 + 2\log_a x$          (b)   $y = 2\log_a x + 4\log_a y$

         (c)   $y = \log_a 7 + \log_a x - \log_a 9 - \log_a y$      (d)   $y = 5\log_a x - 3\log_a y$

**11.46.**    (a)   $y = 4\ln x + 2\ln y - 5\ln z$      (b)   $y = \frac{1}{4}\ln x$

         (c)   $y = \frac{1}{3}\ln x - \frac{1}{2}\ln y$      (d)   $y = \frac{1}{2}(7\ln x - 5\ln y)$

**11.47.**    (a)   $y = a^{5x+8y}$      (b)   $y = a^{2x-7y}$      (c)   $y = a^{6xy}$      (d)   $y = a^{(7/3)x}$

**11.48.**    (a)   $y = e^{-3x} = \dfrac{1}{e^{3x}}$      (b)   $y = e^{4x-5y}$

         (c)   $y = e^{2.5x}$      (d)   $y = e^{7x+4y}$

**11.49.**    (a)   $f' = -9e^{-9x}$      (b)   $y' = 15x^2 e^{5x^3}$      (c)   $f'(x) = -42xe^{8-3x^2}$

         (d)   $y' = 72x^2 e^{4x^3-17}$      (e)   $f' = -x^5 e^{(1/3)x^6}$      (f)   $y' = \frac{1}{2}xe^{5-x^2}$

**11.50.**    (a)   $y' = \dfrac{4x}{2x^2 - 5}$      (b)   $f'(x) = \dfrac{1}{x}$      (c)   $y' = \dfrac{5}{x}$

         (d)   $f' = \dfrac{3x^2 + 8x - 9}{x^3 + 4x^2 - 9x + 16}$      (e)   $y' = \dfrac{12}{x}$      (f)   $f'(x) = \dfrac{30x^2}{5x^3 - 4}$

**11.51.**    (a)   $f'(x) = -15e^{-15x}$          (b)   $y' = 28xe^{4x} + 7e^{4x}$

         (c)   $f' = (9e^{3x} + 6e^{2x})(e^{3x} + e^{2x})^2$      (d)   $y' = \dfrac{12xe^{6x} + 4e^{6x}}{(2x + 1)^2}$

         (e)   $f'(x) = \dfrac{6}{3x - 7}$      (f)   $y' = 8x(3 + 2\ln x^3)$

         (g)   $f' = \dfrac{8\ln 5x^4}{x}$      (h)   $y' = \dfrac{e^x[\ln x - (1/x)]}{\ln^2 x}$

**11.52.**    (a)   \$1469.33      (b)   \$1480.24      (c)   \$1485.95      (d)   \$1491.82

**11.53.**    (a)   \$3156.19      (b)   \$3166.93      (c)   \$3172.46      (d)   \$3178.12

**11.54.**    (a)   \$986.91      (b)   \$1006.10      (c)   \$1016.40      (d)   \$1027.22

**11.55.**    (a)   \$19,925.63      (b)   \$20,223.70      (c)   \$20,381.03      (d)   \$20,544.33

**11.56.**     8.836 years

**11.57.**     12.748 years

**11.58.**     (*a*)  $2332.54        (*b*)  $2290.56        (*c*)  $2268.85        (*d*)  $2246.64

**11.59.**     (*a*)  $6581.62        (*b*)  $6565.97        (*c*)  $6557.97        (*d*)  $6549.85

**11.60.**     (*a*)  $9433.96        (*b*)  $8396.19        (*c*)  $7472.58        (*d*)  $5583.95

**11.61.**     (*a*)  $9335.11        (*b*)  $8135.01        (*c*)  $7089.19        (*d*)  $5025.66

**11.62.**     (*a*)  $S = 23.2e^{0.055t}$        (*b*)  $G = 0.055$ or 5.5 percent

**11.63.**     (*a*)  $\text{TR} = 345e^{0.0725t}$        (*b*)  $G = 7.25$ percent

**11.64.**     (*a*)  $S = 8.9e^{0.11t}$        (*b*)  $G = 11$ percent

**11.65.**     50.63 million

**11.66.**     $62.45 million

**11.67.**     $G = 6.5$ percent

**11.68.**     $2.42 million

# Chapter 12

## Integral Calculus

### 12.1 INTEGRATION

Chapters 9 and 10 were devoted to differential calculus, which measures the rate of change of functions. Differentiation, we learned, is the process of finding the derivative $F'(x)$ of a function $F(x)$. Frequently in business and economics, however, we know the rate of change of a function $F'(x)$ and need to find the original function $F(x)$. Reversing the process of differentiation and finding the original function from the derivative is called *antidifferentiation* or *integration*, and the original function $F(x)$ is called the *antiderivative* or *integral* of $F'(x)$.

**EXAMPLE 1.** Letting $f(x) = F'(x)$ for simplicity, the antiderivative of $f(x)$ is expressed mathematically as

$$\int f(x)\, dx = F(x) + c$$

Here the left-hand side of the equation is read "the *indefinite integral* of $f$ of $x$ with respect to $x$." The symbol $\int$ is an *integral sign*, $f(x)$ is the *integrand*, and $c$ is the *constant of integration*, which will be explained in Example 3.

### 12.2 RULES FOR INDEFINITE INTEGRALS

The following rules for indefinite integrals are obtained by reversing the corresponding rules of differentiation. Their accuracy is easily checked since the derivative of the integral must equal the integrand. The rules are illustrated in Example 2 and Problems 12.1 to 12.3

**Rule 1.** The integral of a constant $k$ is

$$\int k\, dx = kx + c$$

**Rule 2.** The integral of 1, written simply as $dx$, not 1 $dx$, is

$$\int dx = x + c$$

**Rule 3.** The integral of a power function $x^n$, where $n \neq -1$, is given by the *power rule*:

$$\int x^n\, dx = \frac{1}{n+1} x^{n+1} + c \qquad (n \neq -1)$$

**Rule 4.** The integral of $x^{-1} \left( \text{or } \dfrac{1}{x} \right)$ is

$$\int x^{-1}\, dx = \ln x + c \qquad (x > 0)$$

The condition $x > 0$ is needed because only positive numbers have logarithms. For negative numbers,

$$\int x^{-1}\, dx = \ln |x| + c \qquad (x \neq 0)$$

**Rule 5.** The integral of a natural exponential function is

$$\int e^{kx}\, dx = \frac{1}{k} e^{kx} + c$$

**Rule 6.** The integral of a constant times a function equals the constant times the integral of the function.

$$\int kf(x)\, dx = k \int f(x)\, dx$$

**Rule 7.** The integral of the sum or difference of two or more functions equals the sum or difference of their integrals.

$$\int [f(x) \pm g(x)]\, dx = \int f(x)\, dx \pm \int g(x)\, dx$$

**Rule 8.** The integral of the negative of a function equals the negative of the integral of the function.

$$\int -f(x)\, dx = - \int f(x)\, dx$$

**EXAMPLE 2.**    The rules for indefinite integrals are illustrated below. Check each answer on your own by making sure that the derivative of the antiderivative equals the integrand.

(a)
$$\int 9\, dx = 9x + c \qquad\qquad\qquad\qquad\qquad\qquad \text{(Rule 1)}$$

(b)
$$\int x^4\, dx = \frac{1}{4+1} x^{4+1} + c = \frac{1}{5} x^5 + c \qquad\qquad \text{(Rule 3)}$$

(c)
$$\int 4x^3\, dx = 4 \int x^3\, dx \qquad\qquad\qquad\qquad \text{(Rule 6)}$$

$$= 4\left( \frac{1}{4} x^4 + c_1 \right) \qquad\qquad\qquad \text{(Rule 3)}$$

$$= x^4 + c$$

where $c_1$ and $c$ are arbitrary constants and $4c_1 = c$. Since $c$ is an arbitrary constant, it can be ignored in the preliminary calculations and included only in the final solution.

(d)
$$\int (2x^2 - x + 1)\, dx = 2 \int x^2\, dx - \int x\, dx + \int dx \qquad \text{(Rules 6, 7, 8)}$$

$$= 2\left( \frac{1}{3} x^3 \right) - \frac{1}{2} x^2 + x + c \qquad\qquad \text{(Rules 2 and 3)}$$

$$= \frac{2}{3} x^3 - \frac{1}{2} x^2 + x + c$$

(e)
$$\int 12x^{-1}\, dx = 12 \int x^{-1}\, dx \qquad\qquad\qquad \text{(Rule 6)}$$

$$= 12 \ln |x| + c \qquad\qquad\qquad\qquad \text{(Rule 4)}$$

(f)
$$\int x^{-4} = \frac{1}{-4+1} x^{-4+1} = -\frac{1}{3} x^{-3} + c \qquad \text{(Rule 3)}$$

(g)
$$\int 20e^{-4x}\, dx = 20 \cdot \left( -\frac{1}{4} \cdot e^{-4x} \right) = -5e^{-4x} + c \qquad \text{(Rule 5)}$$

**EXAMPLE 3.**    From the rules of differentiation, we know that functions which differ only by a constant $k$ have the same derivative. The function $F(x) = -\frac{1}{2}x + k$, for instance, has the same derivative, $F'(x) = f(x) = -\frac{1}{2}$, for any infinite number of possible values for $k$. If the process is reversed, it is clear that $\int -\left( \frac{1}{2} \right) dx$ must be the antiderivative for an infinite number of functions differing from each other only by a constant. The constant of integration $c$, then, serves to represent the value of any constant that was part of the original function but precluded from the derivative by the rules of differentiation. The graph of an indefinite integral $\int f(x)\, dx = F(x) + c$, where $c$ is unspecified, is a family of curves parallel in the sense that the slope of the tangent to any of them at $x$ is $f(x)$. Specifying $c$ specifies the curve; changing $c$ shifts the curve. This is illustrated in Fig. 12-1 for the indefinite integral $\int -\frac{1}{2}\, dx = -\frac{1}{2}x + c$ where $c = 5, 3, 1,$ and $-1$, respectively.

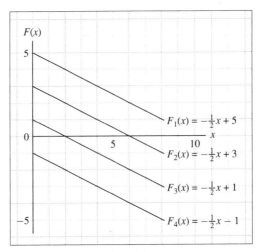

**Fig. 12-1**

## 12.3   AREA UNDER A CURVE

There is no geometric formula for the area under an irregularly shaped curve, such as $y = f(x)$ between $x = a$ and $x = b$ in Fig. 12-2(a). An approximation of the area may be obtained, however, by subdividing the interval $[a, b]$ into $n$ subintervals and creating rectangles such that the height of each, for example, is equal to the largest value of the function in the subinterval, as in Fig. 12-2(b). Then the sum of the areas of the rectangles $\sum_{i=1}^{n} [f(x_i)\Delta x_i]$, called a *Riemann sum*, will approximate, but overestimate, the actual area under the curve. The smaller each subinterval ($\Delta x_i$), the more rectangles are created and the closer the combined area of the rectangles $\sum_{i=1}^{n} [f(x_i)\Delta x_i]$ approaches the actual area under the curve. If the number of subintervals is increased so that $n \to \infty$, each subinterval becomes infinitesimal ($\Delta x_i = dx_i = dx$) and the area $A$ under the curve can be expressed mathematically as

$$A = \lim_{n \to \infty} \sum_{i=1}^{n} f(x_i)\Delta x_i$$

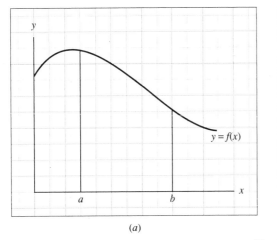

(a)

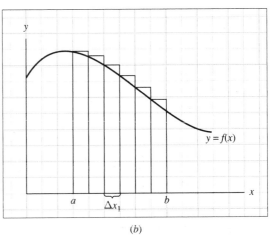

(b)

**Fig. 12-2**

## 12.4   THE DEFINITE INTEGRAL

The area under a graph of a continuous function such as that in Fig. 12-2 can be expressed more succinctly as the *definite integral* of $f(x)$ over the interval $a$ to $b$. Put mathematically,

$$\int_a^b f(x)\,dx = \lim_{n \to \infty} \sum_{i=1}^n f(x_i) \Delta x_i$$

Here the left-hand side is read "the integral from $a$ to $b$ of $f$ of $x$ $dx$," and $a$ is called the *lower limit* of integration and $b$, the *upper limit* of integration. Unlike the indefinite integral, which is a set of functions containing all the antiderivatives of $f(x)$, as explained in Example 3, the definite integral is a real number which can be evaluated, using the fundamental theorem of calculus which is the topic of the next section.

## 12.5   THE FUNDAMENTAL THEOREM OF CALCULUS

The *fundamental theorem of calculus* says simply that the numerical value of the definite integral of a continuous function $f(x)$ over the interval from $a$ to $b$ is given by the antiderivative $F(x) + c$ evaluated at the upper limit of integration $b$, minus the same antiderivative $F(x) + c$ evaluated at the lower limit of integration $a$. With $c$ common to both, the constant of integration is eliminated in the subtraction. Expressed mathematically,

$$\int_a^b f(x)\,dx = F(x)\Big|_a^b = F(b) - F(a)$$

where the symbol $|_a^b$, $]_a^b$, or $[\cdots]_a^b$ indicates that $b$ and $a$ are to be substituted successively for $x$. See Examples 4 and 5 and Problems 12.4 and 12.5.

**EXAMPLE 4.**    The definite integrals given below

$$(a)\ \int_2^6 8x\,dx \qquad\qquad (b)\ \int_1^3 (6x^2 + 5)\,dx$$

are evaluated as follows:

(a)
$$\int_2^6 8x\,dx = 4x^2\Big|_2^6 = 4(6)^2 - 4(2)^2 = 128$$

(b)
$$\int_1^3 (6x^2 + 5)\,dx = \left[2x^3 + 5x\right]_1^3 = \left[2(3)^3 + 5(3)\right] - \left[2(1)^3 + 5(1)\right] = 69 - 7 = 62$$

**EXAMPLE 5.**    The definite integral is used below to determine the area under the curve in Fig. 12-3 over the interval 0 to 5 as follows:

$$A = \int_0^5 10x\,dx = 5x^2\Big|_0^5 = 5(5)^2 - 5(0)^2 = 125 - 0 = 125$$

The answer can easily be checked by using the geometric formula $A = \frac{1}{2}hw$, where $h =$ height and $w =$ width.

$$A = \tfrac{1}{2}hw = \tfrac{1}{2}yx = \tfrac{1}{2}(50)(5) = 125$$

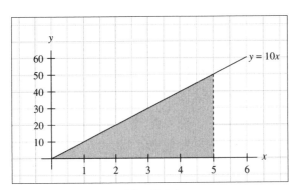

**Fig. 12-3**

## 12.6   PROPERTIES OF DEFINITE INTEGRALS

1.   Reversing the order of the limits of integration changes the sign of the definite integral.

$$\int_a^b f(x)\,dx = -\int_b^a f(x)\,dx \qquad (12.1)$$

2.   If the upper limit of integration equals the lower limit of integration, the value of the definite integral is zero.

$$\int_a^a f(x)\,dx = F(a) - F(a) = 0 \qquad (12.2)$$

3.   The definite integral can be expressed as the sum of component subintegrals.

$$\int_a^c f(x)\,dx = \int_a^b f(x)\,dx + \int_b^c f(x)\,dx \qquad a \le b \le c \qquad (12.3)$$

4.   The sum or difference of two definite integrals with identical limits of integration is equal to the definite integral of the sum or difference of the two functions.

$$\int_a^b f(x)\,dx \pm \int_a^b g(x)\,dx = \int_a^b [f(x) \pm g(x)]\,dx \qquad (12.4)$$

5.   The definite integral of a constant times a function is equal to the constant times the definite integral of the function.

$$\int_a^b k f(x)\,dx = k \int_a^b f(x)\,dx \qquad (12.5)$$

See Example 6 and Problems 12.6 and 12.7.

**EXAMPLE 6.**   The following definite integrals are evaluated to illustrate a sampling of the properties of integrals presented above.

1.   $\displaystyle\int_1^2 3x^4\,dx = -\int_2^1 3x^4\,dx$                                                    (Rule 1)

$$\int_1^2 3x^4\,dx = \tfrac{3}{5}x^5\Big|_1^2 = \tfrac{3}{5}(2)^5 - \tfrac{3}{5}(1)^5 = 18.6$$

$$-\int_2^1 3x^4\,dx = -\tfrac{3}{5}x^5\Big|_2^1 = -\left[\tfrac{3}{5}(1)^5 - \tfrac{3}{5}(2)^5\right] = -(-18.6) = 18.6$$

2.  $\displaystyle\int_4^4 (3x^2 + 4x)\,dx = 0$                                                      (Rule 2)

$$\int_4^4 (3x^2 + 4x)\,dx = (x^3 + 2x^2)\Big|_4^4$$

$$= \left[(4)^3 + 2(4)^2\right] - \left[(4)^3 + 2(4)^2\right] = 0$$

3.  $\displaystyle\int_0^5 8x\,dx = \int_0^2 8x\,dx + \int_2^5 8x\,dx$                               (Rule 3)

$$\int_0^5 8x\,dx = 4x^2\Big|_0^5 = 4(5)^2 - 4(0)^2 = 100$$

$$\int_0^2 8x\,dx = 4x^2\Big|_0^2 = 4(2)^2 - 4(0)^2 = 16$$

$$\int_2^5 8x\,dx = 4x^2\Big|_2^5 = 4(5)^2 - 4(2)^2 = 84$$

And                                                    $16 + 84 = 100$

## 12.7  AREA BETWEEN CURVES

The area of a region between two or more curves can be evaluated by applying Rule 4 of the properties of definite integrals outlined above. The procedure is demonstrated in Example 7 and treated in Problems 12.8 to 12.15.

**EXAMPLE 7.**    Using the properties of integrals, the area $A$ of the region between two functions such as $y_1 = -0.25x + 12$ and $y_2 = -x^2 + 6x$ from $x = 0$ to $x = 6$, is found as follows.

(a)   Draw a rough sketch of the graph of the functions and shade in the desired area as in Fig. 12-4.

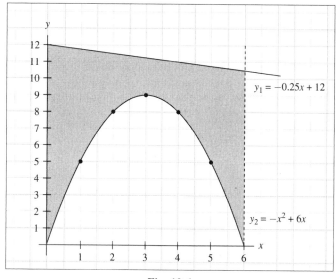

**Fig. 12-4**

(b)   Note the relationship between the curves. Since $y_1$ lies above $y_2$ throughout the indicated region, the desired region is simply the area under $y_1$ minus the area under $y_2$ between $x = 0$ and $x = 6$. Hence,

$$A = \int_0^6 (-0.25x + 12)\,dx - \int_0^6 (-x^2 + 6x)\,dx$$

Using (12.4),

$$A = \int_0^6 \left[(-0.25x + 12) - (-x^2 + 6x)\right] dx$$

$$A = \int_0^6 (x^2 - 6.25x + 12)\,dx$$

$$A = \left(\frac{1}{3}x^3 - 3.125x^2 + 12x\right)\Big|_0^6 = 31.5 - 0 = 31.5$$

See also Problems 12.8 to 12.15.

## 12.8   INTEGRATION BY SUBSTITUTION

Integration of a product or quotient of two differentiable functions of $x$, such as

$$\int 20x^4(x^5 + 7)\,dx$$

is not possible in terms of the simple rules developed earlier in the chapter. If the integrand can be expressed as a *constant* multiple of another function $u$ and its derivative $du/dx$, however, integration by substitution is possible. By expressing the integrand $f(x)$ as a function of $u$ and its derivative $du/dx$, and integrating with respect to $x$, we obtain

$$\int f(x)\,dx = \int \left(u\frac{du}{dx}\right) dx$$

$$\int f(x)\,dx = \int u\,du = F(u) + c$$

*Integration by substitution* reverses the operation of the chain rule and the generalized power function rule in differential calculus. See Examples 8 and 9 and Problems 12.16 to 12.23.

**EXAMPLE 8.**   Integration by substitution is used in the steps below to determine the indefinite integral

$$\int 20x^4(x^5 + 7)\,dx$$

1.   Check to be sure that the integrand can be converted to a product of another function $u$ and its derivative $du/dx$, times a *constant* multiple. (a) Let $u$ equal the function in which the independent variable is raised to the higher power in terms of absolute value; here $u = x^5 + 7$. (b) Differentiate $u$: $du/dx = 5x^4$. (c) Solve algebraically for $dx$ : $dx = du/5x^4$ and (d) substitute $u$ for $x^5 + 7$ and $du/5x^4$ for $dx$ in the original integrand:

$$\int 20x^4(x^5 + 7)\,dx = \int 20x^4 \cdot u \cdot \frac{du}{5x^4} = \int 4u\,du = 4\int u\,du$$

where 4 is a *constant* multiple of $u$.

2.   Integrate with respect to $u$.

$$4\int u\,du = 4\left(\frac{1}{2}u^2\right) = 2u^2 + c$$

3.   Convert back to the terms of the original problem by substituting $x^5 + 7$ for $u$.

$$\int 20x^4(x^5 + 7)\,dx = 2u^2 + c = 2(x^5 + 7)^2 + c$$

4.  Check the answer by differentiating with the generalized power function rule or the chain rule.

$$\frac{d}{dx}\left[2(x^5+7)^2+c\right]=4(x^5+7)(5x^4)=20x^4(x^5+7)$$

**EXAMPLE 9.**   Determine the integral $\int 3x(x+6)^2\,dx$. Let $u=x+6$. Then $du/dx=1$ and $dx=du/1=du$. Substitute $u$ for $x+6$ and $du$ for $dx$ in the original integrand.

$$\int 3x(x+6)^2\,dx=\int 3xu^2\,du=3\int xu^2\,du$$

Since $x$ is a *variable* multiple which cannot be factored out, the original integrand cannot be converted into a *constant* multiple of $u\,du/dx$. Hence integration by substitution will not work. Integration by parts (Section 12.9), however, may help.

## 12.9  INTEGRATION BY PARTS

If an integrand is a product or quotient of differentiable functions of $x$ and cannot be expressed as a constant multiple of $u\,du/dx$, integration by parts may prove helpful. *Integration by parts* is derived by reversing the process of differentiating a product. From the product rule in Section 9.7.5,

$$\frac{d}{dx}[f(x)\cdot g(x)]=f(x)\cdot g'(x)+g(x)\cdot f'(x)$$

Taking the integral of the derivative gives

$$f(x)\cdot g(x)=\int\left[f(x)\cdot g'(x)\right]\,dx+\int\left[g(x)\cdot f'(x)\right]\,dx$$

Then solving algebraically for the first integral on the right-hand side,

$$\int\left[f(x)\cdot g'(x)\right]\,dx=f(x)\cdot g(x)-\int\left[g(x)\cdot f'(x)\right]\,dx \qquad (12.6)$$

See Examples 10 and 11 and Problems 12.24 to 12.27.

**EXAMPLE 10.**   Integration by parts is used below to determine

$$\int 3x(x+6)^2\,dx$$

1.  Separate the integrand into two parts amenable to the formula in (12.6). As a general rule, consider first the simpler function for $f(x)$ and the more complicated function for $g'(x)$. By letting $f(x)=3x$ and $g'(x)=(x+6)^2$, then $f'(x)=3$ and $g(x)=\int(x+6)^2\,dx$, which can be integrated using the simple power rule (Rule 3) from Section 12.2:

$$g(x)=\frac{1}{3}(x+6)^3+c_1$$

2.  Substitute the values for $f(x)$, $f'(x)$, and $g(x)$ in (12.6); and note that $g'(x)$ is not used in the formula.

$$\int 3x(x+6)^2\,dx=f(x)\cdot g(x)-\int[g(x)\cdot f'(x)]\,dx$$

$$=3x\left[\frac{1}{3}(x+6)^3+c_1\right]-\int\left[\frac{1}{3}(x+6)^3+c_1\right](3)\,dx$$

$$=x(x+6)^3+3c_1x-\int\left[(x+6)^3+3c_1\right]\,dx$$

3.  Use Rule 3 to compute the final integral and substitute.

$$\int 3x(x+6)^2 \, dx = x(x+6)^3 + 3c_1 x - \frac{1}{4}(x+6)^4 - 3c_1 x + c$$

$$= x(x+6)^3 - \frac{1}{4}(x+6)^4 + c$$

Note that the $c_1$ term does not appear in the final solution. Since this is common to integration by parts, $c_1$ will henceforth be assumed equal to 0 and not formally included in future problem solving.

4.  Check the answer by letting $y(x) = x(x+6)^3 - \frac{1}{4}(x+6)^4 + c$ and using the product and generalized power function rules to differentiate the answer.

$$y'(x) = \left[x \cdot 3(x+6)^2 + (x+6)^3 \cdot 1\right] - (x+6)^3$$

$$y'(x) = 3x(x+6)^2$$

**EXAMPLE 11.**    The integral $\int 5xe^{x-9} \, dx$ is determined as follows: Let $f(x) = 5x$ and $g'(x) = e^{x-9}$; then $f'(x) = 5$ and, by Rule 5, $g(x) = \int e^{x-9} \, dx = e^{x-9}$. Substituting in (12.6),

$$\int 5xe^{x-9} \, dx = f(x) \cdot g(x) - \int [g(x) \cdot f'(x)] \, dx$$

$$= 5x \cdot e^{x-9} - \int e^{x-9} \cdot 5 \, dx = 5xe^{x-9} - 5 \int e^{x-9} \, dx$$

Applying Rule 5 again, remembering the constant of integration,

$$\int 5xe^{x-9} \, dx = 5xe^{x-9} - 5e^{x-9} + c$$

Then, letting $y(x) = 5xe^{x-9} - 5e^{x-9} + c$ and checking the answer,

$$y'(x) = 5x \cdot e^{x-9} + e^{x-9} \cdot 5 - 5e^{x-9} = 5xe^{x-9}$$

## 12.10   PRESENT VALUE OF A CASH FLOW

In equation (11.18) we saw that the present value $P$ of a sum of money $A$ to be received in the future, when interest is compounded continuously, is $P = Ae^{-rt}$. The present value of a continuous stream of income at the constant rate of $A(t)$ dollars per year, therefore, is simply the integral.

$$P_n = \int_0^n Ae^{-rt} \, dt = A \int_0^n e^{-rt} \, dt = A\left[-\frac{1}{r}e^{-rt}\right]_0^n = -\frac{A}{r}\left(e^{-rt}\right)\Big|_0^n$$

$$P_n = -\frac{A}{r}\left(e^{-rn} - e^{-r(0)}\right) = -\frac{A}{r}\left(e^{-rn} - 1\right)$$

$$P_n = \frac{A}{r}\left(1 - e^{-rn}\right) \tag{12.7}$$

**EXAMPLE 12.**    (a) The present value of a continuous stream of income at the constant rate of \$10,000 a year for 5 years when discounted continuously at 8 percent is calculated below, using (12.7), as is (b) the present value of \$3,000 at 6 percent for 4 years.

(a)                           $$P_n = \frac{10,000}{0.08}\left(1 - e^{-0.08(5)}\right) = 125,000(1 - e^{-0.4})$$

$$P_n = 125,000(1 - 0.6703201) = \$41,209.99$$

(b)                           $$P_n = \frac{3000}{0.06}\left(1 - e^{-0.06(4)}\right) = 50,000(1 - e^{-0.24})$$

$$P_n = 50,000(1 - 0.7866279) = \$10,668.61$$

## 12.11 CONSUMERS' AND PRODUCERS' SURPLUS

A demand function $P = f(Q)$, as in Fig. 12-5(a), represents the different prices consumers are willing to pay for different quantities of a good. If equilibrium in the market occurs at $(Q_0, P_0)$, with all consumers paying the same price, the consumers who would have bought the good at a higher price benefit. Total benefit to consumers, called consumers' surplus, is depicted by the shaded area. Mathematically,

$$\text{Consumers' surplus} = \int_0^{Q_0} f(Q)\, dQ - Q_0 P_0 \qquad (12.8)$$

A supply function $P = g(Q)$, as in Fig. 12-5(b), represents the prices at which producers will supply different quantities of a good. If market equilibrium occurs at $(Q_0, P_0)$, producers willing to supply at prices lower than $P_0$ benefit. Total gain to producers is termed producers' surplus and is designated by the shaded area. Mathematically,

$$\text{Producers' surplus} = Q_0 P_0 - \int_0^{Q_0} g(Q)\, dQ \qquad (12.9)$$

See Examples 13 and 14 and Problems 12.40 and 12.41.

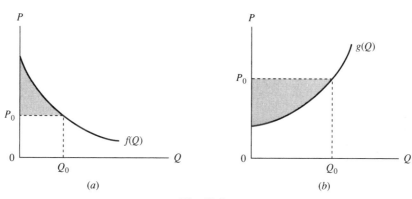

**Fig. 12-5**

**EXAMPLE 13.** Given the demand function $P = 110 - Q^2$ and assuming that at market equilibrium $Q_0 = 9$ and $P_0 = 29$, the consumers' surplus is estimated as follows, using (12.8):

$$\text{Consumers' surplus} = \int_0^9 (110 - Q^2)\, dQ - (9)(29)$$

$$= \left[110Q - \tfrac{1}{3}Q^3\right]_0^9 - 261$$

$$= (990 - 243) - (0) - 261 = 486$$

**EXAMPLE 14.**    Given the supply function $P = (Q + 6)^2$ and assuming that at market equilibrium $Q_0 = 3$ and $P_0 = 81$, the producers' surplus is estimated as follows, using $(12.9)$:

$$\text{Producers' surplus} = (3)(81) - \int_0^3 (Q + 6)^2 \, dQ$$

$$= 243 - \left[ \tfrac{1}{3}(Q + 6)^3 \right]_0^3$$

$$= 243 - \tfrac{1}{3} \left[ (9)^3 - (6)^3 \right] = 72$$

For a variety of different applications, see Problems 12.28 to 12.41.

# Solved Problems

## INDEFINITE INTEGRALS

**12.1.**    Find the following indefinite integrals. Check the answers on your own by making sure that the derivative of the antiderivative equals the integrand.

(a)    $\int 8 \, dx$

$$\int 8 \, dx = 8x + c \qquad\qquad\qquad \text{(Rule 1)}$$

(b)    $\int -16 \, dx$

$$\int -16 \, dx = - \int 16 \, dx = -16x + c \qquad\qquad \text{(Rules 1 and 8)}$$

(c)    $\int x^5 \, dx$

$$\int x^5 \, dx = \frac{1}{5 + 1} x^{5+1} + c = \frac{1}{6} x^6 + c \qquad\qquad \text{(Rule 3)}$$

(d)    $\int \frac{1}{x^4} \, dx$

$$\int \frac{1}{x^4} \, dx = \int x^{-4} \, dx = \left( \frac{1}{-4 + 1} \right) x^{-4+1} + c = -\frac{1}{3} x^{-3} + c \qquad \text{(Rule 3)}$$

(e)    $\int 21x^6 \, dx$

$$\int 21x^6 \, dx = 21 \int x^6 \, dx = 21 \cdot \frac{1}{7} \cdot x^7 + c = 3x^7 + c \qquad \text{(Rules 3 and 6)}$$

(f)    $\int (35x^4 - 8x^3) \, dx$

$$\int (35x^4 - 8x^3) \, dx = \frac{35}{5} x^5 - \frac{8}{4} x^4 + c = 7x^5 - 2x^4 + c \qquad \text{(Rules 3,6,7)}$$

(g)    $\int -\sqrt{x} \, dx$

$$\int -\sqrt{x} \, dx = - \int x^{1/2} \, dx = \frac{-1}{(3/2)} x^{3/2} + c = -\frac{2}{3} x^{3/2} + c \qquad \text{(Rules 3 and 8)}$$

(h) $\displaystyle \int 6e^{-1.5t}\, dt$

$$\int 6e^{-1.5t}\, dt = 6\left(\frac{1}{-1.5}\right)e^{-1.5t} + c = -4e^{-1.5t} + c \qquad \text{(Rules 5 and 6)}$$

(i) $\displaystyle \int 20e^{1.25t}\, dt$

$$\int 20e^{1.25t}\, dt = 20\left(\frac{1}{1.25}\right)e^{1.25t} + c = 16e^{1.25t} + c \qquad \text{(Rules 5 and 6)}$$

(j) $\displaystyle \int \frac{2}{x}\, dx$

$$\int \frac{2}{x}\, dx = \int 2x^{-1}\, dx = 2\ln|x| + c = \ln x^2 + c \qquad \text{(Rules 4 and 6)}$$

**12.2.** Determine the following indefinite integrals.

(a) $\displaystyle \int 15e^{t/4}\, dt$

$$\int 15e^{t/4}\, dt = \int 15e^{(1/4)t}\, dt = 15\left[\frac{1}{(1/4)}\right]e^{(1/4)t} + c = 60e^{t/4} + c$$

(b) $\displaystyle \int x^{2/3}\, dx$

$$\int x^{2/3}\, dx = \frac{1}{(5/3)}x^{5/3} + c = \frac{3}{5}x^{5/3} + c$$

(c) $\displaystyle \int \frac{dx}{\sqrt{x}}$

$$\int \frac{dx}{\sqrt{x}} = \int x^{-(1/2)}\, dx = \frac{1}{(1/2)}x^{1/2} + c = 2x^{1/2} + c = 2\sqrt{x} + c$$

(d) $\displaystyle \int \frac{9}{t^4}\, dt$

$$\int \frac{9}{t^4}\, dt = \int 9t^{-4}\, dt = 9\left(\frac{1}{-3}\right)t^{-3} + c = -3t^{-3} + c$$

(e) $\displaystyle \int \sqrt{x+7}\, dx$

$$\int \sqrt{x+7}\, dx = \int (x+7)^{1/2}\, dx = \frac{1}{(3/2)}(x+7)^{3/2} + c$$
$$= \frac{2}{3}(x+7)^{3/2} + c$$

(f) $\displaystyle \int \frac{1}{x+8}\, dx$

$$\int \frac{1}{x+8}\, dx = \int (x+8)^{-1}\, dx = \ln|x+8| + c$$

(g) $\displaystyle \int 4(x-18)^{-2}\, dx$

$$\int 4(x-18)^{-2}\, dx = 4\left(\frac{1}{-1}\right)(x-18)^{-1} + c = \frac{-4}{x-18} + c$$

(h) $\int (e^3 + 5t^4 + 3e^{-4t}) \, dt$, where $e^3$ is a constant.

$$\int (e^3 + 5t^4 + 3e^{-4t}) \, dt = e^3 t + t^5 - \frac{3}{4} e^{-4t} + c$$

**12.3.** Find the antiderivative for each of the following, given an *initial condition* $F(0) = k$ (a constant), or a *boundary condition* $F(a) = k$ $(a \neq 0)$.

(a) $\int (25x^{1/4} + 16x^{1/3}) \, dx$, given the initial condition $F(0) = 19$.

$$\int (25x^{1/4} + 16x^{1/3}) \, dx = 25 \left(\frac{4}{5}\right) x^{5/4} + 16 \left(\frac{3}{4}\right) x^{4/3} + c$$

$$= 20x^{5/4} + 12x^{4/3} + c$$

At $F(0) = 19$,                              $19 = 20(0)^{5/4} + 12(0)^{4/3} + c$

$c = 19$, and                              $F(x) = 20x^{5/4} + 12x^{4/3} + 19$

(b) $\int (16e^{2t} + 15e^{-3t}) \, dt$, given $F(0) = 9$.

$$\int (16e^{2t} + 15e^{-3t}) \, dt = 8e^{2t} - 5e^{-3t} + c$$

At $F(0) = 9$,                              $9 = 8e^{2(0)} - 5e^{-3(0)} + c$

With $e^0 = 1$,                              $9 = 8 - 5 + c \qquad c = 6$

Thus,                              $F(t) = 8e^{2t} - 5e^{-3t} + 6$

(c) $\int (4x^{-1} + 5x^{-2}) \, dx$, given the boundary condition $F(1) = 3$.

$$\int (4x^{-1} + 5x^{-2}) \, dx = 4 \ln |x| - 5x^{-1} + c = 4 \ln x - \frac{5}{x} + c$$

At $F(1) = 3$,                              $3 = 4 \ln 1 - \frac{5}{1} + c$

With $\ln 1 = 0$,                              $3 = -5 + c \qquad c = 8$

and                              $F(x) = 4 \ln x - \frac{5}{x} + 8$

## DEFINITE INTEGRALS

**12.4.** Evaluate the following definite integrals.

(a) $\int_2^5 3x^2 \, dx$

$$\int_2^5 3x^2 \, dx = x^3 \Big|_2^5 = (5)^3 - (2)^3 = 125 - 8 = 117$$

(b) $\int_1^3 -36x^{-3} \, dx$

$$\int_1^3 -36x^{-3} \, dx = 18x^{-2} \Big|_1^3 = \frac{18}{(3)^2} - \frac{18}{(1)^2} = 2 - 18 = -16$$

(c) $\displaystyle\int_9^{49} \frac{dx}{\sqrt{x}}$

$$\int_9^{49} \frac{dx}{\sqrt{x}} = \int_9^{49} x^{-1/2}\, dx = 2x^{1/2}\Big|_9^{49} = 2\sqrt{49} - 2\sqrt{9} = 8$$

(d) $\displaystyle\int_1^8 3x^{-1}\, dx$

$$\int_1^8 3x^{-1}\, dx = 3\ln x\Big|_1^8 = 3\ln 8 - 3\ln 1 = 3\ln 8 - 3(0) = 3\ln 8$$

(e) $\displaystyle\int_4^9 12\sqrt{x}\, dx$

$$\int_4^9 12\sqrt{x}\, dx = \int_4^9 12x^{1/2}\, dx = 8x^{3/2}\Big|_4^9$$
$$= 8(9)^{3/2} - 8(4)^{3/2} = 8(27) - 8(8) = 152$$

**12.5.** Evaluate the definite integrals given below.

(a) $\displaystyle\int_1^2 (9x^2 + 4x)\, dx$

$$\int_1^2 (9x^2 + 4x)\, dx = \left[3x^3 + 2x^2\right]_1^2$$
$$= \left(3(2)^3 + 2(2)^2\right) - \left(3(1)^3 + 2(1)^2\right)$$
$$= 32 - 5 = 27$$

(b) $\displaystyle\int_1^8 (8x^{1/3} + 4x^{-1/3})\, dx$

$$\int_1^8 (8x^{1/3} + 4x^{-1/3})\, dx = \left[6x^{4/3} + 6x^{2/3}\right]_1^8$$
$$= 6\left[(8)^{4/3} + (8)^{2/3}\right] - 6\left[(1)^{4/3} + (1)^{2/3}\right]$$
$$= 120 - 12 = 108$$

(c) $\displaystyle\int_0^1 e^{t/3}\, dt$

$$\int_0^1 e^{t/3}\, dt = 3e^{t/3}\Big|_0^1 = 3e^{1/3} - 3e^0 = 3(e^{1/3} - 1)$$

(d) $\displaystyle\int_0^1 20e^{-5t}\, dt$

$$\int_0^1 20e^{-5t}\, dt = -4e^{-5t}\Big|_0^1 = -4e^{-5} - (-4e^0) = 4(1 - e^{-5})$$

(e) $\displaystyle\int_0^2 (5 + x)^2\, dx$

$$\int_0^2 (5 + x)^2\, dx = \tfrac{1}{3}(5 + x)^3\Big|_0^2 = \tfrac{1}{3}\left[(7)^3 - (5)^3\right] = \tfrac{1}{3}(218) \approx 72.67$$

**PROPERTIES OF DEFINITE INTEGRALS**

**12.6.** Show $\int_{-3}^{3} (6x^2 + 18x)\,dx = \int_{-3}^{0} (6x^2 + 18x)\,dx + \int_{0}^{3} (6x^2 + 18x)\,dx.$

$$\int_{-3}^{3} (6x^2 + 18x)\,dx = (2x^3 + 9x^2)\Big|_{-3}^{3} = 135 - 27 = 108$$

$$\int_{-3}^{0} (6x^2 + 18x)\,dx = (2x^3 + 9x^2)\Big|_{-3}^{0} = 0 - 27 = -27$$

$$\int_{0}^{3} (6x^2 + 18x)\,dx = (2x^3 + 9x^2)\Big|_{0}^{3} = 135 - 0 = 135$$

Checking this answer,

$$-27 + 135 = 108$$

**12.7.** Show $\int_{0}^{25} (5x + x^{-1/2})\,dx = \int_{0}^{9} (5x + x^{-1/2})\,dx + \int_{9}^{16} (5x + x^{-1/2})\,dx + \int_{16}^{25} (5x + x^{-1/2})\,dx.$

$$\int_{0}^{25} (5x + x^{-1/2})\,dx = (2.5x^2 + 2x^{1/2})\Big|_{0}^{25} = 1562.5 + 10 = 1572.5$$

$$\int_{0}^{9} (5x + x^{-1/2})\,dx = (2.5x^2 + 2x^{1/2})\Big|_{0}^{9} = 202.5 + 6 = 208.5$$

$$\int_{9}^{16} (5x + x^{-1/2})\,dx = (2.5x^2 + 2x^{1/2})\Big|_{9}^{16} = 648 - 208.5 = 439.5$$

$$\int_{16}^{25} (5x + x^{-1/2})\,dx = (2.5x^2 + 2x^{1/2})\Big|_{16}^{25} = 1572.5 - 648 = 924.5$$

Checking this answer,

$$208.5 + 439.5 + 924.5 = 1572.5$$

**AREA BETWEEN CURVES**

**12.8.** (*a*) Draw graphs for the following functions, and (*b*) evaluate the area between the curves over the stated interval:

$$y_1 = 8 - x^2 \quad \text{and} \quad y_2 = -x + 6 \quad \text{from} \quad x = -1 \quad \text{to} \quad x = 2$$

(*a*)   See Fig. 12-6.

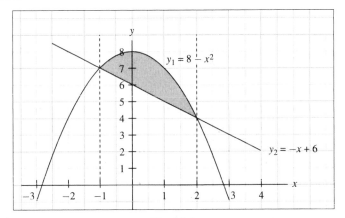

**Fig. 12-6**

(b)  From Fig. 12-6, the desired area is the area under the curve specified by $y_1 = 8 - x^2$ from $x = -1$ to $x = 2$ minus the area under the curve specified by $y_2 = -x + 6$ from $x = -1$ to $x = 2$. Using the properties of definite integrals,

$$A = \int_{-1}^{2} (8 - x^2)\, dx - \int_{-1}^{2} (-x + 6)\, dx$$

From (12.4),      $$A = \int_{-1}^{2} \left[ (8 - x^2) - (-x + 6) \right] dx$$

$$A = \int_{-1}^{2} (-x^2 + x + 2)\, dx = \left[ -\tfrac{1}{3}x^3 + \tfrac{1}{2}x^2 + 2x \right]_{-1}^{2}$$

$$A = 3\tfrac{1}{3} - \left( -1\tfrac{1}{6} \right) = 4\tfrac{1}{2}$$

**12.9.**  Redo Problem 12.8, given

$$y_1 = 6 \qquad y_2 = x^2 - 3 \qquad \text{from} \quad x = -3 \quad \text{to} \quad x = 3$$

(a)  See Fig. 12-7.

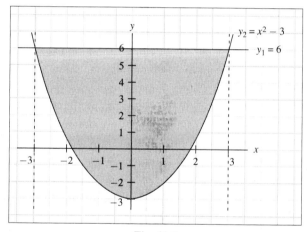

**Fig. 12-7**

(b)      $$A = \int_{-3}^{3} 6\, dx - \int_{-3}^{3} (x^2 - 3)\, dx = \int_{-3}^{3} (-x^2 + 9)\, dx$$

$$A = \left[ -\tfrac{1}{3}x^3 + 9x \right]_{-3}^{3} = 18 - (-18) = 36$$

**12.10.**  Redo Problem 12.8, given

$$y_1 = 5x + 20 \qquad y_2 = 30 - 5x^2 \qquad \text{from} \quad x = -2 \quad \text{to} \quad x = 1$$

(*a*)    See Fig. 12-8.

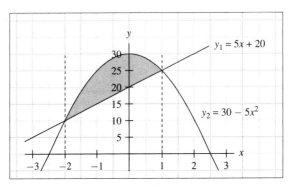

**Fig. 12-8**

(*b*)
$$A = \int_{-2}^{1} (30 - 5x^2)\,dx - \int_{-2}^{1} (5x + 20)\,dx$$

$$A = \int_{-2}^{1} (-5x^2 - 5x + 10)\,dx$$

$$A = \left(-\tfrac{5}{3}x^3 - \tfrac{5}{2}x^2 + 10x\right)\bigg|_{-2}^{1} = \tfrac{35}{6} - \left(-\tfrac{50}{3}\right) = \tfrac{135}{6} = 22.5$$

**12.11.** Redo Problem 12.8, given

$$y_1 = 4x^2 - 8x + 7 \qquad y_2 = -5x^2 + 10x - 3 \qquad \text{from} \quad x = 0 \quad \text{to} \quad x = 2$$

(*a*)    See Fig. 12-9.

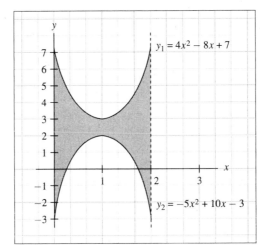

**Fig. 12-9**

(b)
$$A = \int_0^2 (4x^2 - 8x + 7)\, dx - \int_0^2 (-5x^2 + 10x - 3)\, dx$$

$$A = \int_0^2 (9x^2 - 18x + 10)\, dx$$

$$A = (3x^3 - 9x^2 + 10x)\Big|_0^2 = 8 - 0 = 8$$

**12.12.** (a) Draw graphs for the following functions and (b) evaluate the area between the curves over the stated interval:

$$y_1 = x \qquad y_2 = 3 \qquad \text{from} \quad x = 0 \quad \text{to} \quad x = 5$$

Notice the shift in the relative position of the curves at the point of intersection.

(a)    See Fig. 12-10.

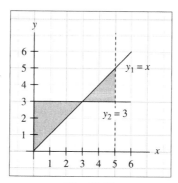

**Fig. 12-10**

(b)
$$A = \int_0^3 (3 - x)\, dx + \int_3^5 (x - 3)\, dx$$

$$A = \left(3x - \tfrac{1}{2}x^2\right)\Big|_0^3 + \left(\tfrac{1}{2}x^2 - 3x\right)\Big|_3^5$$

$$A = 4\tfrac{1}{2} + 2 = 6\tfrac{1}{2}$$

**12.13.** Redo Problem 12.12, given

$$y_1 = x + 2 \qquad y_2 = 12 - 1.5x \qquad \text{from} \quad x = 0 \quad \text{to} \quad x = 6$$

(a)    See Fig. 12-11.

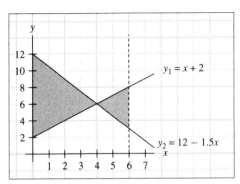

**Fig. 12-11**

(b)

$$A = \int_0^4 [(12 - 1.5x) - (x + 2)] \, dx + \int_4^6 [(x + 2) - (12 - 1.5x)] \, dx$$

$$A = \int_0^4 (-2.5x + 10) \, dx + \int_4^6 (2.5x - 10) \, dx$$

$$A = (-1.25x^2 + 10x)\Big|_0^4 + (1.25x^2 - 10x)\Big|_4^6 = 20 + 5 = 25$$

**12.14.** Redo Problem 12.12, given

$$y_1 = 6x \qquad y_2 = 2x^2 - 8x + 12 \qquad \text{from} \quad x = 0 \quad \text{to} \quad x = 3$$

(a)    See Fig. 12-12.

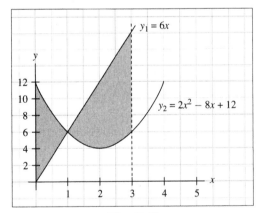

**Fig. 12-12**

(b)

$$A = \int_0^1 \left[ (2x^2 - 8x + 12) - (6x) \right] dx + \int_1^3 \left[ (6x) - (2x^2 - 8x + 12) \right] dx$$

$$A = \int_0^1 (2x^2 - 14x + 12) \, dx + \int_1^3 (-2x^2 + 14x - 12) \, dx$$

$$A = \left( \tfrac{2}{3}x^3 - 7x^2 + 12x \right)\Big|_0^1 + \left( -\tfrac{2}{3}x^3 + 7x^2 - 12x \right)\Big|_1^3 = 5\tfrac{2}{3} + 14\tfrac{2}{3} = 20\tfrac{1}{3}$$

**12.15.** Redo Problem 12.12, given

$$y_1 = -x^2 + 10 \qquad y_2 = -x^2 + 4x + 2 \qquad \text{from} \quad x = 0 \quad \text{to} \quad x = 3$$

(a)   See Fig. 12-13.

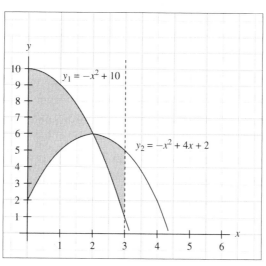

**Fig. 12-13**

(b)
$$A = \int_0^2 \left[(-x^2 + 10) - (-x^2 + 4x + 2)\right] dx + \int_2^3 \left[(-x^2 + 4x + 2) - (-x^2 + 10)\right] dx$$

$$A = \int_0^2 (-4x + 8) \, dx + \int_2^3 (4x - 8) \, dx$$

$$A = (-2x^2 + 8x)\Big|_0^2 + (2x^2 - 8x)\Big|_2^3$$

$$A = 8 + 2 = 10$$

## INTEGRATION BY SUBSTITUTION

**12.16.** Use integration by substitution to determine the indefinite integral for the following function involving a product. Check the answer on your own.

$$\int 60x^2 (x^3 + 5)^4 \, dx$$

Selecting as $u$ the function in which the independent variable is raised to the higher power in absolute value, let $u = x^3 + 5$. Then $du/dx = 3x^2$ and, solving algebraically for $dx$, $dx = du/3x^2$. Substituting these values in the original integrand to reduce it to a function of $u \, du/dx$,

$$\int 60x^2 (x^3 + 5)^4 \, dx = \int 60x^2 \cdot u^4 \cdot \frac{du}{3x^2} = \int 20u^4 \, du$$

Integrating with respect to $u$ with the simple power rule,

$$\int 20u^4 \, du = 20 \cdot \frac{1}{5} u^5 + c = 4u^5 + c$$

Then substituting $x^3 + 5$ for $u$,

$$\int 60x^2(x^3+5)^4 \, dx = 4u^5 + c = 4(x^3+5)^5 + c$$

**12.17.** Redo Problem 12.16, given

$$\int 84x(7x^2-4)^3 \, dx$$

Let $u = 7x^2 - 4$; then $du/dx = 14x$ and $dx = du/14x$. Substituting in the original integrand,

$$\int 84x(7x^2-4)^3 \, dx = \int 84x \cdot u^3 \cdot \frac{du}{14x} = \int 6u^3 \, du$$

Integrating,

$$\int 6u^3 \, du = 6 \cdot \frac{1}{4}u^4 + c = 1.5u^4 + c$$

Substituting,

$$\int 84x(7x^2-4)^3 dx = 1.5u^4 + c = 1.5(7x^2-4)^4 + c$$

**12.18.** Use integration by substitution to determine the indefinite integral for the following function involving a quotient.

$$\int \frac{72x}{(9x^2+2)^5} \, dx$$

Let $u = 9x^2 + 2$; then $du/dx = 18x$ and $dx = du/18x$. Substituting,

$$\int \frac{72x}{(9x^2+2)^5} \, dx = \int 72x \cdot \frac{1}{u^5} \cdot \frac{du}{18x} = \int \frac{4}{u^5} \, du = \int 4u^{-5} \, du$$

Integrating,

$$\int 4u^{-5} \, du = 4\left(-\frac{1}{4}\right)u^{-4} + c = -u^{-4} + c$$

Substituting,

$$\int \frac{72x}{(9x^2+2)^5} \, dx = -u^{-4} + c = -(9x^2+2)^{-4} + c$$

$$= \frac{-1}{(9x^2+2)^4} + c$$

**12.19.** Redo Problem 12.18, given

$$\int \frac{10x^2}{5x^3-8} \, dx$$

Letting $u = 5x^3 - 8$, then $du/dx = 15x^2$ and $dx = du/15x^2$. Substituting,

$$\int \frac{10x^2}{5x^3-8} \, dx = \int 10x^2 \cdot \frac{1}{u} \cdot \frac{du}{15x^2} = \int \frac{2}{3} \cdot \frac{1}{u} du = \frac{2}{3} \int u^{-1} \, du$$

Integrating,

$$\frac{2}{3}\int u^{-1} \, du = \frac{2}{3}\ln|u| + c$$

Substituting,

$$\int \frac{10x^2}{5x^3-8} \, dx = \frac{2}{3}\ln|u| + c = \frac{2}{3}\ln\left|5x^3-8\right| + c$$

**12.20.** Use integration by substitution to determine the indefinite integral for the following function involving a radical.

$$\int 72x^2 \sqrt{8x^3 + 5}\, dx$$

Let $u = 8x^3 + 5$, then $du/dx = 24x^2$ and $dx = du/24x^2$. Substituting in the original integrand,

$$\int 72x^2 \sqrt{8x^3 + 5}\, dx = \int 72x^2 \cdot \sqrt{u} \cdot \frac{du}{24x^2} = 3 \int u^{1/2}\, du$$

Integrating

$$3 \int u^{1/2}\, du = 3 \cdot \frac{2}{3} u^{3/2} + c = 2u^{3/2} + c$$

Substituting,

$$\int 72x^2 \sqrt{8x^3 + 5}\, dx = 2u^{3/2} + c = 2(8x^3 + 5)^{3/2} + c$$

$$= 2 \left( \sqrt{8x^3 + 5} \right)^3 + c$$

**12.21.** Redo Problem 12.20, given

$$\int \frac{15x^4}{\sqrt{6x^5 - 34}}\, dx$$

Letting $u = 6x^5 - 34$; then $du/dx = 30x^4$ and $dx = du/30x^4$. Substituting,

$$\int \frac{15x^4}{\sqrt{6x^5 - 34}}\, dx = \int 15x^4 \cdot \frac{1}{\sqrt{u}} \cdot \frac{du}{30x^4} = \frac{1}{2} \int u^{-1/2} du$$

Integrating

$$\frac{1}{2} \int u^{-1/2}\, du = \frac{1}{2} \cdot 2u^{1/2} + c = u^{1/2} + c$$

Substituting,

$$\int \frac{15x^4}{\sqrt{6x^5 - 34}}\, dx = u^{1/2} + c = (6x^5 - 34)^{1/2} + c$$

$$= \sqrt{6x^5 - 34} + c$$

**12.22.** Use integration by substitution to find the indefinite integral for the following natural exponential function.

$$\int 6xe^{3x^2}\, dx$$

Let $u = 3x^2$; then $du/dx = 6x$ and $dx = du/6x$. Substituting,

$$\int 6xe^{3x^2}\, dx = \int 6x \cdot e^u \cdot \frac{du}{6x} = \int e^u\, du$$

Integrating,

$$\int e^u\, du = e^u + c$$

Substituting $3x^2$ for $u$,

$$\int 6xe^{3x^2}\, dx = e^u + c = e^{3x^2} + c$$

**12.23.** Redo Problem 12.22, given

$$\int 36x^5 e^{2x^6}\, dx$$

Letting $u = 2x^6$; then $du/dx = 12x^5$, $dx = du/12x^5$, and

$$\int 36x^5 e^{2x^6}\, dx = \int 36x^5 \cdot e^u \cdot \frac{du}{12x^5} = 3 \int e^u\, du$$

Integrating

$$3 \int e^u\, du = 3e^u + c$$

Substituting,

$$\int 36x^5 e^{2x^6}\, dx = 3e^u + c = 3e^{2x^6} + c$$

## INTEGRATION BY PARTS

**12.24.** Use integration by parts to find the indefinite integral for each of the following functions involving a product. Check your answers on your own, recalling that the product rule is needed to check an answer obtained through integration by parts.

(a)

$$\int x(x-8)^3\, dx$$

Picking the simpler function for $f(x)$ and the more complicated function for $g'(x)$, let $f(x) = x$ and $g'(x) = (x-8)^3$. Then $f'(x) = 1$ and $g(x) = \int (x-8)^3\, dx = \frac{1}{4}(x-8)^4 + c_1$, where $c_1$ will drop out and can be ignored, as explained in Example 10. Now substituting the relevant values in (12.6),

$$\int x(x-8)^3\, dx = f(x) \cdot g(x) - \int \left[ g(x) \cdot f'(x) \right]\, dx$$

$$= x \cdot \frac{1}{4}(x-8)^4 - \int \left[ \frac{1}{4}(x-8)^4 \cdot (1) \right]\, dx$$

$$= \frac{x}{4}(x-8)^4 - \int \frac{1}{4}(x-8)^4\, dx$$

Then substituting $\int \frac{1}{4}(x-8)^4\, dx = \frac{1}{20}(x-8)^5 + c$ above,

$$\int x(x-8)^3\, dx = \frac{x}{4}(x-8)^4 - \frac{1}{20}(x-8)^5 + c$$

(b)

$$\int 56x(x+11)^6\, dx$$

Letting $f(x) = 56x$ and $g'(x) = (x+11)^6$, then $f'(x) = 56$ and $g(x) = \frac{1}{7}(x+11)^7$. Substituting in (12.6),

$$\int 56x(x+11)^6\, dx = 56x \cdot \frac{1}{7}(x+11)^7 - \int \left[ \frac{1}{7}(x+11)^7 \cdot 56 \right]\, dx$$

$$= 8x(x+11)^7 - \int 8(x+11)^7\, dx$$

Integrating once again,

$$\int 56x(x+11)^6\, dx = 8x(x+11)^7 - (x+11)^8 + c$$

**12.25.** Use integration by parts to find the indefinite integral for each of the following functions involving a quotient.

(a)
$$\int \frac{24x}{(x+7)^4}\, dx$$

Let $f(x) = 24x$ and $g'(x) = (x+7)^{-4}$; then $f'(x) = 24$ and $g(x) = \int (x+7)^{-4}\, dx = -\tfrac{1}{3}(x+7)^{-3}$. Substituting,

$$\int \frac{24x}{(x+7)^4}\, dx = f(x) \cdot g(x) - \int \left[ g(x) \cdot f'(x) \right] dx$$

$$= 24x \left[ -\frac{1}{3}(x+7)^{-3} \right] - \int \left[ -\frac{1}{3}(x+7)^{-3} \cdot 24 \right] dx$$

$$= -8x(x+7)^{-3} + 8 \int (x+7)^{-3}\, dx$$

Substituting $8 \int (x+7)^{-3}\, dx = -4(x+7)^{-2} + c$ above,

$$\int \frac{24x}{(x+7)^4}\, dx = -8x(x+7)^{-3} - 4(x+7)^{-2} + c$$

$$= \frac{-8x}{(x+7)^3} - \frac{4}{(x+7)^2} + c$$

(b)
$$\int -\frac{22x}{(x+4)^3}\, dx$$

Letting $f(x) = -22x$ and $g'(x) = (x+4)^{-3}$, then $f'(x) = -22$ and $g(x) = -\tfrac{1}{2}(x+4)^{-2}$. Substituting in (12.6),

$$\int \frac{-22x}{(x+4)^3}\, dx = -22x \left[ -\frac{1}{2}(x+4)^{-2} \right] - \int \left[ -\frac{1}{2}(x+4)^{-2} \cdot (-22) \right] dx$$

$$= 11x(x+4)^{-2} - \int 11(x+4)^{-2}\, dx$$

Integrating again,

$$\int \frac{-22x}{(x+4)^3}\, dx = 11x(x+4)^{-2} + 11(x+4)^{-1} + c$$

**12.26.** Use integration by parts to find the indefinite integral for each of the following functions involving a radical.

(a)
$$\int 30x\sqrt{9+x}\, dx$$

Let $f(x) = 30x$ and $g'(x) = (9+x)^{1/2}$; then $f'(x) = 30$ and $g(x) = \int (9+x)^{1/2}\, dx = \tfrac{2}{3}(9+x)^{3/2}$. Substituting,

$$\int 30x\sqrt{9+x}\, dx = f(x) \cdot g(x) - \int \left[ g(x) \cdot f'(x) \right] dx$$

$$= 30x \cdot \frac{2}{3}(9+x)^{3/2} - \int \left[ \frac{2}{3}(9+x)^{3/2} \cdot 30 \right] dx$$

$$= 20x(9+x)^{3/2} - 20 \int (9+x)^{3/2}\, dx$$

Substituting $-20 \int (9+x)^{3/2} \, dx = -8(9+x)^{5/2} + c$ above,

$$\int 30x\sqrt{9+x} \, dx = 20x(9+x)^{3/2} - 8(9+x)^{5/2} + c$$

(b)
$$\int \frac{9x}{\sqrt{x+23}} \, dx$$

Letting $f(x) = 9x$ and $g'(x) = (x+23)^{-1/2}$, then $f'(x) = 9$ and $g(x) = 2(x+23)^{1/2}$. Substituting in (12.6),

$$\int \frac{9x}{\sqrt{x+23}} \, dx = 9x \cdot 2(x+23)^{1/2} - \int \left[ 2(x+23)^{1/2} \cdot 9 \right] dx$$

$$= 18x(x+23)^{1/2} - \int 18(x+23)^{1/2} \, dx$$

Integrating again,

$$\int \frac{9x}{\sqrt{x+23}} \, dx = 18x(x+23)^{1/2} - 12(x+23)^{3/2} + c$$

**12.27.** Use integration by parts to find the indefinite integral for each of the following natural exponential functions.

(a)
$$\int (x+5)e^x \, dx$$

Let $f(x) = x + 5$ and $g'(x) = e^x$; then $f'(x) = 1$ and $g(x) = \int e^x \, dx = e^x$. Substituting,

$$\int (x+5)e^x \, dx = (x+5)e^x - \int e^x \, dx$$

Substituting $\int e^x \, dx = e^x + c$ above,

$$\int (x+5)e^x \, dx = (x+5)e^x - e^x + c$$

(b)
$$\int 24x^2 e^{6x} \, dx$$

Let $f(x) = 24x^2$ and $g'(x) = e^{6x}$. Then $f'(x) = 48x$ and $g(x) = \frac{1}{6}e^{6x}$. Substituting in (12.6),

$$\int 24x^2 e^{6x} \, dx = 24x^2 \cdot \frac{1}{6}e^{6x} - \int \left[ \frac{1}{6}e^{6x} \cdot 48x \right] dx$$

$$= 4x^2 e^{6x} - \int 8x e^{6x} \, dx$$

Using integration by parts again,

$$\int 24x^2 e^{6x} \, dx = 4x^2 e^{6x} - \left[ 8x \cdot \frac{1}{6}e^{6x} - \int \left( \frac{1}{6}e^{6x} \cdot 8 \right) dx \right]$$

$$= 4x^2 e^{6x} - \frac{4}{3}x e^{6x} + \int \frac{4}{3}e^{6x} \, dx$$

$$= 4x^2 e^{6x} - \frac{4}{3}x e^{6x} + \frac{2}{9}e^{6x} + c$$

## PRACTICAL APPLICATIONS

**12.28.** A firm's marginal cost function is $MC = x^2 - 6x + 125$, where $x$ is the number of units produced. Fixed costs are \$280. Find the total cost TC of producing $x$ units.

$$TC = \int MC\, dx = \int (x^2 - 6x + 125)\, dx = \tfrac{1}{3}x^3 - 3x^2 + 125x + c$$

Substituting the fixed cost information: $TC(0) = 280$,

$$280 = \tfrac{1}{3}(0)^3 - 3(0)^2 + 125(0) + c \qquad c = 280$$

Thus $$TC = \tfrac{1}{3}x^3 - 3x^2 + 125x + 280$$

Note that in integrating a marginal cost function to obtain a total cost function, the fixed cost FC will always be the constant of integration.

**12.29.** Find the total cost function TC given the marginal cost function $MC = x^2 - 5x + 89$ and fixed costs of \$225.

$$TC = \int (x^2 - 5x + 89)\, dx = \tfrac{1}{3}x^3 - 2.5x^2 + 89x + c$$

Substituting $FC = 225$ for $c$,

$$TC = \tfrac{1}{3}x^3 - 2.5x^2 + 89x + 225$$

**12.30.** A producer's marginal cost is $MC = \tfrac{1}{8}x^2 - x + 320$. What is the total cost TC of producing 2 extra units if 6 units are currently being produced?

$$TC(8) - TC(6) = \int_6^8 MC\, dx = \int_6^8 \left(\tfrac{1}{8}x^2 - x + 320\right) dx$$

$$= \left[\tfrac{1}{24}x^3 - \tfrac{1}{2}x^2 + 320x\right]_6^8 = 2549\tfrac{1}{3} - 1911 = 638\tfrac{1}{3}$$

**12.31.** A firm's marginal revenue is $MR = 425 - 0.5x - 0.15x^2$. Find the total revenue TR.

$$TR = \int MR\, dx = \int (425 - 0.5x - 0.15x^2)\, dx$$

$$TR = 425x - 0.25x^2 - 0.05x^3 + c$$

Since there is no revenue when there are no sales, $TR(0) = 0$. In integrating a MR function to find a TR function, therefore, the constant of integration $c$ always equals zero. So

$$TR = 425x - 0.25x^2 - 0.05x^3$$

**12.32.** Find TR, given $MR = 185 - 0.4x - 0.3x^2$.

$$TR = \int (185 - 0.4x - 0.3x^2)\, dx = 185x - 0.2x^2 - 0.1x^3$$

**12.33.** Find the additional total revenue derived from increasing daily sales from 5 to 9 units, given $MR = 270 - 8x$.

$$TR(9) - TR(5) = \int_5^9 MR\, dx = \int_5^9 (270 - 8x)\, dx$$

$$= \left[270x - 4x^2\right]_5^9 = 2106 - 1250 = 856$$

**12.34.** Find the additional total revenue from increasing sales from 100 to 250 units for the producer in pure competition for whom $MR = 15$.

$$TR(250) - TR(100) = \int_{100}^{250} MR\, dx = \int_{100}^{250} 15\, dx$$

$$= \left[15x\right]_{100}^{250} = 3750 - 1500 = 2250$$

**12.35.** A manufacturer's marginal profit is $\pi' = -3x^2 + 80x + 140$. Find the additional profit $\pi$ earned by increasing production from 2 units to 4 units.

$$\pi(4) - \pi(2) = \int_2^4 (-3x^2 + 80x + 140)\, dx$$

$$= \left(-x^3 + 40x^2 + 140x\right)_2^4 = 704$$

**12.36.** Maintenance costs $M(t)$ in a factory increase as plant and equipment get older. Given the rate of increase in maintenance costs in dollars per year $M'(t) = 75t^2 + 9000$, where $t$ is the number of years, find the total maintenance costs of the factory from 4 to 6 years.

$$M(6) - M(4) = \int_4^6 (75t^2 + 9000)\, dt$$

$$= \left(25t^3 + 9000t\right)_4^6 = 21,800$$

**12.37.** A car depreciates rapidly in value in its first few years and more slowly in later years. Given the rate at which the value of a car depreciates over the years, $V'(t) = 300(t - 8)$ for $0 \le t \le 8$, and a sticker price of \$12,000, find (a) the value of the car $V(t)$, (b) the total amount by which the car depreciates in the first 4 years, and (c) the total amount by which it depreciates in the next 4 years. (d) Use the answer in (a) evaluated at $t = 4$ to check your answer in (b).

(a)
$$V(t) = \int 300(t - 8)\, dt = 150t^2 - 2400t + c$$

With $V(0) = 12,000$,        $V(t) = 150t^2 - 2400t + 12,000$        (12.10)

(b)
$$V(4) - V(0) = \int_0^4 300(t - 8)\, dt$$

$$= \left(150t^2 - 2400t\right)_0^4 = 2400 - 9600 = -\$7200$$

The value of the car decreases, that is, depreciates, by \$7200 in the first 4 years.

(c)
$$V(8) - V(4) = \int_4^8 300(t - 8)\, dt$$

$$= (150t^2 - 2400t)\Big|_4^8 = -\$2400$$

The car depreciates by \$2400 in the next 4 years.

(d) Evaluating (12.10) at $t = 4$,

$$V(4) = 150(4)^2 - 2400(4) + 12,000 = 4800$$

$$V(0) - V(4) = 12,000 - 4800 = \$7200$$

The car has depreciated by \$7200 as was found in (b).

**12.38.** Find the present value of $7500 to be paid each year for 4 years when the interest rate is 10 percent compounded continuously.

From (12.7),
$$P = \frac{A}{r}\left(1 - e^{-rn}\right)$$

Substituting,
$$P = \frac{7500}{0.1}\left(1 - e^{-(0.1)(4)}\right)$$
$$P = 75{,}000(0.32968) = \$24{,}726$$

**12.39.** Find the present value of $15,000 to be paid each year for 8 years when the interest rate is 7.5 percent compounded continuously.

Substituting in (12.7),

$$P = \frac{15{,}000}{0.075}\left(1 - e^{-(0.075)(8)}\right)$$
$$= 200{,}000(0.45119) = \$90{,}238$$

**12.40.** Find the consumers' surplus CS for each of the following demand curves at the level indicated.

(a) $P = 375 - 3Q^2$; $Q_0 = 10$, $P_0 = 75$

From (12.8),
$$CS = \int_0^{Q_0} f(Q)\,dQ - Q_0 P_0$$
$$CS = \int_0^{10} (375 - 3Q^2)\,dQ - 10(75)$$
$$CS = (375Q - Q^3)\Big|_0^{10} - 750$$
$$CS = 3750 - 1000 - 750 = 2000$$

(b) $P = \frac{350}{Q+5}$; $Q_0 = 20$, $P_0 = 12$

$$CS = \int_0^{20} \left[350(Q+5)^{-1}\right] dQ - 12(20)$$
$$CS = [350\ln(Q+5)]_0^{20} - 240$$
$$CS = [350(3.21888 - 1.60944)] - 240 \approx 323.30$$

**12.41.** Find the producers' surplus PS for each of the following supply curves at the level indicated.

(a) $P = Q^2 + 4Q + 60$; $Q_0 = 5$, $P_0 = 85$

From (12.9),  $$PS = Q_0 P_0 - \int_0^{Q_0} g(Q)\,dQ$$
$$PS = 5(85) - \int_0^5 (Q^2 + 4Q + 60)\,dQ$$
$$PS = 425 - \left(\tfrac{1}{3}Q^3 + 2Q^2 + 60Q\right)\Big|_0^5 = 425 - 391.67 = 33.33$$

(b)   $P = 5 + \frac{1}{4}\sqrt{Q}$; $Q_0 = 144$,  $P_0 = 8$

$$PS = 144(8) - \int_0^{144} \left(5 + \frac{1}{4}Q^{1/2}\right) dQ$$

$$PS = 1152 - \left[5Q + \frac{1}{6}Q^{3/2}\right]_0^{144} = 1152 - 720 - 288 = 144$$

# Supplementary Problems

## INDEFINITE INTEGRALS

**12.42.**     Find the following indefinite integrals:

(a)   $\int (24x^5 + 35x^4 - 64x^3)\, dx$      (b)   $\int (8x^{-5} - x^{-3} + 5x^{-2})\, dx$

(c)   $\int 21x^{3/4}\, dx$      (d)   $\int 8x^{-1/3}\, dx$

(e)   $\int 6\sqrt{x}\, dx$      (f)   $\int 9\sqrt{x - 13}\, dx$

(g)   $\int 28e^{-1.75t}\, dt$      (h)   $\int \dfrac{5}{x - 6}\, dx$

**12.43.**     Find the indefinite integrals for each of the following, given an initial condition or a boundary condition.

(a)   $\int (16x^3 - 21x^{2/5})\, dx$, given $F(0) = 13$.      (b)   $\int (18x^5 + 4x^{-3})\, dx$, given $F(1) = 8$.

(c)   $\int (3x^8 - 2x^{-4})\, dx$, given $F(1) = -4$.      (d)   $\int (28e^{4t} - 18e^{-6t})\, dt$, $F(0) = 2$.

## DEFINITE INTEGRALS

**12.44.**     Evaluate each of the following definite integrals:

(a)   $\int_1^4 9x^2\, dx$      (b)   $\int_5^{15} -60x^{-2}\, dx$      (c)   $\int_4^{64} 2.5x^{-1/2}\, dx$

(d)   $\int_1^7 8x^{-1}\, dx$      (e)   $\int_8^{27} \dfrac{dx}{\sqrt[3]{x}}$      (f)   $\int_0^2 3e^{t/5}\, dt$

(g)   $\int_{-0.5}^0 -42e^{-6t}\, dt$      (h)   $\int_{16}^{36} 15\sqrt{x}\, dx$

**12.45.**     Evaluate each of the following definite integrals:

(a)   $\int_2^5 (8x + 7)\, dx$      (b)   $\int_2^4 (15x^2 - 6x)\, dx$

(c)   $\int_8^{125} (4x^{-1/3} - x^{-2/3})\, dx$      (d)   $\int_8^{12} 20(x - 7)^3\, dx$

(e)   $\int_{-2}^2 (6x^2 + 10x + 3)\, dx$      (f)   $\int_0^{0.25} (12e^{-3t} - 8e^{-2t})\, dt$

## AREA BETWEEN CURVES

**12.46.**     Evaluate the area between the curves over the stated interval for each of the following sets of functions, after having drawn the curves on your own:

(a)   $y_1 = 2x^2 - 8$ and $y_2 = -2x + 4$ from $x = -3$ to $x = 2$

(b)   $y_1 = -4x^2 + 64$ and $y_2 = 4x + 16$ from $x = -4$ to $x = 3$

(c)   $y_1 = 7x^2 - 14x + 13$ and $y_2 = -3x^2 + 6x - 2$ from $x = 0$ to $x = 2$

(d)   $y_1 = -3x^2 + 24x - 2$ and $y_2 = -17x + 68$ from $x = 0$ to $x = 4$

## INTEGRATION BY SUBSTITUTION

**12.47.** Use integration by substitution to determine the indefinite integral for each of the following functions:

(a) $\int 252x^2(7x^3 - 12)^2\, dx$

(b) $\int 90x^3(3x^4 + 19)^4\, dx$

(c) $\int 672x(8x^2 - 9)^5\, dx$

(d) $\int \dfrac{120x^3}{(5x^4 - 18)^4}\, dx$

(e) $\int \dfrac{-576x^2}{(8x + 13)^5}\, dx$

(f) $\int \dfrac{30x^4}{6x^5 - 21}\, dx$

(g) $\int \dfrac{18x^3}{\sqrt{9x^4 - 5}}\, dx$

(h) $\int 99x^2\sqrt{22x^3 + 19}\, dx$

(i) $\int 96x^2 e^{4x^3}\, dx$

(j) $\int 120x^3 e^{-6x^4}\, dx$

## INTEGRATION BY PARTS

**12.48.** Use integration by parts to determine each of the following indefinite integrals:

(a) $\int 20x(x + 9)^4\, dx$

(b) $\int 42x(x - 3)^5\, dx$

(c) $\int \dfrac{36x}{(x + 8)^5}\, dx$

(d) $\int \dfrac{-18x}{(x - 7)^3}\, dx$

(e) $\int \dfrac{6x}{\sqrt{x + 13}}\, dx$

(f) $\int 45x\sqrt{x + 21}\, dx$

(g) $\int 7xe^{x-3}\, dx$

(h) $\int 32x^2 e^{4x}\, dx$

**12.49.** A firm's marginal cost function is $MC = x^2 - 11x + 385$. Its fixed costs are \$450. Find the total cost TC function.

**12.50.** Find a firm's total cost of producing 3 extra units as a firm moves from a production level of 4 units to 7 units, given $MC = 3x^2 - 4x + 525$.

**12.51.** Find a firm's total revenue TR function, given the marginal revenue function $MR = -0.24x^2 - 1.5x + 660$.

**12.52.** Estimate the additional total revenue received from increasing sales from 10 to 12 units, given $MR = 104 - 6x$.

**12.53.** A car worth \$6000 depreciates over the years at the rate $V'(t) = 250(t - 6)$ for $0 \le t \le 6$. Find the total amount by which the car depreciates in (a) the first 3 years and (b) the last three years.

**12.54.** Find the consumers' surplus at $x_0 = 5$ and $p_0 = 125$, given the demand function $p = -6x^2 + 275$.

**12.55.** Find the producers' surplus at $x_0 = 6$ and $p_0 = 83.4$, given the supply function $p = 0.15x^2 + 8x + 30$.

# Answers to Supplementary Problems

**12.42.**
(a) $F(x) = 4x^6 + 7x^5 - 16x^4 + c$
(b) $F(x) = -2x^{-4} + \frac{1}{2}x^{-2} - 5x^{-1} + c$
(c) $F(x) = 12x^{7/4} + c$
(d) $F(x) = 12x^{2/3} + c$
(e) $F(x) = 4x^{3/2} + c$
(f) $F(x) = 6(x - 13)^{3/2} + c$
(g) $F(t) = -16e^{-1.75t} + c$
(h) $F(x) = 5\ln|x - 6| + c$

**12.43.**
(a) $F(x) = 4x^4 - 15x^{7/5} + 13$
(b) $F(x) = 3x^6 - 2x^{-2} + 7$
(c) $F(x) = \frac{1}{3}x^9 + \frac{2}{3}x^{-3} - 5$
(d) $F(t) = 7e^{4t} + 3e^{-6t} - 8$

**12.44.**  (a)  $F(x) = 3x^3 \Big|_1^4 = 189$       (b)  $F(x) = \dfrac{60}{x}\Big|_5^{15} = -8$

(c)  $F(x) = 5x^{1/2}\Big|_4^{64} = 30$       (d)  $F(x) = 8\ln x \Big|_1^7 = 8\ln 7$

(e)  $F(x) = \frac{3}{2}x^{2/3}\Big|_8^{27} = 7.5$       (f)  $F(x) = 15e^{t/5}\Big|_0^2 = 15(e^{2/5} - 1)$

(g)  $F(x) = 7e^{-6t}\Big|_{-0.5}^0 = 7(1 - e^3)$       (h)  $F(x) = 10x^{3/2}\Big|_{16}^{36} = 1520$

**12.45.**  (a)  $F(x) = (4x^2 + 7x)\Big|_2^5 = 105$       (b)  $F(x) = (5x^3 - 3x^2)\Big|_2^4 = 244$

(c)  $F(x) = (6x^{2/3} - 3x^{1/3})\Big|_8^{125} = 117$       (d)  $F(x) = 5(x - 7)^4\Big|_8^{12} = 3120$

(e)  $F(x) = (2x^3 + 5x^2 + 3x)\Big|_{-2}^2 = 44$       (f)  $F(x) = 4(e^{-2t} - e^{-3t})\Big|_0^{0.25} = 0.5366564$

**12.46.**  (a)  $F(x) = \left(-\frac{2}{3}x^3 - x^2 + 12x\right)\Big|_{-3}^2 = 41.67$       (b)  $F(x) = \left(-\frac{4}{3}x^3 - 2x^2 + 48x\right)\Big|_{-4}^3 = 228.67$

(c)  $F(x) = \left(3\frac{1}{3}x^3 - 10x^2 + 15x\right)\Big|_0^2 = 16.67$

(d)  $F(x) = (x^3 - 20.5x^2 + 70x)\Big|_0^2 + (-x^3 + 20.5x^2 - 70x)\Big|_2^4 = 116$

**12.47.**  (a)  $F(x) = 4(7x^3 - 12)^3 + c$       (b)  $F(x) = 1.5(3x^4 + 19)^5 + c$

(c)  $F(x) = 7(8x^2 - 9)^6 + c$       (d)  $F(x) = -2(5x^4 - 18)^{-3} + c$

(e)  $F(x) = 6(8x^3 + 13)^{-4} + c$       (f)  $F(x) = \ln\left|6x^5 - 21\right| + c$

(g)  $F(x) = \sqrt{9x^4 - 5} + c$       (h)  $F(x) = (22x^3 + 19)^{3/2} + c$

(i)  $F(x) = 8e^{4x^3} + c$       (j)  $F(x) = -5e^{-6x^4} + c$

**12.48.**  (a)  $F(x) = 4x(x + 9)^5 - \frac{2}{3}(x + 9)^6 + c$       (b)  $F(x) = 7x(x - 3)^6 - (x - 3)^7 + c$

(c)  $F(x) = -9x(x + 8)^{-4} - 3(x + 8)^{-3} + c$       (d)  $F(x) = 9x(x - 7)^{-2} + 9(x - 7)^{-1} + c$

(e)  $F(x) = 12x(x + 13)^{1/2} - 8(x + 13)^{3/2} + c$       (f)  $F(x) = 30x(x + 21)^{3/2} - 12(x + 21)^{5/2} + c$

(g)  $F(x) = 7xe^{x-3} - 7e^{x-3} + c$       (h)  $F(x) = 8x^2e^{4x} - 4xe^{4x} + e^{4x} + c$

**12.49.**  $\text{TC} = \frac{1}{3}x^3 - 5.5x^2 + 385x + 450$

**12.50.**  $\text{TC} = (x^3 - 2x^2 + 525x)\Big|_4^7 = 1788$

**12.51.**  $\text{TR} = -0.08x^3 - 0.75x^2 + 660x$

**12.52.**  $\text{TR} = (104x - 3x^2)\Big|_{10}^{12} = 76$

**12.53.**  (a)  $(125t^2 - 1500t)\Big|_0^3 = -\$3375$       (b)  $(125t^2 - 1500t)\Big|_3^6 = -\$1125$

**12.54.**  $\text{CS} = (-2x^3 + 275x)\Big|_0^5 - 625 = 500$

**12.55.**  $\text{PS} = 500.4 - (0.05x^3 + 4x^2 + 30x)\Big|_0^6 = 165.6$

# Chapter 13

# Calculus of Multivariable Functions

## 13.1 FUNCTIONS OF SEVERAL INDEPENDENT VARIABLES

Study of the derivative in Chapters 9 and 10 was limited to functions of a single independent variable such as $y = f(x)$. Since many economic activities involve more than one independent variable, however, we now consider functions of two or more independent variables. Definitionally, $z = f(x, y)$ is a *function of two independent variables* if there exists one and only one value of $z$ in the range of $f$ for each ordered pair of real numbers $(x, y)$ in the domain of $f$. By convention, $z$ is the *dependent variable*; $x$ and $y$, the *independent variables*.

**EXAMPLE 1.** A firm produces one good $x$ for which the cost function is $C(x) = 125 + 7x$ and another good $y$ for which the cost function is $C(y) = 305 + 9y$. By adding individual costs, the total cost to the firm can be expressed simply as

$$C(x, y) = 125 + 7x + 305 + 9y$$

$$C(x, y) = 430 + 7x + 9y$$

Other examples of multivariable or multivariate functions include

$$f(x, y) = x^2 + 5xy + y^2$$

$$Q(K, L) = -2K^2 + 78KL - 3L^2$$

**EXAMPLE 2.** Multivariable functions can be evaluated for specific values of $x$ and $y$, such as $x = 2, y = 4$, by replacing $x$ and $y$ with the desired values. Using the functions from Example 1,

$$C(2, 4) = 430 + 7(2) + 9(4) = 480$$

$$f(2, 4) = (2)^2 + 5(2)(4) + (4)^2 = 60$$

$$Q(2, 4) = -2(2)^2 + 78(2)(4) - 3(4)^2 = 568$$

## 13.2 PARTIAL DERIVATIVES

Given a multivariate function such as $z = f(x, y)$, to measure the effect of a change in one of the independent variables ($x$ or $y$) on the dependent variable ($z$), we need a partial derivative. There is a distinct partial derivative for each of the independent variables. The *partial derivative of $z$ with respect to $x$* measures the instantaneous rate of change of $z$ with respect to $x$ while $y$ is held constant. Written $\partial z/\partial x$, $\partial f/\partial x$, $f_x(x, y)$, $f_x$, $z_x$, $f_1$, or $z_1$, it is defined as

$$\frac{\partial z}{\partial x} = \lim_{\Delta x \to 0} \frac{f(x + \Delta x, y) - f(x, y)}{\Delta x} \tag{13.1a}$$

Similarly, the *partial derivative of $z$ with respect to $y$* measures the rate of change of $z$ with respect to $y$ while $x$ is held constant. Written $\partial z/\partial y$, $\partial f/\partial y$, $f_y(x, y)$, $f_y$, $z_y$, $f_2$, or $z_2$, it is defined as

$$\frac{\partial z}{\partial y} = \lim_{\Delta y \to 0} \frac{f(x, y + \Delta y) - f(x, y)}{\Delta y} \tag{13.1b}$$

To find the partial derivative of a function with respect to one of the independent variables, simply treat the other independent variable as a constant and follow the ordinary rules of differentiation. See Examples 3 and 4 and Problems 13.1 to 13.8.

**EXAMPLE 3.**   The partial derivatives of a multivariable function such as $z = 8x^3 y^5$ are found as follows.

(a)   When differentiating with respect to $x$, treat the $y$ term as a constant by mentally bracketing it with the coefficient:

$$z = \left[8y^5\right] \cdot x^3$$

Then take the derivative of the $x$ term, holding the $y$ term constant,

$$\frac{\partial z}{\partial x} = z_x = \left[8y^5\right] \cdot \frac{d}{dx}(x^3)$$

$$= \left[8y^5\right] \cdot 3x^2$$

Recalling that a multiplicative constant remains in the process of differentiation, simply multiply and rearrange terms to obtain

$$\frac{\partial z}{\partial x} = z_x = 24x^2 y^5$$

(b)   When differentiating with respect to $y$, treat the $x$ term as a constant by bracketing it with the coefficient; then take the derivative with respect to $y$:

$$z = \left[8x^3\right] \cdot y^5$$

$$\frac{\partial z}{\partial y} = z_y = \left[8x^3\right] \cdot \frac{d}{dy}(y^5)$$

$$= \left[8x^3\right] \cdot 5y^4 = 40x^3 y^4$$

**EXAMPLE 4.**   To find the partial derivatives for

$$z = 12x^4 - 10x^2 y^3 + 15y^6$$

(a)   When differentiating with respect to $x$, mentally bracket off all $y$ terms to remember to treat them as constants:

$$z = 12x^4 - \left[10y^3\right]x^2 + \left[15y^6\right]$$

then take the derivative with respect to $x$, remembering that in differentiation multiplicative constants remain but additive constants drop out because the derivative of a constant is zero.

$$\frac{\partial z}{\partial x} = \frac{d}{dx}(12x^4) - \left[10y^3\right] \cdot \frac{d}{dx}(x^2) + \frac{d}{dx}\left[15y^6\right]$$

$$= 48x^3 - \left[10y^3\right] \cdot 2x + 0$$

$$= 48x^3 - 20xy^3$$

(b)   When differentiating with respect to $y$, block off all the $x$ terms and differentiate with respect to $y$.

$$z = \left[12x^4\right] - \left[10x^2\right]y^3 + 15y^6$$

$$\frac{\partial z}{\partial y} = \frac{d}{dy}\left[12x^4\right] - \left[10x^2\right] \cdot \frac{d}{dy}(y^3) + \frac{d}{dy}(15y^6)$$

$$= 0 - \left[10x^2\right] \cdot 3y^2 + 90y^5$$

$$= -30x^2 y^2 + 90y^5$$

See Problem 13.1.

## 13.3   RULES OF PARTIAL DIFFERENTIATION

The rules of partial differentiation have the same basic structure as the rules of ordinary differentiation presented in Section 9.7. A few key rules are given below, illustrated in Examples 5 to 9, treated in Problems 13.2 to 13.8, and verified in Problem 13.45.

### 13.3.1   Product Rule

Given $z = g(x, y) \cdot h(x, y)$,

$$\frac{\partial z}{\partial x} = g(x, y) \cdot \frac{\partial h}{\partial x} + h(x, y) \cdot \frac{\partial g}{\partial x} \qquad (13.2a)$$

$$\frac{\partial z}{\partial y} = g(x, y) \cdot \frac{\partial h}{\partial y} + h(x, y) \cdot \frac{\partial g}{\partial y} \qquad (13.2b)$$

**EXAMPLE 5.**   Given $z = (4x + 9)(8x + 5y)$, by the product rule,

$$\frac{\partial z}{\partial x} = (4x + 9)(8) + (8x + 5y)(4) = 64x + 72 + 20y$$

$$\frac{\partial z}{\partial y} = (4x + 9)(5) + (8x + 5y)(0) = 20x + 45$$

### 13.3.2   Quotient Rule

Given $z = \dfrac{g(x, y)}{h(x, y)} \qquad h(x, y) \neq 0,$

$$\frac{\partial z}{\partial x} = \frac{h(x, y) \cdot \partial g/\partial x - g(x, y) \cdot \partial h/\partial x}{[h(x, y)]^2} \qquad (13.3a)$$

$$\frac{\partial z}{\partial y} = \frac{h(x, y) \cdot \partial g/\partial y - g(x, y) \cdot \partial h/\partial y}{[h(x, y)]^2} \qquad (13.3b)$$

**EXAMPLE 6.**   Given $z = (2x + 9y)/(8x + 7y)$, by the quotient rule,

$$\frac{\partial z}{\partial x} = \frac{(8x + 7y)(2) - (2x + 9y)(8)}{(8x + 7y)^2}$$

$$\frac{\partial z}{\partial x} = \frac{16x + 14y - 16x - 72y}{(8x + 7y)^2} = \frac{-58y}{(8x + 7y)^2}$$

$$\frac{\partial z}{\partial y} = \frac{(8x + 7y)(9) - (2x + 9y)(7)}{(8x + 7y)^2}$$

$$\frac{\partial z}{\partial y} = \frac{72x + 63y - 14x - 63y}{(8x + 7y)^2} = \frac{58x}{(8x + 7y)^2}$$

### 13.3.3   Generalized Power Function Rule

Given $z = [g(x, y)]^n$,

$$\frac{\partial z}{\partial x} = n\,[g(x, y)]^{n-1} \cdot \frac{\partial g}{\partial x} \qquad (13.4a)$$

$$\frac{\partial z}{\partial y} = n\,[g(x, y)]^{n-1} \cdot \frac{\partial g}{\partial y} \qquad (13.4b)$$

**EXAMPLE 7.**   Given $z = \left(4x^2 + 9y^3\right)^5$, by the generalized power function rule,

$$\frac{\partial z}{\partial x} = 5\left(4x^2 + 9y^3\right)^4 \cdot (8x) = 40x\left(4x^2 + 9y^3\right)^4$$

$$\frac{\partial z}{\partial y} = 5\left(4x^2 + 9y^3\right)^4 \cdot (27y^2) = 135y^2\left(4x^2 + 9y^3\right)^4$$

### 13.3.4   Natural Exponential Function Rule

Given $z = e^{g(x,y)}$,

$$\frac{\partial z}{\partial x} = e^{g(x,y)} \cdot \frac{\partial g}{\partial x} \tag{13.5a}$$

$$\frac{\partial z}{\partial y} = e^{g(x,y)} \cdot \frac{\partial g}{\partial y} \tag{13.5b}$$

**EXAMPLE 8.**   Given $z = e^{4x^2y^3}$, by the natural exponential function rule,

$$\frac{\partial z}{\partial x} = e^{4x^2y^3} \cdot 8xy^3 = 8xy^3 e^{4x^2y^3}$$

$$\frac{\partial z}{\partial y} = e^{4x^2y^3} \cdot 12x^2y^2 = 12x^2y^2 e^{4x^2y^3}$$

### 13.3.5   Natural Logarithmic Function Rule

Given $z = \ln|g(x, y)|$,

$$\frac{\partial z}{\partial x} = \frac{1}{g(x, y)} \cdot \frac{\partial g}{\partial x} \tag{13.6a}$$

$$\frac{\partial z}{\partial y} = \frac{1}{g(x, y)} \cdot \frac{\partial g}{\partial y} \tag{13.6b}$$

**EXAMPLE 9.**   Given $z = \ln\left|3x^3 + 4y^2\right|$, by the natural logarithmic function rule,

$$\frac{\partial z}{\partial x} = \frac{1}{3x^3 + 4y^2} \cdot 9x^2 = \frac{9x^2}{3x^3 + 4y^2} \qquad\qquad \frac{\partial z}{\partial y} = \frac{1}{3x^3 + 4y^2} \cdot 8y = \frac{8y}{3x^3 + 4y^2}$$

## 13.4   SECOND-ORDER PARTIAL DERIVATIVES

Given a function $z = f(x, y)$, *the second-order (direct) partial derivative* signifies that the function has been differentiated partially with respect to one of the independent variables twice while the other independent variable has been held constant:

$$f_{xx} = (f_x)_x = \frac{\partial}{\partial x}\left(\frac{\partial z}{\partial x}\right) = \frac{\partial^2 z}{\partial x^2} \qquad\qquad f_{yy} = (f_y)_y = \frac{\partial}{\partial y}\left(\frac{\partial z}{\partial y}\right) = \frac{\partial^2 z}{\partial y^2}$$

In effect, $f_{xx}$ measures the rate of change of the first-order partial derivative $f_x$ with respect to $x$ while $y$ is held constant. $f_{yy}$ is exactly parallel. See Problems 13.9 and 13.10.

The *cross (or mixed) partial derivatives*, $f_{xy}$ and $f_{yx}$, indicate that the primitive function has been first partially differentiated with respect to one independent variable and then that partial derivative has in turn been partially differentiated with respect to the other independent variable:

$$f_{xy} = (f_x)_y = \frac{\partial}{\partial y}\left(\frac{\partial z}{\partial x}\right) = \frac{\partial^2 z}{\partial y\,\partial x} \qquad\qquad f_{yx} = (f_y)_x = \frac{\partial}{\partial x}\left(\frac{\partial z}{\partial y}\right) = \frac{\partial^2 z}{\partial x\,\partial y}$$

In brief, a cross partial measures the rate of change of a first-order partial derivative with respect to the other independent variable. Notice how the order of independent variables changes in the different forms of notation. See Problems 13.11 and 13.12.

**EXAMPLE 10.**   The (a) first, (b) second, and (c) cross partial derivatives are taken as shown below for

$$z = 4x^5 + 7xy + 8y^4$$

(a)
$$z_x = \frac{\partial z}{\partial x} = 20x^4 + 7y \qquad z_y = \frac{\partial z}{\partial y} = 7x + 32y^3$$

(b)
$$z_{xx} = \frac{\partial^2 z}{\partial x^2} = 80x^3 \qquad z_{yy} = \frac{\partial^2 z}{\partial y^2} = 96y^2$$

(c)
$$z_{xy} = \frac{\partial^2 z}{\partial y\, \partial x} = \frac{\partial}{\partial y}\left(\frac{\partial z}{\partial x}\right) = \frac{\partial}{\partial y}(20x^4 + 7y) = 7$$

$$z_{yx} = \frac{\partial^2 z}{\partial x\, \partial y} = \frac{\partial}{\partial x}\left(\frac{\partial z}{\partial y}\right) = \frac{\partial}{\partial x}(7x + 32y^3) = 7$$

**EXAMPLE 11.**   The (a) first, (b) second, and (c) cross partial derivatives are found and evaluated below at $x = 1$, $y = 2$ for

$$z = 2x^3 y^4$$

(a)
$$z_x = 6x^2 y^4 \qquad\qquad\qquad z_y = 8x^3 y^3$$
$$z_x(1, 2) = 6(1)^2(2)^4 = 96 \qquad\qquad z_y(1, 2) = 8(1)^3(2)^3 = 64$$

(b)
$$z_{xx} = 12xy^4 \qquad\qquad\qquad z_{yy} = 24x^3 y^2$$
$$z_{xx}(1, 2) = 12(1)(2)^4 = 192 \qquad\qquad z_{yy}(1, 2) = 24(1)^3(2)^2 = 96$$

(c)
$$z_{xy} = \frac{\partial}{\partial y}(6x^2 y^4) = 24x^2 y^3 \qquad\qquad z_{yx} = \frac{\partial}{\partial x}(8x^3 y^3) = 24x^2 y^3$$
$$z_{xy}(1, 2) = 24(1)^2(2)^3 = 192 \qquad\qquad z_{yx}(1, 2) = 24(1)^2(2)^3 = 192$$

By *Young's theorem*, if both cross partial derivatives are continuous, they will be identical. See Problems 13.11 and 13.12.

## 13.5   OPTIMIZATION OF MULTIVARIABLE FUNCTIONS

For a multivariable function such as $z = f(x, y)$ to be at a relative minimum or maximum, three conditions must be met:

1.  The first-order partial derivatives must equal zero simultaneously. This indicates that at the given point $(a, b)$, called a *critical point*, the function is neither increasing nor decreasing with respect to the principal axes but is at a relative plateau.
2.  The second-order direct partial derivatives, when evaluated at the critical point $(a, b)$, must both be positive for a minimum and negative for a maximum. This ensures that from a relative plateau at $(a, b)$ the function is moving upward in relation to the principal axes in the case of a minimum, and downward in relation to the principal axes in the case of a maximum.
3.  The product of the second-order direct partials evaluated at the critical point must exceed the product of the cross partials evaluated at the critical point. In sum, as seen in Fig. 13-1, when evaluated at a critical point $(a, b)$, the following values apply for a relative maximum and a relative minimum.

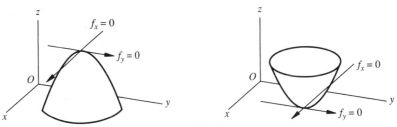

**Fig. 13-1**

| Relative Maximum | Relative Minimum |
|---|---|
| $f_x = f_y = 0$ | $f_x = f_y = 0$ |
| $f_{xx}, f_{yy} < 0$ | $f_{xx}, f_{yy} > 0$ |
| $f_{xx} \cdot f_{yy} > (f_{xy})^2$ | $f_{xx} \cdot f_{yy} > (f_{xy})^2$ |

The following observations are important to note.

1) Since, by Young's theorem, $f_{xy} = f_{yx}$, $f_{xy} \cdot f_{yx} = (f_{xy})^2$. Step 3 may also be written $f_{xx} \cdot f_{yy} - (f_{xy})^2 > 0$.

2) If $f_{xx} \cdot f_{yy} < (f_{xy})^2$, (a) when $f_{xx}$ and $f_{yy}$ have the same signs, the function is at an *inflection point*; (b) when $f_{xx}$ and $f_{yy}$ have different signs, the function is at a *saddle point*, as seen in Fig. 13-2, where the function is at a minimum when viewed from one axis, here the $x$ axis, but at a maximum when viewed from the other axis, here the $y$ axis.

3) If $f_{xx} \cdot f_{yy} = (f_{xy})^2$, the test is inconclusive.

See Example 12 and Problems 13.13 to 13.20; for inflection points, see Problems 13.15, 13.18, and 13.19; for saddle points, see Problems 13.16, 13.17, and 13.20.

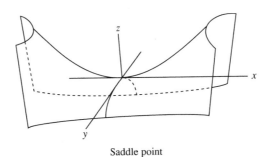

Saddle point
**Fig. 13-2**

**EXAMPLE 12.** (a) Find the critical points and (b) test to see if the function is at a relative maximum or minimum, given

$$z = 3x^3 + 2y^3 + 9x^2 - 12y^2 - 72x - 126y + 19$$

(a) Take the first-order partial derivatives, set them equal to zero, and solve for $x$ and $y$:

$$z_x = 9x^2 + 18x - 72 = 0 \qquad\qquad z_y = 6y^2 - 24y - 126 = 0 \qquad\qquad (13.7)$$

$$9(x^2 + 2x - 8) = 0 \qquad\qquad 6(y^2 - 4y - 21) = 0$$

$$9(x - 2)(x + 4) = 0 \qquad\qquad 6(y + 3)(y - 7) = 0$$

$$x = 2 \quad x = -4 \qquad\qquad y = -3 \quad y = 7$$

There are, thus, four distinct sets of critical points: $(2, -3)$, $(2, 7)$, $(-4, -3)$, and $(-4, 7)$.

(b) Take the second-order direct partials from (13.7), evaluate them at each of the critical points, and check the signs:

$$z_{xx} = 18x + 18 \qquad\qquad z_{yy} = 12y - 24$$

(1)  $z_{xx}(2, -3) = 18(2) + 18 = 54 > 0$  $\qquad z_{yy}(2, -3) = 12(-3) - 24 = -60 < 0$

(2)  $z_{xx}(2, 7) = 18(2) + 18 = 54 > 0$  $\qquad z_{yy}(2, 7) = 12(7) - 24 = 60 > 0$

(3)  $z_{xx}(-4, -3) = 18(-4) + 18 = -54 < 0$  $\qquad z_{yy}(-4, -3) = 12(-3) - 24 = -60 < 0$

(4)  $z_{xx}(-4, 7) = 18(-4) + 18 = -54 < 0$  $\qquad z_{yy}(-4, 7) = 12(7) - 24 = 60 > 0$

Since there are different signs for each of the second direct partials in (1) and (4), the function cannot be at a relative maximum or minimum at $(2, -3)$ or $(-4, 7)$. When $f_{xx}$ and $f_{yy}$ are of different signs, $f_{xx} \cdot f_{yy}$ cannot be greater than $(f_{xy})^2$, and the function is at a saddle point. With both signs of the second direct partials positive in (2) and negative in (3), the function *may be* at a relative minimum at $(2, 7)$ and at a relative maximum at $(-4, -3)$, but the third condition must be tested first to ensure against the possibility of an inflection point.

(c) From (13.7), take the cross partial derivatives and check to make sure that $z_{xx}(a, b) \cdot z_{yy}(a, b) > \left[z_{xy}(a, b)\right]^2$.

$$z_{xy} = 0 \qquad z_{yx} = 0$$

$$z_{xx}(a, b) \cdot z_{yy}(a, b) > \left[z_{xy}(a, b)\right]^2$$

From (2),  $\qquad\qquad (54) \quad \cdot \quad (60) \quad > (0)^2$

$$3240 > 0$$

From (3),  $\qquad\qquad (-54) \quad \cdot \quad (-60) \quad > (0)^2$

$$3240 > 0$$

Thus all three conditions are met for a relative minimum at $(2, 7)$ and a relative maximum at $(-4, -3)$. Earlier we found that the function was at a saddle point at $(2, -3)$ and $(-4, 7)$. For examples of inflection points, see Problems 13.15, 13.18, and 13.19.

## 13.6  CONSTRAINED OPTIMIZATION WITH LAGRANGE MULTIPLIERS

Differential calculus can also be used to maximize or minimize a function subject to constraint. Given a function $f(x, y)$ subject to a constraint $g(x, y) = k$ (a constant), a new function $F$ can be formed by (1) setting the constraint equal to zero, (2) multiplying it by $\lambda$ (the *Lagrange multiplier*), and (3) adding the product to the original function:

$$F(x, y, \lambda) = f(x, y) + \lambda\, [k - g(x, y)] \tag{13.8}$$

Here $F(x, y, \lambda)$ is called the *Lagrangian function*, $f(x, y)$ is the original or *objective function*, and $k - g(x, y)$ is the *constraint set equal to zero*. Since the constraint is always rewritten such that it equals zero, the product $\lambda[k - g(x, y)]$ also equals zero and the addition of the term does not change the value of the objective function. Critical values $x_0$, $y_0$, and $\lambda_0$, at which the function is optimized, are found by taking the partial derivatives of $F$ with respect to *all three* independent variables, setting them equal to zero, and solving simultaneously:

$$F_x(x, y, \lambda) = 0$$

$$F_y(x, y, \lambda) = 0$$

$$F_\lambda(x, y, \lambda) = 0$$

*Note*: Since in (*13.8*) above $\lambda[k - g(x, y)] = 0 = \lambda[g(x, y) - k]$, either form can be added to *or* subtracted from the objective function without changing the critical values of $x$ and $y$. Only the sign of $\lambda$ will be affected. The Lagrange multiplier $\lambda$ approximates the effect on the objective function of a small change in the constant of the constraint. See Example 13. For an economic interpretation of $\lambda$, see Dowling, *Schaum's Outline of Introduction to Mathematical Economics*, Section 5.6 and Problems 5.12 to 5.14. For the second-order conditions for constrained optimization, see Dowling, *ibid.*, Section 12.5, and Problems 12.23 to 12.32.

**EXAMPLE 13.**    Use the Lagrange multiplier method to optimize the function

$$z = 6x^2 + 5xy + 2y^2$$

subject to the constraint $2x + y = 96$.

1.   Start by setting the constraint equal to zero,

$$96 - 2x - y = 0$$

Multiply it by $\lambda$ and add it to the objective function to form the Lagrangian function $Z$.

$$Z = 6x^2 + 5xy + 2y^2 + \lambda(96 - 2x - y) \tag{13.9}$$

2.   Take the first-order partials, set them equal to zero, and solve simultaneously.

$$Z_x = 12x + 5y - 2\lambda = 0 \tag{13.10}$$

$$Z_y = 5x + 4y - \lambda = 0 \tag{13.11}$$

$$Z_\lambda = 96 - 2x - y = 0 \tag{13.12}$$

Eliminate $\lambda$ by subtracting 2 times (*13.11*) from (*13.10*).

$$2x - 3y = 0 \qquad x = 1.5y$$

Substitute $x = 1.5y$ in (*13.12*), thus making use of all three equations.

$$96 - 2(1.5y) - y = 0 \qquad y_0 = 24$$

Next substitute $y_0 = 24$ back in (*13.12*) to find $x_0$.

$$96 - 2x - 24 = 0 \qquad x_0 = 36$$

To find $\lambda$, substitute $x_0 = 36$, $y_0 = 24$ in (*13.11*) or (*13.10*).

$$5(36) + 4(24) - \lambda = 0 \qquad \lambda_0 = 276$$

Last, substitute the critical values in (*13.9*),

$$Z = 6(36)^2 + 5(36)(24) + 2(24)^2 + (276)[96 - 2(36) - 24]$$

$$Z = 6(1296) + 5(864) + 2(576) + 276(0) = 13{,}248$$

Note above that with the constraint equal to zero at the critical values the Lagrangian function $Z$ exactly equals the objective function $z$. Note, too, that with $\lambda_0 = 276$, an increase in the constant of the constraint from 96 to 97 would lead to an increase in $Z$ of approximately 276. See Problems 13.21 to 13.24.

## 13.7   INCOME DETERMINATION MULTIPLIERS

Partial differentiation is used to derive the various multipliers of an income determination model. Given

$$Y = C + I \qquad C = C_0 + bY_d \qquad T = T_0 + tY \qquad Y_d = Y - T \qquad I = I_0$$

we found in equation (*4.52*) of Problem 4.19 that

$$Y_e = \frac{1}{1 - b + bt}(C_0 + I_0 - bT_0) \tag{13.13}$$

Taking the partial derivative of ($13.13$) with respect to any of the independent variables or parameters gives the multiplier for that variable or parameter. Thus, recalling that the fraction $1/(1 - b + bt)$ multiplies in turn each of the terms within the parentheses, the *autonomous consumption multiplier* is

$$\frac{\partial Y_e}{\partial C_0} = \frac{1}{1 - b + bt}$$

The *autonomous tax multiplier* is

$$\frac{\partial Y_e}{\partial T_0} = \frac{-b}{1 - b + bt}$$

and, using the quotient rule, the *multiplier for a change in the tax rate t* is

$$\frac{\partial Y_e}{\partial t} = \frac{(1 - b + bt)(0) - (C_0 + I_0 - bT_0)(b)}{(1 - b + bt)^2} = \frac{-b(C_0 + I_0 - bT_0)}{(1 - b + bt)^2}$$

Then making use of ($13.13$) to simplify,

$$\frac{\partial Y_e}{\partial t} = \frac{-bY_e}{(1 - b + bt)}$$

For $\partial Y_e/\partial b$, the *multiplier for a change in the MPC*, see Problems 13.25 and 13.26.

## 13.8  OPTIMIZING MULTIVARIABLE FUNCTIONS IN BUSINESS AND ECONOMICS

In monopolistic competition where producers have control over prices, companies frequently manufacture more than one line of the same product: assorted brand names of toothpaste, soap, and deodorants; various lines of women's apparel, from designer clothes to off-the-rack imitations; and different qualities of audio and visual electronic equipment. Find the prices a monopolistic firm should charge for its two products in order to maximize total profits when the demand for its products is given by

$$P_1 = 105 - 3Q_1 \tag{13.14}$$

$$P_2 = 86 - 2Q_2 \tag{13.15}$$

and the joint total cost function is

$$\text{TC} = 2Q_1^2 + 4Q_1Q_2 + Q_2^2$$

First find the total profit function, $\pi = \text{TR} - \text{TC}$, where $\text{TR} = P_1Q_1 + P_2Q_2$. Substituting the given information, and remembering to subtract each term in the total cost function,

$$\pi = (105 - 3Q_1)Q_1 + (86 - 2Q_2)Q_2 - (2Q_1^2 + 4Q_1Q_2 + Q_2^2)$$

$$\pi = 105Q_1 - 5Q_1^2 + 86Q_2 - 3Q_2^2 - 4Q_1Q_2$$

Next optimize the function,

$$\pi_1 = \frac{\partial \pi}{\partial Q_1} = 105 - 10Q_1 - 4Q_2 = 0$$

$$\pi_2 = \frac{\partial \pi}{\partial Q_2} = 86 - 4Q_1 - 6Q_2 = 0$$

and solve simultaneously for the critical values,

$$\overline{Q}_1 = 6.5 \qquad \overline{Q}_2 = 10$$

Check the second-order conditions.

$$\pi_{11} = -10 \qquad \pi_{22} = -6 \qquad \pi_{12} = -4 = \pi_{21}$$

With both second-order direct partials negative and $\pi_{11}\pi_{22} > (\pi_{12})^2$, the function is maximized at the critical values.

Last, substitute $\overline{Q}_1 = 6.5$ and $\overline{Q}_2 = 10$ in (13.14) and (13.15) respectively to find the profit-maximizing prices.

$$\overline{P}_1 = 105 - 3(6.5) = 85.5$$

$$\overline{P}_2 = 86 - 2(10) = 66$$

and $\pi = 105(6.5) - 5(6.5)^2 + 86(10) - 3(10)^2 - 4(6.5)(10) = 771.25$. See Problems 13.27 to 13.32.

## 13.9   CONSTRAINED OPTIMIZATION OF MULTIVARIABLE ECONOMIC FUNCTIONS

People in business and economics frequently need to optimize functions under specific constraints. At times they may wish to maximize utility subject to a budget constraint; at other times they may be called on to minimize costs subject to some minimal requirement of output such as a production quota. Assume a firm has pledged to donate to charity a combination of its two products $x$ and $y$ with a total market value of \$140. Given $P_x = \$2$ and $P_y = \$3$, minimize the cost of fulfilling the pledge under the constraint, when unrestrained total costs are

$$c = 18x^2 - 6xy + 20.5y^2$$

Set up the Lagrange function incorporating the constraint $2x + 3y = 140$.

$$C = 18x^2 - 6xy + 20.5y^2 + \lambda(140 - 2x - 3y)$$

Take the partial derivative with respect to all three variables and solve simultaneously for the critical values.

$$C_x = 36x - 6y - 2\lambda = 0 \qquad (13.16)$$

$$C_y = -6x + 41y - 3\lambda = 0 \qquad (13.17)$$

$$C_\lambda = 140 - 2x - 3y = 0 \qquad (13.18)$$

Subtracting 1.5 times (13.16) from (13.17) to eliminate $\lambda$,

$$-60x + 50y = 0 \qquad y = 1.2x$$

Substituting in (13.18),

$$140 - 2x - 3(1.2x) = 0 \qquad x_0 = 25$$

From which we find by successive substitution, $y_0 = 30$, $\lambda_0 = 360$. See also Problems 13.33 to 13.38.

## 13.10   CONSTRAINED OPTIMIZATION OF COBB-DOUGLAS PRODUCTION FUNCTIONS

A function much used in economic analysis for its favorable properties is the *Cobb-Douglas production function*:

$$q = AK^\alpha L^\beta \qquad (A > 0; \quad 0 < \alpha, \beta < 1)$$

where $q$ is the quantity of output in physical units, $K$ the quantity of capital, and $L$ the quantity of labor. Here $\alpha$ is the *output elasticity of capital*, which measures the percentage change in $q$ for a 1 percent change in $K$ while $L$ is held constant; $\beta$, the *output elasticity of labor*, is exactly parallel; and $A$ is an *efficiency parameter* reflecting the level of technology. A *strict* Cobb-Douglas function, in which $\alpha + \beta = 1$, exhibits constant returns to scale and is often written $q = AK^\alpha L^{1-\alpha}$ $(0 < \alpha < 1)$.

A *generalized* Cobb-Douglas function, in which $\alpha + \beta \neq 1$, exhibits *increasing returns to scale* if $\alpha + \beta > 1$, and *decreasing returns to scale* if $\alpha + \beta < 1$.

Given a firm operating at constant returns to scale whose (strict) Cobb-Douglas production function is

$$q = 3K^{0.2}L^{0.8}$$

and whose budgetary parameters are

$$P_K = 9 \qquad P_L = 4 \qquad B = 450$$

the Lagrange function to be maximized is

$$Q = 3K^{0.2}L^{0.8} + \lambda(450 - 9K - 4L)$$

Taking the partial derivatives for optimization,

$$Q_K = 0.6K^{-0.8}L^{0.8} - 9\lambda = 0 \tag{13.19}$$

$$Q_L = 2.4K^{0.2}L^{-0.2} - 4\lambda = 0 \tag{13.20}$$

$$Q_\lambda = 450 - 9K - 4L = 0 \tag{13.21}$$

Rearranging, then dividing (13.19) by (13.20) to eliminate $\lambda$,

$$\frac{0.6K^{-0.8}L^{0.8}}{2.4K^{0.2}L^{-0.2}} = \frac{9\lambda}{4\lambda}$$

Remembering to subtract exponents in division,

$$0.25K^{-1}L^1 = 2.25$$

$$\frac{L}{K} = 9 \qquad L = 9K$$

Substituting $L = 9K$ in (13.21),

$$450 - 9K - 4(9K) = 0 \qquad K_0 = 10$$

Finally by substituting $K_0 = 10$ in (13.21),

$$L_0 = 90$$

See also Problems 13.39 to 13.42. For constrained optimization of a constant elasticity of substitution (CES) production function, see Dowling, *Schaum's Outline of Introduction to Mathematical Economics*, Section 6.11 and Problems 6.47, 6.48, and 6.63 to 6.73.

## 13.11  IMPLICIT AND INVERSE FUNCTION RULES (OPTIONAL)

As discussed in Section 4.9, functions of the form $y = f(x)$ express $y$ explicitly in terms of $x$ and are called *explicit functions*. Functions of the form $f(x, y) = 0$ do not express $y$ explicitly in terms of $x$ and are called *implicit functions*. With the help of the implicit function rule, however, it is possible to find the derivative $dy/dx$ of an implicit function without first solving for $y$ in terms of $x$. The *implicit function rule* states that if an implicit function $f(x, y)$ exists and $f_y \neq 0$ at the point around which the implicit function is defined, then

$$\frac{dy}{dx} = -\frac{f_x}{f_y} \tag{13.22}$$

Given a function $y = f(x)$, an *inverse function* $x = f^{-1}(y)$ exists if each value of $y$ yields one and only one value of $x$. Assuming that the inverse function exists, and most economic functions have inverse functions, the *inverse function rule* states that the derivative of the inverse function is the reciprocal of the derivative of the original function. Thus, if $Q = f(P)$ is the original function, the derivative of the original function is $dQ/dP$, the derivative of the inverse function $[P = f^{-1}(Q)]$ is $dP/dQ$, and

$$\frac{dP}{dQ} = \frac{1}{dQ/dP} \qquad \text{provided} \qquad \frac{dQ}{dP} \neq 0 \qquad\qquad (13.23)$$

See Examples 14 and 15 and Problems 13.43 and 13.44.

**EXAMPLE 14.** Find the derivative $dy/dx$ for each of the following implicit functions:

(a)
$$9x^2 - y = 0$$

From (13.22),
$$\frac{dy}{dx} = -\frac{f_x}{f_y}$$

Here $f_x = 18x$ and $f_y = -1$. Substituting above,

$$\frac{dy}{dx} = -\frac{18x}{(-1)} = 18x$$

Here the function has deliberately been kept simple so the answer can easily be checked by solving for $y$ in terms of $x$ and then taking the derivative directly. Since $y = 9x^2$, $dy/dx = 18x$.

(b)
$$5x^4 - 3y^5 - 49 = 0$$

$$\frac{dy}{dx} = -\frac{f_x}{f_y} = -\frac{20x^3}{-15y^4} = \frac{4x^3}{3y^4}$$

Note that the derivative $dy/dx$ is always the *negative* of the *reciprocal* of the corresponding partials: $-f_x/f_y$.

**EXAMPLE 15.** Find the derivative for the inverse of each of the following functions:

(a)
$$Q = 94 - 3P$$

From (13.23),
$$\frac{dP}{dQ} = \frac{1}{dQ/dP}$$

Here $dQ/dP = -3$. So

$$\frac{dP}{dQ} = \frac{1}{-3} = -\frac{1}{3}$$

(b)
$$Q = 14 + 4P^3$$

$$\frac{dP}{dQ} = \frac{1}{dQ/dP} = \frac{1}{12P^2} \qquad (P \neq 0)$$

# Solved Problems

## FIRST-ORDER PARTIAL DERIVATIVES

**13.1.** Find the first-order partial derivatives for each of the following functions.

(a)    $z = 7x^2 - 13xy + 5y^2$            (b)    $z = 5x^4 + 3x^2y^5 - 9y^3$

$\qquad\qquad z_x = 14x - 13y$                   $z_x = 20x^3 + 6xy^5$

$\qquad\qquad z_y = -13x + 10y$               $z_y = 15x^2y^4 - 27y^2$

(c)  $z = 8w^3 + 6wx + 4x^2 - 9xy - 5y^2$          (d)  $z = 3w^4 + 7wxy - 4x^2 + y^3$

$z_w = 24w^2 + 6x$                                      $z_w = 12w^3 + 7xy$

$z_x = 6w + 8x - 9y$                                    $z_x = 7wy - 8x$

$z_y = -9x - 10y$                                       $z_y = 7wx + 3y^2$

**13.2.** Use the product rule from equation (*13.2*) to find the first-order partial derivatives for each of the following functions:

(a)                                $z = (8x + 15y)(12x - 7y)$

$z_x = (8x + 15y)(12) + (12x - 7y)(8)$          $z_y = (8x + 15y)(-7) + (12x - 7y)(15)$

$z_x = 96x + 180y + 96x - 56y$                   $z_y = -56x - 105y + 180x - 105y$

$z_x = 192x + 124y$                              $z_y = 124x - 210y$

(b)                                $z = (4x^2 - 5y)(3x + 2y^3)$

$z_x = (4x^2 - 5y)(3) + (3x + 2y^3)(8x)$         $z_y = (4x^2 - 5y)(6y^2) + (3x + 2y^3)(-5)$

$z_x = 12x^2 - 15y + 24x^2 + 16xy^3$             $z_y = 24x^2y^2 - 30y^3 - 15x - 10y^3$

$z_x = 36x^2 - 15y + 16xy^3$                     $z_y = 24x^2y^2 - 15x - 40y^3$

(c)                        $z = (5w - 3x + 8y)(7w^2 + 9x^4 - 2y^5)$

$$z_w = (5w - 3x + 8y)(14w) + (7w^2 + 9x^4 - 2y^5)(5)$$

$$z_w = 105w^2 - 42wx + 112wy + 45x^4 - 10y^5$$

$$z_x = (5w - 3x + 8y)(36x^3) + (7w^2 + 9x^4 - 2y^5)(-3)$$

$$z_x = 180wx^3 - 135x^4 + 288x^3y - 21w^2 + 6y^5$$

$$z_y = (5w - 3x + 8y)(-10y^4) + (7w^2 + 9x^4 - 2y^5)(8)$$

$$z_y = -50wy^4 + 30xy^4 - 96y^5 + 56w^2 + 72x^4$$

**13.3.** Use the quotient rule from equation (*13.3*) to find the first-order partials of each of the following functions:

(a)                                $z = \dfrac{14x}{9x - 4y}$

$$z_x = \frac{(9x - 4y)(14) - (14x)(9)}{(9x - 4y)^2} = \frac{-56y}{(9x - 4y)^2}$$

$$z_y = \frac{(9x - 4y)(0) - (14x)(-4)}{(9x - 4y)^2} = \frac{56x}{(9x - 4y)^2}$$

(b)                                $z = \dfrac{4x^2 + 3y^3}{5x - 2y}$

$$z_x = \frac{(5x - 2y)(8x) - (4x^2 + 3y^3)(5)}{(5x - 2y)^2} = \frac{20x^2 - 16xy - 15y^3}{(5x - 2y)^2}$$

$$z_y = \frac{(5x - 2y)(9y^2) - (4x^2 + 3y^3)(-2)}{(5x - 2y)^2} = \frac{45xy^2 - 12y^3 + 8x^2}{(5x - 2y)^2}$$

(c)
$$z = \frac{4w + 9x - 2y}{7w - 2x + 3y}$$

$$z_w = \frac{(7w - 2x + 3y)(4) - (4w + 9x - 2y)(7)}{(7w - 2x + 3y)^2} = \frac{-71x + 26y}{(7w - 2x + 3y)^2}$$

$$z_x = \frac{(7w - 2x + 3y)(9) - (4w + 9x - 2y)(-2)}{(7w - 2x + 3y)^2} = \frac{71w + 23y}{(7w - 2x + 3y)^2}$$

$$z_y = \frac{(7w - 2x + 3y)(-2) - (4w + 9x - 2y)(3)}{(7w - 2x + 3y)^2} = \frac{-26w - 23x}{(7w - 2x + 3y)^2}$$

**13.4.** Find the first-order partial derivatives for each of the following functions by using the generalized power function rule from Equation (*13.4*):

(a)
$$z = (6x - 7y)^3$$

$$z_x = 3(6x - 7y)^2(6) = 18(6x - 7y)^2$$

$$z_y = 3(6x - 7y)^2(-7) = -21(6x - 7y)^2$$

(b)
$$z = (2x^4 + 9y^2)^5$$

$$z_x = 5(2x^4 + 9y^2)^4(8x^3) = 40x^3(2x^4 + 9y^2)^4$$

$$z_y = 5(2x^4 + 9y^2)^4(18y) = 90y(2x^4 + 9y^2)^4$$

(c)
$$z = (4w^3 - 7x^5 + 8y^2)^4$$

$$z_w = 4(4w^3 - 7x^5 + 8y^2)^3(12w^2) = 48w^2(4w^3 - 7x^5 + 8y^2)^3$$

$$z_x = 4(4w^3 - 7x^5 + 8y^2)^3(-35x^4) = -140x^4(4w^3 - 7x^5 + 8y^2)^3$$

$$z_y = 4(4w^3 - 7x^5 + 8y^2)^3(16y) = 64y(4w^3 - 7x^5 + 8y^2)^3$$

**13.5.** Find the first-order partial derivatives for the following natural exponential functions, using Equation (*13.5*).

(a)
$$z = 9e^{4xy}$$

$$z_x = 9e^{4xy} \cdot 4y = 36ye^{4xy} \qquad z_y = 9e^{4xy} \cdot 4x = 36xe^{4xy}$$

(b)
$$z = e^{(2x+7y)}$$

$$z_x = e^{(2x+7y)} \cdot 2 = 2e^{(2x+7y)} \qquad z_y = e^{(2x+7y)} \cdot 7 = 7e^{(2x+7y)}$$

(c)
$$z = 25e^{(5w-8x+3y)}$$

$$z_w = 25e^{(5w-8x+3y)} \cdot 5 = 125e^{(5w-8x+3y)}$$

$$z_x = 25e^{(5w-8x+3y)} \cdot (-8) = -200e^{(5w-8x+3y)}$$

$$z_y = 25e^{(5w-8x+3y)} \cdot 3 = 75e^{(5w-8x+3y)}$$

**13.6.** Use Equation (*13.6*) to find the first-order partial derivatives for the following natural logarithmic functions:

(a)
$$z = \ln|4x + 9y|$$

$$z_x = \frac{1}{4x + 9y} \cdot 4 = \frac{4}{4x + 9y}$$

$$z_y = \frac{1}{4x + 9y} \cdot 9 = \frac{9}{4x + 9y}$$

(b)
$$z = \ln \left| x^3 + y^2 \right|$$

$$z_x = \frac{1}{x^3 + y^2} \cdot 3x^2 = \frac{3x^2}{x^3 + y^2}$$

$$z_y = \frac{1}{x^3 + y^2} \cdot 2y = \frac{2y}{x^3 + y^2}$$

(c)
$$z = \ln \left| w^4 x^5 y^3 \right|$$

$$z_w = \frac{1}{w^4 x^5 y^3} \cdot 4w^3 x^5 y^3 = \frac{4}{w} \qquad\qquad z_x = \frac{1}{w^4 x^5 y^3} \cdot 5w^4 x^4 y^3 = \frac{5}{x}$$

$$z_y = \frac{1}{w^4 x^5 y^3} \cdot 3w^4 x^5 y^2 = \frac{3}{y}$$

**13.7.** Use whatever combination of rules is necessary to find the first-order partials for each of the following functions:

(a)
$$z = \frac{(8x + 15y)^4}{3x + 7y}$$

Using the quotient rule and the generalized power function rule,

$$z_x = \frac{(3x + 7y)\left[4(8x + 15y)^3(8)\right] - (8x + 15y)^4(3)}{(3x + 7y)^2} = \frac{(96x + 224y)(8x + 15y)^3 - 3(8x + 15y)^4}{(3x + 7y)^2}$$

$$z_y = \frac{(3x + 7y)\left[4(8x + 15y)^3(15)\right] - (8x + 15y)^4(7)}{(3x + 7y)^2} = \frac{(180x + 420y)(8x + 15y)^3 - 7(8x + 15y)^4}{(3x + 7y)^2}$$

(b)
$$z = \frac{(9x + 4y)(3x - 8y)}{5x - 2y}$$

Using the quotient rule and the product rule,

$$z_x = \frac{(5x - 2y)\left[(9x + 4y)(3) + (3x - 8y)(9)\right] - (9x + 4y)(3x - 8y)(5)}{(5x - 2y)^2}$$

$$z_x = \frac{(5x - 2y)(27x + 12y + 27x - 72y) - (9x + 4y)(15x - 40y)}{(5x - 2y)^2}$$

$$z_x = \frac{135x^2 - 108xy + 280y^2}{(5x - 2y)^2}$$

$$z_y = \frac{(5x - 2y)\left[(9x + 4y)(-8) + (3x - 8y)(4)\right] - (9x + 4y)(3x - 8y)(-2)}{(5x - 2y)^2}$$

$$z_y = \frac{(5x - 2y)(-72x - 32y + 12x - 32y) - (9x + 4y)(-6x + 16y)}{(5x - 2y)^2}$$

$$z_y = \frac{-246x^2 - 320xy + 64y^2}{(5x - 2y)^2}$$

(c)
$$z = (3x - 5y)^3(6x + 7y)$$

Using the product rule and the generalized power function rule,

$$z_x = (3x - 5y)^3(6) + (6x + 7y)\left[3(3x - 5y)^2(3)\right] = 6(3x - 5y)^3 + (54x + 63y)(3x - 5y)^2$$

$$z_y = (3x - 5y)^3(7) + (6x + 7y)\left[3(3x - 5y)^2(-5)\right] = 7(3x - 5y)^3 - (90x + 105y)(3x - 5y)^2$$

$(d)$
$$z = \left(\frac{2x + 9y}{5x + 4y}\right)^2$$

Using the generalized power function rule and the quotient rule,

$$z_x = 2\left(\frac{2x + 9y}{5x + 4y}\right)\left[\frac{(5x + 4y)(2) - (2x + 9y)(5)}{(5x + 4y)^2}\right] = \frac{4x + 18y}{5x + 4y}\left[\frac{-37y}{(5x + 4y)^2}\right]$$

$$z_x = \frac{-(148xy + 666y^2)}{(5x + 4y)^3}$$

$$z_y = 2\left(\frac{2x + 9y}{5x + 4y}\right)\left[\frac{(5x + 4y)(9) - (2x + 9y)(4)}{(5x + 4y)^2}\right] = \frac{4x + 18y}{5x + 4y}\left[\frac{37x}{(5x + 4y)^2}\right]$$

$$z_y = \frac{666xy + 148x^2}{(5x + 4y)^3}$$

**13.8.** Redo Problem 13.7 for the following natural exponential and logarithmic functions:

$(a)$
$$z = 7x^3 e^{5xy}$$

Using the product rule and natural exponential function rule,

$$z_x = 7x^3 \cdot 5ye^{5xy} + e^{5xy} \cdot 21x^2 \qquad z_y = 7x^3 \cdot 5xe^{5xy} + e^{5xy} \cdot 0$$

$$z_x = 7x^2 e^{5xy}(5xy + 3) \qquad z_y = 35x^4 e^{5xy}$$

$(b)$
$$z = \ln|3x + 8y| \cdot e^{2xy}$$

By the product, logarithmic, and exponential function rules,

$$z_x = \ln|3x + 8y| \cdot 2ye^{2xy} + e^{2xy}\left[\frac{1}{3x + 8y} \cdot 3\right] = e^{2xy}\left[2y(\ln|3x + 8y|) + \frac{3}{3x + 8y}\right]$$

$$z_y = \ln|3x + 8y| \cdot 2xe^{2xy} + e^{2xy}\left(\frac{1}{3x + 8y} \cdot 8\right) = e^{2xy}\left[2x(\ln|3x + 8y|) + \frac{8}{3x + 8y}\right]$$

$(c)$
$$z = \frac{6xy}{e^{5x+2}}$$

By the quotient and natural exponential function rule,

$$z_x = \frac{e^{5x+2}(6y) - 6xy(5e^{5x+2})}{(e^{5x+2})^2} \qquad\qquad z_y = \frac{e^{5x+2}(6x) - 6xy(0)}{(e^{5x+2})^2}$$

$$z_x = \frac{6ye^{5x+2}(1 - 5x)}{(e^{5x+2})^2} \qquad\qquad z_y = \frac{6xe^{5x+2}}{(e^{5x+2})^2}$$

$$z_x = \frac{6y(1 - 5x)}{e^{5x+2}} \qquad\qquad z_y = \frac{6x}{e^{5x+2}}$$

$(d)$
$$z = 6xye^{-(5x+2)}$$

This is the same function as in $(c)$. By the product and natural exponential function rules,

$$z_x = 6xy\left(-5e^{-(5x+2)}\right) + e^{-(5x+2)}(6y) \qquad z_y = 6xy(0) + e^{-(5x+2)}(6x)$$

$$z_x = 6ye^{-(5x+2)}(-5x + 1) \qquad\qquad z_y = 6xe^{-(5x+2)}$$

$$z_x = \frac{6y(1 - 5x)}{e^{5x+2}} \qquad\qquad z_y = \frac{6x}{e^{5x+2}}$$

## SECOND-ORDER PARTIAL DERIVATIVES

**13.9.** Find the second-order direct partial derivatives $z_{xx}$ and $z_{yy}$ for each of the following functions:

(a)
$$z = 6x^4 - 17xy + 4y^5$$

$$z_x = 24x^3 - 17y \qquad\qquad z_y = -17x + 20y^4$$

$$z_{xx} = 72x^2 \qquad\qquad z_{yy} = 80y^3$$

(b)
$$z = 7x^6 - 2x^3y^2 + 9xy^4 - 13y^5$$

$$z_x = 42x^5 - 6x^2y^2 + 9y^4 \qquad\qquad z_y = -4x^3y + 36xy^3 - 65y^4$$

$$z_{xx} = 210x^4 - 12xy^2 \qquad\qquad z_{yy} = -4x^3 + 108xy^2 - 260y^3$$

(c)
$$z = 5w^3x^6y^4$$

$$z_w = 15w^2x^6y^4 \qquad z_x = 30w^3x^5y^4 \qquad z_y = 20w^3x^6y^3$$

$$z_{ww} = 30wx^6y^4 \qquad z_{xx} = 150w^3x^4y^4 \qquad z_{yy} = 60w^3x^6y^2$$

(d)
$$z = (4x - 7y)^3$$

$$z_x = 3(4x - 7y)^2(4) \qquad\qquad z_y = 3(4x - 7y)^2(-7)$$

$$= 12(4x - 7y)^2 \qquad\qquad = -21(4x - 7y)^2$$

$$z_{xx} = 24(4x - 7y)(4) \qquad\qquad z_{yy} = -42(4x - 7y)(-7)$$

$$= 96(4x - 7y) \qquad\qquad = 294(4x - 7y)$$

**13.10.** Find the second-order direct partial derivatives $z_{xx}$ and $z_{yy}$ for each of the following Cobb-Douglas functions:

(a)
$$z = x^{0.1}y^{0.9}$$

$$z_x = 0.1x^{-0.9}y^{0.9} \qquad\qquad z_y = 0.9x^{0.1}y^{-0.1}$$

$$z_{xx} = -0.09x^{-1.9}y^{0.9} \qquad\qquad z_{yy} = -0.09x^{0.1}y^{-1.1}$$

(b)
$$z = 2x^{0.6}y^{0.3}$$

$$z_x = 1.2x^{-0.4}y^{0.3} \qquad\qquad z_y = 0.6x^{0.6}y^{-0.7}$$

$$z_{xx} = -0.48x^{-1.4}y^{0.3} \qquad\qquad z_{yy} = -0.42x^{0.6}y^{-1.7}$$

(c)
$$z = 3x^{0.4}y^{0.5}$$

$$z_x = 1.2x^{-0.6}y^{0.5} \qquad\qquad z_y = 1.5x^{0.4}y^{-0.5}$$

$$z_{xx} = -0.72x^{-1.6}y^{0.5} \qquad\qquad z_{yy} = -0.75x^{0.4}y^{-1.5}$$

**13.11.** Find the cross partial derivatives $z_{xy}$ and $z_{yx}$ for each of the following functions:

(a)
$$z = 7x^4 - 15xy + 2y^5$$

$$z_x = 28x^3 - 15y \qquad\qquad z_y = -15x + 10y^4$$

$$z_{xy} = -15 \qquad\qquad z_{yx} = -15$$

(b)
$$z = 8x^3 - 11x^6y^4 - 6y^2$$

$$z_x = 24x^2 - 66x^5y^4 \qquad\qquad z_y = -44x^6y^3 - 12y$$

$$z_{xy} = -264x^5y^3 \qquad\qquad z_{yx} = -264x^5y^3$$

(c)
$$z = 10(9x - 4y)^5$$

$$z_x = 50(9x - 4y)^4(9) \qquad\qquad z_y = 50(9x - 4y)^4(-4)$$

$$z_x = 450(9x - 4y)^4 \qquad\qquad z_y = -200(9x - 4y)^4$$

$$z_{xy} = 1800(9x - 4y)^3(-4) \qquad\qquad z_{yx} = -800(9x - 4y)^3(9)$$

$$z_{xy} = -7200(9x - 4y)^3 \qquad\qquad z_{yx} = -7200(9x - 4y)^3$$

(d)
$$z = w^3 x^{-5} y^4$$

$$z_w = 3w^2 x^{-5} y^4 \qquad z_x = -5w^3 x^{-6} y^4 \qquad z_y = 4w^3 x^{-5} y^3$$

$$z_{wx} = -15w^2 x^{-6} y^4 \qquad z_{xw} = -15w^2 x^{-6} y^4 \qquad z_{yw} = 12w^2 x^{-5} y^3$$

$$z_{wy} = 12w^2 x^{-5} y^3 \qquad z_{xy} = -20w^3 x^{-6} y^3 \qquad z_{yx} = -20w^3 x^{-6} y^3$$

Note how by Young's theorem $z_{xy} = z_{yx}$ in parts (a) to (c) and in part (d) $z_{wx} = z_{xw}$, $z_{wy} = z_{yw}$, and $z_{xy} = z_{yx}$.

**13.12.** Find the cross partial derivatives for each of the following Cobb-Douglas functions.

(a)
$$z = 1.25x^{0.4} y^{0.6}$$

$$z_x = 0.5x^{-0.6} y^{0.6} \qquad\qquad z_y = 0.75x^{0.4} y^{-0.4}$$

$$z_{xy} = 0.3x^{-0.6} y^{-0.4} \qquad\qquad z_{yx} = 0.3x^{-0.6} y^{-0.4}$$

(b)
$$z = 1.4x^{0.2} y^{0.7}$$

$$z_x = 0.28x^{-0.8} y^{0.7} \qquad\qquad z_y = 0.98x^{0.2} y^{-0.3}$$

$$z_{xy} = 0.196x^{-0.8} y^{-0.3} \qquad\qquad z_{yx} = 0.196x^{-0.8} y^{-0.3}$$

(c)
$$Q = 2.5K^{0.3} L^{0.9}$$

$$Q_K = 0.75K^{-0.7} L^{0.9} \qquad\qquad Q_L = 2.25K^{0.3} L^{-0.1}$$

$$Q_{KL} = 0.675K^{-0.7} L^{-0.1} \qquad\qquad Q_{LK} = 0.675K^{-0.7} L^{-0.1}$$

## OPTIMIZATION OF MULTIVARIABLE FUNCTIONS

**13.13.** For the following quadratic function, (1) find the critical points where the function may be at an optimum and (2) determine whether at these points the function is at a relative maximum, relative minimum, inflection point, or saddle point.

$$z = 5x^2 - 8x - 2xy - 6y + 4y^2 + 27 \qquad\qquad (13.24)$$

(1)  Take the first-order partial derivatives, set them equal to zero, and solve simultaneously using the methods of Section 13.5.

$$z_x = 10x - 8 - 2y = 0 \qquad\qquad (13.25)$$

$$z_y = -2x - 6 + 8y = 0 \qquad\qquad (13.26)$$

$$x = 1 \qquad y = 1 \qquad (1, 1) \text{ critical point}$$

(2)  Take the second-order direct partial derivatives from (13.25) and (13.26), evaluate them at the critical point, and check the signs.

$$z_{xx} = 10 \qquad\qquad z_{yy} = 8$$

$$z_{xx}(1, 1) = 10 > 0 \qquad z_{yy}(1, 1) = 8 > 0$$

With both second-order direct partial derivatives positive, the function is possibly at a relative minimum. Now take the cross partial derivative from (13.25) or (13.26),

$$z_{xy} = -2 = z_{yx}$$

evaluate it at the critical point,

$$z_{xy}(1, 1) = -2 = z_{yx}(1, 1)$$

and test the third condition to be sure $z_{xx} \cdot z_{yy} > (z_{xy})^2$. Here

$$10 \cdot 8 > (-2)^2$$

With $z_{xx}, z_{yy} > 0$ and $z_{xx}z_{yy} > (z_{xy})^2$, the function is indeed at a relative minimum at $(1, 1)$. In $(13.24)$, $z = 20$.

**13.14.** Redo Problem 13.13, given

$$f(x, y) = -7x^2 + 88x - 6xy + 42y - 2y^2 + 4$$

(1)   Take the first partials, set them equal to zero, and solve.

$$f_x = -14x + 88 - 6y = 0 \qquad\qquad (13.27)$$

$$f_y = -6x + 42 - 4y = 0 \qquad\qquad (13.28)$$

$$x = 5 \qquad y = 3 \qquad (5, 3) \text{ critical point}$$

(2)   Take the second direct partials, evaluate them at the critical point, and check their signs.

$$f_{xx} = -14 \qquad\qquad\qquad f_{yy} = -4$$

$$f_{xx}(5, 3) = -14 < 0 \qquad\qquad f_{yy}(5, 3) = -4 < 0$$

Take the cross partial from $(13.27)$ or $(13.28)$ and evaluate it at the critical point.

$$f_{xy} = -6 = f_{yx}$$

$$f_{xy}(5, 3) = -6 = f_{yx}(5, 3)$$

Then check the third condition $f_{xx}f_{yy} > (f_{xy})^2$. Here

$$(-14) \cdot (-4) > (-6)^2$$

With $f_{xx}, f_{yy} < 0$ and $f_{xx}f_{yy} > (f_{xy})^2$, we are sure the function is at a relative maximum at $(5, 3)$.

**13.15.** Redo Problem 13.13, given

$$z = 4x^2 + 128x - 12xy + 96y + 3y^2 + 17$$

(1)

$$z_x = 8x + 128 - 12y = 0$$

$$z_y = -12x + 96 + 6y = 0$$

$$x = 20 \qquad y = 24 \qquad (20, 24) \text{ critical point}$$

(2)   Test the second direct partials at the critical point.

$$z_{xx} = 8 \qquad\qquad\qquad z_{yy} = 6$$

$$z_{xx}(20, 24) = 8 > 0 \qquad z_{yy}(20, 24) = 6 > 0$$

With $z_{xx}$ and $z_{yy} > 0$ at the critical point, the function may be at a relative minimum. Check the cross partials,

$$z_{xy} = -12 = z_{yx}$$

$$z_{xy}(20, 24) = -12 = z_{yx}(20, 24)$$

and test the third condition. Here

$$8 \cdot 6 \ngtr (-12)^2$$

With $z_{xx}$ and $z_{yy}$ of the same sign and $z_{xx}z_{yy} < (z_{xy})^2$, the function is at an inflection point at $(20, 24)$.

**13.16.** Redo Problem 13.13, given

$$f(x, y) = 6x^2 - 108x + 4xy + 12y - 2y^2 - 19$$

(1)
$$f_x = 12x - 108 + 4y = 0$$

$$f_y = 4x + 12 - 4y = 0$$

$$x = 6 \quad y = 9 \quad (6, 9) \text{ critical point}$$

(2)
$$f_{xx} = 12 \qquad f_{yy} = -4$$

$$f_{xx}(6, 9) = 12 > 0 \qquad f_{yy}(6, 9) = -4 < 0$$

Taking the cross partials,

$$f_{xy} = 4 = f_{yx}$$

$$f_{xy}(6, 9) = 4 = f_{yx}(6, 9)$$

and testing the third condition,

$$(12) \cdot (-4) \not> (4)^2$$

With $f_{xx}$ and $f_{yy}$ of different signs and $f_{xx}f_{yy} < (f_{xy})^2$, the function is at a saddle point. Whenever $f_{xx}$ and $f_{yy}$ are of different signs, $f_{xx}f_{yy}$ cannot be greater than $(f_{xy})^2$, and the function will be at a saddle point.

**13.17.** Given the following cubic function, (1) find the critical points and (2) determine whether at these points the function is at a relative maximum, relative minimum, inflection point, or saddle point.

$$z(x, y) = 2y^3 - 15x^2 + 60x - 384y + 95$$

(1)   Take the first-order partials and set them equal to zero.

$$z_x = -30x + 60 = 0 \tag{13.29}$$

$$z_y = 6y^2 - 384 = 0 \tag{13.30}$$

Solve for the critical points.

$$-30x + 60 = 0 \qquad\qquad 6y^2 - 384 = 0$$

$$x = 2 \qquad\qquad\qquad y^2 = 64$$

$$y = \pm 8$$

$$(2, 8), \qquad (2, -8) \quad \text{critical points}$$

(2)   From (13.29) and (13.30), take the second direct partials,

$$z_{xx} = -30 \qquad z_{yy} = 12y$$

evaluate them at the critical points and note the signs.

$$z_{xx}(2, 8) = -30 < 0 \qquad z_{yy}(2, 8) = 12(8) = 96 > 0$$

$$z_{xx}(2, -8) = -30 < 0 \qquad z_{yy}(2, -8) = 12(-8) = -96 < 0$$

Then take the cross partial from (13.29) or (13.30),

$$z_{xy} = 0 = z_{yx}$$

evaluate it at the critical points and test the third condition.

$$z_{xx}(a, b) \cdot z_{yy}(a, b) > \left[z_{xy}(a, b)\right]^2 \; A$$

At (2, 8),                                $-30 \cdot \quad 96 \quad < 0$

At (2, -8),                               $-30 \cdot \quad -96 \quad > 0$

With $f_{xx}$ and $f_{yy}$ of different signs and $f_{xx}f_{yy} < (f_{xy})^2$ at $(2, 8)$, $f(2, 8)$ is a saddle point. With $f_{xx}, f_{yy} < 0$ and $f_{xx}f_{yy} > (f_{xy})^2$ at $(2, -8)$, $f(2, -8)$ is a relative maximum.

**13.18.** Redo Problem 13.17, given

$$f(x, y) = 4x^3 - 60xy + 5y^2 + 297$$

(1)   Set the first-order partial derivatives equal to zero,

$$f_x = 12x^2 - 60y = 0 \tag{13.31}$$

$$f_y = 10y - 60x = 0 \tag{13.32}$$

and solve for the critical values:

$$60y = 12x^2 \qquad\qquad 10y = 60x$$

$$y = \tfrac{1}{5}x^2 \qquad\qquad\quad y = 6x \tag{13.33}$$

Setting $y$ equal to $y$,                          $\tfrac{1}{5}x^2 = 6x$

$$x^2 - 30x = 0$$

$$x(x - 30) = 0$$

$$x = 0 \qquad x = 30$$

Substituting $x = 0$ and $x = 30$ in $y = 6x$ from (13.33),

$$y = 6(0) = 0$$

$$y = 6(30) = 180$$

Therefore,                         $(0, 0)$      $(30, 180)$      critical points

(2)   Take the second direct partials from (13.31) and (13.32),

$$f_{xx} = 24x \qquad f_{yy} = 10$$

evaluate them at the critical points and note the signs.

$$f_{xx}(0, 0) = 24(0) \;\; = 0 \qquad\qquad f_{yy}(0, 0) \;= 10 > 0$$

$$f_{xx}(30, 180) = 24(30) = 720 > 0 \qquad f_{yy}(30, 180) = 10 > 0$$

Then take the cross partial from (13.31) or (13.32),

$$f_{xy} = -60 = f_{yx}$$

evaluate it at the critical points and test the third condition:

$$f_{xx}(a, b) \cdot f_{yy}(a, b) > \left[f_{xy}(a, b)\right]^2$$

At (0, 0),                              $0 \cdot \quad 10 \quad < (-60)^2$

At (30, 180),                          $720 \cdot \quad 10 \quad > (-60)^2$

$$7200 > 3600$$

With $f_{xx}$ and $f_{yy}$ of the same sign and $f_{xx}f_{yy} < (f_{xy})^2$ at $(0, 0)$, $f(0, 0)$ is an inflection point. With $f_{xx}, f_{yy} > 0$ and $f_{xx}f_{yy} > (f_{xy})^2$ at $(30, 180)$, $f(30, 180)$ is a relative minimum.

**13.19.** Redo Problem 13.17, given

$$z = 8x^3 + 96xy - 8y^3$$

(1)

$$z_x = 24x^2 + 96y = 0 \qquad (13.34)$$

$$z_y = 96x - 24y^2 = 0 \qquad (13.35)$$

From (13.34),

$$96y = -24x^2$$

$$y = -\tfrac{1}{4}x^2$$

Substituting $y = -\tfrac{1}{4}x^2$ in (13.35) and solving algebraically,

$$96x - 24\left(-\tfrac{1}{4}x^2\right)^2 = 0$$

$$96x - \tfrac{3}{2}x^4 = 0$$

$$3x(64 - x^3) = 0$$

$$3x = 0 \qquad \text{or} \qquad 64 - x^3 = 0$$

$$x = 0 \qquad\qquad\qquad x^3 = 64$$

$$x = 4$$

Then substituting these values in (13.34), we find that if $x = 0$, $y = 0$, and if $x = 4$, $y = -4$. Therefore,

$$(0, 0) \qquad \text{and} \qquad (4, -4) \qquad \text{critical points}$$

(2) Test the second-order conditions from (13.34) and (13.35).

$$z_{xx} = 48x \qquad\qquad\qquad z_{yy} = -48y$$

$$z_{xx}(0, 0) = 48(0) = 0 \qquad\qquad z_{yy}(0, 0) = -48(0) = 0$$

$$z_{xx}(4, -4) = 48(4) = 192 > 0 \qquad z_{yy}(4, -4) = -48(-4) = 192 > 0$$

$$z_{xy} = 96 = z_{yx}$$

$$z_{xx}(a, b) \cdot z_{yy}(a, b) > \left[z_{xy}(a, b)\right]^2$$

At $(0, 0)$,          $0 \cdot\quad 0 \quad < (96)^2$

At $(4, -4)$,       $192 \cdot\quad 192 \quad > (96)^2$

With $z_{xx}$ and $z_{yy}$ of the same sign at $(0, 0)$ and $z_{xx}z_{yy} < (z_{xy})^2$, the function is at an inflection point at $(0, 0)$. With $z_{xx}$ and $z_{yy} > 0$ and $z_{xx}z_{yy} > (z_{xy})^2$ at $(4, -4)$, the function is at a relative minimum at $(4, -4)$.

**13.20.** Redo Problem 13.17, given

$$f(x, y) = 3x^3 - 9x^2 + 2y^3 + 24y^2 - 432x - 54y + 127$$

(1)

$$f_x = 9x^2 - 18x - 432 = 0 \qquad f_y = 6y^2 + 48y - 54 = 0 \qquad (13.36)$$

$$9(x^2 - 2x - 48) = 0 \qquad\qquad 6(y^2 + 8y - 9) = 0$$

$$(x + 6)(x - 8) = 0 \qquad\qquad (y - 1)(y + 9) = 0$$

$$x = -6 \quad\;\; x = 8 \qquad\quad y = 1 \qquad\quad y = -9$$

Hence $(-6, 1)$, $(-6, -9)$, $(8, 1)$, and $(8, -9)$ are critical points.

(2)  Test the second direct partials at each of the critical points. From ($13.36$),

$$f_{xx} = 18x - 18 \qquad f_{yy} = 12y + 48$$

(i) $\qquad f_{xx}(-6, 1) = -126 < 0 \qquad f_{yy}(-6, 1) = 60 > 0$

(ii) $\qquad f_{xx}(-6, -9) = -126 < 0 \qquad f_{yy}(-6, -9) = -60 < 0$

(iii) $\qquad f_{xx}(8, 1) = 126 > 0 \qquad f_{yy}(8, 1) = 60 > 0$

(iv) $\qquad f_{xx}(8, -9) = 126 > 0 \qquad f_{yy}(8, -9) = -60 < 0$

With different signs in (i) and (iv), $(-6, 1)$ and $(8, -9)$ can be ignored, if desired, as saddle points. Now take the cross partials from ($13.36$) and test the third condition.

$$f_{xy} = 0 = f_{yx}$$
$$f_{xx}(a, b) \cdot f_{yy}(a, b) > \left[ f_{xy}(a, b) \right]^2$$

From (ii), $\qquad\qquad (-126) \cdot (-60) \quad > (0)^2$

From (iii), $\qquad\qquad (126) \cdot (60) \quad > (0)^2$

The function is at a relative maximum at $(-6, -9)$, at a relative minimum at $(8, 1)$, and at saddle points at $(-6, 1)$ and $(8, -9)$.

## CONSTRAINED OPTIMIZATION AND LAGRANGE MULTIPLIERS

**13.21.** Use the Lagrange multiplier method to find the critical values at which the following function is optimized subject to the given constraint.

$$z = 5x^2 - 2xy + 8y^2, \qquad \text{subject to} \qquad x + y = 60$$

(1)  Set the constraint equal to zero, multiply it by $\lambda$ and add it to the objective function to obtain

$$Z = 5x^2 - 2xy + 8y^2 + \lambda(60 - x - y)$$

The first-order conditions are

$$Z_x = 10x - 2y - \lambda = 0 \qquad\qquad (13.37)$$

$$Z_y = -2x + 16y - \lambda = 0 \qquad\qquad (13.38)$$

$$Z_\lambda = 60 - x - y = 0 \qquad\qquad (13.39)$$

Subtracting ($13.38$) from ($13.37$) to eliminate $\lambda$,

$$12x - 18y = 0 \qquad x = 1.5y$$

Substituting $x = 1.5y$ in ($13.39$) and rearranging,

$$1.5y + y = 60 \qquad y_0 = 24$$

Next substituting $y_0 = 24$ in ($13.39$) and rearranging,

$$x + 24 = 60 \qquad x_0 = 36$$

Finally, substituting $x_0 = 36$ and $y_0 = 24$ in ($13.37$) or ($13.38$), we find $\lambda_0 = 312$. To see how matrix algebra can also be used to solve the first-order conditions for the critical values, see Problem 6.13 where Cramer's rule is used to find the same solution. In order to test the second-order conditions and determine whether the function is at a relative maximum or relative minimum, the bordered Hessian is necessary. See Dowling, *Schaum's Outline of Introduction to Mathematical Economics*, Section 12.5.

**13.22.** Use the Lagrange multiplier method to optimize the following function subject to the given constraint:

$$f(x, y) = 4x^2 - 6xy + 9y^2, \qquad \text{subject to} \qquad 2x + y = 104$$

The Lagrangian function is

$$F = 4x^2 - 6xy + 9y^2 + \lambda(104 - 2x - y)$$

The first-order conditions are

$$F_x = 8x - 6y - 2\lambda = 0 \qquad (13.40)$$

$$F_y = -6x + 18y - \lambda = 0 \qquad (13.41)$$

$$F_\lambda = 104 - 2x - y = 0 \qquad (13.42)$$

Subtracting 2 times (13.41) from (13.40) to eliminate $\lambda$,

$$20x - 42y = 0 \qquad x = 2.1y$$

Substituting $x = 2.1y$ in (13.42) and rearranging,

$$2(2.1y) + y = 104 \qquad y_0 = 20$$

Then substituting $y_0 = 20$ in (13.42) and rearranging,

$$2x + 20 = 104 \qquad x_0 = 42$$

Last, substituting $x_0 = 42$ and $y_0 = 20$ in (13.40) or (13.41), we see $\lambda_0 = 108$.

**13.23.** Optimize the following function subject to the constraint:

$$z = 120x - 4x^2 + 2xy - 3y^2 + 96y - 102, \qquad \text{subject to} \qquad x + 3y = 69$$

Incorporating the constraint and optimizing,

$$Z = 120x - 4x^2 + 2xy - 3y^2 + 96y - 102 + \lambda(69 - x - 3y)$$

$$Z_x = 120 - 8x + 2y - \lambda = 0 \qquad (13.43)$$

$$Z_y = 96 + 2x - 6y - 3\lambda = 0 \qquad (13.44)$$

$$Z_\lambda = 69 - x - 3y = 0 \qquad (13.45)$$

Subtracting 3 times (13.43) from (13.44) to eliminate $\lambda$,

$$-264 + 26x - 12y = 0 \qquad (13.46)$$

Subtracting 4 times (13.45) from (13.46) to eliminate $y$,

$$-540 + 30x = 0$$

$$x_0 = 18$$

Substituting $x_0 = 18$ in (13.45) to find $y$,

$$y_0 = 17$$

Then substituting $x_0 = 18$ and $y_0 = 17$ in (13.43) or (13.44), $\lambda_0 = 10$.

**13.24.** Optimize the following function subject to the constraint:

$$f(x, y) = 120x - 2x^2 - xy - 3y^2 + 160y + 7, \qquad \text{subject to} \qquad 3x + y = 480$$

Setting up the Lagrangian function and optimizing,

$$F = 120x - 2x^2 - xy - 3y^2 + 160y + 7 + \lambda(480 - 3x - y)$$

$$F_x = 120 - 4x - y - 3\lambda = 0 \qquad (13.47)$$

$$F_y = 160 - x - 6y - \lambda = 0 \qquad (13.48)$$

$$F_\lambda = 480 - 3x - y = 0 \qquad (13.49)$$

Subtracting 3 times (13.48) from (13.47) to eliminate $\lambda$,

$$-360 - x + 17y = 0 \qquad (13.50)$$

Subtracting 3 times ($13.50$) from ($13.49$) to eliminate $x$,

$$1560 - 52y = 0$$

$$y_0 = 30$$

Substituting $y_0 = 30$ in ($13.49$) to find $x$,

$$x_0 = 150$$

Then substituting $x_0 = 150$ and $y_0 = 30$ in ($13.47$) or ($13.48$), $\lambda_0 = -170$. With $\lambda$ negative, $f(x, y)$ is at a relative minimum.

## INCOME DETERMINATION MULTIPLIERS

**13.25.** Given the reduced form equation in ($4.53$) from Problem 4.20,

$$Y_e = \frac{1}{1 - b + z}(C_0 + I_0 + G_0 + X_0 - Z_0) \qquad (13.51)$$

find (1) the autonomous investment $I_0$ multiplier, (2) the autonomous import $Z_0$ multiplier, and (3) the multiplier for the marginal propensity to consume $b$.

(1)   To find any of the different multipliers, simply take the partial derivative of $Y_e$ with respect to the desired variable or parameter. Recalling that the fraction multiplies each term in the parentheses separately,

$$\frac{\partial Y_e}{\partial I_0} = \frac{1}{1 - b + z}$$

(2)

$$\frac{\partial Y_e}{\partial Z_0} = \frac{-1}{1 - b + z}$$

(3)   Using the quotient rule,

$$\frac{\partial Y_e}{\partial b} = \frac{(1 - b + z)(0) - (C_0 + I_0 + G_0 + X_0 - Z_0)(-1)}{(1 - b + z)^2}$$

$$\frac{\partial Y_e}{\partial b} = \frac{1(C_0 + I_0 + G_0 + X_0 - Z_0)}{(1 - b + z)^2} = \frac{1}{(1 - b + z)}\left(\frac{C_0 + I_0 + G_0 + X_0 - Z_0}{(1 - b + z)}\right)$$

And making use of ($13.51$) to simplify,

$$\frac{\partial Y_e}{\partial b} = \frac{Y_e}{(1 - b + z)}$$

**13.26.** Given the reduced form equation in ($4.52$) from Problem 4.19,

$$Y_e = \frac{1}{1 - b + bt}(C_0 + I_0 - bT_0) \qquad (13.52)$$

find the multiplier for the marginal propensity to consume $b$.

Using the quotient rule,

$$\frac{\partial Y_e}{\partial b} = \frac{(1 - b + bt)(-T_0) - (C_0 + I_0 - bT_0)(-1 + t)}{(1 - b + bt)^2}$$

Multiplying through by the negatives and then rearranging,

$$\frac{\partial Y_e}{\partial b} = \frac{-(1 - b + bt)T_0 + (1 - t)(C_0 + I_0 - bT_0)}{(1 - b + bt)^2}$$

$$\frac{\partial Y_e}{\partial b} = \frac{-(1 - b + bt)T_0 - t(C_0 + I_0 - bT_0) + (C_0 + I_0 - bT_0)}{(1 - b + bt)^2}$$

Recalling that the denominator divides each of the three terms in the numerator separately, and making use of (13.52) to simplify,

$$\frac{\partial Y_e}{\partial b} = \frac{-T_0}{1 - b + bt} - \frac{tY_e}{1 - b + bt} + \frac{Y_e}{1 - b + bt}$$

Rearranging and simplifying once again,

$$\frac{\partial Y_e}{\partial b} = \frac{1}{1 - b + bt}[Y_e - (T_0 + tY_e)] = \frac{1}{1 - b + bt}(Y_e - T_e)$$

But $Y - T = Y_d$ or disposable income, so

$$\frac{\partial Y_e}{\partial b} = \frac{1}{1 - b + bt}Y_d$$

## OPTIMIZATION OF BUSINESS AND ECONOMIC FUNCTIONS

**13.27.** For a firm producing two goods $x$ and $y$ (*a*) find the levels of $x$ and $y$ at which profit is maximized, (*b*) test the second-order conditions, and (*c*) evaluate the function at the critical values $x_0$ and $y_0$, given the following profit function:

$$\pi = 120x - 2x^2 - 3xy - 3y^2 + 165y - 250$$

(*a*)    Taking the first partials and setting them equal to zero,

$$\pi_x = 120 - 4x - 3y = 0 \tag{13.53}$$

$$\pi_y = 165 - 3x - 6y = 0 \tag{13.54}$$

Subtracting 2 times (13.53) from (13.54) to eliminate $y$,

$$-75 + 5x = 0$$

$$x_0 = 15$$

Substituting $x_0 = 15$ in (13.53) or (13.54) to find $y_0$,

$$y_0 = 20$$

(*b*)    Taking the second partials from (13.53) and (13.54),

$$\pi_{xx} = -4 \qquad \pi_{yy} = -6 \qquad \pi_{xy} = -3$$

With $\pi_{xx}$ and $\pi_{yy}$ both negative and $\pi_{xx}\pi_{yy} > (\pi_{xy})^2$, $\pi$ is maximized.

(*c*)    Substituting $x_0 = 15$ and $y_0 = 20$ in the original function,

$$\pi = 2300$$

**13.28.** Redo Problem 13.27, given

$$\pi = 324x - 5x^2 - 2xy - 4y^2 + 308y - 1240$$

(*a*)

$$\pi_x = 324 - 10x - 2y = 0 \tag{13.55}$$

$$\pi_y = 308 - 2x - 8y = 0 \tag{13.56}$$

Subtracting 4 times (13.55) from (13.56) to eliminate $y$,

$$-988 + 38x = 0$$

$$x_0 = 26$$

Substituting $x_0 = 26$ in (13.55) or (13.56) to find $y_0$,

$$y_0 = 32$$

(b)
$$\pi_{xx} = -10 \qquad \pi_{yy} = -8 \qquad \pi_{xy} = -2$$

With $\pi_{xx}$ and $\pi_{yy}$ both negative and $\pi_{xx}\pi_{yy} > (\pi_{xy})^2$, $\pi$ is maximized.

(c)
$$\pi = 7900$$

**13.29.** A monopolist produces two goods $x$ and $y$ for which the demand functions are

$$P_x = 315 - 4x \tag{13.57}$$

$$P_y = 260 - 3y \tag{13.58}$$

and the joint total cost function is

$$c = 2x^2 + 3xy + y^2 + 400$$

Find (a) the profit-maximizing level of output for each product, (b) the profit-maximizing price for each product, and (c) the maximum profit.

(a)   Given that $\pi = $ TR $-$ TC and here TR $= P_x \cdot x + P_y \cdot y$,

$$\pi = P_x \cdot x + P_y \cdot y - c$$

Substituting from the given information and optimizing,

$$\pi = (315 - 4x)x + (260 - 3y)y - (2x^2 + 3xy + y^2 + 400)$$
$$\pi = 315x - 6x^2 - 3xy - 4y^2 + 260y - 400$$

$$\pi_x = 315 - 12x - 3y = 0 \tag{13.59}$$

$$\pi_y = 260 - 3x - 8y = 0 \tag{13.60}$$

Subtracting 4 times (13.60) from (13.59) to eliminate $x$,

$$-725 + 29y = 0$$

$$y_0 = 25$$

Substituting $y_0 = 25$ in (13.59) or (13.60) to find $x_0$,

$$x_0 = 20$$

Testing the second-order conditions to be sure of a maximum,

$$\pi_{xx} = -12 \qquad \pi_{yy} = -8 \qquad \pi_{xy} = -3$$

With $\pi_{xx}$ and $\pi_{yy}$ both negative and $\pi_{xx}\pi_{yy} > (\pi_{xy})^2$, $\pi$ is maximized.

(b)   Substituting $x_0 = 20$ and $y_0 = 25$ in (13.57) and (13.58),

$$P_x = 315 - 4(20) = 235 \qquad P_y = 260 - 3(25) = 185$$

(c)
$$\pi = 6000$$

**13.30.** Find the profit-maximizing level of (a) output, (b) price, and (c) profit for the monopolist with the demand functions

$$P_x = 92 - 2x \tag{13.61}$$

$$P_y = 176 - 5y \tag{13.62}$$

and the cost function $c = 3x^2 + xy + 2y^2 + 424$.

(a)   Substituting into the profit function and optimizing,

$$\pi = (92 - 2x)x + (176 - 5y)y - (3x^2 + xy + 2y^2 + 424)$$

$$\pi = 92x - 5x^2 - xy - 7y^2 + 176y - 424$$

$$\pi_x = 92 - 10x - y = 0 \qquad\qquad (13.63)$$

$$\pi_y = 176 - x - 14y = 0 \qquad\qquad (13.64)$$

Subtracting 10 times (13.64) from (13.63) to eliminate $x$,

$$-1668 + 139y = 0$$

$$y_0 = 12 \qquad x_0 = 8$$

Testing the second-order conditions,

$$\pi_{xx} = -10 \qquad \pi_{yy} = -14 \qquad \pi_{xy} = -1$$

With $\pi_{xx}$ and $\pi_{yy}$ both negative and $\pi_{xx}\pi_{yy} > (\pi_{xy})^2$, $\pi$ is maximized.

(b)   Substituting $x_0 = 8$ and $y_0 = 12$ in (13.61) and (13.62),

$$P_x = 76 \qquad P_y = 116$$

(c)                                   $$\pi = 1000$$

**13.31.** Find the profit-maximizing level of (a) output, (b) price, and (c) profit for the monopolist with the demand functions

$$P_1 = 138 - 1.5Q_1 \qquad\qquad (13.65)$$

$$P_2 = 202 - 3.5Q_2 \qquad\qquad (13.66)$$

and total cost function $c = 4Q_1^2 + 2Q_1Q_2 + 3Q_2^2 + 204$.

(a)   Setting up the profit function and optimizing,

$$\pi = (138 - 1.5Q_1)Q_1 + (202 - 3.5Q_2)Q_2 - (4Q_1^2 + 2Q_1Q_2 + 3Q_2^2 + 204)$$

$$\pi = 138Q_1 - 5.5Q_1^2 - 2Q_1Q_2 - 6.5Q_2^2 + 202Q_2 - 204$$

$$\pi_1 = 138 - 11Q_1 - 2Q_2 = 0 \qquad\qquad (13.67)$$

$$\pi_2 = 202 - 2Q_1 - 13Q_2 = 0 \qquad\qquad (13.68)$$

Subtracting 5.5 times (13.68) from (13.67) to eliminate $Q_1$,

$$-973 + 69.5Q_2 = 0$$

$$\overline{Q}_2 = 14 \qquad \overline{Q}_1 = 10$$

Checking the second-order conditions,

$$\pi_{11} = -11 \qquad \pi_{22} = -13 \qquad \pi_{12} = -2$$

With $\pi_{11}, \pi_{22} < 0$ and $\pi_{11}\pi_{22} > (\pi_{12})^2$, $\pi$ is maximized.

(b)                   $$P_1 = 138 - 1.5(10) = 123 \qquad P_2 = 202 - 3.5(14) = 153$$

(c)                                   $$\pi = 1900$$

**13.32.** Find the profit-maximizing level of (*a*) output, (*b*) price, and (*c*) profit for the monopolist with the demand functions

$$P_1 = 920 - 10Q_1 \qquad\qquad (13.69)$$

$$P_2 = 950 - 6Q_2 \qquad\qquad (13.70)$$

and total cost function $c = 2Q_1^2 + 5Q_1Q_2 + 4Q_2^2 + 1800$.

(*a*)   Setting up the profit function and optimizing,

$$\pi = (920 - 10Q_1)Q_1 + (950 - 6Q_2)Q_2 - (2Q_1^2 + 5Q_1Q_2 + 4Q_2^2 + 1800)$$

$$\pi = 920Q_1 - 12Q_1^2 - 5Q_1Q_2 - 10Q_2^2 + 950Q_2 - 1800$$

$$\pi_1 = 920 - 24Q_1 - 5Q_2 = 0 \qquad\qquad (13.71)$$

$$\pi_2 = 950 - 5Q_1 - 20Q_2 = 0 \qquad\qquad (13.72)$$

Subtracting 4 times (*13.71*) from (*13.72*) to eliminate $Q_2$,

$$-2730 + 91Q_1 = 0$$

$$\overline{Q}_1 = 30 \qquad \overline{Q}_2 = 40$$

Checking the second-order conditions,

$$\pi_{11} = -24 \qquad \pi_{22} = -20 \qquad \pi_{12} = -5$$

With $\pi_{11}, \pi_{22} < 0$ and $\pi_{11}\pi_{22} > (\pi_{12})^2$, $\pi$ is maximized.

(*b*) $$P_1 = 920 - 10(30) = 620 \qquad P_2 = 950 - 6(40) = 710$$

(*c*) $$\pi = 31,000$$

## ECONOMIC APPLICATIONS OF THE LAGRANGE MULTIPLIER METHOD

**13.33.** A women's clothing manufacturer is under contract to deliver 77 units of skirts *x* and slacks *y* in any combination of its choosing. (*a*) Find the combination that minimizes the cost of fulfilling the contract given the firm's total cost function

$$c = 7x^2 - 2xy + 5y^2 + 64$$

(*b*) What would happen to costs if the constant of the constraint were increased or decreased by one unit?

(*a*)   First forming the constraint,

$$x + y = 77$$

then setting up the Lagrangian function and optimizing,

$$C = 7x^2 - 2xy + 5y^2 + 64 + \lambda(77 - x - y)$$

$$C_x = 14x - 2y - \lambda = 0 \qquad\qquad (13.73)$$

$$C_y = -2x + 10y - \lambda = 0 \qquad\qquad (13.74)$$

$$C_\lambda = 77 - x - y = 0 \qquad\qquad (13.75)$$

Subtracting (*13.74*) from (*13.73*) to eliminate $\lambda$,

$$16x - 12y = 0$$

$$y = \tfrac{4}{3}x$$

Substituting in (13.75),

$$77 - x - \left(\tfrac{4}{3}x\right) = 0$$

$$x_0 = 33$$

Substituting $x_0 = 33$ in (13.75), $y_0 = 44$. Then substituting $x_0 = 33$ and $y_0 = 44$ in (13.73) or (13.74), $\lambda_0 = 374$.

(b)   With $\lambda_0 = 374$, if the constant of the constraint were increased by one unit to 78, total cost would increase by approximately \$374. And if the constant of the constraint were decreased by one unit to 76, total cost would decrease by approximately \$374. To see the first-order conditions of this problem solved with matrix algebra, see Problem 6.14.

**13.34.** A shoe manufacturer is under contract to deliver 36 pairs of loafers $x$ and oxfords $y$ in any combination it wants. (a) Find the least-cost combination that fulfills the contract if

$$c = 2x^2 - xy + 3y^2 + 18$$

(b) What happens to $C$ if an extra pair of shoes is added to the contract?

(a)   Setting up the Lagrangian function and optimizing,

$$C = 2x^2 - xy + 3y^2 + 18 + \lambda(36 - x - y)$$

$$C_x = 4x - y - \lambda = 0 \tag{13.76}$$

$$C_y = -x + 6y - \lambda = 0 \tag{13.77}$$

$$C_\lambda = 36 - x - y = 0 \tag{13.78}$$

Subtracting (13.77) from (13.76) to eliminate $\lambda$,

$$5x - 7y = 0$$

$$y = \tfrac{5}{7}x$$

Substituting in (13.78),

$$36 - x - \left(\tfrac{5}{7}x\right) = 0$$

$$x_0 = 21$$

and

$$y_0 = 15 \qquad \lambda_0 = 69$$

(b)   With $\lambda_0 = 69$, $C$ would increase by approximately \$69.

**13.35.** (a) Find the output mix that will maximize profits for a firm with a maximum joint output capacity of 22 when

$$\pi = 160x - 4x^2 - xy - 3y^2 + 210y$$

(b) Estimate the effect on profits if output capacity is expanded by one unit.

(a)   Incorporating the constraint and optimizing,

$$\Pi = 160x - 4x^2 - xy - 3y^2 + 210y + \lambda(22 - x - y)$$

$$\Pi_x = 160 - 8x - y - \lambda = 0 \tag{13.79}$$

$$\Pi_y = 210 - x - 6y - \lambda = 0 \tag{13.80}$$

$$\Pi_\lambda = 22 - x - y = 0 \tag{13.81}$$

Subtracting $(13.80)$ from $(13.79)$ to eliminate $\lambda$,

$$-50 - 7x + 5y = 0 \qquad (13.82)$$

Adding 5 times $(13.81)$ to $(13.82)$ to eliminate $y$,

$$60 - 12x = 0$$

$$x_0 = 5 \qquad y_0 = 17 \qquad \lambda_0 = 103$$

(b)   With $\lambda_0 = 103$, $\pi$ would increase by approximately \$103.

**13.36.** A department store has found that its value of sales $z$ depends on the number of advertisements in circulars $x$ and in newspapers $y$, given by

$$z = 420x - 2x^2 - 3xy - 5y^2 + 640y + 1725$$

If the price per ad is \$1 in circulars and \$4 in newspapers, and the advertising budget is \$180, (a) find the number of ads in circulars and newspapers that will maximize sales subject to the budget constraint. (b) Estimate the effect on sales of a \$1 increase in the advertising budget.

(a)   Setting up the Lagrangian function,

$$Z = 420x - 2x^2 - 3xy - 5y^2 + 640y + 1725 + \lambda(180 - x - 4y)$$

$$Z_x = 420 - 4x - 3y - \lambda = 0 \qquad (13.83)$$

$$Z_y = 640 - 3x - 10y - 4\lambda = 0 \qquad (13.84)$$

$$Z_\lambda = 180 - x - 4y = 0 \qquad (13.85)$$

Subtracting 4 times $(13.83)$ from $(13.84)$ to eliminate $\lambda$,

$$-1040 + 13x + 2y = 0 \qquad (13.86)$$

Adding 2 times $(13.86)$ to $(13.85)$ to eliminate $y$,

$$-1900 + 25x = 0$$

$$x_0 = 76 \qquad y_0 = 26 \qquad \lambda_0 = 38$$

(b)   With $\lambda_0 = 38$, sales would increase by \$38.

**13.37.** Because of reduced warehouse space from current renovations, a firm with the normal profit function

$$\pi = 160x - 3x^2 - xy - 2y^2 + 240y - 665$$

has had storage space temporarily reduced to 40 square feet ($\text{ft}^2$). If $x$ requires 1 $\text{ft}^2$ of space and $y$ requires 2 $\text{ft}^2$, (a) find the critical values for $x$ and $y$ that will maximize profits subject to the temporary storage constraint. (b) Estimate the effect on profits of a 1-$\text{ft}^2$ increase in storage space.

(a)   Incorporating the constraint and taking the first partials,

$$\Pi = 160x - 3x^2 - xy - 2y^2 + 240y - 665 + \lambda(40 - x - 2y)$$

$$\Pi_x = 160 - 6x - y - \lambda = 0 \qquad (13.87)$$

$$\Pi_y = 240 - x - 4y - 2\lambda = 0 \qquad (13.88)$$

$$\Pi_\lambda = 40 - x - 2y = 0 \qquad (13.89)$$

Subtracting 2 times $(13.87)$ from $(13.88)$ to eliminate $\lambda$,

$$-80 + 11x - 2y = 0 \qquad (13.90)$$

Subtracting $(13.90)$ from $(13.89)$ to eliminate $y$,

$$120 - 12x = 0$$

$$x_0 = 10 \qquad y_0 = 15 \qquad \lambda_0 = 85$$

(*b*)    With $\lambda_0 = 85$, profits would increase by approximately \$85.

**13.38.**  A monopolistic producer of TVs ($Q_1$) and VCRs ($Q_2$) in a small developing country has a maximum joint capacity of 40 items a day. Find the profit-maximizing level of output subject to the production constraint, given the demand functions

$$P_1 = 820 - 6Q_1$$
$$P_2 = 980 - 8Q_2$$

and total cost function $c = 2Q_1^2 + Q_1Q_2 + 4Q_2^2 + 225$.

First setting up the profit function,

$$\pi = (820 - 6Q_1)Q_1 + (980 - 8Q_2)Q_2 - (2Q_1^2 + Q_1Q_2 + 4Q_2^2 + 225)$$

$$\pi = 820Q_1 - 8Q_1^2 - Q_1Q_2 - 12Q_2^2 + 980Q_2 - 225$$

Then forming the Lagrangian function and optimizing,

$$\Pi = 820Q_1 - 8Q_1^2 - Q_1Q_2 - 12Q_2^2 + 980Q_2 - 225 + \lambda(40 - Q_1 - Q_2)$$

$$\Pi_1 = 820 - 16Q_1 - Q_2 - \lambda = 0 \tag{13.91}$$

$$\Pi_2 = 980 - Q_1 - 24Q_2 - \lambda = 0 \tag{13.92}$$

$$\Pi_\lambda = 40 - Q_1 - Q_2 = 0 \tag{13.93}$$

Subtracting (*13.92*) from (*13.91*) to eliminate $\lambda$,

$$-160 - 15Q_1 + 23Q_2 = 0 \tag{13.94}$$

Subtracting 15 times (*13.93*) from (*13.94*) to eliminate $Q_1$,

$$-760 + 38Q_2 = 0$$

$$\overline{Q}_2 = 20 \qquad \overline{Q}_1 = 20 \qquad \overline{\lambda} = 480$$

## OPTIMIZING COBB-DOUGLAS FUNCTIONS UNDER CONSTRAINTS

**13.39.**  Maximize output for a firm operating at constant returns to scale for which the strict Cobb-Douglas production function is

$$q = 1.5K^{0.6}L^{0.4}$$

when $P_K = \$6$, $P_L = \$2$, and the production budget is \$960.

Setting up the Lagrangian function for constrained optimization and setting the first partials equal to zero,

$$Q = 1.5K^{0.6}L^{0.4} + \lambda(960 - 6K - 2L)$$

$$Q_K = 0.9K^{-0.4}L^{0.4} - 6\lambda = 0 \tag{13.95}$$

$$Q_L = 0.6K^{0.6}L^{-0.6} - 2\lambda = 0 \tag{13.96}$$

$$Q_\lambda = 960 - 6K - 2L = 0 \tag{13.97}$$

Rearranging, then dividing (*13.95*) by (*13.96*) to eliminate $\lambda$,

$$\frac{0.9K^{-0.4}L^{0.4}}{0.6K^{0.6}L^{-0.6}} = \frac{6\lambda}{2\lambda}$$

Remembering to subtract exponents in division,

$$1.5K^{-1}L^1 = 3$$

$$\frac{L}{K} = 2 \qquad L = 2K$$

Substituting $L = 2K$ in ($13.97$),

$$960 - 6K - 2(2K) = 0 \qquad K_0 = 96$$

Finally by substituting $K_0 = 96$ in ($13.97$),

$$L_0 = 192$$

**13.40.** Production for a firm with decreasing returns to scale is given by the generalized Cobb-Douglas production function

$$q = K^{0.3} L^{0.5}$$

If $P_K = \$12$, $P_L = \$8$, and the production budget is $\$1280$, maximize output subject to the budget constraint.

Setting up the Lagrangian function,

$$Q = K^{0.3} L^{0.5} + \lambda(1280 - 12K - 8L)$$

$$Q_K = 0.3K^{-0.7} L^{0.5} - 12\lambda = 0 \qquad\qquad (13.98)$$

$$Q_L = 0.5K^{0.3} L^{-0.5} - 8\lambda = 0 \qquad\qquad (13.99)$$

$$Q_\lambda = 1280 - 12K - 8L = 0 \qquad\qquad (13.100)$$

Rearranging, then dividing ($13.98$) by ($13.99$) to eliminate $\lambda$,

$$\frac{0.3K^{-0.7} L^{0.5}}{0.5K^{0.3} L^{-0.5}} = \frac{12\lambda}{8\lambda}$$

$$0.6\frac{L}{K} = 1.5 \qquad L = 2.5K$$

Substituting $L = 2.5K$ in ($13.100$),

$$K_0 = 40 \qquad L_0 = 100$$

**13.41.** Maximize utility for the consumer whose utility function is

$$u = x^{0.8} y^{0.1}$$

when $P_x = \$16$, $P_y = \$4$, and her budget $B$ is $\$864$.

Setting up the Lagrangian function and optimizing,

$$U = x^{0.8} y^{0.1} + \lambda(864 - 16x - 4y)$$

$$U_x = 0.8x^{-0.2} y^{0.1} - 16\lambda = 0 \qquad\qquad (13.101)$$

$$U_y = 0.1x^{0.8} y^{-0.9} - 4\lambda = 0 \qquad\qquad (13.102)$$

$$U_\lambda = 864 - 16x - 4y = 0 \qquad\qquad (13.103)$$

Rearranging, then dividing ($13.101$) by ($13.102$) to eliminate $\lambda$,

$$\frac{0.8x^{-0.2} y^{0.1}}{0.1x^{0.8} y^{-0.9}} = \frac{16\lambda}{4\lambda}$$

$$8\frac{y}{x} = 4 \qquad y = \frac{1}{2}x$$

Substituting $y = \frac{1}{2}x$ in ($13.103$),

$$x_0 = 48 \qquad y_0 = 24$$

**13.42.** Maximize utility for the consumer, given

$$u = x^{0.2}y^{0.7}$$

subject to $P_x = 3$, $P_y = 5$, and $B = 405$.

Forming the Lagrangian function and optimizing,

$$U = x^{0.2}y^{0.7} + \lambda(405 - 3x - 5y)$$

$$U_x = 0.2x^{-0.8}y^{0.7} - 3\lambda = 0 \qquad (13.104)$$

$$U_y = 0.7x^{0.2}y^{-0.3} - 5\lambda = 0 \qquad (13.105)$$

$$U_\lambda = 405 - 3x - 5y = 0 \qquad (13.106)$$

Dividing (13.104) by (13.105) to eliminate $\lambda$,

$$\frac{0.2x^{-0.8}y^{0.7}}{0.7x^{0.2}y^{-0.3}} = \frac{3\lambda}{5\lambda}$$

$$\frac{0.2y}{0.7x} = 0.6 \qquad y = 2.1x$$

Substituting in (13.106),

$$x_0 = 30 \qquad y_0 = 63$$

## IMPLICIT AND INVERSE FUNCTION RULES

**13.43.** Find the derivative $dy/dx$ for each of the following implicit functions.

(a) $7x^6 + 4y^5 - 96 = 0$

From the implicit function rule in (13.22),

$$\frac{dy}{dx} = -\frac{f_x}{f_y} \qquad (f_y \neq 0)$$

Here $f_x = 42x^5$ and $f_y = 20y^4$. Substituting above,

$$\frac{dy}{dx} = -\frac{42x^5}{20y^4} = -\frac{21x^5}{10y^4}$$

(b) $6x^2 - 7xy + 2y^2 = 81$

Here $f_x = 12x - 7y$, $f_y = 4y - 7x$. Substituting in the formula,

$$\frac{dy}{dx} = -\frac{f_x}{f_y} = -\left(\frac{12x - 7y}{4y - 7x}\right) = \frac{7y - 12x}{4y - 7x}$$

(c) $5x^3 + 8y^4 - 3y^3 - 109 = 0$

$$\frac{dy}{dx} = -\frac{f_x}{f_y} = -\frac{15x^2}{32y^3 - 9y^2}$$

(d) $9x^5 - 2x^2y^2 + 7y^3 = 224$

$$\frac{dy}{dx} = -\frac{f_x}{f_y} = -\frac{45x^4 - 4xy^2}{21y^2 - 4x^2y}$$

(e) $(13x - 7)^3 + 15y^5 = 1029$

Using the generalized power function rule for $f_x$,

$$\frac{dy}{dx} = -\frac{f_x}{f_y} = -\frac{39(13x - 7)^2}{75y^4}$$

(f)  $(4x^4 + 9y^2)^3 = 1645$

Using the generalized power function rule for $f_x$ and $f_y$,

$$\frac{dy}{dx} = -\frac{f_x}{f_y} = -\frac{48x^3(4x^4 + 9y^2)^2}{54y(4x^4 + 9y^2)^2} = -\frac{8x^3}{9y}$$

(g)  $\sqrt{2x^3 + 3y^5} = 98$

Recalling that $\sqrt{2x^3 + 3y^5} = (2x^3 + 3y^5)^{1/2}$,

$$\frac{dy}{dx} = -\frac{f_x}{f_y} = -\frac{3x^2(2x^3 + 3y^5)^{-1/2}}{7.5y^4(2x^3 + 3y^5)^{-1/2}} = -\frac{x^2}{2.5y^4}$$

**13.44.** Find the derivative of the inverse function $dP/dQ$ for each of the following explicit functions of $Q$.

(a)  $Q = 496 - 5P$

From the inverse function rule in (13.23),

$$\frac{dP}{dQ} = \frac{1}{dQ/dP} \qquad \left(\frac{dQ}{dP} \neq 0\right)$$

Substituting $dQ/dP = -5$ in the formula,

$$\frac{dP}{dQ} = -\frac{1}{5}$$

(b)  $Q = 87 - 0.4P$

$$\frac{dP}{dQ} = \frac{1}{-0.4} = -2.5$$

(c)  $Q = 13 + 3P^2$

$$\frac{dP}{dQ} = \frac{1}{6P} \qquad (P \neq 0)$$

(d)  $Q = 246 - P^2 + 3P$

$$\frac{dP}{dQ} = \frac{1}{3 - 2P} \qquad (P \neq 1.5)$$

## VERIFICATION OF THE RULES OF PARTIAL DIFFERENTIATION

**13.45.** For each of the following functions, use (1) the definition in (13.1a) to find $\partial z/\partial x$, and (2) the definition in (13.1b) to find $\partial z/\partial y$ in order to confirm the rules of differentiation.

(a)  $z = 94 + 5x - 8y$

(1) From (13.1a),
$$\frac{\partial z}{\partial x} = \lim_{\Delta x \to 0} \frac{f(x + \Delta x, y) - f(x, y)}{\Delta x}$$

Substituting,
$$\frac{\partial z}{\partial x} = \lim_{\Delta x \to 0} \frac{[94 + 5(x + \Delta x) - 8y] - (94 + 5x - 8y)}{\Delta x}$$

$$\frac{\partial z}{\partial x} = \lim_{\Delta x \to 0} \frac{94 + 5x + 5\Delta x - 8y - 94 - 5x + 8y}{\Delta x}$$

$$\frac{\partial z}{\partial x} = \lim_{\Delta x \to 0} \frac{5\Delta x}{\Delta x} = \lim_{\Delta x \to 0} 5 = 5$$

(2) From (13.1b),
$$\frac{\partial z}{\partial y} = \lim_{\Delta y \to 0} \frac{f(x, y + \Delta y) - f(x, y)}{\Delta y}$$

Substituting,
$$\frac{\partial z}{\partial y} = \lim_{\Delta y \to 0} \frac{[94 + 5x - 8(y + \Delta y)] - (94 + 5x - 8y)}{\Delta y}$$

$$\frac{\partial z}{\partial y} = \lim_{\Delta y \to 0} \frac{94 + 5x - 8y - 8\Delta y - 94 - 5x + 8y}{\Delta y}$$

$$\frac{\partial z}{\partial y} = \lim_{\Delta y \to 0} \frac{-8\Delta y}{\Delta y} = \lim_{\Delta y \to 0} (-8) = -8$$

(b)   $z = 16x - 9xy + 13y$

(1)
$$\frac{\partial z}{\partial x} = \lim_{\Delta x \to 0} \frac{[16(x + \Delta x) - 9(x + \Delta x)y + 13y] - (16x - 9xy + 13y)}{\Delta x}$$

$$\frac{\partial z}{\partial x} = \lim_{\Delta x \to 0} \frac{16x + 16\Delta x - 9xy - 9y\,\Delta x + 13y - 16x + 9xy - 13y}{\Delta x}$$

$$\frac{\partial z}{\partial x} = \lim_{\Delta x \to 0} \frac{16\Delta x - 9y\,\Delta x}{\Delta x}$$

$$\frac{\partial z}{\partial x} = \lim_{\Delta x \to 0} (16 - 9y) = 16 - 9y$$

(2)
$$\frac{\partial z}{\partial y} = \lim_{\Delta y \to 0} \frac{[16x - 9x(y + \Delta y) + 13(y + \Delta y)] - (16x - 9xy + 13y)}{\Delta y}$$

$$\frac{\partial z}{\partial y} = \lim_{\Delta y \to 0} \frac{16x - 9xy - 9x\,\Delta y + 13y + 13\Delta y - 16x + 9xy - 13y}{\Delta y}$$

$$\frac{\partial z}{\partial y} = \lim_{\Delta y \to 0} \frac{-9x\,\Delta y + 13\Delta y}{\Delta y} = \lim_{\Delta y \to 0} (-9x + 13) = -9x + 13$$

(c)   $z = 7x^2 y$

(1)
$$\frac{\partial z}{\partial x} = \lim_{\Delta x \to 0} \frac{\left[7(x + \Delta x)^2 y\right] - (7x^2 y)}{\Delta x}$$

$$\frac{\partial z}{\partial x} = \lim_{\Delta x \to 0} \frac{7x^2 y + 14xy\,\Delta x + 7y(\Delta x)^2 - 7x^2 y}{\Delta x}$$

$$\frac{\partial z}{\partial x} = \lim_{\Delta x \to 0} \frac{14xy\,\Delta x + 7y(\Delta x)^2}{\Delta x}$$

$$\frac{\partial z}{\partial x} = \lim_{\Delta x \to 0} (14xy + 7y\,\Delta x) = 14xy$$

(2)
$$\frac{\partial z}{\partial y} = \lim_{\Delta y \to 0} \frac{\left[7x^2(y + \Delta y)\right] - (7x^2 y)}{\Delta y}$$

$$\frac{\partial z}{\partial y} = \lim_{\Delta y \to 0} \frac{7x^2 y + 7x^2\,\Delta y - 7x^2 y}{\Delta y}$$

$$\frac{\partial z}{\partial y} = \lim_{\Delta y \to 0} \frac{7x^2\,\Delta y}{\Delta y} = \lim_{\Delta y \to 0} 7x^2 = 7x^2$$

(d)   $z = 3x^2 y^2$

(1)
$$\frac{\partial z}{\partial x} = \lim_{\Delta x \to 0} \frac{\left[3(x + \Delta x)^2 y^2\right] - (3x^2 y^2)}{\Delta x}$$

$$\frac{\partial z}{\partial x} = \lim_{\Delta x \to 0} \frac{6y^2 x \,\Delta x + 3y^2 (\Delta x)^2}{\Delta x}$$

$$\frac{\partial z}{\partial x} = \lim_{\Delta x \to 0} (6xy^2 + 3y^2 \,\Delta x) = 6xy^2$$

(2)

$$\frac{\partial z}{\partial y} = \lim_{\Delta y \to 0} \frac{\left[3x^2 (y + \Delta y)^2\right] - (3x^2 y^2)}{\Delta y}$$

$$\frac{\partial z}{\partial y} = \lim_{\Delta y \to 0} \frac{3x^2 y^2 + 6x^2 y \,\Delta y + 3x^2 (\Delta y)^2 - 3x^2 y^2}{\Delta y}$$

$$\frac{\partial z}{\partial y} = \lim_{\Delta y \to 0} \frac{6x^2 y \,\Delta y + 3x^2 (\Delta y)^2}{\Delta y} = \lim_{\Delta y \to 0} (6x^2 y + 3x^2 \,\Delta y) = 6x^2 y$$

# Supplementary Problems

## PARTIAL DERIVATIVES

**13.46.** Find the first-order partial derivatives for each of the following functions:

(a)  $z = 15x^2 + 23xy - 14y^2$    (b) $z = 6x^5 - 13x^2 y^3 + 8y^4$

(c)  $z = 7w^3 - 2w^2 x + 16x^2 + 11x^3 y^4 + 20y^4$

**13.47.** Use the product rule to find the first-order partial derivatives for each of the following functions:

(a)  $z = (8x - 5y)(3x + 4y)$    (b) $z = (2x^3 + 5y)(6x^2 - 8y^2)$

(c)  $z = (7w^2 - 4x^3 - 10y^4)(3w^2 + 8x^2 - 2y^3)$

**13.48.** Use the quotient rule to find the first-order partial derivatives for each of the following functions:

(a) $z = \dfrac{8x - 7y}{18x}$  (b) $z = \dfrac{27y}{12x + 5y}$

(c) $z = \dfrac{5x^3 - 9y^2}{4x + 3y}$  (d) $z = \dfrac{6w - 4x - 9y}{8w - 3x + 2y}$

**13.49.** Use the generalized power function rule to find the first-order partial derivatives for each of the following functions:

(a) $z = (12x - 13y)^5$  (b) $z = (3x^2 + 5y^3)^4$
(c) $z = (14x^3 + 9y^2)^{1/2}$  (d) $z = (5x^4 - 2xy + 4y^2)^{-3}$
(e) $z = (3x^2 + 8xy + 5y^2)^{-1/2}$  (f) $z = (6w^2 - 7x^4 - 9y^5)^3$

**13.50.** Use the natural exponential rule to find the first-order partial derivatives for each of the following functions:

(a) $z = 6e^{3xy}$  (b) $z = -15e^{-3x^2 y^2}$
(c) $z = 7e^{(2x+5y)}$  (d) $z = e^{w^2 x^3 y^2}$

**13.51.** Use the natural logarithmic rule to find the first-order partial derivatives for each of the following functions:

(a) $z = \ln|8x + 11y|$  (b) $z = \ln\left|5x^3 + 4y^2\right|$
(c) $z = \ln\left|6x^4 y^5\right|$  (d) $z = \ln\left|7w^3 + 4x^2 + 5y^4\right|$

**13.52.** Use whatever combination of rules is necessary to find the first-order partials for each of the following functions:

(a) $z = (9x + 2y)(3x - 7y)^3$

(b) $z = \dfrac{(5x + 8y)^5}{4x - 3y}$

(c) $z = \dfrac{(2x - 9y)(8x - 5y)}{3x + 11y}$

(d) $z = 6x^2 y^3 e^{5xy}$

## SECOND-ORDER PARTIAL DERIVATIVES

**13.53.** Find all the second-order partial derivatives for each of the following functions:

(a) $z = 7x^3 - 4xy + 12y^4$

(b) $z = 9x^5 + 6x^2 y^4 - 3x^3 y^5 - 13y^3$

(c) $z = 16x^3 y^2$

(d) $z = (5x + 12y)^4$

(e) $z = x^{0.6} y^{0.4}$

(f) $z = 8x^{0.5} y^{0.7}$

## OPTIMIZATION OF MULTIVARIABLE FUNCTIONS

**13.54.** Find the critical values at which each of the following quadratic functions is optimized and test the second-order conditions to see if the function is at a relative maximum, relative minimum, inflection point, or saddle point.

(a)  $z = -8x^2 + 148x - 4xy + 118y - 5y^2 + 6$

(b)  $z = 6x^2 - 51x - 9xy + 182y - 11y^2 - 17$

(c)  $z = 7x^2 - 78x - 6xy - 48y + 4y^2 - 23$

(d)  $z = -3x^2 - 6x + 14xy - 158y - 2y^2 + 37$

**13.55.** Find the critical values at which each of the following cubic functions is optimized and test the second-order conditions to see if the function is at a relative maximum, relative minimum, inflection point, or saddle point.

(a)  $z = 5x^3 + 7y^2 - 240x - 154y + 61$

(b)  $z = -9x^2 - 4y^3 + 324x + 108y - 37$

(c)  $z = 8x^2 - 6y^3 + 144xy - 323$

(d)  $z = 6x^3 + 6y^3 + 54xy - 195$

(e)  $z = 4x^3 + 5y^3 + 18x^2 - 52.5y^2 - 336x - 270y + 189$

(f)  $z = 3x^3 - 7y^3 + 22.5x^2 - 10.5y^2 - 216x + 630y - 231$

## CONSTRAINED OPTIMIZATION

**13.56.** Find the critical values at which each of the following functions is optimized subject to the given constraints:

(a)  $z = 4x^2 - 5xy + 6y^2$, subject to $x + y = 30$

(b)  $z = -7x^2 + 6xy + -9y^2$, subject to $2x + y = 165$

(c)  $z = 8x^2 - 70x - 4xy - 50y + 5y^2$, subject to $x + y = 35$

(d)  $z = -3x^2 + 40x + 8xy + 288y - 10y^2$, subject to $x + 2y = 58$

**OPTIMIZATION OF BUSINESS AND ECONOMICS FUNCTIONS**

**13.57.**    Find the critical values at which each firm's profit $\pi$ is maximized, given the following profit functions:

(a)   $\pi = 422x - 4x^2 - 7xy - 6y^2 + 522y - 167$

(b)   $\pi = 496x - 9x^2 - 11xy - 7y^2 + 405y - 235$

(c)   $\pi = 406x - 5x^2 - 8xy - 9y^2 + 522y - 237$

(d)   $\pi = 685x - 12x^2 - 13xy - 8y^2 + 595y - 305$

**13.58.**    Find the critical values at which each firm's profit $\pi$ is maximized, given the following demand functions and total cost functions:

(a)   $P_x = 712 - 5x,\ P_y = 976 - 9y,\ C = 3x^2 + 4xy + 2y^2 + 375$

(b)   $P_x = 1173 - 9x,\ P_y = 1200 - 11y,\ C = 3x^2 + 5xy + 4y^2 + 222$

(c)   $P_x = 1532 - 13x,\ P_y = 1396 - 8y,\ C = 4x^2 + 7xy + 5y^2 + 624$

(d)   $P_x = 2501 - 15x,\ P_y = 1648 - 14y,\ C = 5x^2 + 11xy + 3y^2 + 535$

**13.59.**    Find the critical values at which each firm's total cost $C$ is minimized, given the production quotas that must be fulfilled:

(a)   $C = 8x^2 - 3xy + 11y^2 + 1055$, subject to $x + y = 88$

(b)   $C = 12x^2 - 4xy + 7y^2 + 945$, subject to $x + y = 69$

(c)   $C = 9x^2 - 6xy + 13y^2 + 775$, subject to $x + 4y = 362$

(d)   $C = 14x^2 - 9xy + 16y^2 + 835$, subject to $3x + y = 148$

**13.60.**    Find the critical values at which each firm's profit $\pi$ is maximized, given the production limitations in parentheses that cannot be exceeded:

(a)   $\pi = 145x - 5x^2 - 2xy - 8y^2 + 201y - 325$     $(x + y = 37)$

(b)   $\pi = 440x - 7x^2 - 3xy - 4y^2 + 445y - 295$     $(x + y = 49)$

(c)   $\pi = 210x - 11x^2 - 5xy - 9y^2 + 182y - 435$     $(2x + y = 36)$

(d)   $\pi = 494x - 6x^2 - 2xy - 15y^2 + 240y - 565$     $(x + 3y = 81)$

# Answers to Supplementary Problems

**13.46.**    (a)   $z_x = 30x + 23y,\qquad z_y = 23x - 28y$

(b)   $z_x = 30x^4 - 26xy^3,\qquad z_y = -39x^2y^2 + 32y^3$

(c)   $z_w = 21w^2 - 4wx,\qquad z_x = -2w^2 + 32x + 33x^2y^4,\qquad z_y = 44x^3y^3 + 80y^3$

**13.47.**    (a)   $z_x = 48x + 17y,\qquad z_y = 17x - 40y$

(b)   $z_x = 60x^4 - 48x^2y^2 + 60xy,\qquad z_y = -32x^3y + 30x^2 - 120y^2$

(c)   $z_w = 84w^3 + 112wx^2 - 28wy^3 - 24wx^3 - 60wy^4$

$z_x = 112w^2x - 36w^2x^2 - 160x^4 + 24x^2y^3 - 160xy^4$

$z_y = -42w^2y^2 + 24x^3y^2 - 120w^2y^3 - 320x^2y^3 + 140y^6$

**13.48.**    (a)  $z_x = \dfrac{126y}{(18x)^2}$                                           $z_y = \dfrac{-7}{18x}$

(b)  $z_x = \dfrac{-324y}{(12x + 5y)^2}$                                    $z_y = \dfrac{324x}{(12x + 5y)^2}$

(c)  $z_x = \dfrac{40x^3 + 45x^2y + 36y^2}{(4x + 3y)^2}$                     $z_y = \dfrac{-15x^3 - 72xy - 27y^2}{(4x + 3y)^2}$

(d)  $z_w = \dfrac{14x + 84y}{(8w - 3x + 2y)^2}$                             $z_x = \dfrac{-14w - 35y}{(8w - 3x + 2y)^2}$

$$z_y = \dfrac{-84w + 35x}{(8w - 3x + 2y)^2}$$

**13.49.**    (a)  $z_x = 60(12x - 13y)^4$                                   $z_y = -65(12x - 13y)^4$

(b)  $z_x = 24x(3x^2 + 5y^3)^3$                                             $z_y = 60y^2(3x^2 + 5y^3)^3$

(c)  $z_x = 21x^2(14x^3 + 9y^2)^{-1/2}$                                     $z_y = 9y(14x^3 + 9y^2)^{-1/2}$

(d)  $z_x = (-60x^3 + 6y)(5x^4 - 2xy + 4y^2)^{-4}$                          $z_y = (6x - 24y)(5x^4 - 2xy + 4y^2)^{-4}$

(e)  $z_x = (-3x - 4y)(3x^2 + 8xy + 5y^2)^{-3/2}$                           $z_y = (-4x - 5y)(3x^2 + 8xy + 5y^2)^{-3/2}$

(f)  $z_w = 36w(6w^2 - 7x^4 - 9y^5)^2$                                      $z_x = -84x^3(6w^2 - 7x^4 - 9y^5)^2$

$$z_y = -135y^4(6w^2 - 7x^4 - 9y^5)^2$$

**13.50.**    (a)  $z_x = 18ye^{3xy}$                                       $z_y = 18xe^{3xy}$

(b)  $z_x = 90xy^2e^{-3x^2y^2}$                                             $z_y = 90x^2ye^{-3x^2y^2}$

(c)  $z_x = 14e^{(2x+5y)}$                                                  $z_y = 35e^{(2x+5y)}$

(d)  $z_w = 2wx^3y^2e^{w^2x^3y^2}$      $z_x = 3w^2x^2y^2e^{w^2x^3y^2}$      $z_y = 2w^2x^3ye^{w^2x^3y^2}$

**13.51.**    (a)  $z_x = \dfrac{8}{8x + 11y}$                              $z_y = \dfrac{11}{8x + 11y}$

(b)  $z_x = \dfrac{15x^2}{5x^3 + 4y^2}$                                     $z_y = \dfrac{8y}{5x^3 + 4y^2}$

(c)  $z_x = \dfrac{4}{x}$                                                   $z_y = \dfrac{5}{y}$

(d)  $z_w = \dfrac{21w^2}{7w^3 + 4x^2 + 5y^4}$                              $z_x = \dfrac{8x}{7w^3 + 4x^2 + 5y^4}$

$$z_y = \dfrac{20y^3}{7w^3 + 4x^2 + 5y^4}$$

**13.52.**    (a)   $z_x = 9(3x - 7y)^3 + (81x + 18y)(3x - 7y)^2$    $z_y = 2(3x - 7y)^3 - (189x + 42y)(3x - 7y)^2$

(b)   $z_x = \dfrac{-4(5x + 8y)^5 + (100x - 75y)(5x + 8y)^4}{(4x - 3y)^2}$    $z_y = \dfrac{3(5x + 8y)^5 + (160x - 120y)(5x + 8y)^4}{(4x - 3y)^2}$

(c)   $z_x = \dfrac{48x^2 + 353xy - 1037y^2}{(3x + 11y)^2}$    $z_y = \dfrac{-422x^2 + 270xy + 495y^2}{(3x + 11y)^2}$

(d)   $z_x = 30x^2y^4e^{5xy} + 12xy^3e^{5xy}$    $z_y = 30x^3y^3e^{5xy} + 18x^2y^2e^{5xy}$

**13.53.**    (a)   $z_{xx} = 42x$    $z_{yy} = 144y^2$    $z_{xy} = -4 = z_{yx}$

(b)   $z_{xx} = 180x^3 + 12y^4 - 18xy^5$    $z_{yy} = 72x^2y^2 - 60x^3y^3 - 78y$    $z_{xy} = 48xy^3 - 45x^2y^4 = z_{yx}$

(c)   $z_{xx} = 96xy^2$    $z_{yy} = 32x^3$    $z_{xy} = 96x^2y = z_{yx}$

(d)   $z_{xx} = 300(5x + 12y)^2$    $z_{yy} = 1728(5x + 12y)^2$    $z_{xy} = 720(5x + 12y)^2 = z_{yx}$

(e)   $z_{xx} = -0.24x^{-1.4}y^{0.4}$    $z_{yy} = -0.24x^{0.6}y^{-1.6}$    $z_{xy} = 0.24x^{-0.4}y^{-0.6} = z_{yx}$

(f)   $z_{xx} = -2x^{-1.5}y^{0.7}$    $z_{yy} = -1.68x^{0.5}y^{-1.3}$    $z_{xy} = 2.8x^{-0.5}y^{-0.3} = z_{yx}$

**13.54.**    (a)   $(7, 9)$, relative maximum    (b)   $(8, 5)$, saddle point

(c)   $(12, 15)$, relative minimum    (d)   $(13, 6)$, inflection point

**13.55.**    (a)   $(-4, 11)$, saddle point; $(4, 11)$, relative minimum

(b)   $(18, 3)$, relative maximum; $(18, -3)$, saddle point

(c)   $(0, 0)$, inflection point; $(648, -72)$, relative minimum

(d)   $(0, 0)$, inflection point; $(-3, -3)$, relative maximum

(e)   $(-7, -2)$, relative maximum; $(-7, 9)$, saddle point; $(4, -2)$, saddle point; $(4, 9)$, relative minimum

(f)   $(-8, -6)$, saddle point; $(-8, 5)$, relative maximum; $(3, -6)$, relative minimum; $(3, 5)$, saddle point

**13.56.**    (a) $(17, 13)$    (b) $(63, 39)$    (c) $(15, 20)$    (d) $(22, 18)$

**13.57.**    (a) $(30, 26)$    (b) $(19, 14)$    (c) $(27, 17)$    (d) $(15, 25)$

**13.58.**    (a) $(35, 38)$    (b) $(42, 33)$    (c) $(36, 44)$    (d) $(54, 31)$

**13.59.**    (a) $(50, 38)$    (b) $(27, 42)$    (c) $(50, 78)$    (d) $(42, 22)$

**13.60.**    (a) $(21, 16)$    (b) $(15, 34)$    (c) $(13, 10)$    (d) $(45, 12)$

# Chapter 14

## Sequences and Series

### 14.1  SEQUENCES

A sequence is a set of numbers that follow an order or rule. For example, the sequence

$$1, 4, 9, 16, \ldots, n^2, \ldots$$

represents the squared natural numbers. Each element within this sequence can be represented mathematically as $a_n = n^2$. Here, $a_n$ is an element, and $n$ denotes the index and order in the sequence.

A second example is the series of numbers known as the **Fibonacci sequence:**

$$0, 1, 1, 2, 3, 5, 8, 13$$

The element $a_n$ is the sum of the two previous elements, where $8 = 5 + 3$, $5 = 3 + 2$, $2 = 1 + 1$, except for the first and second terms, which are 0 and 1 ($= 0 + 1$), respectively. The rule of the Fibonacci sequence is summarized then as

$$\begin{cases} a_1 = 0 \\ a_2 = 1 \\ a_n = a_{n-1} + a_{n-2} \text{ if } n > 2 \end{cases}$$

The representation of the set of elements in the sequences is given by $\{a_n\}$, where $\{a_n\} = \{n^2\}_{n=1}^{j}$. The elements included in $\{a_n\}$ follow the rule of the sequence, which is $n^2$, given that the (integer) range of $n$ goes from 1 to $j$. Thus, there are $j$ elements.

**EXAMPLE 1.**  Different rules may be given to represent a sequence, as shown in the examples below for illustration, in which the sequences are expanded.

(a)  $\{a_n\} = \{n+1\}_{i=1}^{4}$

$$a_1 = 1 + 1 = 2 \qquad\qquad a_2 = 2 + 1 = 3$$

$$a_3 = 3 + 1 = 4 \qquad\qquad a_4 = 4 + 1 = 5$$

$$\{a_n\} = \{2, 3, 4, 5\}$$

(b)  $\{b_n\} = \{n^2 + 3n\}_{j=1}^{3}$

$$b_1 = 1^2 + 3(1) = 4 \qquad\qquad b_2 = 2^2 + 3(2) = 10$$

$$b_3 = 3^2 + 3(3) = 18$$

$$\{b_n\} = \{4, 10, 18\}$$

376

(c)   $\{b_n\} = \left\{\dfrac{(-1)^{n+1}}{n}\right\}_{j=1}^{5}$

$$b_1 = \frac{(-1)^2}{1} = 1 \qquad\qquad b_2 = \frac{(-1)^3}{2} = -\frac{1}{2}$$

$$b_3 = \frac{(-1)^4}{3} = \frac{1}{3} \qquad\qquad b_4 = \frac{(-1)^5}{4} = -\frac{1}{4}$$

$$b_5 = \frac{(-1)^6}{5} = \frac{1}{5}$$

$$\{b_n\} = \left\{1,\ -\frac{1}{2},\ \frac{1}{3},\ -\frac{1}{4},\ \frac{1}{5}\right\}$$

## 14.2   REPRESENTATION OF ELEMENTS

The notation of sequences is also useful to identify relations and apply operations on tables or matrices. For these cases, $\{a_n\}$ represents the elements following a specific rule. The subscript may include multiple indices (usually two to indicate row and column), as presented in Examples 2 and 3.

**EXAMPLE 2.**   Given the matrix.

$$A = \begin{pmatrix} 4 & 5 & 1 \\ 0 & -2 & 3 \\ -1 & 0 & 8 \end{pmatrix}$$

Identify the elements given the rules:

(a)   $\{a_n\} = \{b_{i,2}\}_{i=1}^{3}$

$$a_1 = b_{1,2} = 5 \qquad\qquad a_2 = b_{2,2} = -2$$

$$a_3 = b_{3,2} = 0$$

$$\{a_n\} = \{5, -2,\ 0\} \qquad\qquad \text{These are the elements of the second column.}$$

(b)   $\{a_n\} = \{b_{ij}\}_{i=2,j=1}^{3,2}$

$$a_1 = b_{2,1} = \ \ 0 \qquad\qquad a_2 = b_{2,2} = -2$$

$$a_3 = b_{3,1} = -1 \qquad\qquad a_4 = b_{3,2} = 0$$

$$\{a_n\} = \{0, -2, -1,\ \ 0\}$$

(c)   $\{a_n\} = \{c_{3j}\}_{j=2}^{3}$

$$a_1 = c_{3,2} = \ \ 0 \qquad\qquad a_2 = c_{3,3} = 8$$

$$\{a_n\} = \{0,\ \ 8\} \qquad\qquad \text{These are the last two elements of the third row.}$$

**EXAMPLE 3.**   Given the $5 \times 8$ table below,

| 5  | 7   | 1 | 12 | 9  | 2  | 4  | 6  |
|----|-----|---|----|----|----|----|----|
| 8  | 15  | 4 | 14 | 6  | 1  | 5  | 15 |
| 11 | 4   | 5 | 11 | 3  | 8  | 9  | 12 |
| 14 | −10 | 6 | 2  | 1  | 5  | −7 | 9  |
| 17 | 7   | 7 | 8  | −9 | 13 | 17 | 7  |

identify the elements and operations within elements given the rules:

(a)   $\{a_n\} = \{b_{i,j}\}_{i=3,\,j=4}^{4,7}$

$$a_1 = b_{3,4} = \ 11 \qquad\qquad a_2 = b_{3,5} = \ 3$$

$$a_3 = b_{3,6} = \ 8 \qquad\qquad a_4 = b_{3,7} = \ 9$$

$$a_5 = b_{4,4} = \ 2 \qquad\qquad a_6 = b_{4,5} = \ 1$$

$$a_7 = b_{4,6} = \ 5 \qquad\qquad a_8 = b_{4,7} = -7$$

$$\{a_n\} = \{11, \ 3, \ 8, \ 9, \ 2, \ 1, \ 5, \ -7\}$$

(b)   $\{a_n\} = \{b_{ii}\}_{i=2}^{4}$

$$a_1 = b_{2,2} = \ 15 \qquad\qquad a_2 = b_{3,3} = 5$$

$$a_3 = b_{4,4} = \ 2$$

$$\{a_n\} = \{15, \ 5, \ 2\} \qquad\qquad \text{These are diagonal elements.}$$

(c)   $\{a_n\} = \{b_{2j} + b_{2,j+1}\}_{j=1}^{2}$

$$a_1 = b_{2,1} + b_{2,2} = 8 + 15 = 22$$

$$a_2 = b_{2,2} + b_{2,3} = 15 + 4 = 19$$

$$\{a_n\} = \{22, \ 19\}$$

(d)   $\{a_n\} = \{b_{1k} - b_{k,1}\}_{j=2}^{3}$

$$a_1 = b_{1,2} - b_{2,1} = 7 - 8 = -1$$

$$a_2 = b_{1,3} - b_{3,1} = 1 - 11 = -10$$

$$\{a_n\} = \{-1, -10\}$$

## 14.3   SERIES AND SUMMATIONS

A series $s_n$ is the summation of the elements within a sequence. Here $n$ indicates the last element to be summed. For example, if the sequence $\{a_n\}$ has as elements 3, 4, 7, 11, 15, 21, 28, the series of the first three terms can be represented as $s_3$:

$$s_3 = a_1 + a_2 + a_3$$
$$s_3 = 3 + 4 + 7 = 14$$

In a similar manner, $s_5$ represents the addition of the first five elements:

$$s_5 = a_1 + a_2 + a_3 + a_4 + a_5$$

$$s_5 = 3 + 4 + 7 + 11 + 15 = 37$$

Because $s_n$ can add any number of elements within the sequence, $s_n$ is also referred as a **partial sum** operation.

The partial sum may involve a large number of elements; thus, the representation of the series requires the use of the **summation operator** ($\Sigma$, which is capital letter sigma). Consider the sequence $\{x_i : i = 1, 2, 3, \ldots, n\}$, in which index $i$ ranges from 1 to a natural integer $n$. The series or the sum of these numbers $s_n$ can be represented as

$$\sum_{i=1}^{n} x_i = x_1 + x_2 + x_3 + \ldots + x_n = s_n$$

Examples 4 and 5 provide an example of how to use the summation notation.

**EXAMPLE 4.**   Determine the partial sum $s_5$ given the following sequence $\{a_n\} = \{i\}_{i=1}^{6}$.
Expand the sequence: $\{a_n\} = 1, 2, 3, 4, 5, 6$

$$s_5 = \sum_{i=1}^{5} a_i = a_1 + a_2 + a_3 + a_4 + a_5 = 1 + 2 + 3 + 4 + 5$$

$$s_5 = \sum_{i=1}^{5} a_i = 15$$

**EXAMPLE 5.**   Determine the partial sum $s_4$ given the following sequence $\{a_n\} = \{j^2 - 2\}_{j=1}^{6}$.
Expand the sequence: $\{a_n\} = -1, 2, 7, 14, 23, 34$

$$s_4 = \sum_{i=1}^{4} a_i = a_1 + a_2 + a_3 + a_4 = -1 + 2 + 7 + 14$$

$$s_5 = \sum_{i=1}^{4} a_i = 22$$

In most cases, the series is directly defined, including any sequence operation in the summation notation. Examples 6, 7, and 8 illustrate this notion. Example 9 shows how it can be used when working with tables.

**EXAMPLE 6.**   Given the sequence $\{x_i = -2, 4, 1, \text{and } 8\}$, the partial sum of their squared values can be represented as follows:

$$\sum_{i=1}^{4} x_i^2 = x_1^2 + x_2^2 + x_3^2 + x_4^2 = (-2)^2 + (4)^2 + (1)^2 + (8)^2$$

$$\sum_{i=1}^{4} x_i^2 = 4 + 16 + 1 + 64 = 85$$

**EXAMPLE 7.** Compute the following summation:

$$\sum_{i=1}^{5}\frac{i}{2i-1}=\frac{1}{2-1}+\frac{2}{4-1}+\frac{3}{6-1}+\frac{4}{8-1}+\frac{5}{10-1}$$

$$\sum_{i=1}^{4}x_i^2=\frac{1}{2}+\frac{2}{3}+\frac{3}{5}+\frac{4}{7}+\frac{5}{9}\approx2.8937$$

**EXAMPLE 8.** Expand the partial sum:

$$\sum_{i=1}^{4}\frac{(-1)^i(x^i)}{3i+1}=\frac{-x}{3+1}+\frac{x^2}{6+1}+\frac{-x^3}{9+1}+\frac{x^4}{12+1}$$

$$\sum_{i=1}^{4}\frac{(-1)^i(x^i)}{3i+1}=-\frac{x}{4}+\frac{x^2}{7}-\frac{x^3}{10}+\frac{x^4}{13}$$

**EXAMPLE 9.** Given the table below,

| $i$ | $x_i$ |
|---|---|
| 1 | 3 |
| 2 | 6 |
| 3 | 9 |
| 4 | 7 |
| 5 | −1 |
| 6 | −3 |
| 7 | 5 |

The series

$$\sum_{i=1}^{5}(x_i-2)$$

can be computed by first expanding the elements:

$$\sum_{i=1}^{5}(x_i-2)=(3-2)+(6-2)+(9-2)+(7-2)+(-1-2)$$

$$\sum_{i=1}^{5}(x_i-2)=1+4+7+5-3=14$$

## 14.4   PROPERTY OF SUMMATIONS

Summations hold the properties listed below.

### 14.4.1   Sum of Monomials

Given $c$ is a constant real number and $\{x_i : 1, 2, 3, \ldots, n\}$, the constant coefficient of the operation can be moved out of the summation.

$$\sum_{i=1}^{n} cx_i = c\sum_{i=1}^{n} x_i$$

**EXAMPLE 10.**

$$\sum_{i=1}^{3} 5x_i = 5\sum_{i=1}^{3} x_i = 5(x_1 + x_2 + x_3)$$

$$\sum_{i=1}^{4} 2i^2 = 2\sum_{i=1}^{4} i^2 = 2(1^2 + 2^2 + 3^2 + 4^2) = 60$$

### 14.4.2   Sum of a Constant

Given $c$ is a constant real number and $\{x_i : 1, 2, 3, \ldots, n\}$, the sum of constant terms is the multiplication of the constant times the number of repetitions.

$$\sum_{i=1}^{n} c = cn$$

**EXAMPLE 11.**

$$\sum_{i=1}^{3} 7 = (7 + 7 + 7) = 7(3) = 21$$

### 14.4.3   Summation of Two Sequences

Considering $\{x_i\}$ and $\{y_i\}$ are defined sequences, and assuming $a$ and $b$ are constants, the summation of two sequences can be separated into monomials.

$$\sum_{i=1}^{n} (ax_i \pm by_i) = a\sum_{i=1}^{n} x_i + b\sum_{i=1}^{n} y_i$$

**EXAMPLE 12.**

$$\sum_{i=1}^{4} (2x_i + 3y_i) = 2\sum_{i=1}^{4} x_i + 3\sum_{i=1}^{4} y_i$$

$$\sum_{i=1}^{7} (9x_i - x_i^2) = 9\sum_{i=1}^{7} x_i - \sum_{i=1}^{7} x_i^2$$

### 14.4.4  Split of a Summation

Given $c$ is a constant real number $\{x_i : 1, 2, 3, \ldots, n\}$, and $m \leq n$, the summation can be split into two or more summation terms.

$$\sum_{i=1}^{n} cx_i = \sum_{j=1}^{m} cx_j + \sum_{k=m+1}^{n} cx_k$$

**EXAMPLE 13.**

$$\sum_{i=1}^{4} 7x_i = \sum_{j=1}^{2} 7x_j + \sum_{k=3}^{4} 7x_k$$

$$\sum_{i=1}^{5} 2i^2 = 2\sum_{j=1}^{3} j^2 + 2\sum_{i=3}^{4} k^2 = 2(1^2 + 2^2 + 3^2) + 2(4^2 + 5^2) = 28 + 82 = 110$$

## 14.5  SPECIAL FORMULAS OF SUMMATIONS

Some summations have well-established formulations, among them:

1.  Sum of the first $n$ consecutive natural numbers:

$$\sum_{i=1}^{n} i = \frac{n(n+1)}{2}$$

2.  Sum of the first $n$ squares natural numbers:

$$\sum_{i=1}^{n} i^2 = \frac{n(n+1)(2n+1)}{6}$$

3.  Sum of the first $n$ cubic numbers:

$$\sum_{i=1}^{n} i^3 = \left[\frac{n(n+1)}{2}\right]^2$$

**EXAMPLE 14.**   Determine the following summation using the formulas presented in section 14.5:

$$\sum_{i=1}^{4} (2i + i^2) = 2\sum_{i=1}^{4} i + \sum_{i=1}^{4} i^2 = 2\left[\frac{4(4+1)}{2}\right] + \left[\frac{4(4+1)(2 \cdot 4+1)}{6}\right]$$

$$= 4(5) + \frac{4(5)(9)}{6} = 20 + 30 = 50$$

**EXAMPLE 15.**   Compute the summation below:

$$\sum_{i=1}^{10} (5 - 3i)^2 = \sum_{i=1}^{10} (25 - 30i + 9i^2) = \sum_{i=1}^{10} 25 - 30\sum_{i=1}^{10} i + 9\sum_{i=1}^{10} i^2$$

$$= 25(10) - 30\left[\frac{10(11)}{2}\right] + 9\left[\frac{10(11)(21)}{6}\right]$$

$$= 250 - 1650 + 3465 = 2065$$

## 14.6   ECONOMICS APPLICATIONS: MEAN AND VARIANCE

In economics, sample data is collected for analysis and inference, usually from a whole population. Two statistics are often used to describe sample data: the mean and the variance.

### 14.6.1   Mean

The **mean** (denoted as $\overline{X}$) is the average of all $n$ elements or observations:

$$\overline{X} = \frac{\sum_{i=1}^{n} x_i}{n}$$

**EXAMPLE 16.**   Compute the mean for the data below:

| $i$ | $x_i$ |
|-----|-------|
| 1 | 3 |
| 2 | 6 |
| 3 | 9 |
| 4 | 7 |
| 5 | −1 |
| 6 | −3 |
| 7 | 5 |

$$\overline{X} = \frac{3+6+9+7-1-3-5}{7} = \frac{26}{7} = 3.7143$$

### 14.6.2   Standard Deviation

The **variance** (denoted as $S_x^2$) is the squared average distance of all observations from the mean:

$$S_x^2 = \frac{\sum_i \left(x_i - \overline{X}\right)^2}{n-1}$$

The **standard deviation** ($S_x$) is the squared root of the variance:

$$S_x = \sqrt{S_x^2} = \sqrt{\frac{\sum_i \left(x_i - \overline{X}\right)^2}{n-1}}$$

**EXAMPLE 17.**   Compute the variance and standard deviation of the data below:

| $i$ | $x_i$ |
|-----|-------|
| 1 | 8 |
| 2 | 9 |
| 3 | 7 |
| 4 | 16 |

First, compute the mean:

$$\bar{X} = \frac{8+9+7+16}{4} = \frac{40}{4} = 10$$

Thus, the variance is

$$S_x^2 = \frac{\sum_i (x_i - 10)^2}{4-1} = \frac{(8-10)^2 + (9-10)^2 + (7-10)^2 + (16-10)^2}{3} = \frac{4+1+9+36}{3}$$

$$S_x^2 = \frac{50}{3} = 16.66\overline{6}$$

The standard deviation is

$$S_x^2 = \sqrt{S_x^2} = \sqrt{16.66\overline{6}} = 4.0825$$

## 14.7   INFINITE SERIES

In cases where there is no upper limit, in which $n$ tends to infinity ($n \to \infty$), the sequence is said to be an **infinite sequence**. A sequence converges to a number when the sequence approaches a single value as $n$ increases. For example, the sequence

$$\left\{ 1, \frac{1}{2}, \frac{1}{3}, \frac{1}{4} \cdots \frac{1}{n}, \cdots \right\}$$

approaches to zero as $n$ becomes larger.

An infinite series is the summation of the elements in an infinite sequence. Thus,

$$\sum_{i=1}^{\infty} x_i = x_1 + x_2 + x_3 + \cdots + x_n + \cdots$$

If this partial sum approaches a single number $L$, then the infinite series is said to *converge* on $L$. The most prominent example is the **infinite geometric series**, defined as

$$\sum_{i=0}^{\infty} r^i = 1 + r + r^2 + r^3 + \cdots + r^n + \cdots$$

here $r$ is a real number and known as the **common ratio**. Note that the index starts at 0, not at 1. If $|r| < 1$, then the series converges to a specific ratio:

$$\sum_{i=0}^{\infty} r^i = \frac{1}{1-r}, \qquad |r| < 1$$

otherwise, if $|r| \geq 1$, then the series does not approach a specific number; in other words, the series **diverges**.

**EXAMPLE 18.**   Determine if the geometric series converges or diverges. If it converges, find the numerical value of the sum:

$$s_n = \sum_{i=0}^{\infty} \left( \frac{1}{2} \right)^i$$

Here, $r = \frac{1}{2} < 1$; therefore, the geometric series converges. The solution is

$$\sum_{i=0}^{\infty} \left(\frac{1}{2}\right)^i = 1 + \frac{1}{2} + \frac{1}{4} + \frac{1}{8} + \cdots = \frac{1}{1 - 1/2} = \frac{1}{1/2}$$

$$s_n = \sum_{i=0}^{\infty} \left(\frac{1}{2}\right)^i = 2$$

## 14.8   FINANCE APPLICATIONS: NET PRESENT VALUE

In section 11.8, interest compounding was established, in which the value of $A$ at the end of $t$ years was equal to

$$A_t = P(1+r)^t \tag{11.7}$$

In finance, $A$ is known as the future value of the cash flow. Solve for $P$:

$$P = \frac{A_t}{(1+r)^t}$$

$P$ is defined as the present value of the cash flow, which is the value of money at the present time.

In a project, it is common to have cash flows in multiple years or periods. In finance, the **net present value** (**NPV**) is used to evaluate the combination of cash flows within a project. Thus, $NPV$ is the sum of all the present value of the cash flows during the life span $n$ of the evaluated project, including any initial investments.

$$NPV = \sum_{t=0}^{n} P_t = P_0 + P_1 + P_2 + \cdots P_n$$

When you replace $P$ in terms of $A$, $NPV$ can also be seen as the discounted sum of all future cash flows:

$$NPV = \sum_{t=0}^{n} \frac{A_t}{(1+r)^t} = A_0 + \frac{A_1}{1+r} + \frac{A_2}{(1+r)^2} + \cdots + \frac{A_n}{(1+r)^n}$$

In finance, if $NPV > 0$ at the interest rate $r$, then the project results in a positive net cash flow for the firm, and therefore it is convenient to accept the project. Conversely, if $NPV \leq 0$, then the project should be rejected.

**EXAMPLE 19.**   Calculate the net present value of the project given the cash flows below and determine if the project must be approved or rejected. The discount rate is 10%.

|                  | Year |      |      |      |      |
|------------------|------|------|------|------|------|
|                  | 0    | 1    | 2    | 3    | 4    |
| Future value $A$ | $-1000$ | 2000 | 800 | 400 | 300 |

$$NPV = -1000 + \frac{2000}{1+0.1} + \frac{800}{(1+0.1)^2} + \frac{400}{(1+0.1)^3} + \frac{300}{(1+0.1)^4}$$

$$NPV = -1000 + 1818.18 + 661.16 + 300.53 + 204.90$$

$$NPV = \ 1984.77 > 0 \qquad \text{Accept the project as } NPV > 0.$$

# Solved Problems

## SEQUENCES

**14.1.** Describe the elements of

(a)   $\{a_n\} = \left\{n - \dfrac{1}{n}\right\}_{i=1}^{3}$

$\qquad\qquad a_1 = 1 - 1/1 = 0$ $\qquad\qquad\qquad\qquad a_2 = 2 - 1/2 = 3/2$

$\qquad\qquad a_3 = 3 - 1/3 = 8/3$

$\qquad\{a_n\} = \left\{0,\ \dfrac{3}{2},\ \dfrac{8}{3}\right\}$

(b)   $\{b_n\} = \{n^2 + n^{1/2}\}_{i=1}^{4}$

$\qquad\qquad b_1 = 1^2 + (1)^{1/2} = 2$ $\qquad\qquad\qquad b_2 = 2^2 + (2)^{1/2} = 5.41$

$\qquad\qquad b_3 = 3^2 + (3)^{1/2} = 10.73$ $\qquad\qquad b_4 = 4^2 + (4)^{1/2} = 18$

$\qquad\{b_n\} = \{2,\ 5.41,\ 10.73,\ 18\}$

**14.2.** Given the formula below for the $n$th term in the sequence $a_n$, find the values for $a_1$, $a_3$, and $a_5$.

(a)   $a_n = n - \dfrac{1}{2n-1}$

$\qquad\qquad a_1 = 1 - \dfrac{1}{2(1)-1} = 1 - \dfrac{1}{1} = \dfrac{1}{2}$ $\qquad\qquad a_3 = 3 - \dfrac{1}{2(3)-1} = 3 - \dfrac{1}{5} = \dfrac{14}{15}$

$\qquad\qquad a_5 = 5 - \dfrac{1}{2(5)-1} = 5 - \dfrac{1}{9} = \dfrac{44}{9}$

(b)   $a_n = 4 - (-2)^n$

$\qquad\qquad a_1 = 4 - (-2)^1 = 4 + 2 = 6$ $\qquad\qquad a_3 = 4 - (-2)^3 = 4 + 8 = 12$

$\qquad\qquad a_5 = 4 - (-2)^5 = 4 + 32 = 36$

(c)   $a_n = \dfrac{9 - n^2}{2^n}$

$\qquad\qquad a_1 = \dfrac{9 - 1^2}{2^1} = \dfrac{8}{2} = 4$ $\qquad\qquad\qquad a_3 = \dfrac{9 - 3^2}{2^3} = \dfrac{0}{8} = 0$

$\qquad\qquad a_5 = \dfrac{9 - 5^2}{2^5} = \dfrac{-16}{32} = -\dfrac{1}{2}$

**14.3.** Find the first four terms of the sequence $\{a_n\}$ given the rule:

(a) $\begin{cases} a_1 = 8 \\ a_n = 5a_{n-1} \qquad \text{if } n \geq 2 \end{cases}$

$\qquad a_1 = 8 \qquad\qquad\qquad\qquad a_2 = 5(a_1) = 5(8) = 40$

$\qquad\qquad a_3 = 5(a_2) = 5(40) = 200 \qquad\qquad a_4 = 5(a_3) = 5(200) = 1000$

(b) $\begin{cases} a_1 = 3 \\ a_2 = 5 \\ a_n = 2a_{n-1} - a_{n-2} \qquad \text{if } n \geq 2 \end{cases}$

$\qquad a_1 = 3 \qquad\qquad\qquad\qquad a_2 = 5$

$\qquad\qquad a_3 = 2(a_2) - a_1 = 2(5) - 3 = 12 \qquad\qquad a_4 = 2(a_3) - a_2 = 2(12) - 5 = 19$

## SEQUENCES AND SUMMATION

**14.4.** Given the sequence $\{x_n\} = \{8, 64, 27, 1, 125\}$, compute:

$$\sum_{i=1}^{5} (1 + \sqrt[3]{x_i})$$

The summation has five elements ($i = 1$ to $5$). Expand the operation.

$$\sum_{i=1}^{5} (1 + \sqrt[3]{x_i}) = (4 + \sqrt[3]{8}) + (4 + \sqrt[3]{64}) + (4 + \sqrt[3]{27}) + (4 + \sqrt[3]{1}) + (4 + \sqrt[3]{125})$$

$$\sum_{i=1}^{5} (1 + \sqrt[3]{x_i}) = 6 + 8 + 7 + 5 + 9 = 35$$

**14.5.** Find the value of the sum:

$$\sum_{i=1}^{4} (8 + 2^i)$$

Here, the index is located in the exponent (in $2^i$), which will be the term that will vary in the summation.

$$\sum_{i=1}^{4} (8 + 2^i) = (8 + 2^1) + (8 + 2^2) + (8 + 2^3) + (8 + 2^4)$$

$$\sum_{i=1}^{4} (8 + 2^i) = 10 + 12 + 16 + 24 = 62$$

**14.6.** Calculate the partial sum:

$$\sum_{k=1}^{5} \frac{3^k}{k^2+1}$$

The index is located in two components, $3^k$ and $k^2$.

$$\sum_{k=1}^{5} \frac{3^k}{k^2+1} = \frac{3^1}{1^2+1} + \frac{3^2}{2^2+1} + \frac{3^3}{3^2+1} + \frac{3^4}{4^2+1} + \frac{3^5}{5^2+1}$$

$$\sum_{k=1}^{5} \frac{3^k}{k^2+1} = \frac{3}{2} + \frac{9}{5} + \frac{27}{10} + \frac{81}{17} + \frac{243}{26} = 20.11$$

**14.7.** Calculate the partial sum:

$$\sum_{i=1}^{3} \frac{1}{i^{i+1}-2i}$$

The index is located in the denominator,

$$\sum_{i=1}^{3} \frac{1}{i^{i+1}-2i} = \frac{1}{1^2-2(1)} + \frac{1}{2^3-2(2)} + \frac{1}{3^4-2(3)}$$

$$\sum_{i=1}^{3} \frac{1}{i^{i+1}-2i} = -1 + \frac{1}{4} + \frac{1}{75} = -0.7367$$

**14.8.** Given the table below,

| $k$ | $x_k$ |
|-----|-------|
| 1 | 2 |
| 2 | 1 |
| 3 | 3 |
| 4 | −1 |

Calculate:

$$\sum_{k=1}^{4} \frac{2x_k}{k+1}$$

The index is located in both the numerator and denominator,

$$\sum_{k=1}^{4} \frac{2x_k}{k+1} = \frac{2x_1}{1+1} + \frac{2x_2}{2+1} + \frac{2x_3}{3+1} + \frac{2x_4}{4+1} = \frac{2(1)}{1+1} + \frac{2(2)}{2+1} + \frac{2(3)}{3+1} + \frac{2(4)}{4+1}$$

$$\sum_{k=1}^{4} \frac{2x_k}{k+1} = \frac{2}{2} + \frac{4}{3} + \frac{6}{4} + \frac{8}{5} = 5.4\overline{3}$$

**14.9.** Given the table below,

| $k$ | $x_k$ |
|-----|-------|
| 1   | 1     |
| 2   | 9     |
| 5   | 16    |

Calculate:

$$\sum_{k=1}^{3} (-1)^k (\sqrt{k} - 1)$$

The index is located in two terms,

$$\sum_{k=1}^{5} (-1)^k (\sqrt{k} - 1) = (-1)^1 (\sqrt{1} - 1) + (-1)^2 (\sqrt{9} - 1) + (-1)^3 (\sqrt{16} - 1)$$

$$\sum_{k=1}^{5} (-1)^k (\sqrt{k} - 1) = (-1)(0) + (1)(2) + (-1)(3)$$

$$\sum_{k=1}^{5} (-1)^k (\sqrt{k} - 1) = 0 + 2 - 3 = -1$$

## APPLICATION OF PROPERTIES AND FORMULAS OF SUMMATIONS

**14.10.** Compute:

$$\sum_{k=1}^{5} (k^2 - 3)$$

Split the summation.

$$\sum_{i=1}^{5} (k^2 - 3) = \sum_{i=1}^{5} k^2 - \sum_{i=1}^{5} 3$$

Use the sum of squared numbers in the first operator and sum of constants in the second one.

$$\sum_{i=1}^{5} (k^2 - 3) = \frac{5(5+1)(2 \cdot 5 + 1)}{6} - 3(5) = \frac{5(6)(11)}{6} - 15$$

$$\sum_{i=1}^{5} (k^2 - 3) = 55 - 15 = 40$$

**14.11.** Compute:

$$\sum_{k=1}^{100} (2k^2 + 4k)$$

Split the summation and take constants out of the operators:

$$\sum_{k=1}^{100} (2k^2 + 4k) = \sum_{k=1}^{100} (2k^2) + \sum_{k=1}^{100} (4k) = 2\sum_{k=1}^{100} k^2 + 4\sum_{k=1}^{100} k$$

Use the formulas of summations:

$$\sum_{k=1}^{100} (2k^2 + 4k) = 2\left[\frac{100(100+1)(2\cdot 100 + 1)}{6}\right] + 4\left[\frac{100(100+1)}{2}\right]$$

$$\sum_{k=1}^{100} (2k^2 + 4k) = 2\left[\frac{100(101)(201)}{6}\right] + 4\left[\frac{100(101)}{2}\right] = 2[338350] + 4[5050]$$

$$\sum_{k=1}^{100} (2k^2 + 4k) = 676700 + 20200 = 696900$$

**14.12.** Given the table below,

| $k$ | $x_k$ |
|-----|-------|
| 1 | 2 |
| 2 | 1 |
| 3 | 3 |
| 4 | 5 |

Calculate:

$$\sum_{k=1}^{4} (2x_k + 3)^2$$

Expand the polynomial in the summation:

$$\sum_{k=1}^{4} (2x_k + 3)^2 = \sum_{k=1}^{4} (4x_k^2 + 12x_k + 9)$$

Split the summation:

$$= 4\sum_{k=1}^{4} x_k^2 + 12\sum_{k=1}^{4} x_k + \sum_{k=1}^{4} 9$$

Expand the terms:

$$= 4(2^2 + 1^2 + 3^2 + 5^2) + 12(2+1+3+5) + 9(4)$$

$$\sum_{k=1}^{4} (2x_k + 3)^2 = 4(39) + 12(11) + 36 = 324$$

**14.13.** Given the table below,

| $k$ | $x_k$ |
|-----|-------|
| 1 | 27 |
| 2 | 729 |
| 3 | −1 |

Calculate:

$$\sum_{k=1}^{3} (\sqrt[3]{x_k} + k)$$

Split the summation:

$$\sum_{k=1}^{3} (\sqrt[3]{x_k} + k) = \sum_{k=1}^{3} \sqrt[3]{x_k} + \sum_{k=1}^{3} k$$

Expand the first term and use the summation formula for the second term:

$$\sum_{k=1}^{3} \left(\sqrt[3]{x_k} + k\right) = (\sqrt[3]{27} + \sqrt[3]{729} + \sqrt[3]{-1}) + (1+2+3)$$

$$\sum_{k=1}^{3} \left(\sqrt[3]{x_k} + k\right) = (\sqrt[3]{27} + \sqrt[3]{729} + \sqrt[3]{-1}) + \frac{3(4)}{2}$$

$$\sum_{k=1}^{3} \left(\sqrt[3]{x_k} + k\right) = (3 + 9 - 1) + \frac{3(4)}{2}$$

$$\sum_{k=1}^{3} \left(\sqrt[3]{x_k} + k\right) = 11 + 6 = 17$$

**14.14.** Given the table below,

| $k$ | $x_k$ |
|-----|-------|
| 1 | 0.6 |
| 2 | −0.4 |
| 3 | −0.1 |
| 4 | 0.8 |

Calculate:

$$\sum_{k=1}^{4}\left(10x_k - \frac{1}{2}k^3\right)$$

Split the summation:

$$\sum_{k=1}^{4}\left(10x_k - \frac{1}{2}k^3\right) = 10\sum_{k=1}^{4}x_k - \frac{1}{2}\sum_{k=1}^{4}k^3$$

Expand the first term and use the summation formula for the second term:

$$\sum_{k=1}^{4}\left(10x_k - \frac{1}{2}k^3\right) = 10(0.6 - 0.4 - 0.1 + 0.8) - \frac{1}{2}(1^3 + 2^3 + 3^3 + 4^3)$$

$$= 10(0.6 - 0.4 - 0.1 + 0.8) - \frac{1}{2}\left[\frac{4(4+1)}{2}\right]^2$$

$$= 10(0.9) - \frac{1}{2}\left[\frac{4(5)}{2}\right]^2 = 9 - \frac{1}{2}(100)$$

$$\sum_{k=1}^{4}\left(10x_k - \frac{1}{2}k^3\right) = 9 - 50 = -41$$

**14.15.** Given the table below,

| $k$ | $x_k$ |
|-----|-------|
| 1   | 8     |
| 2   | 9     |
| 3   | 17    |

Calculate:

$$\sum_{k=1}^{2}(x_k + 3x_{i+1})$$

Split the summation:

$$\sum_{k=1}^{2}(x_k + 3x_{i+1}) = \sum_{k=1}^{2}x_k + 3\sum_{k=1}^{2}x_{i+1}$$

Expand carefully both terms:

$$\sum_{k=1}^{2}(x_k + 3x_{i+1}) = (x_1 + x_2) + 3(x_2 + x_3)$$

$$= (8 + 9) + 3(9 + 17)$$

$$\sum_{k=1}^{2}(x_k + 3x_{i+1}) = (17) + 3(26) = 95$$

**14.16.** Given the table below,

| $k$ | $x_k$ |
|-----|-------|
| 1 | 6 |
| 2 | 8 |
| 3 | 10 |
| 4 | 13 |

Show that

$$\sum_{k=1}^{4} x_k = \sum_{j=0}^{3} x_{j+1}$$

Find the sum of the operator in the left-hand side of the equation.

$$\sum_{k=1}^{3} x_k = x_1 + x_2 + x_3 + x_4 = 6 + 8 + 10 + 13 = 37$$

Find the sum of the right operator.

$$\sum_{j=0}^{3} x_{j+1} = x_{0+1} + x_{1+1} + x_{2+1} + x_{3+1}$$

$$\sum_{j=0}^{3} x_{j+1} = x_1 + x_2 + x_3 + x_4$$

$$\sum_{j=0}^{3} x_{j+1} = 6 + 8 + 10 + 13 = 37$$

## MEAN AND STANDARD DEVIATION

**14.17.** Compute the mean and standard deviation of the data below:

| $i$ | $x_i$ |
|-----|-------|
| 1 | 30 |
| 2 | 14 |
| 3 | 17 |
| 4 | 35 |
| 5 | 24 |

Compute the mean:

$$\bar{X} = \frac{1}{5} \sum_{i=1}^{5} x_i = \frac{30 + 14 + 17 + 35 + 24}{5} = \frac{120}{7} = 24$$

To determine the standard deviation, first compute the variance:

$$S_x^2 = \frac{1}{5-1} \sum_{i=1}^{5} (x_i - \bar{X})^2$$

$$S_x^2 = \frac{(30-24)^2 + (14-24)^2 + (17-24)^2 + (35-24)^2 + (24-24)^2}{5-1}$$

$$= \frac{(6)^2 + (-10)^2 + (-7)^2 + (11)^2 + (0)^2}{5-1} = \frac{36+100+49+121+0}{4}$$

$$S_x^2 = \frac{306}{4} = 76.5$$

The standard deviation is the square root of the variance,

$$S_x = \sqrt{S_x^2} = \sqrt{76.5} = 8.7464$$

**14.18.** Compute the mean and standard deviation of the data below:

| $i$ | $x_i$ |
|-----|-------|
| 1   | 11    |
| 2   | −21   |
| 3   | 34    |
| 4   | −12   |

Compute the mean:

$$\bar{X} = \frac{1}{4} \sum_{i=1}^{4} x_i = \frac{11-21+34-14}{4} = \frac{12}{4} = 3$$

To determine the standard deviation, first compute the variance:

$$S_x^2 = \frac{1}{4-1} \sum_{i=1}^{4} (x_i - \bar{X})^2$$

$$S_x^2 = \frac{(11-3)^2 + (-21-3)^2 + (34-3)^2 + (-12-3)^2}{4-1}$$

$$= \frac{(8)^2 + (-24)^2 + (31)^2 + (-15)^2}{4-1} = \frac{64+576+961+225}{3}$$

$$S_x^2 = \frac{1826}{3} = 608.67$$

The standard deviation is the square root of the variance,

$$S_x = \sqrt{S_x^2} = \sqrt{608.67} = 24.6711$$

## INFINITE GEOMETRIC SERIES

**14.19.** Determine if the geometric series converges or diverges. If it converges, find the numerical value of the sum:

$$s_n = \sum_{i=0}^{\infty} 12\left(\frac{3}{4}\right)^i$$

Here, $r = \frac{3}{4} < 1$; therefore, the geometric series converges. The solution is

$$\sum_{i=0}^{\infty} 12\left(\frac{3}{4}\right)^i = 12\sum_{i=0}^{\infty} \left(\frac{3}{4}\right)^i = 12\left[1 + \frac{3}{4} + \left(\frac{3}{4}\right)^2 + \left(\frac{3}{4}\right)^3 + \cdots\right]$$

$$= 12\left[1 + \frac{3}{4} + \frac{9}{16} + \frac{27}{64} + \cdots\right]$$

$$= 12\left[\frac{1}{1 - \frac{3}{4}}\right] = 12\left[\frac{1}{\frac{1}{4}}\right]$$

$$\sum_{i=0}^{\infty} 12\left(\frac{3}{4}\right)^i = 12(4) = 48$$

**14.20.** Determine if the geometric series converges or diverges. If it converges, find the numerical value of the sum:

$$s_n = \sum_{i=2}^{\infty} \left(\frac{1}{3}\right)^i$$

Here, $r = 1/3 < 1$; therefore, the geometric series converges. However, note that the index starts at $i = 2$.

$$\sum_{i=2}^{\infty} \left(\frac{1}{3}\right)^i = \left(\frac{1}{3}\right)^2 + \left(\frac{1}{3}\right)^3 + \left(\frac{1}{3}\right)^4 + \cdots$$

In order to use the geometric formula, complete the series by adding and subtracting the first two terms needed to have the index start at $i = 0$, which are 1 and 1/3:

$$= \left[\left(\frac{1}{3}\right)^2 + \left(\frac{1}{3}\right)^3 + \left(\frac{1}{3}\right)^4 + \cdots\right] + \left(1 + \frac{1}{3}\right) - \left(1 + \frac{1}{3}\right)$$

Reorganize terms:

$$= \left[1 + \frac{1}{3} + \left(\frac{1}{3}\right)^2 + \left(\frac{1}{3}\right)^3 + \left(\frac{1}{3}\right)^4 + \cdots\right] - \left(1 + \frac{1}{3}\right)$$

Use the formula:

$$\sum_{i=2}^{\infty} \left(\frac{1}{3}\right)^i = \left[\frac{1}{1 - 1/3}\right] - \frac{4}{3}$$

$$\sum_{i=2}^{\infty}\left(\frac{1}{3}\right)^{i} = \left[\frac{1}{2/3}\right] - \frac{4}{3} = \frac{3}{2} - \frac{4}{3}$$

$$\sum_{i=2}^{\infty}\left(\frac{1}{3}\right)^{i} = \frac{1}{6}$$

**14.21.** Determine if the geometric series converges or diverges. If it converges, find the numerical value of the sum:

$$s_n = \sum_{k=0}^{\infty}\left(\frac{1}{2^k} + \frac{2}{3^k}\right)$$

First split the series and take constant coefficients out.

$$s_n = \sum_{k=0}^{\infty}\left(\frac{1}{2^k} + \frac{2}{3^k}\right) = \sum_{k=0}^{\infty}\left(\frac{1}{2}\right)^{k} + 2\sum_{k=0}^{\infty}\left(\frac{1}{3}\right)^{k}$$

The ratios 1/2 and 1/3 are both less than 1, and each component converges, so the geometric series converges.

$$s_n = \sum_{k=0}^{\infty}\left(\frac{1}{2^k} + \frac{2}{3^k}\right) = \left(1 + \frac{1}{2} + \frac{1}{4} + \frac{1}{8} + \cdots\right) + 2\left(1 + \frac{1}{3} + \frac{1}{9} + \cdots\right)$$

Use the formula and note that for the first term, $r_1 = 1/2$ and for the second term, $r_2 = 1/3$.

$$s_n = \sum_{k=0}^{\infty}\left(\frac{1}{2^k} + \frac{2}{3^k}\right) = \frac{1}{1-1/2} + 2\left(\frac{1}{1-1/3}\right)$$

$$= \frac{1}{1/2} + 2\left(\frac{1}{2/3}\right)$$

$$= 2 + 2\left(\frac{3}{2}\right) = 2 + 3 = 5$$

$$s_n = \sum_{k=0}^{\infty}\left(\frac{1}{2^k} + \frac{2}{3^k}\right) = 2 + 3 = 5$$

## NET PRESENT VALUE

**14.22.** Compute the net present value of the project given the cash flows below and determine if the project must be approved or rejected. The discount rate is 20%.

| | Year | | | | | | |
|---|---|---|---|---|---|---|---|
| | 0 | 1 | 2 | 3 | 4 | 5 | 6 |
| Future value $A$ | −3000 | −1500 | 500 | 500 | 500 | 2000 | 3000 |

Use the NPV formula:

$$NPV = -3000 + \frac{-1500}{1+0.2} + \frac{500}{(1+0.2)^2} + \frac{500}{(1+0.2)^3} + \frac{500}{(1+0.2)^4} + \frac{2000}{(1+0.2)^5} + \frac{3000}{(1+0.2)^6}$$

Compute and round up to two decimals:

$$NPV = -3000 - 1250 + 347.22 + 289.35 + 241.13 + 803.75 + 1004.69$$
$$NPV = -1563.85 < 0 \qquad \text{Reject the project as } NPV < 0.$$

**14.23.** Compute the net present value of the project given the cash flows below and determine if the project must be approved or rejected. The discount rate is 12%.

| | | | | Year | | |
|---|---|---|---|---|---|---|
| | 0 | 1 | 2 | 3 | 4 | 5 |
| Future value $A$ | −2000 | 0 | 0 | 0 | 7000 | −2000 |

Note that in years 1 to 3 there is no cash flow. Use the NPV formula:

$$NPV = -2000 + \frac{7000}{(1+0.12)^4} + \frac{-2000}{(1+0.12)^5}$$

Compute and round up to two decimals:

$$NPV = -2000 + 4448.63 - 1134.85$$
$$NPV = 1313.77 > 0 \qquad \text{Accept the project as } NPV > 0.$$

# Supplementary Problems

## SEQUENCES

**14.24.** Describe the elements of

(a)  $\{a_n\} = \{n^2 - n\}_{i=1}^5$

(b)  $\{b_n\} = \{4 + \sqrt[3]{n}\}_{i=1}^3$

**14.25.** Write a formula for the $n$th term of the sequence.

(a)  $-1, 4, -9, 16, \ldots$

(b)  $\dfrac{1}{2}, \dfrac{2}{3}, \dfrac{3}{4}, \dfrac{4}{5}, \ldots$

(c)  $\sqrt[3]{4 + \sqrt{1}}, \sqrt[3]{9 + \sqrt{2}}, \sqrt[3]{16 + \sqrt{3}}, \sqrt[3]{25 + \sqrt{4}}, \ldots$

**14.26.** Find the fifth term of the sequence given the rules below:

(a)  $\begin{cases} a_1 = 2 \\ a_n = a_{n-1} + 1 & \text{if } n \geq 2 \end{cases}$

(b)  $\begin{cases} a_1 = 2 \\ a_2 = 5 \\ a_3 = a_{n-1} + a_{n-2} & \text{if } n \geq 3 \end{cases}$

## SEQUENCES AND SUMMATION

**14.27.** Given the sequence $\{x_n\} = \{-1, 3, -2, 0, 21, 7\}$, compute:

$$\sum_{i=4}^{5} (1 + x_{i-1})$$

**14.28.** Given the sequence $\{x_n\} = \{4, 1, -2, 10, 8, 13, 1, 9\}$, compute:

$$\sum_{j=0}^{3} (x_{j+2} + x_{j+1})$$

**14.29.** Calculate the partial sum:

$$\sum_{k=0}^{3} \frac{(-1)^k}{k+1}$$

Use the table below for Problems 14.30 and 14.31.

| $k$ | $x_k$ |
|-----|-------|
| 1   | 7     |
| 2   | 12    |
| 3   | -3    |
| 4   | 2     |

**14.30.** Calculate:

$$\sum_{k=0}^{3} \frac{x_{k+1} - k}{4}$$

**14.31.** Calculate:

$$\sum_{j=2}^{4} x_j + \sum_{i=0}^{2} x_{i+1}$$

## APPLICATION OF PROPERTIES AND FORMULAS OF SUMMATIONS

**14.32.** Compute:

$$\sum_{k=1}^{30} 3k - k^2$$

**14.33.** Compute:

$$\sum_{k=1}^{40} (4k^3 - 2k^2)$$

Use the table below for Problems 14.34 and 14.35.

| $k$ | $x_k$ |
|---|---|
| 1 | −1 |
| 2 | 9 |
| 3 | 7 |
| 4 | −6 |
| 5 | 4 |
| 6 | −1 |

**14.34.** Calculate:

$$\sum_{k=0}^{3} (5x_{k+1} - (k+1)^2)$$

**14.35.** Calculate:

$$\sum_{k=4}^{6} (\sqrt{|x_k|} - k)$$

## MEAN AND STANDARD DEVIATION

**14.36.** Compute the mean and standard deviation of the data table used in Problem 14.30.

**14.37.** Find the mean and variance of the data table used in Problem 14.33.

## INFINITE GEOMETRIC SERIES

**14.38.** Determine if the geometric series converges or diverges. If it converges, find the numerical value of the sum:

$$s_n = \sum_{i=0}^{\infty} \left(\frac{4}{3}\right)^i$$

**14.39.** Determine if the geometric series converges or diverges. If it converges, find the numerical value of the sum:

$$s_n = \sum_{i=1}^{\infty} \left(-\frac{2}{5}\right)^i$$

## NET PRESENT VALUE

**14.40.** Compute the net present value of the project given the cash flows below and determine if the project must be approved or rejected. The discount rate is 14%.

| | Year | | | | | | |
|---|---|---|---|---|---|---|---|
| | 0 | 1 | 2 | 3 | 4 | 5 | 6 |
| Future value $A$ | −8000 | −2500 | 5000 | 5000 | 5000 | 5000 | 8000 |

**14.41.** Compute the net present value of the project given the cash flows below and determine if the project must be approved or rejected. The discount rate is 12%.

|               | Year     |      |      |      |       |
| ------------- | -------- | ---- | ---- | ---- | ----- |
|               | 0        | 1    | 2    | 3    | 4     |
| Future value $A$ | $-10000$ | 900  | 200  | 9000 | 32000 |

# Answers to Supplementary Problems

**14.24.** (*a*)   $\{a_n\} = \{0, 2, 6, 12, 20\}$

       (*b*)   $\{b_n\} = \{5,\ 4+\sqrt[3]{2},\ 4+\sqrt[3]{3}\}$

**14.25.** (*a*)   $a_n = (-1)^n (n)^2$

       (*b*)   $a_n = \dfrac{n}{n+1}$

       (*c*)   $a_n = \sqrt[3]{(n+1)^2 + \sqrt{n}}$

**14.26.** (*a*)   $a_5 = 6$

       (*b*)   $a_5 = 19$

**14.27.** 0

**14.28.** 30

**14.29.** 0.583

**14.30.** 3

**14.31.** 27

**14.32.** $-8060$

**14.33.** $-41000$

**14.34.** 31

**14.35.** $-9.55$

**14.36.** Mean = 4.5, Standard deviation = 6.45497

**14.37.** Mean = 2, Variance = 32

**14.38.** Geometric series diverges as $r = 4/3 > 1$

**14.39.** Series converges to $-0.28571$

**14.40.** 6231.15

**14.41.** 17705.61

# Excel Practice

## Microsoft Excel™ Practice A: Linear Programming

Linear programming is a useful tool in economics for optimization problems such as maximizing profits or minimizing costs. Chapters 7 and 8 provide the foundation on how to solve these models when there are two decision variables. In particular, the graphical approach and the simplex algorithm were introduced. However, the computation becomes much more complex if more than two decision variables are added and there are more than three active constraints. For these cases, Microsoft Excel™ provides a useful package tool called *Solver*™, which is commonly used in what-if analysis and simulations that involve determining conditions to maximize or minimize a formula or function.

It is important to note that *Solver*™ must first be installed, but this add-in can be downloaded and installed from the Excel™ add-ins, depending on the version of the software and operating system that is used. Once installed, *Solver*™ will appear as a default application. For example, in Microsoft Office 2019™, the installed add-in can be found in the *Data* tab, with the following icon ?₊ Solver .

Examples A.1. to A.4. presented below show step-by-step how to use the *Solver*™ add-in to solve linear program models (specifically maximization and minimization problems, which were seen in Chapters 7 and 8, but with more than two decision variables and more than three constraints).

**MAXIMIZATION PROBLEMS**

**A.1.**  Maximize $\qquad\qquad\qquad Z = 15x + 10y + 20w$

subject to

$$5x + 2y + 3w \le 55 \qquad\qquad \text{(Constraint } A\text{)}$$
$$x + y + 2w \le 26 \qquad\qquad \text{(Constraint } B\text{)}$$
$$3x + y + w \le 30 \qquad\qquad \text{(Constraint } C\text{)}$$
$$x + y + 3w \le 25 \qquad\qquad \text{(Constraint } D\text{)}$$
$$4x + 2y + 5w \le 57 \qquad\qquad \text{(Constraint } E\text{)}$$
$$x, y, w \ge 0$$

*(1)*  Write the linear programming model in Microsoft Excel™, in the specific cell positions presented in Fig. A-1. Each variable and coefficient is separated in different cells.

*Objective function*: Rows 2 and 3 describe the objective variable.

- Cell B3: Identifies whether the objective is to maximize or minimize. In this case it's *to maximize*.

- Cell C3: Provides the variable name.

- Cell D2: *Objective* label.

- Cell D3: Objective formula of this maximization problem. This cell is highlighted in *orange* as *Solver*™ will optimize it.

*Declaration of decision variables*: Rows 4, 5, and 6 provide details on the decision variables and their respective objective coefficient.

- Cell B5: *Variables* title.

- Cells C5 to E5: Label each decision variable (in this case are $x$, $y$, and $w$, respectively).

- Cell B6: *Obj. Coefficients* title.
- Cells C6 to E6: Write the objective coefficients of each of the variables. For example, C6 is 15, which is the objective coefficient of *x*, which is found in the objective formula $Z = 15x + 10y + 20w$. In the same manner, cell D6 value is 10 (coefficient of *y*) and cell E6 is 20 (coefficient of *w*).
- Cell B7: *Optimal values* title.
- Cells C7 to E7: These cells are highlighted in *yellow*, but no information is written inside. Instead, *Solver*™ will provide their optimal values when running the simulation.

*Declaration of the constraint*: Rows 9 to 20 offer details on the structure to declare the constraints.

- Cell B9: *Subject to* label.
- Cell D10: *Coefficients of constraints* label.
- Cells B12 to B20: Write the labels to identify which row will represent each constraint (for example, constraint C will be located in row 16; therefore, cell B16 is labeled *Constraint C*).
- Column C: Declare the first decision variable (in this case *x*) and its coefficient for each structural constraint (A to E). In order, these are 5 (from constraint *A*) and written in cell C12, 1 (from constraint *B*), 3 (from constraint *C*), 1 (from constraint *D*), 4 (from constraint *E*).
- Column D: Declare the second decision variable (*y*) together with its coefficient for each structural constraint.
- Column E: Write the third decision variable (*w*) and each coefficient of *w* for each constraint.

|  | A | B | C | D | E | F | G | H |
|---|---|---|---|---|---|---|---|---|
| 1 | | | | | | | | |
| 2 | | | | Objective | | | | |
| 3 | | Maximize | Z | | | | | |
| 4 | | | | | | | | |
| 5 | | Variables | x | y | w | | | |
| 6 | | Obj. Coefficients | 15 | 10 | 20 | | | |
| 7 | | Optimal values | | | | | | |
| 8 | | | | | | | | |
| 9 | | Subject to: | | | | | | |
| 10 | | | Coefficients of constraints | | | | | |
| 11 | | | x | y | w | Use | Sign | Limit |
| 12 | | Constraint A: | 5 | 2 | 3 | | <= | 55 |
| 13 | | | | | | | | |
| 14 | | Constraint B: | 1 | 1 | 2 | | <= | 26 |
| 15 | | | | | | | | |
| 16 | | Constraint C: | 3 | 1 | 1 | | <= | 30 |
| 17 | | | | | | | | |
| 18 | | Constraint D: | 1 | 1 | 3 | | <= | 25 |
| 19 | | | | | | | | |
| 20 | | Constraint E: | 4 | 2 | 5 | | <= | 57 |
| 21 | | | | | | | | |

**Fig. A-1**

- Column F: Declare the formula of the left-hand side of each constraint. *Use* (label of cell F11) refers to how much of the endowment (right-hand side constant) is used by the model. Thus, cells F12 to F20 will be highlighted (in *light blue*) as equations are later inserted.
- Column G: The signs of each structural constraint will be written here (in strict order).

- Column F: The *limit* label (written as title in cell F11) refers to the maximum value that each constraint can use, which is the right-hand side of each equation. For example, for constraint C: $3x + y + w \leq 30$, the limit is 30, and this is inserted into cell F16. This procedure is the same for all other structural constraints.

*(2)* Write the objective function in cell D3. For this maximization problem, this is:

$$Z = 15x + 10y + 20w$$

Note that the formulation:

First multiplies each objective coefficient with its respective optimal value (for example 15, which is the objective coefficient of $x$ and located in cell C6, with the currently empty cell D6 that represents the optimal value of $x$);

Second adds all partial products.

Microsoft Excel™ provides a function to do this process: *sumproduct*, which as its name suggests, first takes the product of each pair of cells and then adds all these partial products. Thus, insert in the objective cell D3 the formula:

= SUMPRODUCT(C6:E6, C7:E7)

This function selects first the objective coefficients and then the optimal coefficients. This is illustrated in Fig. A-2.

| SUM | | ⟩ | × ✓ $f_x$ | =SUMPRODUCT(C6:E6,C7:E7) | |
|---|---|---|---|---|---|
| | A | B | C | D | E |
| 1 | | | | | |
| 2 | | | | **Objective** | |
| 3 | | **Maximize** | =SUMPRODUCT(C6:E6,C7:E7) | | |
| 4 | | | | | |
| 5 | | **Variables** | x | y | w |
| 6 | | Obj. Coefficients | 15 | 10 | 20 |
| 7 | | Optimal values | | | |

**Fig. A-2**

Once the formula is written, if the cell D3 is selected, notice that cells C6:E6 will be highlighted in light blue, whereas cells C7:D7 have a red border. This is to identify which cells are being used by the formula. For now, 0 appears in cell D3 as there are no current values in cells C7:E7.

*(3)* Write the left-hand side of the constraint *A* equation in cell F12. For this maximization problem, constraint A is:

$$5x + 2y + 3w \leq 55 \qquad \text{(Constraint } A\text{)}$$

The formulation is similar to the objective; thus, the Excel function *sumproduct* will be used. In cell F12, write:

= SUMPRODUCT(C7:E7, C12:E12)

This formula will multiply each decision variable (located in cells C7 to E7) with its respective coefficient in constraint A (in cells C12 to E12), and then it will add up these results. In other words, it is representing this operation: $5x + 2y + 3w$. Note that these groups of cells will be highlighted when writing the formula as it appears in Fig. A-3.

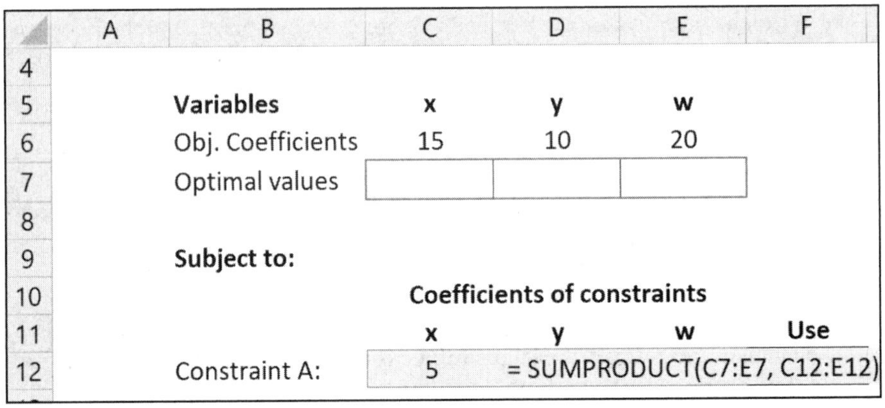

**Fig. A-3**

For now, cell F2 shows a zero.

(4) Repeat operation in (2) for each constraint. The following table describes each equation, Excel function, and cell location.

| Constraint | Left-Hand Side | Cell Location | Excel Function |
|---|---|---|---|
| B | $x + y + 2w$ | F14 | =SUMPRODUCT(C7:E7,C14:E14) |
| C | $3x + y + w$ | F16 | =SUMPRODUCT(C7:E7,C16:E16) |
| D | $x + y + 3w$ | F18 | =SUMPRODUCT(C7:E7,C18:E18) |
| E | $4x + 2y + 5w$ | F20 | =SUMPRODUCT(C7:E7,C20:E20) |

The template will look as shown in Fig. A-4.

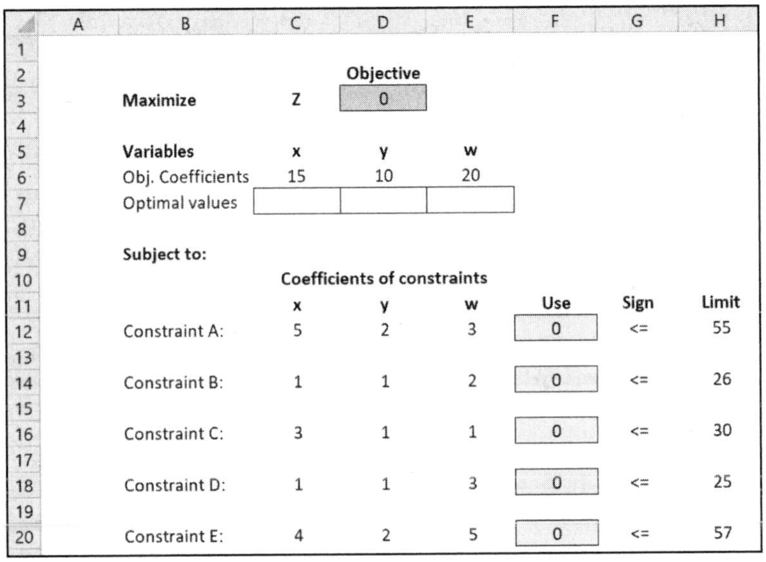

**Fig. A-4**

(5) For MS Office 2019™, go to the Data tab → Analyze section → select *Solver*™, as shown in Fig. A-5.

**Fig. A-5**

For other operating systems or versions of MS Excel™, please look at their official web page.

(6) Once *Solver*™ is selected, the menu presented in Fig. A-6 will show.

**Fig. A-6**

For other operating systems or versions of Microsoft Excel™, please look at their official web page.

(7) First, select the solving method. In this case, it is the Simplex algorithm (called here *Simplex LP*), which is shown in Fig. A-7.

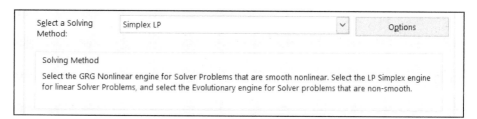

**Fig. A-7**

(8)   Second, establish the objective function and specify the following (illustrated in Fig. A-8):

   - *Set Objective*: Write the objective cell D3 (it usually appears as $D$3) or select it by using this
     button: ⬆. This is the objective value for Z that was highlighted in orange.

   - *To*: Because this is a maximization problem, select Max.

   - *By Changing Variables Cells*: Select or write the cells $C$7:$E$7, which are the cells highlighted
     in yellow and represent the optimal values for *x*, *y*, and *w*.

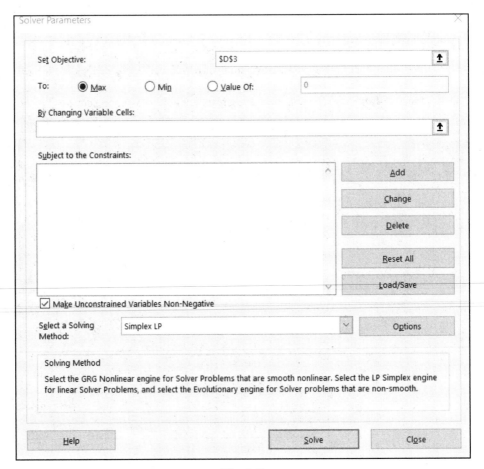

**Fig. A-8**

(9)   Third, establish the constraints:

   - *Subject to constraints*: Click the *Add* button and a menu called *Add constraint* will appear,
     as shown in Fig. A-9.

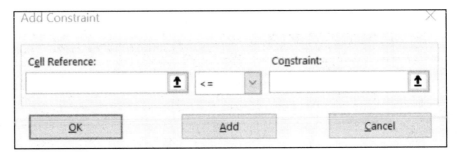

**Fig. A-9**

In this menu, we will insert each of the constraints.

For constraint A, recall that:

$$5x + 2y + 3w \leq 55 \qquad\qquad\qquad\qquad \text{(Constraint } A\text{)}$$

    I. In *Cell Reference*, select the cell $F$12 that indicates *Use* for that constraint (which is the left-hand side of the operation).

    II. Choose the correct sign (in this case <= from the drop-down box).

    III. Select cell $H$12 that represents the *Limit* of constraint *A* (right-hand side of the equation). The template of the menu should look like Fig. A-10.

**Fig. A-10**

Then, click the *Add* button to continue adding each of the other four constraints. The operation will be pretty similar for constraints *B* to *E*, and it is described in the table below.

| Constraint | Cell Location | Sign | Constraint |
|:---:|:---:|:---:|:---:|
| B | $F$14 | <= | =$H$12 |
| C | $F$16 | <= | =$H$14 |
| D | $F$18 | <= | =$H$16 |
| E | $F$20 | <= | =$H$18 |

Once all constraints are inserted, click the *OK* button. These restrictions will appear in the menu of *Solver Parameters* (as shown in Fig. A-11).

*(10)* Specify that decision variables are nonnegative by checking the box "*make unconstrained variables non-negative.*" The final result looks like Fig. A-11.

*(11)* Click the *Solve* button. A menu that indicates the *Solver™* results will appear (Fig. A-12).

As shown in Fig. A-12, *Solver™* found a solution. Before clicking *OK*, select in *Reports* the option *Sensitivity*, which will provide the shadow prices of the linear programming problem. Then, click *OK*.

Notice that cells C3 (optimal value of the objective), C7 to E7 (optimal values for the decision variables), and F12 to F20 (which represent the use in the constraints) are now filled out, as illustrated in Fig. A-13.

*(12)* *Interpretation of results*

Figure A-13 shows the following:

- Cell C3: The **optimal value** of **Z** is 258.33.

- Cells C7 to E7: The **optimal values for the decision variables** are $x^* = 1.67$, $y^* = 23.3$, and $w^* = 0$.

- Cells F12: Constraint *A* **used** 55 units, which is the limit imposed on the right-hand side; therefore, this is an **active** constraint that delimits the optimal solution.

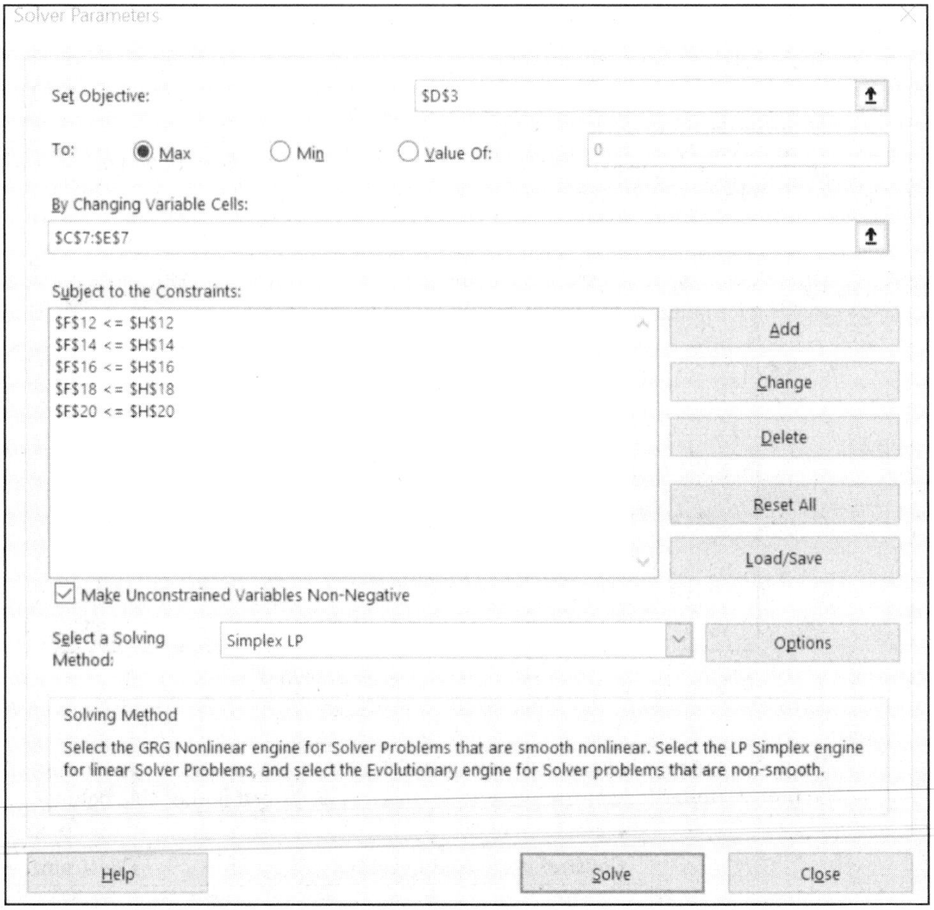

**Fig. A-11**

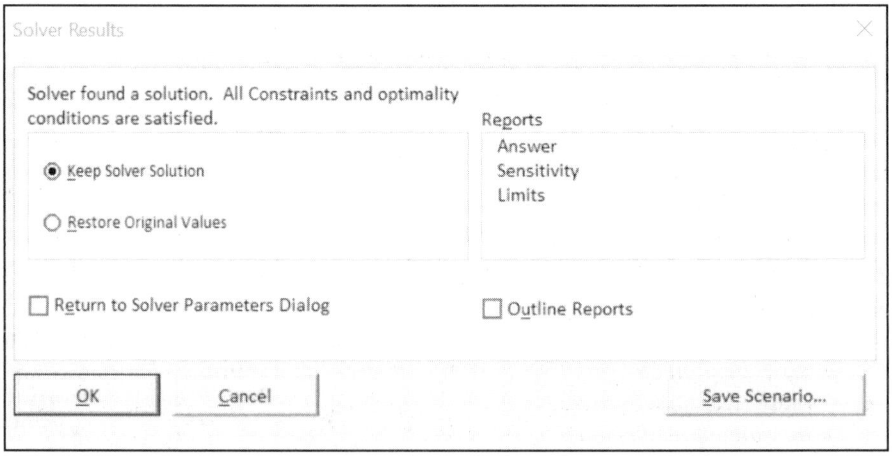

**Fig. A-12**

|  | A | B | C | D | E | F | G | H |
|---|---|---|---|---|---|---|---|---|
| 1 | | | | | | | | |
| 2 | | | | **Objective** | | | | |
| 3 | | **Maximize** | Z | 258.3333333 | | | | |
| 4 | | | | | | | | |
| 5 | | **Variables** | **x** | **y** | **w** | | | |
| 6 | | Obj. Coefficients | 15 | 10 | 20 | | | |
| 7 | | Optimal values | 1.666667 | 23.33333 | 0 | | | |
| 8 | | | | | | | | |
| 9 | | **Subject to:** | | | | | | |
| 10 | | | | **Coefficients of constraints** | | | | |
| 11 | | | **x** | **y** | **w** | **Use** | **Sign** | **Limit** |
| 12 | | Constraint A: | 5 | 2 | 3 | 55 | <= | 55 |
| 13 | | | | | | | | |
| 14 | | Constraint B: | 1 | 1 | 2 | 25 | <= | 26 |
| 15 | | | | | | | | |
| 16 | | Constraint C: | 3 | 1 | 1 | 28.33333 | <= | 30 |
| 17 | | | | | | | | |
| 18 | | Constraint D: | 1 | 1 | 3 | 25 | <= | 25 |
| 19 | | | | | | | | |
| 20 | | Constraint E: | 4 | 2 | 5 | 53.33333 | <= | 57 |

**Fig. A-13**

- Cells F14: Constraint *B* **used** 25 units, which is lower than the limit (26); therefore, this is not an active constraint.

- Cells F16: Constraint *C* **used** 28.33 units, which is lower than the limit (30); therefore, this is also not an active constraint.

- Cells F18: Constraint *D* **used** 25 units, which is the limit imposed on the right-hand side; therefore, this is an **active** constraint that delimits the optimal solution.

- Cells F20: Constraint *E* **used** 53.33 units, which is lower than the limit (57); therefore, this is not an active constraint.

*(13) Sensitivity Analysis*

The *Solver Results* dialog box (in Fig. A-12) offers the option to collect a report on the Sensitivity Analysis, which appears as an additional spreadsheet tab called "Sensitivity Report 1." Figure A-14 displays this report.

This report shows the following details.

Rows 7 to 11 discusses sensitivity on the objective function:

- Cells D9 to D11: The optimal values for *x*, *y*, and *w,* which was 1.67, 23.3 and 0, respectively.

- Cells F9 and F11: These show the objective coefficients for each decision variable.

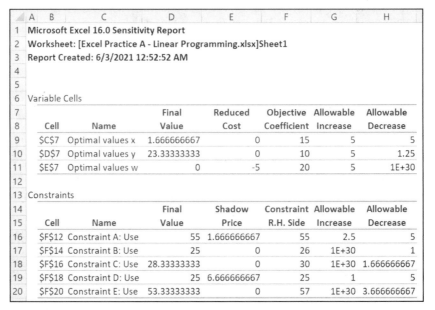

| | A B | C | D | E | F | G | H |
|---|---|---|---|---|---|---|---|
| 1 | Microsoft Excel 16.0 Sensitivity Report | | | | | | |
| 2 | Worksheet: [Excel Practice A - Linear Programming.xlsx]Sheet1 | | | | | | |
| 3 | Report Created: 6/3/2021 12:52:52 AM | | | | | | |
| 4 | | | | | | | |
| 5 | | | | | | | |
| 6 | Variable Cells | | | | | | |
| 7 | | | Final | Reduced | Objective | Allowable | Allowable |
| 8 | Cell | Name | Value | Cost | Coefficient | Increase | Decrease |
| 9 | $C$7 | Optimal values x | 1.666666667 | 0 | 15 | 5 | 5 |
| 10 | $D$7 | Optimal values y | 23.33333333 | 0 | 10 | 5 | 1.25 |
| 11 | $E$7 | Optimal values w | 0 | -5 | 20 | 5 | 1E+30 |
| 12 | | | | | | | |
| 13 | Constraints | | | | | | |
| 14 | | | Final | Shadow | Constraint | Allowable | Allowable |
| 15 | Cell | Name | Value | Price | R.H. Side | Increase | Decrease |
| 16 | $F$12 | Constraint A: Use | 55 | 1.666666667 | 55 | 2.5 | 5 |
| 17 | $F$14 | Constraint B: Use | 25 | 0 | 26 | 1E+30 | 1 |
| 18 | $F$16 | Constraint C: Use | 28.33333333 | 0 | 30 | 1E+30 | 1.666666667 |
| 19 | $F$18 | Constraint D: Use | 25 | 6.666666667 | 25 | 1 | 5 |
| 20 | $F$20 | Constraint E: Use | 53.33333333 | 0 | 57 | 1E+30 | 3.666666667 |

**Fig. A-14**

- Cells G9 and G11 (**Allowable Increase**) and H9 to H11 (**Allowable Decrease**): These numbers show by how much the **objective coefficient may increase** or **decrease** but still have the optimal solution combination. For example, for $y$, the objective coefficient is 10, but if the coefficient increases by less than or equal to 5 (so up to 15) or decreases by less than or equal to 1.25 (so up to 8.5), the optimal combination ($x$ and $y$ are nonzero, whereas $w$ is zero) will stay the same and the same active constraints (in this case A and D) will remain as the delimiting constraints. If the change is greater than indicated, then the optimal combination will change, and the active constraints pair will be different.

Rows 13 to 20 report the sensitivity on the constraints:

- Cells D16 to D20 describe the value of each constraint (limit) that was used (called *Use* in the spreadsheet). These values are the ones that also appeared in the optimal solution. For example, Constraint $C$ shows that the limit was 28.33.

- Cells F16 to F20 show the current right-hand side value of each constraint. For example, Constraint $C$ shows that the limit is 30.

- Cells E16 to E20 report the **shadow prices**, which are the **optimal values of the dual** or **marginal values** of the **resources (constraints)**. Recall that the shadow price is greater than zero only if the active constraint is active. Thus, the shadow price for constraint $A$ (in cell E16) is 1.67, which in this case is active (this can be seen as the Use = Constraint Right-Hand Side). This **shadow price indicates that if the right-hand side of active constraint** A **increases by 1 unit** (from 55 to 56), **then the objective function will increase by** 1.67 units. The shadow prices for the nonactive constraints, which are B, C, and E, are zero, which means that even if the limit is increased or decreased by 1 unit, this will not affect the objective value or the combination of active constraints. The shadow price of the active constraint D is 6.67, which means that increasing constraint D by 1 unit (from 25 to 26) will increase the objective function by 6.67.

## SUMMARY RESULTS

The following table provides the summary of the most important results.

| Variable | Cell Location | Optimal Values | Interpretation or Comments |
|---|---|---|---|
| **Objective Function—Main Sheet** | | | |
| Objective variable $Z$ | D3 | 258.33 | |
| Decision variable $x$ | C7 | 1.67 | |
| Decision variable $y$ | D7 | 23 | |
| Decision variable $w$ | E7 | 0 | |
| **Constraints—Main Sheet** | | | |
| Use of constraint $A$ | F12 | 55 | Limit: 55; this is an active constraint |
| Use of constraint $B$ | F14 | 25 | Limit: 26; this constraint is not active |
| Use of constraint $C$ | F16 | 28.33 | Limit: 55; this constraint is not active |
| Use of constraint $D$ | F18 | 25 | Limit: 25; this is an active constraint |
| Use of constraint $E$ | F20 | 53.33 | Limit: 57; this constraint is not active |
| **Shadow Prices—Sensitivity Report** | | | |
| Shadow price of constraint $A$ | E16 | 1.67 | Active: 1 more unit increases objective by 1.67 |
| Shadow price of constraint $B$ | E17 | 0 | Not active: 1 more unit does not change objective |
| Shadow price of constraint $C$ | E18 | 0 | Not active: 1 more unit does not change objective |
| Shadow price of constraint $D$ | E19 | 6.67 | Active: 1 unit increases objective by 1.67 |
| Shadow price of constraint $E$ | E20 | 0 | Not active: 1 more unit does not change objective |

**A.2.**  Maximize

$$Z = 5x + 2y + 3u + 8v$$

subject to

$$40x + 20y + 15u + 50v \leq 130 \qquad \text{(Constraint } A)$$
$$3x + 2y \qquad\qquad\quad \leq 6 \qquad \text{(Constraint } B)$$
$$2x + 2y + 3u + 4v \leq 10 \qquad \text{(Constraint } C)$$
$$2x + 2y + 5u + 3v \leq 40 \qquad \text{(Constraint } D)$$
$$1u + 5v \leq 8 \qquad \text{(Constraint } E)$$
$$x, y, w \geq 0$$

Solve the problem and interpret the objective coefficients and shadow prices.

*(1)*  Write the maximization problem in Microsoft Excel™, in the cells shown in the template provided in Fig. A-15. Note that some decision variables are not present in each constraint; for those cases, just leave it blank. As in the previous case, highlight in orange the cell C3 (in which the objective function will be declared), in yellow cells C7 to F7, which will be the optimal values for the decision variables (in this case $x$, $y$, $u$, and $v$), and in light blue the cells G12 to G20 (which represent the resource use for each constraint).

*(2)*  Write the objective formula using the Excel™ function *sumproduct*: by selecting the objective coefficients (C6:F6) and the optimal value cells (C7:F7). Thus, write in cell D3

=SUMPRODUCT(C6:F6,C7:F7)

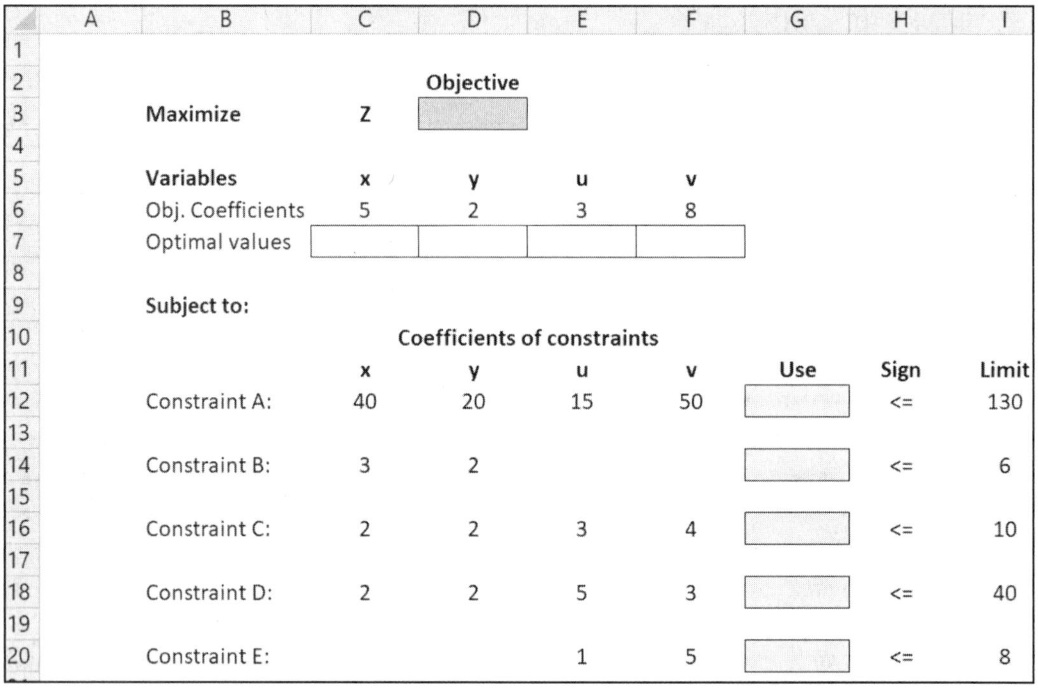

| | A | B | C | D | E | F | G | H | I |
|---|---|---|---|---|---|---|---|---|---|
| 1 | | | | | | | | | |
| 2 | | | | Objective | | | | | |
| 3 | | Maximize | Z | | | | | | |
| 4 | | | | | | | | | |
| 5 | | Variables | x | y | u | v | | | |
| 6 | | Obj. Coefficients | 5 | 2 | 3 | 8 | | | |
| 7 | | Optimal values | | | | | | | |
| 8 | | | | | | | | | |
| 9 | | Subject to: | | | | | | | |
| 10 | | | | Coefficients of constraints | | | | | |
| 11 | | | x | y | u | v | Use | Sign | Limit |
| 12 | | Constraint A: | 40 | 20 | 15 | 50 | | <= | 130 |
| 13 | | | | | | | | | |
| 14 | | Constraint B: | 3 | 2 | | | | <= | 6 |
| 15 | | | | | | | | | |
| 16 | | Constraint C: | 2 | 2 | 3 | 4 | | <= | 10 |
| 17 | | | | | | | | | |
| 18 | | Constraint D: | 2 | 2 | 5 | 3 | | <= | 40 |
| 19 | | | | | | | | | |
| 20 | | Constraint E: | | | 1 | 5 | | <= | 8 |

**Fig. A-15**

(3) Declare the functions for each *Use* cell. For constraint *A*, this is the multiplication of each decision variable (cells C7:F7) by their respective coefficients in constraint *A* (cells C12: F12) and then adding each partial sum. Thus, write in *Use* (cell G12) for the first constraint:

=SUMPRODUCT(C7:F7,C12:F12)

Repeat a similar pattern for each constraint. The specific formulations are shown in the table below and in Fig. A-16.

| Formulation | Cell Location | Formula | Mathematical Representation |
|---|---|---|---|
| **Objective Function** | | | |
| Optimal value of $Z$ | C3 | =SUMPRODUCT(C6:F6,C7:F7) | $Z = 5x + 2y + 3u + 8v$ |
| **Constraints** | | | |
| Use of constraint $A$ | G12 | =SUMPRODUCT(C7:F7,C12:F12) | $40x + 20y + 15u + 50v \le 130$ |
| Use of constraint $B$ | G14 | =SUMPRODUCT(C7:F7,C14:F14) | $3x + 2y \le 6$ |
| Use of constraint $C$ | G16 | =SUMPRODUCT(C7:F7,C16:F16) | $2x + 2y + 3u + 4v \le 10$ |
| Use of constraint $D$ | G18 | =SUMPRODUCT(C7:F7,C18:F18) | $2x + 2y + 5u + 3v \le 40$ |
| Use of constraint $E$ | G20 | =SUMPRODUCT(C7:F7,C20:F20) | $1u + 5v \le 8$ |

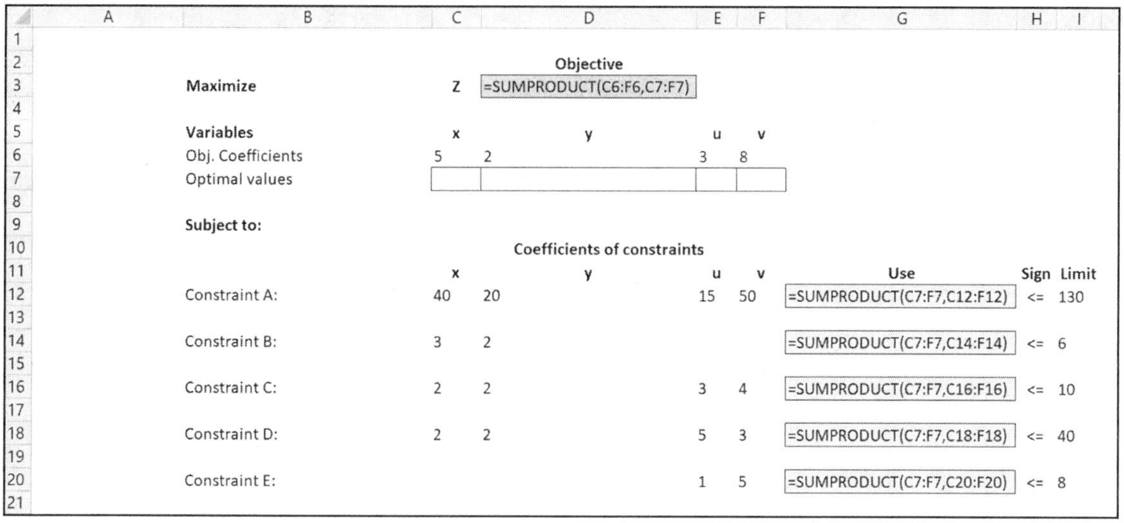

**Fig. A-16**

(4) Open the *Solver*™ menu, select *Simplex LP* as the solving method, and then declare the functions as shown in the table below.

| Label | Statement | Mathematical Representation |
|---|---|---|
| **Objective Function** | | |
| Set objective | **$D$3** | This is the cell of the objective function |
| To | Max | This maximizes the objective |
| By changing variables | $C$7:$F$7 | Decision variables |
| **Subject to Constraints** | | |
| Constraint *A* | $G$12<=$I$12 | Use of A ≤ Limit on right-hand side of *A* |
| Constraint *B* | $G$14<=$I$14 | Use of B ≤ Limit on right-hand side of *B* |
| Constraint *C* | $G$16<=$I$16 | Use of C ≤ Limit on right-hand side of *C* |
| Constraint *D* | $G$18<=$I$18 | Use of D ≤ Limit on right-hand side of *D* |
| Constraint *E* | $G$20<=$I$20 | Use of E ≤ Limit on right-hand side of *E* |

To add the constraints, click the *Add* button.

Finally, click the option "*make unconstrained variables non-negative*," which is the nonnegativity constraint. All these actions are illustrated in Fig. A-17. Then click the *Solve* button.

(5) The Solver Results dialog box shows that there is an optimal solution. Selet Keep Solver Solution, then select the Sensitivity Report and click OK (as in Fig. A-18).

(6) Figure A-19 displays the optimal values of the decision and objective variables, as well as the use of each constraint. Figure A-20 shows the sensitivity analysis, and in particular, the shadow prices.

**Fig. A-17**

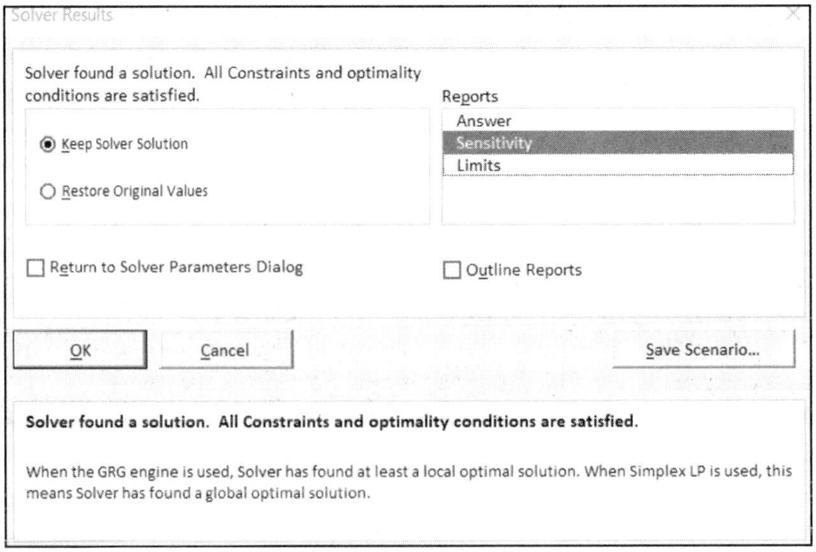

**Fig. A-18**

| | A | B | C | D | E | F | G | H | I |
|---|---|---|---|---|---|---|---|---|---|
| 1 | | | | | | | | | |
| 2 | | | | | **Objective** | | | | |
| 3 | | **Maximize** | | Z | 19.48717949 | | | | |
| 4 | | | | | | | | | |
| 5 | | **Variables** | | x | y | u | v | | |
| 6 | | Obj. Coefficients | | 5 | 2 | 3 | 8 | | |
| 7 | | Optimal values | | 1.179487 | 0 | 0.564103 | 1.487179 | | |
| 8 | | | | | | | | | |
| 9 | | **Subject to:** | | | | | | | |
| 10 | | | | | **Coefficients of constraints** | | | | |
| 11 | | | | x | y | u | v | **Use** | **Sign** | **Limit** |
| 12 | | Constraint A: | | 40 | 20 | 15 | 50 | 130 | <= | 130 |
| 13 | | | | | | | | | |
| 14 | | Constraint B: | | 3 | 2 | | | 3.538461538 | <= | 6 |
| 15 | | | | | | | | | |
| 16 | | Constraint C: | | 2 | 2 | 3 | 4 | 10 | <= | 10 |
| 17 | | | | | | | | | |
| 18 | | Constraint D: | | 2 | 2 | 5 | 3 | 9.641025641 | <= | 40 |
| 19 | | | | | | | | | |
| 20 | | Constraint E: | | | | 1 | 5 | 8 | <= | 8 |

**Fig. A-19**

| | A B | C | D | E | F | G | H |
|---|---|---|---|---|---|---|---|
| 6 | **Variable Cells** | | | | | | |
| 7 | | | **Final** | **Reduced** | **Objective** | **Allowable** | **Allowable** |
| 8 | **Cell** | **Name** | **Value** | **Cost** | **Coefficient** | **Increase** | **Decrease** |
| 9 | $C$7 | Optimal values x | 1.17948718 | 0 | 5 | 1 | 2.05882353 |
| 10 | $D$7 | Optimal values y | 0 | -0.8974359 | 2 | 0.8974359 | 1E+30 |
| 11 | $E$7 | Optimal values u | 0.56410256 | 0 | 3 | 1.5 | 0.775 |
| 12 | $F$7 | Optimal values v | 1.48717949 | 0 | 8 | 3.875 | 1 |
| 13 | | | | | | | |
| 14 | **Constraints** | | | | | | |
| 15 | | | **Final** | **Shadow** | **Constraint** | **Allowable** | **Allowable** |
| 16 | **Cell** | **Name** | **Value** | **Price** | **R.H. Side** | **Increase** | **Decrease** |
| 17 | $G$12 | Constraint A: Use | 130 | 0.105128205 | 130 | 22 | 41.8181818 |
| 18 | $G$14 | Constraint B: Use | 3.53846154 | 0 | 6 | 1E+30 | 2.46153846 |
| 19 | $G$16 | Constraint C: Use | 10 | 0.397435897 | 10 | 14.2650602 | 1.1 |
| 20 | $G$18 | Constraint D: Use | 9.64102564 | 0 | 40 | 1E+30 | 30.3589744 |
| 21 | $G$20 | Constraint E: Use | 8 | 0.230769231 | 8 | 3.66666667 | 3.55555556 |

**Fig. A-20**

The table below summarizes the results, including the interpretation of the major outcomes.

| Variable | Cell Location | Optimal Values | Interpretation or Comments |
|---|---|---|---|
| **Objective Function—Main Sheet** | | | |
| Objective variable $Z$ | D3 | 19.49 | |
| Decision variable $x$ | C7 | 1.18 | |
| Decision variable $y$ | D7 | 0 | |
| Decision variable $u$ | E7 | 0.56 | |
| Decision variable $v$ | F7 | 1.49 | |
| **Constraints—Main Sheet** | | | |
| Use of constraint $A$ | G12 | 130 | Limit: 130; this is an active constraint |
| Use of constraint $B$ | G14 | 3.54 | Limit: 6; this constraint is not active |
| Use of constraint $C$ | G16 | 10 | Limit: 10; this is an active constraint |
| Use of constraint $D$ | G18 | 9.64 | Limit: 40; this constraint is not active |
| Use of constraint $E$ | G20 | 8 | Limit: 8; this is an active constraint |
| **Shadow Prices—Sensitivity Report** | | | |
| Shadow price of constraint $A$ | E16 | 1.11 | Active: 1 more unit increases objective by 1.11 |
| Shadow price of constraint $B$ | E17 | 0 | Not active: 1 more unit does not change objective |
| Shadow price of constraint $C$ | E18 | 0.40 | Active: 1 unit increases objective by 0.40 |
| Shadow price of constraint $D$ | E19 | 0 | Not active: 1 more unit does not change objective |
| Shadow price of constraint $E$ | E20 | 0.23 | Active: 1 unit increases objective by 0.23 |

## MINIMIZATION PROBLEMS

**A.3.**  Minimize

$$Z = 30x + 40y + 80w$$

subject to

$$4x + 2y + 5w \geq 70 \qquad \text{(Constraint } A)$$
$$x + y + w \geq 50 \qquad \text{(Constraint } B)$$
$$5y + 2w \geq 40 \qquad \text{(Constraint } C)$$
$$2x + 2y \geq 8 \qquad \text{(Constraint } D)$$
$$x, y, w \geq 0$$

Solve the problem and interpret the objective coefficients and shadow prices.

(1)  Write the minimization problem in MS Excel™ and fill out the cells as shown in the template in Fig. A-21. Note that cell B3 says "minimize" to properly reflect this problem. There are three decision variables, $x$, $y$, and $w$, located in columns C, D, and E. Highlight the cells as shown below.

(2)  Write the objective formula using the Excel™ function *sumproduct*: by selecting the objective coefficients (C6:E6) and the optimal value cells (C7:E7). Thus, write in cell D3

$$=\text{SUMPRODUCT(C6:E6,C7:E7)}$$

(3)  Declare the functions for each *Use* cell. For constraint $A$, Use is the sum of the product between decision variables (cells C7:E7) by their respective coefficients (cells C12:E12). Thus, write in *Use* (cell G12) for the first constraint

$$=\text{SUMPRODUCT(C7:E7,C12:E12)}$$

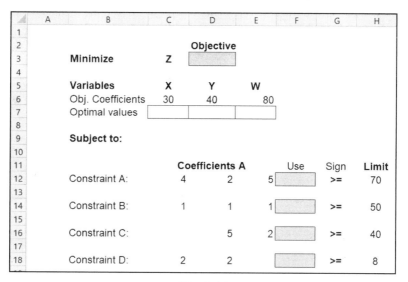

**Fig. A-21**

Repeat a similar pattern for each constraint. The specific formulations are shown in the table below and in Fig. A-22.

| Formulation | Cell Location | Formula | Mathematical Representation |
|---|---|---|---|
| **Objective Function** | | | |
| Optimal value of $Z$ | C3 | =SUMPRODUCT(C6:E6,C7:E7) | $Z = 30x + 40y + 80w$ |
| **Constraints** | | | |
| Use of constraint $A$ | F12 | =SUMPRODUCT(C7:E7,C12:E12) | $4x + 2y + 5w \geq 70$ |
| Use of constraint $B$ | F14 | =SUMPRODUCT(C7:E7,C14:E14) | $x + y + w \geq 50$ |
| Use of constraint $C$ | F16 | =SUMPRODUCT(C7:E7,C16:E16) | $5y + 2w \geq 40$ |
| Use of constraint $D$ | F18 | =SUMPRODUCT(C7:E7,C18:E18) | $2x + 2y \geq 8$ |

**Fig. A-22**

An important detail to remember is the sign. Note that for minimization problems, the primal form requires that the signs in the constraint must be greater than or equal to ($\geq$).

(4)  Open the *Solver*™ menu, select *Simplex LP* as the solving method, then declare the functions as shown in the table below.

| Label | Statement | Mathematical Representation |
|---|---|---|
| **Objective Function** | | |
| Set objective | **$D$3** | This is the cell of the objective function |
| To | Min | This maximizes the objective |
| By changing variables | $C$7:$E$7 | Decision variables |
| **Subject to Constraints** | | |
| Constraint $A$ | $F$12<=$H$12 | Use of $A \le$ Limit on right-hand side of $A$ |
| Constraint $B$ | $F$14<=$H$14 | Use of $B \le$ Limit on right-hand side of $B$ |
| Constraint $C$ | $F$16<=$H$16 | Use of $C \le$ Limit on right-hand side of $C$ |
| Constraint $D$ | $F$20<=$H$20 | Use of $D \le$ Limit on right-hand side of $D$ |

To add the constraints, click the *Add* button. Be careful when selecting the signs.

Finally, click the option "make unconstrained variables non-negative," which is the nonnegativity constraint. All these actions are illustrated in Fig. A-23. Then click the *Solve* button.

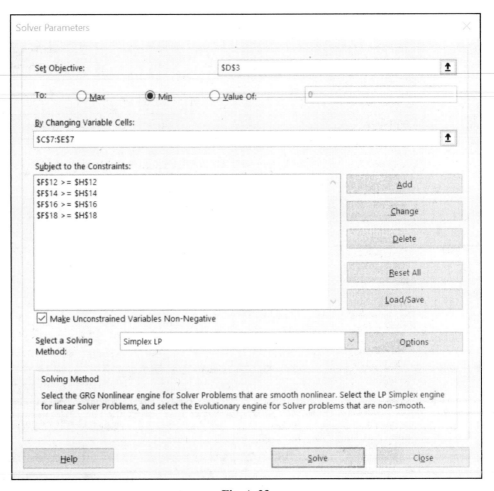

**Fig. A-23**

(5) The Solver Results dialog box shows that there is an optimal solution. Select Keep Solver Solution, then select the Sensitivity Report and click OK (similar to Fig. A-18).

(6) Figure A-24 displays the optimal values of the decision and objective variables, as well as the use of each constraint. Figure A-25 show the sensitivity analysis, and in particular, the shadow prices.

| | A | B | C | D | E | F | G | H |
|---|---|---|---|---|---|---|---|---|
| 1 | | | | | | | | |
| 2 | | | | **Objective** | | | | |
| 3 | | **Minimize** | **Z** | 1580 | | | | |
| 4 | | | | | | | | |
| 5 | | **Variables** | **X** | **Y** | **W** | | | |
| 6 | | Obj. Coefficients | 30 | 40 | 80 | | | |
| 7 | | Optimal values | 42 | 8 | 0 | | | |
| 8 | | | | | | | | |
| 9 | | **Subject to:** | | | | | | |
| 10 | | | | | | | | |
| 11 | | | | **Coefficients A** | | Use | Sign | **Limit** |
| 12 | | Constraint A: | 4 | 2 | 5 | 184 | >= | 70 |
| 13 | | | | | | | | |
| 14 | | Constraint B: | 1 | 1 | 1 | 50 | >= | 50 |
| 15 | | | | | | | | |
| 16 | | Constraint C: | | 5 | 2 | 40 | >= | 40 |
| 17 | | | | | | | | |
| 18 | | Constraint D: | 2 | 2 | | 100 | >= | 8 |

**Fig. A-24**

| | A | B | C | D | E | F | G | H |
|---|---|---|---|---|---|---|---|---|
| 6 | Variable Cells | | | | | | | |
| 7 | | | | Final | Reduced | Objective | Allowable | Allowable |
| 8 | | Cell | Name | Value | Cost | Coefficient | Increase | Decrease |
| 9 | | $C$7 | Optimal values X | 42 | 0 | 30 | 10 | 30 |
| 10 | | $D$7 | Optimal values Y | 8 | 0 | 40 | 115 | 10 |
| 11 | | $E$7 | Optimal values W | 0 | 46 | 80 | 1E+30 | 46 |
| 12 | | | | | | | | |
| 13 | Constraints | | | | | | | |
| 14 | | | | Final | Shadow | Constraint | Allowable | Allowable |
| 15 | | Cell | Name | Value | Price | R.H. Side | Increase | Decrease |
| 16 | | $F$12 | Constraint A: Use | 184 | 0 | 70 | 114 | 1E+30 |
| 17 | | $F$14 | Constraint B: Use | 50 | 30 | 50 | 1E+30 | 28.5 |
| 18 | | $F$16 | Constraint C: Use | 40 | 2 | 40 | 210 | 40 |
| 19 | | $F$18 | Constraint D: Use | 100 | 0 | 8 | 92 | 1E+30 |

**Fig. A-25**

The table below summarizes the results, including the interpretation of the major outcomes.

| Variable | Cell Location | Optimal Values | Interpretation or Comments |
|---|---|---|---|
| **Objective Function—Main Sheet** | | | |
| Objective variable $Z$ | D3 | 1580 | |
| Decision variable $x$ | C7 | 42 | |
| Decision variable $y$ | D7 | 8 | |
| Decision variable $w$ | E7 | 0 | |
| **Constraints—Main Sheet** | | | |
| Use of constraint $A$ | F12 | 184 | Limit: 6, this constraint is not active |
| Use of constraint $B$ | F14 | 50 | Limit: 50, this is an active constraint |
| Use of constraint $C$ | F16 | 40 | Limit: 40, this is an active constraint |
| Use of constraint $D$ | F18 | 100 | Limit: 100, this constraint is not active |
| **Shadow Prices—Sensitivity Report** | | | |
| Shadow price of constraint $A$ | E16 | 0 | Not Active: 1 more unit does not change objective |
| Shadow price of constraint $B$ | E17 | 30 | Active: 1 more unit increases objective by 30 |
| Shadow price of constraint $C$ | E18 | 2 | Active: 1 more unit increases objective by 2 |
| Shadow price of constraint $D$ | E19 | 0 | Not Active: 1 more unit does not change objective |

**A.4.**  Minimize $\qquad\qquad Z = 8x + 5y$

subject to

$$4x + 3y \geq 120 \qquad\qquad \text{(Constraint } A\text{)}$$
$$2x + 3y \geq 80 \qquad\qquad \text{(Constraint } B\text{)}$$
$$2x + \ y \geq 50 \qquad\qquad \text{(Constraint } C\text{)}$$
$$x, y \geq 0$$

Solve the problem and interpret the objective coefficients and shadow prices.

(*1*)  Write the minimization problem as shown in the template in Fig. A-26. There are two decision variables, $x$ and $y$, located in columns C and D. Highlight the cells as shown below.

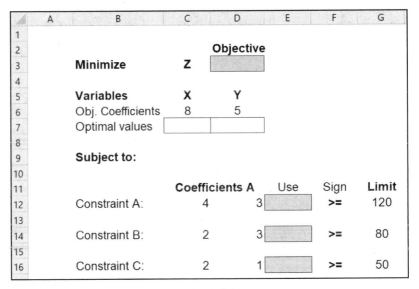

**Fig. A-26**

(2)  Write the objective formula using the Excel™ function *sumproduct*: by selecting the objective coefficients (C6:D6) and the optimal value cells (C7:D7). Thus, write in cell D3

$$=\text{SUMPRODUCT(C6:D6,C7:D7)}$$

(3)  Declare the *sumproduct* function for each *Use* cell. For constraint *A*, this function will use the decision variables (cells C7:D7) and their respect coefficients (cells C12:D12). Thus, write in *Use* (cell E12) for the first constraint

$$=\text{SUMPRODUCT(C7:D7,C12:D12)}$$

Repeat a similar pattern for each constraint, as shown in the table below and in Fig. A-27.

| Formulation | Cell Location | Formula | Mathematical Representation |
|---|---|---|---|
| **Objective Function** | | | |
| Optimal value of $Z$ | C3 | =SUMPRODUCT(C6:D6,C7:D7) | $Z = 8x_1 + 5x_2$ |
| **Constraints** | | | |
| Use of constraint $A$ | E12 | =SUMPRODUCT(C7:D7,C12:D12) | $4x + 3y \geq 120$ |
| Use of constraint $B$ | E14 | =SUMPRODUCT(C7:D7,C14:D14) | $2x + 3y \geq 80$ |
| Use of constraint $C$ | E16 | =SUMPRODUCT(C7:D7,C16:D16) | $2x + y \geq 50$ |

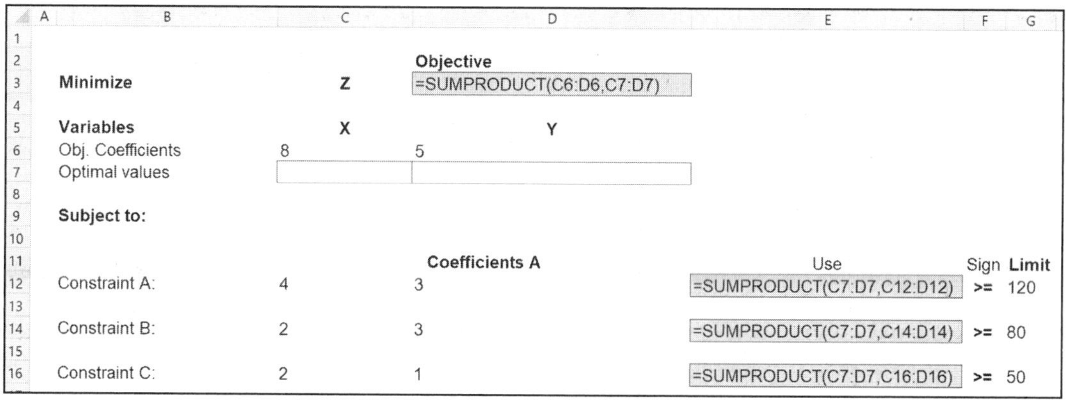

**Fig. A-27**

(4)  Open the *Solver*™ menu, select *Simplex LP* as the solving method, and then declare the functions as shown in the table below.

| Label | Statement | Mathematical Representation |
|---|---|---|
| **Objective Function** | | |
| Set objective | **$D$3** | This is the cell of the objective function |
| To | Min | This maximizes the objective |
| By changing variables | $C$7:$D$7 | Decision variables |
| **Subject to Constraints** | | |
| Constraint $A$ | $E$12<=$G$12 | Use of $A \leq$ Limit in right-hand side of $A$ |
| Constraint $B$ | $E$14<=$G$14 | Use of $B \leq$ Limit in right-hand side of $B$ |
| Constraint $C$ | $E$16<=$G$16 | Use of $C \leq$ Limit in right-hand side of $C$ |

To add the constraints, click the *Add* button. Be careful when selecting the signs.

Finally, click the option "*make unconstrained variables non-negative*," which is the nonnegativity constraint. All these actions are illustrated in Fig. A-28. Then click the *Solve* button.

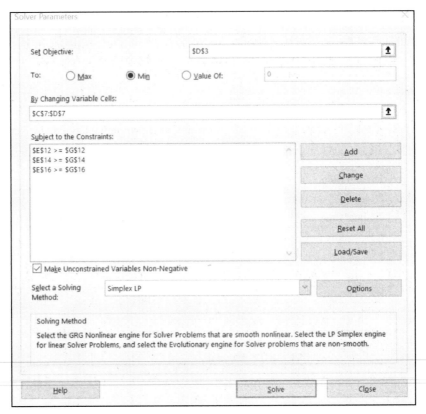

**Fig. A-28**

(*4*) The Solver Results dialog box will show that there is an optimal solution. Select Keep Solver Solution, then select the Sensitivity Report and click OK (similar to Fig. A-18).

(*5*) Figure A-29 displays the optimal values of the decision and objective variables, as well as the use of each constraint. Figure A-30 show the sensitivity analysis, and in particular, the shadow prices.

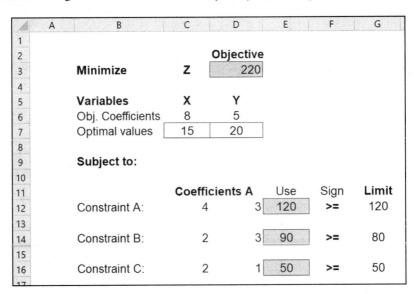

**Fig. A-29**

| | A | B | C | D | E | F | G | H |
|---|---|---|---|---|---|---|---|---|
| 6 | | Variable Cells | | | | | | |
| 7 | | | | Final | Reduced | Objective | Allowable | Allowable |
| 8 | | Cell | Name | Value | Cost | Coefficient | Increase | Decrease |
| 9 | | $C$7 | Optimal values X | 15 | 0 | 8 | 2 | 1.333333333 |
| 10 | | $D$7 | Optimal values Y | 20 | 0 | 5 | 1 | 1 |
| 11 | | | | | | | | |
| 12 | | Constraints | | | | | | |
| 13 | | | | Final | Shadow | Constraint | Allowable | Allowable |
| 14 | | Cell | Name | Value | Price | R.H. Side | Increase | Decrease |
| 15 | | $E$12 | Constraint A: Use | 120 | 1 | 120 | 30 | 5 |
| 16 | | $E$14 | Constraint B: Use | 90 | 0 | 80 | 10 | 1E+30 |
| 17 | | $E$16 | Constraint C: Use | 50 | 2 | 50 | 3.333333333 | 10 |

**Fig. A-30**

The table below summarizes the results, including the interpretation of the major outcomes.

| Variable | Cell Location | Optimal Values | Interpretation or Comments |
|---|---|---|---|
| **Objective Function—Main Sheet** | | | |
| Objective variable $Z$ | D3 | 220 | |
| Decision variable $x$ | C7 | 15 | |
| Decision variable $y$ | D7 | 20 | |
| **Constraints—Main Sheet** | | | |
| Use of constraint $A$ | E12 | 120 | Lower limit: 120; this is an active constraint |
| Use of constraint $B$ | E14 | 90 | Lower limit: 80; this constraint is not active |
| Use of constraint $C$ | E16 | 50 | Lower limit: 50; this is an active constraint |
| **Shadow Prices—Sensitivity Report** | | | |
| Shadow price of constraint $A$ | E16 | 1 | Active: 1 more unit increases objective by 1 |
| Shadow price of constraint $B$ | E17 | 0 | Not active: 1 more unit does not change objective |
| Shadow price of constraint $C$ | E18 | 2 | Active: 1 more unit increases objective by 2 |

# Excel Practice

## Microsoft Excel™ Practice B: Present Value and Net Present Value

Series and sequences are often used in economic and financial analysis to represent formulas that may include hundreds or thousands of observations. In this practice, the focus will be two major tools used in finance, which were introduced in Chapter 14: the present value and the net present value. Examples B.1. and B.2. explain the MS Excel™ formulation for present value. Examples B.3. and B.4. show step-by-step how to compute the net present value formula.

### THE PRESENT VALUE

Present value $P$ is the value of money at the present time, given a specific discount rate; its formula is

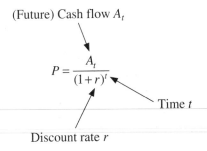

(Future) Cash flow $A_t$

$$P = \frac{A_t}{(1+r)^t}$$

Time $t$

Discount rate $r$

Thus, to calculate the present value of a future investment that will be paid in a determined future year, this depends three components:

- $A_t$: value of the cash flow at the specific period (*future value*)
- $r$: rate of investment (*discount rate*)
- $t$: years of investment (*time*)

Although we can compute the present value by implementing this formula in MS Excel™, the software provides a specific financial formulation called **PV** (present value) with the following syntax:

$$=\textbf{PV}(\text{rate, nper, pmt, [fv], [type]})$$

where

- **rate** is the *discount rate*
- **nper** is the *time* of investment
- **pmt** is any intermediate payment (which for this practice will be zero)
- **[fv]** is the future value of the cash flow, which is typically written with the opposite sign as the present value
- **[type]** is an optional statement that declares when payments are due by default. This is at the end of the period, which will be the case to reflect the present value $P$ formula

Examples B1 and B2 show how to implement the MS Excel™ function to calculate the present value.

**B.1.** Yousef is expecting to receive $1,800 in 5 years. Given that the discount rate is 21% return, calculate the present value of this cash flow.

(*1*) Write each component in a spreadsheet of MS Excel™ as shown in Fig. B-1:

- Row 2 is used to label each title.
- Column B describes each component.
- Column C shows the name of the component that corresponds to the argument in MS Excel™.
- In cell D3, write the discount rate as "21%".
- In cell D4, write the future value as "$1800"; it will automatically format it as a *currency*.
- In cell D5, provide the number of periods by writing "5".
- Column E displays the period type of the cash flow: years, weeks, or months.
- Row 7 is where the present value function will be written.
- Label cell B7 as present value and cell C7 as "pv".

**Fig. B-1**

(*2*) Write the MS Excel™ function for present value in cell D7 (shown in Fig. B-2):

$$=PV(D3,D5,0,-D4)$$

- Declare "=PV("
- Select the interest rate value or write D3 followed by a comma "," to separate the next argument
- Select the number of years or write "D5," also followed by a comma.
- Write "0," as there are no intermediate payments, as Yousef will receive this amount at the end of the 5 years, then write a comma.
- Write a negative sign "-" before selecting the future value cell D4 or write "D4." This is because future value is considered a cash inflow that a person will receive, whereas the present value is a cash outflow (as the person invests).
- Then, write the closing parenthesis ")".

**Fig. B-2**

The final result will appear in cell D7, which is $693.98, as displayed in Fig. B-3.

| | A | B | C | D | E |
|---|---|---|---|---|---|
| 1 | | | | | |
| 2 | | **Component** | **Argument** | **Value** | **Period** |
| 3 | | Discount rate | *rate* | 21% | |
| 4 | | Future value | *fv* | $1,800 | |
| 5 | | Time | *nper* | | 5 years |
| 6 | | | | | |
| 7 | | Present value | *pv* | $693.98 | |

**Fig. B-3**

Thus, the present value is $693.98. Alternatively, you can verify the result by computing the present value manually, and the values must be the same.

**B.2.**   Mirella is receiving an inheritance of $100,000 in 30 years. She wants to know the value of this cash flow today. Given a discount rate is 5% return, calculate the present value.

(*1*)   Write the components in a spreadsheet of MS Excel™, including labels as shown in Fig. B-4:

| | A | B | C | D | E |
|---|---|---|---|---|---|
| 1 | | | | | |
| 2 | | **Component** | **Argument** | **Value** | **Period** |
| 3 | | Discount rate | *rate* | 5% | |
| 4 | | Future value | *fv* | $100,000 | |
| 5 | | Time | *nper* | | 30 years |
| 6 | | | | | |
| 7 | | Present value | *pv* | | |

**Fig. B-4**

(*2*)   Write the MS Excel™ function for present value in cell D7:

**=PV(D3,D5,0,-D4)**

*rate    nper    fv*

The present value of the inheritance (in cell D7) is $23,137.74, as displayed in Fig. B-5:

| | A | B | C | D | E | F |
|---|---|---|---|---|---|---|
| 1 | | | | | | |
| 2 | | **Component** | **Argument** | **Value** | **Period** | |
| 3 | | Discount rate | *rate* | 5% | | |
| 4 | | Future value | *fv* | $100,000 | | |
| 5 | | Time | *nper* | | 30 years | |
| 6 | | | | | | |
| 7 | | Present value | *pv* | $23,137.74 | =PV(D3,D5,0,-D4) | |

**Fig. B-5**

## THE NET PRESENT VALUE

Net present value $NPV$ is the sum of all cash flows discounted at a given rate, as explained in section 14.8; its formula is

$$NPV = \sum_{t=0}^{n} \frac{A_t}{(1+r)^t} = A_0 + \frac{A_1}{1+r} + \frac{A_2}{(1+r)^2} + \cdots + \frac{A_n}{(1+r)^n}$$

Here, $A_t$ represents the cash flow at year $t$ and $r$ is the discounting rate. This computation becomes cumbersome if a project lasts for a long period of time (10 or 20 years). Thus, MS Excel™ provides the NPV function for the task, whose syntax is

**=NPV(rate, value 1, [value 2], ...)**

- *Rate* is the discount rate.
- Value 1 is the cash flow in the first period.
- Value 2 is the cash flow in the second period.
- You can continue adding more cash flows by writing the cell locations or selecting them.

It is important to note that the MS Excel™ NPV function actually represents:

$$NPV = \frac{A_1}{1+r} + \frac{A_2}{(1+r)^2} + \cdots + \frac{A_n}{(1+r)^n}$$

Thus, if there is any cash outflow (expenditure when starting the project or *time* = 0), this has to be inserted manually. The use of this formula is explained in examples B.3. and B.4.

**B.3.** Find the net present value of the following project and determine if it must be accepted or rejected according to the NPV rule. The discount rate is 12%.

| | Year | | | | | |
|---|---|---|---|---|---|---|
| Future value | 0 | 1 | 2 | 3 | 4 | 5 |
| *A* | −2000 | 0 | 0 | 0 | 7000 | −2000 |

(*1*) Write the cash flows in a spreadsheet using the template format in Fig. B-6:

| | A | B | C | D | E | F | G | H |
|---|---|---|---|---|---|---|---|---|
| 1 | | | | | | | | |
| 2 | | Discount rate | 12% | | | | | |
| 3 | | | | | | | | |
| 4 | | | Year | | | | | |
| 5 | | | 0 | 1 | 2 | 3 | 4 | 5 |
| 6 | | Cash Flow | -2000 | 0 | 0 | 0 | 7000 | -2000 |
| 7 | | | | | | | | |
| 8 | | Net Present Value | | | | | | |

**Fig. B-6**

- In row 2, the discount rate is declared. Write "12%" in cell C2.
- Row 4 represents the years of the cash flows. Please note that all consecutive years from 1 to 5 must be declared for the NPV function to work properly.

- Row 5 declares all the cash flows; it is important to write "0" if there is no value for the cash flow. If left empty, the formulation will not work appropriately.

- NPV will be computed in row 8.

(2)   In cell C6, write the following syntax (in Fig. B-7):

$$=NPV(C2,D6:H6)+C6$$

- The first component of the addition is the NPV function; write "=NPV(" select the discount rate (in cell C2), then write a comma, and then select the cash flows starting from year 1 to 5 (cells D6 to H6), and close the statement with ")".

- To include the initial outlet, we manually write "+" and select the cash flow at year 0 (in cell C6).

| | A | B | C | D | E | F | G | H |
|---|---|---|---|---|---|---|---|---|
| 1 | | | | | | | | |
| 2 | | Discount rate | 12% | | | | | |
| 3 | | | | | | | | |
| 4 | | | | | Year | | | |
| 5 | | | 0 | 1 | 2 | 3 | 4 | 5 |
| 6 | | Cash Flow | -2000 | 0 | 0 | 0 | 7000 | -2000 |
| 7 | | | | | | | | |
| 8 | | Net Present Value | $1,313.77 | =NPV(C2,D6:H6)+C6 | | | | |

**Fig. B-7**

The net present value is $1,313.77 (shown in Fig. B-7). Because NPV > 0, the project must be accepted.

You can verify this result by computing the NPV manually, which is explained in Problem 14.23.

**B.4.**   Compute the net present value of the project given the cash flows below and determine if the project must be approved or rejected. The discount rate is 20%.

| | Year | | | | | | |
|---|---|---|---|---|---|---|---|
| Future value | 0 | 1 | 2 | 3 | 4 | 5 | 6 |
| A | −3000 | −1500 | 500 | 500 | 500 | 2000 | 3000 |

(1)   Write the cash flows in a spreadsheet using the template format in Fig. B-8.

| | A | B | C | D | E | F | G | H | I |
|---|---|---|---|---|---|---|---|---|---|
| 1 | | | | | | | | | |
| 2 | | Discount rate | 20% | | | | | | |
| 3 | | | | | | | | | |
| 4 | | | | | Year | | | | |
| 5 | | | 0 | 1 | 2 | 3 | 4 | 5 | 6 |
| 6 | | Cash Flow | -3000 | -1500 | 500 | 500 | 500 | 2000 | 3000 |
| 7 | | | | | | | | | |
| 8 | | Net Present Value | | | | | | | |

**Fig. B-8**

As before, discount rate is in row 2, the years in row 5, and the net present value is computed in row 8.

(2)  In cell C6, write the following syntax (in Fig. B-9):

$$=NPV(C2,D6:H6)+C6$$

- The first part of the summation is the NPV function; write "=NPV(" select the discount rate (in cell C2), write a comma, and then select the cash flows starting from year 1 to 6 (cells D6 to I6) and close the statement with ")."

- Include the initial outlet by manually writing "+" and select the cash flow at year 0 (in cell C6).

| | A | B | C | D | E | F | G | H | I |
|---|---|---|---|---|---|---|---|---|---|
| 1 | | | | | | | | | |
| 2 | | Discount rate | 20% | | | | | | |
| 3 | | | | | | | | | |
| 4 | | | | | | Year | | | |
| 5 | | | 0 | 1 | 2 | 3 | 4 | 5 | 6 |
| 6 | | Cash Flow | -3000 | -1500 | 500 | 500 | 500 | 2000 | 3000 |
| 7 | | | | | | | | | |
| 8 | | Net Present Value | $(1,563.85) | =NPV(C2,D6:I6)+C6 | | | | | |

**Fig. B-9**

The net present value is −$1,563.85 (shown in Fig. B-9). Note that in MS Excel™, the negative values for currencies are represented in parenthesis "( )." Because $NPV < 0$, the project must be rejected.

You can verify this result by computing the NPV manually, as explained in Problem 14.22.

# Additional Practice Problems

# Chapter 1

## THEORETICAL CONCEPTS

**S1.1.** Consider the expression $\sqrt[3]{ax^2}$. Assuming the radicand $a$ is a positive real number, the radical can be expressed as an exponential term. Is this true or false? Why?

**S1.2.** When a polynomial has only two terms, it is called _____ (monomial/binomial).

**S1.3.** Given the polynomial $5x^{0.5}y + 3x^{-2}y^4 - 4z$, what is its degree?

    *a.*  1.5                                 *c.*  1

    *b.*  2                                   *d.*  4

**S1.4.** The term $x^{-1/2}$ can also be expressed as $\sqrt{1/x}$. Is this true or false?

**S1.5.** To add two binomials, the coefficients of the monomials can be different, but the variables and all exponents of these variables must be the same. _____ True _____ False

## EXPONENTS AND FRACTIONS

Use the rules of exponents to simplify these fractions:

**S1.6.**

$$\frac{x^3 \cdot x^8}{x^7} = x^{\boxed{\phantom{xxx}}} = x^{\boxed{\phantom{x}}}$$

**S1.7.**

$$\frac{x^{1/2} \cdot x^{-1/3}}{x^{1/6}} = x^{\boxed{\phantom{xxx}}} = x^{\boxed{\phantom{x}}} = 1$$

## POLYNOMIALS

Multiply these polynomials:

**S1.8.**

$$(4x - 2y)(4x^2 + 3xy + 1)$$

**S1.9.**

$$(3x - 7y)(3 + 6/y)$$

## FACTORING

**S1.10.** Factor the following polynomial and express it as two factors:

$$2x^2 - 5x - 12$$

## ORDER OF MATHEMATICAL OPERATIONS

**S1.11.** Simplify the following operations and express them in exponential terms:

$$\sqrt{\left[\frac{20x^5y^{-3}}{10x^2y^{-1}}\right] \cdot 2z^{\left[3\left(\frac{4}{3}+\frac{3}{4}\right)\right]}}$$

# Chapter 2

## THEORETICAL CONCEPTS

**S2.1.** A constant $k$ can be added on both sides of the equation without affecting the solution of the equation. Is this true or false? Why?

**S2.2.** In the fourth quadrant, the values of $x$ are _____ (positive/negative), whereas $y$ is _____ (positive/negative).

**S2.3.** Which of the following represents the steepest positive slope?

    a.   −2.5                          c.   1.25

    b.   7.5                             d.   −6.3

**S2.4.** In a linear cost function, the intercept represents the _____ (variable/fixed) cost.

**S2.5.** In a linear depreciation, the slope of the equation is negative, because the older the object (larger $x$), the less remaining value $y$.     _____ True   _____ False

## EQUATIONS

Solve the following linear equations by using properties of equality:

**S2.6.**

$$4x + 2 = 2(x-1) + 10$$

$$\boxed{\phantom{xx}}\,x = \boxed{\phantom{xx}}$$

$$x = \boxed{\phantom{xx}}$$

**S2.7.**

$$(1/4)x - 2 = (1/5)x + 1$$

$$\boxed{\phantom{xx}}\,x = \boxed{\phantom{xx}}$$

$$x = \boxed{\phantom{xx}}$$

**S2.8.**

$$\frac{4}{x-2} = \frac{3}{2x}$$

$$\boxed{\phantom{xx}}\,x = \boxed{\phantom{xxxx}}$$

$$x = \boxed{\phantom{xx}}$$

## SLOPES AND INTERCEPTS

**S2.9.** Find the slope of

$$5x + 2y = 20$$

**S2.10.** Find the $x$- and $y$-intercepts of the following equation:

$$4x - 2y = 10$$

**S2.11.** Find the slope-intercept form of

$$3x + 7y = 21$$

## APPLICATIONS OF LINEAR EQUATIONS

**S2.12.** An agricultural firm operates in pure competition, receiving $3 for each bag of potatoes sold. It has a variable cost of $1.2 per item and a fixed cost of $2,000. What is the number of bags the farmer must sell to break even (i.e., profit is neither negative nor positive)?

**S2.13.** Ernesto has a budget of $50 that can be spent for lunch every week. He can buy either spaghetti or salad. The price of spaghetti is $8 per meal and the price of salad is $4 per serving.

   I.   Consider $x$ as the number of spaghetti meals and $y$ as servings of salad per week. Write Ernesto's budget equation in standard form.

$$\boxed{\phantom{00}}\,x+\boxed{\phantom{00}}\,y=\boxed{\phantom{00}}$$

   II.  Graph the budget line for Ernesto's lunch.

# Chapter 3

## THEORETICAL CONCEPTS

**S3.1.** A cubic function may not include a quadratic term in the polynomial.          _____ True _____ False

**S3.2.** The coefficient $a$ of the profit function $\pi = ax^2 + bx + c$ is _____ (positive/negative). Thus, the profit function $\pi$ has a _____ (maximum/minimum) point.

**S3.3.** The vertex of the function $y = x^2 - 4x + 1$ is

   (*a*)  (2, −3)                                      (*c*)  (4, 13)

   (*b*)  (−2, 13)                                     (*d*)  (−3, 2)

**S3.4.** The vertex represents the _____ (maximum/minimum) point for an inverted U–shaped parabola.

**S3.5.** The rule of the function may match different values of $x$ to the same value of $f(x)$.

_____ True _____ False

## THE ALGEBRA OF FUNCTIONS

**S3.6.** Given $F(x) = x^2$ and $G(x) = x - \dfrac{1}{x}$, find $G(F(x))$ and $F(G(x))$

$$G(F(x)) = \boxed{\phantom{xxxxxx}} \qquad F(G(x)) = \boxed{\phantom{xxxxxx}}$$

**S3.7.** Given $F(x) = 2x/(x-2)$ and $G(x) = x^2 - 4$, find $(F \cdot G)(x)$

$$(F \cdot G)(x) = \boxed{\phantom{xxxxxx}}$$

**S3.8.** Given $F(x) = 4/(x-1)$ and $G(x) = 2/(x^2 - 1)$, find $(F - G)(x)$

$$(F - G)(x) = \boxed{\phantom{xxxxxx}}$$

## SOLVING QUADRATIC FUNCTIONS

**S3.9.** Graph the quadratic function indicating its vertex and intercepts:

$$F(x) = x^2 - 3x + 2$$

(a)  Here, $a = \boxed{\phantom{xx}}$, $b = \boxed{\phantom{xx}}$, and $c = \boxed{\phantom{xx}}$. With $a$ ___ 0, the parabola opens ___ (up/down).

(b)  The coordinates of the vertex are ($\boxed{\phantom{xx}}$, $\boxed{\phantom{xx}}$).

(c)  The $x$-coordinates of the $x$-intercepts are found by setting the equation equal to zero. The $x$-intercepts are ($\boxed{\phantom{xx}}$, $\boxed{\phantom{xx}}$) and ($\boxed{\phantom{xx}}$, $\boxed{\phantom{xx}}$).

(d)  The $y$-coordinate of the $y$-intercept is found by setting $\boxed{\phantom{xx}}$ = 0. The $y$-intercept is ($\boxed{\phantom{xx}}$, $\boxed{\phantom{xx}}$).

(e)  The graph of the function $F(x) = x^2 - 3x + 2$ is:

**NONLINEAR FUNCTIONS IN BUSINESS, ECONOMICS, AND FINANCE**

**S3.10.** Given the total revenue $TR = 90x - x^2$ and total cost $TC = 60x + 120$,

(a) Find the profit function $\pi$

$$\pi = \boxed{\phantom{xxxxxxxxxxxxxx}}$$

(b) Determine the maximum level of profit by finding the vertex of the parabola.

The coordinates of the vertex are ( $\boxed{\phantom{xxx}}$ , $\boxed{\phantom{xxx}}$ ).

(c) Compute the break-even points of the profit function by finding the $x$-intercepts.

The break-even points are ( $\boxed{\phantom{xxx}}$ , $\boxed{\phantom{xxx}}$ ) and ( $\boxed{\phantom{xxx}}$ , $\boxed{\phantom{xxx}}$ ).

(d) Graph the profit function, indicating the $x$-intercepts and vertex.

# Chapter 4

**THEORETICAL CONCEPTS**

**S4.1.** The substitution and elimination methods should provide the same solution for any solvable system of simultaneous equations. _____ True _____ False

**S4.2.** A 4×3 system of equations has ___ equations and ___ variables. This system is _____ (exactly/under/over) constrained.

**S4.3.** The slope of the supply equation is _____; this means that a higher price of the good will _____ its quantity supplied.

(*a*) negative, increase           (*c*) negative, decrease

(*b*) positive, increase            (*d*) positive, decrease

**S4.4.** For linear cost and revenue functions, any quantity lower than the break-even point results in profits.

_____ True _____ False

## ELIMINATION AND SUBSTITUTION METHOD

Use the substitution and elimination methods to solve the following system of simultaneous equations and compare their results:

**S4.5.**
$$4x + y = 21 \tag{4.84}$$

$$3x - 2y = 13 \tag{4.85}$$

(*a*) Elimination method

1) Select _____ (*x/y*) to be eliminated and multiply equation _____ by $\boxed{\phantom{xx}}$ and _____ by $\boxed{\phantom{xx}}$.

$$\boxed{\phantom{xxxxxxxxxxxxxxxxxxxxxx}} \tag{4.86}$$

$$\boxed{\phantom{xxxxxxxxxxxxxxxxxxxxxx}} \tag{4.87}$$

2) Add (*4.86*) and (*4.87*) to eliminate _____ (*x/y*). Then, the value of _____ = $\boxed{\phantom{xx}}$.

3) Substituting the value found in step 2) in either (*4.84*) or (*4.85*), the value of ____ = $\boxed{\phantom{xx}}$.

(*b*) Substitution method

1) Solve one of the original equations (*4.84*) for *y*:

$$y = \boxed{\phantom{xxxxxxxxxxxxxxxx}} \tag{4.88}$$

2) Substituting the value *y* in (*4.88*) into the original equation (*4.85*), we find that $x = \boxed{\phantom{xx}}$.

3) Then substituting the value of *x* in (*4.84*) or (*4.85*), we get $y = \boxed{\phantom{xx}}$.

## SUPPLY AND DEMAND

**S4.6.** The market of a product is given by these linear equations:

$$\text{Supply: } -2P + 4Q = -12 \qquad \text{Demand: } 5P + 20Q = 150$$

where $P$ is price and $Q$ is quantity of the product.

I. Find the equilibrium price ($P^*$) and quantity ($Q^*$) sold in this market.

II. Graph the supply and demand of this market, setting price in the $y$-axis and quantity in the $x$-axis.

**BREAK-EVEN ANALYSIS**

**S4.7.** Given the revenue $R(x) = 90x - x^2$ and cost $C(x) = 60x + 120$,

(a) Find the break-even points $x^*$ by setting $R(x) = C(x)$

$$x^* = \boxed{\phantom{xxx}} \text{ and } \boxed{\phantom{xxx}}$$

(b) Graph the revenue $R(x)$ and cost $C(x)$ functions, indicating the points of intersection.

## INCOME DETERMINATION MODEL

**S4.8.** Find the equilibrium level of income $Y_e$, given

$$Y = C + I + (X - Z) \qquad C = C_0 + bY_d \qquad Z = Z_0 + zY_d$$
$$I = I_0 \qquad\qquad\qquad X = X_0 \qquad\qquad Y_d = Y - T$$
$$T = T_0 + tY$$

where $C_0 = 80$, $I_0 = 100$, $X_0 = 90$, $Z_0 = 50$, $T_0 = 10$, $b = 0.8$, $z = 0.1$, $t = 0.2$

(*a*) Find the reduced form of $Y_e$.

(*b*) Determine the numerical value of $Y_e$.

(*c*) Find the effect of the multiplier if an income tax T with a proportional component t is incorporated into the model.

The multiplier is changed from [ ] to [ ]. Numerically,

This _____ (reduces/increases) the size of the multiplier.

# Chapter 5

## THEORETICAL CONCEPTS

**S5.1.** In the addition of two matrices with the same dimension $A$ and $B$, the order the matrices in this addition does not affect the resultant matrix.      _____ True      _____ False

**S5.2.** $A$ is a (4×3) matrix and $B$ is a (3×5) matrix. The matrix multiplication $AB$ _____ (is/is not) conformable. The resultant $AB$ is a ( ___ × _____ ) matrix.

**S5.3.** The transpose of the matrix

$$C = \begin{bmatrix} 0 & 4 & 6 & 7 & 9 \\ 1 & 2 & 3 & 8 & 5 \\ 5 & 3 & 4 & 0 & 2 \end{bmatrix} \quad \text{is} \quad \boxed{\phantom{xxxxxxxxxxxxxxx}}.$$

**S5.4.** In Gaussian elimination, the system of linear equations is converted from augmented to matrix form.

_____ True      _____ False

## MATRIX OPERATIONS

Refer to the following matrices in answering the questions below:

$$A = \begin{bmatrix} 12 & 8 & 3 \\ 10 & 1 & 1 \end{bmatrix} \quad B = \begin{bmatrix} 2 & -10 & 5 \\ -11 & 24 & 4 \end{bmatrix} \quad C = \begin{bmatrix} 9 & 3 \\ 7 & 6 \end{bmatrix} \quad D = \begin{bmatrix} 10 & 0 \\ -1 & 1 \end{bmatrix}$$

**S5.5.** Compute $A + 2B$

**S5.6.** Compute $C - 3D$

**S5.7.** Compute $AB'$.

    *1)*   Find $B'$.

    *2)*   The dimensions of $AB'$ are ____ × ____.

    *3)*   The matrix $AB'$ is:

## GAUSSIAN ELIMINATION

Use Gaussian elimination methods to solve the following system of simultaneous equations:

**S5.8.**
$$4x + y = 21$$
$$3x - 2y = 13$$

**SUPPLY AND DEMAND**

**S5.9.** The market of a product is given by these linear equations:

$$\text{Supply: } -2P + 4Q = -12 \qquad \text{Demand: } 5P + 20Q = 150$$

where $P$ is price and $Q$ is quantity of the product. Find the equilibrium price ($P^*$) and quantity ($Q^*$) sold in this market by using Gaussian elimination methods.

*1)* Compute the matrix form.

*2)* Find the augmented matrix.

*3)* Proceed with the row operations.

*4)* The equilibrium price $P^*$ is _____ and quantity $Q^*$ is _____ .

# Chapter 6

## THEORETICAL CONCEPTS

**S6.1.** The $B$ matrix is nonsingular; therefore, $|B| = 0$. _____ True _____ False

**S6.2.** $A$ is a (4×4) nonsingular matrix; the multiplication $AA^{-1}$ results in a(n) _____.

    (*a*) singular matrix            (*c*) identity matrix

    (*b*) matrix of (1×1) dimension     (*d*) non determined matrix

**S6.3.** The substitution method to solve a system of linear equations provides the same solution as Cramer's rule. _____ True _____ False

**S6.4.** In order to solve a (3×3) system of equations by Cramer's rule (in matrix form $AX = B$), three supplementary matrices must be computed from the original $A$ matrix. _____ True _____ False

## MATRIX DETERMINANTS

Refer to the following matrices in answering the questions below:

$$A = \begin{bmatrix} 9 & 3 \\ 7 & 6 \end{bmatrix} \qquad B = \begin{bmatrix} 1 & 4 & 0 \\ 5 & 2 & 8 \\ 7 & 6 & 3 \end{bmatrix}$$

**S6.5.** Compute $|A|$ and state if matrix $A$ is singular.

**S6.6.** Compute $|B|$ and state if the matrix $B$ is singular.

## SOLVING SYSTEMS OF LINEAR EQUATIONS

Given the following system of simultaneous equations,

$$4x + y = 21$$

$$3x - 2y = 13$$

**S6.7.** Solve the system using inverse matrices.

(a) Find the inverse matrix $A^{-1}$

    1. Sett up the augmented matrix.

    2. Use Gaussian elimination method to obtain the inverse.

        2a. Multiply row 1 by ☐ to obtain 1 in the $a_{11}$ position.

        2b. Subtract ☐ times row 1 from row 2 to clear column 1.

        2c. Multiply row 2 by ☐ to obtain 1 in the $a_{22}$ position.

        2d. Subtract ☐ times row 2 from row 1 to clear column 2.

    3. The identity matrix is now on the left-hand side of the bar.

$$A^{-1} = \boxed{\phantom{xxxxxxxxxxxxx}}$$

(b)  Solve for the unknown variables using the inverse matrix.

**S6.8.**  Solve the system using Cramer's rule.

*1)*  Express the equations in matrix form.

*2)*  Find the determinant of $A$.

*3)*  To solve for $x$, replace column 1 of $A$, which contains the coefficients of $x$, with the column vector of constants $B$, forming a new matrix $A_1$.

Find the determinant of $A_1$.

Use Cramer's rule and find $x$.

4) To solve for $y$, replace column 2 of $A$, which contains the coefficients of $y$, with the column vector of constants $B$, forming a new matrix $A_2$.

Find the determinant of $A_2$.

Use Cramer's rule and find $y$.

## SOLVING SYSTEMS OF LINEAR EQUATIONS

**S6.9.** Solve the following system of equations using Cramer's rule.

$$x_1 + 2x_2 = 8$$
$$5x_1 + 4x_3 = 14$$
$$3x_1 + 2x_2 + x_3 = 13$$

1) Express the equations in matrix form.

*2)* Find the determinant of *A*.

*3)* Solve for $x_1$.

*4)* Solve for $x_2$.

*5)* Solve for $x_3$.

## BUSINESS AND ECONOMIC APPLICATIONS: SUPPLY AND DEMAND

**S6.10.** The market of a product is given by these linear equations:

$$\text{Supply: } -2P + 4Q = -12 \qquad \text{Demand: } 5P + 20Q = 150$$

where $P$ is price and $Q$ is quantity of the product. Find the equilibrium price ($P^*$) and quantity ($Q^*$) sold in this market by using Cramer's rule.

*1)* Express the equations in matrix form.

*2)* Find the determinant of $A$.

*3)* To solve for $P$, replace column 1 of $A$, which contains the coefficients of $P$, with the column vector of constants $B$, forming a new matrix $A_1$.

Find the determinant of $A_1$.

Use Cramer's rule and find $P^*$.

*4)* To solve for $Q$, replace column 2 of $A$, which contains the coefficients of $Q$, with the column vector of constants $B$, forming a new matrix $A_2$.

Find the determinant of $A_2$.

<br><br><br><br>

Use Cramer's rule and find $Q^*$.

<br><br><br><br>

5) The equilibrium price $P^*$ is [      ] and quantity $Q^*$ is [      ].

# Chapter 7

## THEORETICAL CONCEPTS

**S7.1.** Assume that the constrained function can be solved by linear programming and has a unique solution. The solution will be situated in one of the corners of the feasible regions according to the _____ theorem.

**S7.2.** The graph of an equality constraint requires the inequality to be in _____ (standard/slope-intercept) form.

**S7.3.** A "greater than or equal to" inequality may be converted to an equation by adding a _____ (slack/surplus) variable.

**S7.4.** In economics, the decision variables are usually nonnegative numbers. _____ True _____ False

**S7.5.** A linear programming problem has 5 equations and 7 variables (two decision variables $x_1$ and $x_2$, and five slack variables). The calculations necessary to determine the total number of basic solutions is:

(a)  21

(b)  32

(c)  14

(d)  35

## OPTIMIZATION USING GRAPHS

Use graphs to solve the following linear programming problems.

**S7.6.** Maximize $\pi = 12x_1 + 10x_2$

subject to

$$2x_1 + 5x_2 \leq 90 \quad \text{(Constraint } A\text{)}$$
$$14x_1 + 7x_2 \leq 210 \quad \text{(Constraint } B\text{)}$$
$$8x_1 + 10x_2 \leq 200 \quad \text{(Constraint } C\text{)}$$
$$x_1, x_2 \geq 0$$

(1)   For all inequalities constraints, first solve each $x_2$ in terms of $x_1$:

      *1a.*  For constraint $A$:          $x_2 = \boxed{\phantom{xx}} x_1 + 18$

      *1b.*  For constraint $B$:          $x_2 = \quad -2\,x_1 + \boxed{\phantom{xx}}$

      *1c.*  For constraint $C$:          $x_2 = \boxed{\phantom{xx}} x_1 + \boxed{\phantom{xx}}$

The nonnegativity constrains set limit the analysis to the first quadrant.

(2)   Graph all inequalities constraints.

(3)   For the objective function, solve $x_2$ in terms of $x_1$:

      $x_2 = \boxed{\phantom{xx}} x_1 + \pi/10$

The slope is $\boxed{\phantom{xx}}$.

*(4)* Graph the feasible region (be sure to label all intersections) and indicate the point of tangency with the objective function (solution).

The solution is: $x_1^* = \boxed{\phantom{00}}$, $x_2^* = \boxed{\phantom{00}}$, $\pi^* = \boxed{\phantom{00}}$.

**S7.7.** Minimize

$$\pi = 8x_1 + 5x_2$$

subject to

$$4x_1 + 3x_2 \geq 120 \qquad \text{(Constraint } A)$$

$$2x_1 + 3x_2 \geq 80 \qquad \text{(Constraint } B)$$

$$2x_1 + \phantom{3}x_2 \geq 50 \qquad \text{(Constraint } C)$$

$$x_1, x_2 \geq 0$$

*(1)* For all inequalities constraints, first solve each $x_2$ in terms of $x_1$.

*1a.* For constraint $A$: $\quad x_2 = \boxed{\phantom{00}}\; x_1 + 40$

*1b.* For constraint $B$: $\quad x_2 = \boxed{\phantom{00}}\; x_1 + \boxed{\phantom{00}}$

*1c.* For constraint $C$: $\quad x_2 = \quad -2\,x_1 + \boxed{\phantom{00}}$

The nonnegativity constraints set the limit of the analysis to the first quadrant.

(2) Graph all inequalities constraints.

(3) For the objective function, solve $x_2$ in terms of $x_1$.

$$x_2 = \boxed{\phantom{xx}} x_1 + \pi/5$$

The slope is $\boxed{\phantom{xx}}$.

(4) Graph the feasible region (be sure to label all intersections) and indicate the point of tangency with the objective function (solution).

The solution is: $x_1^* = \boxed{\phantom{xx}}$, $x_2^* = \boxed{\phantom{xx}}$, $\pi^* = \boxed{\phantom{xx}}$.

## REPRESENTATION OF BUSINESS AND ECONOMICS PROBLEMS

**S7.8.** A farmer is deciding whether to plant corn and/or soybeans. He has 40 acres of land, capital of $27,000 on hand, and counts on 320 hours of labor throughout the season devoted to each acre. Each acre of corn requires an investment of $900 and 12 hours of labor but provides a net revenue of $640. An acre of soybeans needs $540 of investment and 4 hours of labor and offers a net return of $360. Express the data in equations and inequalities suitable to determine the maximum profit planting corn and/or soybeans.

*(1)*   Represent the data as a linear programming problem and fill the missing values.

Maximize $\pi = \boxed{\phantom{00}} x_1 + 360x_2$

subject to $\boxed{\phantom{00}} x_1 + \boxed{\phantom{00}} x_2 \leq \boxed{\phantom{00}}$   (Land constraint)

$900\, x_1 + \boxed{\phantom{00}} x_2 \leq \boxed{\phantom{00}}$   (Capital constraint)

$\boxed{\phantom{00}} x_1 + \boxed{\phantom{00}} x_2 \leq 320$   (Labor constraint)

$x_1, x_2 \geq 0$

*(2)*   Graph the feasible region and determine the solution.

The solution is: corn = $\boxed{\phantom{00}}$ acres, soybeans = $\boxed{\phantom{00}}$ acres, profit = $\boxed{\phantom{0000}}$.

# Chapter 8

## THEORETICAL CONCEPTS

**S8.1.** The computational method to solve a linear programming model is called the _____ algorithm.

**S8.2.** The displacement ratio is found by multiplying the elements of the constant column by the elements of the pivot column. ____ True ____ False

**S8.3.** The dual of a maximization problem is a _____ (maximization/minimization) problem.

**S8.4.** If the second decision variable $y_2$ of the dual is zero ($y_2 = 0$), what is the value of the slack variable of the second constraint of the primal form of a maximization linear programming problem?

(*a*) Zero                  (*c*) Lower than 0

(*b*) Greater than 0         (*d*) Need more information

**S8.5.** In zero-one programming, the decision variables may only be integer values of 0 or 1.

                                                       ____ True ____ False

## OPTIMIZATION USING THE SIMPLEX ALGORITHM

**S8.6.** Maximize subject to

$$\pi = 12x_1 + 10x_2$$
$$2x_1 + 5x_2 \le 90 \qquad \text{(Constraint } A)$$
$$14x_1 + 7x_2 \le 210 \qquad \text{(Constraint } B)$$
$$8x_1 + 10x_2 \le 200 \qquad \text{(Constraint } C)$$
$$x_1, x_2 \ge 0$$

(*1*) Construct the initial simplex tableau.

    *1a.* Add slack variables to the inequalities to make them equations.

    *1b.* Express the constraints in matrix form.

*1c.* Form the initial tableau.

> [blank box]

The first basic solution is: $x_1 = $ [ ] , $x_2 = $ [ ] , $s_1 = $ [ ] , $s_2 = $ [ ] , $s_3 = $ [ ] , $\pi = $ [ ] .

(2) Change the basis. The negative indicator with the largest absolute value determines the pivot column, in this case _____. The displacement ratios (in order) are ___, _____, and _____. The smallest displacement ratio decides the pivot row. Thus, [ ] becomes the pivot element.

(3) Pivot.

*3a.* Multiply the pivot row by the reciprocal of the pivot element; in other words, multiply row ___ of the initial tableau by [ ] .

> [blank box]

*3b.* Having reduced the pivot element to 1, clear the pivot column.

> [blank box]

The second basic solution is: $x_1 = $ [ ] , $x_2 = $ [ ] , $s_1 = $ [ ] , $s_2 = $ [ ] , $s_3 = $ [ ] , $\pi = $ [ ] .

*3c.* Since _____ in the second column is the only negative indicator, variable _____ is brought to the basis, and column _____ becomes the pivot column. The displacement ratios (in order) are ___, _____, and _____. Thus, [ ] becomes the pivot element. Multiply the pivot row by [ ] .

> [blank box]

*3d.* Having reduced the pivot element to 1, clear the pivot column.

> [blank box]

The solution is: $x_1^* = $ [ ] , $x_2^* = $ [ ] , $\pi^* = $ [ ] .

## THE DUAL

Find the dual of the following problems:

**S8.7.**   Maximize subject to

$$\pi = 12x_1 + 10x_2$$
$$2x_1 + 5x_2 \le 90 \qquad \text{(Constraint A)}$$
$$14x_1 + 7x_2 \le 210 \qquad \text{(Constraint B)}$$
$$8x_1 + 10x_2 \le 200 \qquad \text{(Constraint C)}$$
$$x_1, x_2 \ge 0$$

**S8.8.**   Minimize subject to

$$c = 8x_1 + 5x_2$$
$$4x_1 + 3x_2 \ge 120 \qquad \text{(Constraint A)}$$
$$2x_1 + 3x_2 \ge 80 \qquad \text{(Constraint B)}$$
$$2x_1 + x_2 \ge 50 \qquad \text{(Constraint C)}$$
$$x_1, x_2 \ge 0$$

## OPTIMIZATION WITH THE DUAL

**S8.9.**   Answer the questions below based on the following linear programming problem.

Minimize subject to

$$c = 24y_1 + 60y_2 + 90y_3$$
$$y_1 + 2y_2 + 3y_3 \ge 1 \qquad \text{(Constraint I)}$$
$$y_1 + 5y_2 + 4y_3 \ge 2 \qquad \text{(Constraint II)}$$
$$y_1, y_2, y_3 \ge 0$$

(*1*)   Find the dual of this model and label all decision variables of the dual with the *x* notation followed by subscripts and $\pi$ as the dual optimal variable.

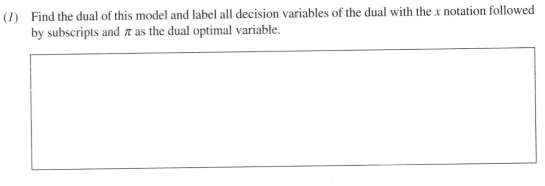

(2) Solve the dual graphically and indicate the value of $\pi^*$.

(3) Write both the primal and the dual including slack (label them as $s$) and surplus variables (label them as $t$).

(4) Find the value of the slack variables ($s_1$, $s_2$, and $s_3$) using the dual equation system.

Which constraints are active in the dual?

Constraint $A$ is _____ (active/not active). The value of the slack/surplus variable = ___.

Constraint $B$ is _____ (active/not active). The value of the slack/surplus variable = ___.

Constraint $C$ is _____ (active/not active). The value of the slack/surplus variable = ___.

(5) Which constraints are active in the dual?

Constraint $I$ is _____ (active/not active). The value of the surplus variable ___ (is/is not) equal to zero.

Constraint $II$ is _____ (active/not active). The value of the surplus variable ___ (is/is not) equal to zero.

(6) Using the information in (4) and (5), solve the primal system of equations in part (3).

The solution is: $y_1^* = \square$, $y_2^* = \square$, $y_3^* = \square$, $c^* = \square$.

# Chapter 9

## THEORETICAL CONCEPTS

**S9.1.** Graphically, the derivative is the line that is tangent to the function at a specific point.

_____ True    _____ False

**S9.2.** The derivative of the function is the infinitesimal change of that function with respect to one of its independent variables.

_____ True    _____ False

**S9.3.** The derivative of a quadratic function is a _____ (constant/linear) function.

**S9.4.** The technique used to take the derivative of a composite function is called the _____ rule.

(*a*)   quotient                      (*c*)   chain

(*b*)   product                       (*d*)   linear function

**S9.5.** The third-order derivative can be obtained by taking twice the derivative of the function.

_____ True    _____ False

## LIMITS AND DIFFERENTIATION

**S9.6.** Find $\lim_{x \to 5} f(x)$, if it exists, given

$$f(x) = \frac{2x - 10}{x^2 - 25} \qquad\qquad (x \neq 5)$$

(*1*)   Plug values of *x* in the function, fill out the table (use three decimal points), and graph it.

| x | f(x) |
|-----|------|
| 3 | |
| 4 | |
| 4.5 | |
| 4.8 | |
| 4.9 | |
| 5.1 | |
| 5.2 | |
| 5.5 | |
| 6 | |

(2)   Solve the function algebraically; be careful when simplifying terms in the denominator.

<br>
<br>
<br>
<br>
<br>

**S9.7.**   Using limits, find the derivative of $f(x) = x^2 - 2x$

(1).   Employ the general formula of the derivative and replace the function.

$$f'(x) = \lim_{\Delta x \to 0} \frac{f(x + \Delta x) - f(x)}{\Delta x} = \boxed{\phantom{xxxxxxxxxxxxxxxxxxxx}}$$

(2)   Simplify the results.

<br>
<br>
<br>

(3)   Divide through by $\Delta x$.

<br>
<br>
<br>

(4)   Take the limit of the simplified expression.

$$f'(x) = \boxed{\phantom{xxxxxxxxxxxxxxxxxx}}$$

## DERIVATIVE RULES

Find the first derivative of the following functions:

**S9.8.**   Use the basic rules to differentiate $f(x) = 12x^{1/3} - 10x^{-1/5} + 8$

$$f'(x) = \boxed{\phantom{xxxxxxxxxxxxxxxxxx}}$$

**S9.9.**  Use the product rule to differentiate  $f(x) = (9x^{1/3} + 4)(10x^{-1/5} + 2x)$

$$f'(x) = \boxed{\phantom{xxxxxxxx}} \cdot \boxed{\phantom{xxxxxxxx}} + \boxed{\phantom{xxxxxxxx}} \cdot \boxed{\phantom{xxxxxxxx}}$$

$\quad\qquad\;\; g(x) \qquad\qquad\; h'(x) \qquad\qquad\; g'(x) \qquad\qquad\; h(x)$

Simplify algebraically.

$$f'(x) = \boxed{\phantom{xxxxxxxxxxxxxxxxx}}$$

**S9.10.** Use the quotient rule to differentiate  $f(x) = \dfrac{4x^{5/4}}{5x - 1}$

$$f'(x) = \boxed{\phantom{xxxxxxxx}} \cdot \boxed{\phantom{xxxxxxxx}} + \boxed{\phantom{xxxxxxxx}} \cdot \boxed{\phantom{xxxxxxxx}}$$

$\qquad\qquad g(x) \qquad\qquad\; h'(x) \qquad\qquad\; g'(x) \qquad\qquad\; h(x)$

$$f'(x) = \; -\; \dfrac{\boxed{\phantom{xxxxxxxxxxxxxxxxx}}}{\boxed{\phantom{xxxxxxxxxxx}}}$$

$\qquad\qquad\qquad\qquad\qquad\qquad [h(x)]^2$

Simplify algebraically.

$$f'(x) = \boxed{\phantom{xxxxxxxxxxxxxxxx}}$$

**S9.11.** Use the generalized power rule to differentiate  $f(x) = 1/\sqrt[3]{4x^2 - 5}$

Define the function in exponential terms.

$$f(x) = (4x^2 - 5)^{\boxed{\phantom{x}}}$$

Take the derivative with respect to $x$.

$$f'(x) = \boxed{\phantom{x}} \cdot \boxed{\phantom{xxxxxxxx}} \cdot \boxed{\phantom{xxxxxxxx}}$$

$\qquad\quad\; n \qquad\quad [g(x)]^{n-1} \qquad\qquad g'(x)$

Simplify algebraically.

$$f'(x) = \boxed{\phantom{xxxxxxxx}}$$

**S9.12.** Use the chain rule to differentiate  $f(x) = (6x^{1/2} + 9x)^8$

Let $\quad y = \boxed{\phantom{xxxxxxxxxxxxx}}$ and $\quad u = \boxed{\phantom{xxxxxxxxxxx}}$

Then $dy/du = \boxed{\phantom{xxxxxxxxxxxxx}}$ and $du/dx = \boxed{\phantom{xxxxxxxxxxx}}$

Substitute in the formula.

$$\frac{dy}{dx} = \boxed{\phantom{xxxxxxx}} \cdot \boxed{\phantom{xxxxxxx}}$$

Replace $u$ in terms of $x$.

$$\frac{dy}{dx} = \boxed{\phantom{xxxxxxxxx}}$$

## HIGHER-ORDER DERIVATIVE

**S9.13.** Given the function $f(x) = (x^2 + 8)^3$, find the second-order derivative.

(1)   Take the first derivative of the function.

(2)   Take the derivative of the first-order derivative to obtain the second-order derivative.

**S9.14.** Given the function $f(x) = x^{-1/4} + x^{1/4}$, find the third-order derivative.

(1)   Take the first derivative of the function.

(2)   Find the second derivative of the function.

(3)   Compute the derivative of the second-order derivative to obtain the third-order derivative.

# Chapter 10

## THEORETICAL CONCEPTS

**S10.1.** A differentiable function $f(x)$ whose second-order derivative evaluated at $x = 5$ is negative can be classified as a(n) _____ function.

    (*a*)   convex                      (*c*)   increasing

    (*b*)   concave                    (*d*)   decreasing

**S10.2.** A point classified as relative extrema can be increasing at that point, but not decreasing.

                                                     ____ True     ____ False

**S10.3.** The inflection point is where the _____ (first/second) derivative of the function equals zero.

**S10.4.** A function is increasing if the first derivative of the function evaluated at $x = a$ _____ zero.

    (*a*)   is greater than       (*b*)   is lower than       (*c*)   equals

**S10.5.** The optimal value of a function can be found by taking the first-order condition.

                                                     ____ True     ____ False

## INCREASING AND DECREASING FUNCTIONS

**S10.6.** Determine whether the function $f(x) = 7x^4 + 3x^2 - x - 1000$ is decreasing or decreasing at the point $x = 3$

    (*1*)   Take the first order derivative.

    (*2*)   Substitute $x = 3$ and classify the point.

## CONCAVITY AND CONVEXITY

**S10.7.** Determine whether the function $y = 5x^5 + 3x^4 - x^2 - 200$ is convex or concave at the point $x = 2$.

    (*1*)   Find the second-order derivative.

(2)   Substitute $x = 2$ and classify the point.

<br><br><br><br>

## SKETCHING THE GRAPHS: RELATIVE EXTREMA AND INFLECTION POINT

**S10.8.**   Graph the following function:  $y = 0.25x^4 - 3x^3 + 10x^2 + 90$

(1)   Find the $y$-intercept.

<br><br><br><br>

(2)   Find the critical values (relative extrema) and classify them as minima or maxima points.

<br><br><br><br><br><br>

(3)   Find the inflection points and indicate if they are decreasing or increasing.

<br><br><br><br><br><br>

(4)   Find the $y$-values for each critical and inflection point.

<br><br><br><br><br><br>

(5)  Using all previous information, draw the curve.

Note: Graph $x$ in the range between $-2$ (where $y = 158$) and $7$ (with $y = 151.25$).

## SKETCHING THE GRAPHS: MARGINAL AND AVERAGE FUNCTIONS IN ECONOMICS

**S10.9.**  Given the total cost function $TC = 2Q^3 - 12Q^2 + 225Q$, create a graph with two panels. (*I*) The first one sketches the total cost curve, indicating the inflection point, and (*II*) the second panel depicts the marginal and average cost curves, indicating their point of intersection and the minimum point of the MC curve.

(1)  Take the first and second derivative of the total cost function.

(2)  Check for (*a*) inflection points and (*b*) concavity, using the second derivative.

(3)  Find the average cost functions and the relative extrema.

(*4*)  Find the marginal cost functions and the relative extrema.

(*5*)  Verify the point of intersection between the average and the marginal cost functions (note $Q > 0$).

(*6*)  Sketch the graph.

## OPTIMIZATION IN BUSINESS AND ECONOMICS

**S10.10.** Find the critical points at which profit $\pi$ is maximized given the total revenue $TR = 470Q - 3Q^2$ and total cost $TC = 32Q + 10,500$

(*1*)  Compute marginal revenue (*MR*) and marginal cost (*MC*) functions.

(*2*)  Equate $MR = MC$ to find $Q^*$.

(*3*)  Verify that $Q^*$ is a relative maximum point.

(*4*)  Compute the maximum profit level $\pi^*$ by establishing $\pi^* = \pi(Q^*)$.

# Chapter 11

## THEORETICAL CONCEPTS

**S11.1.**  The base of the natural logarithm function is *e*.  ____ True  ____ False

**S11.2.**  Any exponential function raised to zero will be equal to one, independently of its base.

____ True  ____ False

**S11.3.**  Logarithmic functions cannot be negative.  ____ True  ____ False

**S11.4.**  Continuous interest compounding can be solved using natural exponential functions.

____ True  ____ False

**S11.5.** The summation of two logarithmic functions can be combined by writing only one time its base and _____ (adding/multiplying) the terms in each logarithm.

## EXPONENTIAL-LOGARITHMIC CONVERSIONS

Change each of the following functions to its corresponding inverse function.

**S11.6.** Convert $y = e^{5x+8}$ into a logarithmic function by solving for $x$.

**S11.7.** Convert $y = \ln(3x - 2)$ into an exponential function by solving for $x$.

## PROPERTIES OF LOGARITHMIC AND EXPONENTIAL FUNCTIONS

Convert the expressions to sums, differences, or products.

**S11.8.** Convert $y = \ln(5x + 4) + \ln(4x - 9)$ into a product.

**S11.9.** Convert $z = \ln\left(\dfrac{4e^2x}{\sqrt[3]{y^2}}\right)$ into additions and subtractions.

**S11.10.** Simplify $y = e^{[\ln(x^2)+x]}$

**DERIVATIVES**

**S11.11.** Differentiate the logarithmic function $y = \ln\left(\dfrac{x^2}{x-2}\right)$

    *(1)*   Use logarithmic properties to convert the expression in subtraction of terms.

    *(2)*   Take the first-order derivative.

**S11.12.** Differentiate the natural exponential function $y = e^{(x^2+5)}$ by combining derivative rules.

**S11.13.** Differentiate the function

$$y = e^{[\ln(x-4)-\ln(x-2)]}$$

    *(1)*   Use logarithmic and exponential properties to convert the expression into a ratio of terms.

    *(2)*   Take the first-order derivative.

# Chapter 12

## THEORETICAL CONCEPTS

**S12.1.** The result for a definite integral is a real number. ____ True ____ False

**S12.2.** The integral of the marginal revenue is the _____

**S12.3.** Integration by parts is the inverse rule of the _____ rule.

(*a*) chain  (*c*) quotient

(*b*) product  (*d*) addition

**S12.4.** The antiderivative is the reverse process of differentiation. ____ True ____ False

**S12.5.** _____ (Definite/Indefinite) integrals may be interpreted as the area between the function and the ____ (*x/y*)-axis.

## DEFINITE AND INDEFINITE INTEGRALS

**S12.6.** Compute the indefinite integral

$$\int (5x^{1/2} - 2x^{-1/3})dx$$



**S12.7.** Find the antiderivative formula, given the boundary condition $F(1) = 8$ and the function:

$$f(x) = 18x^5 + 4x^{-3}$$

(*1*) Compute the indefinite integral.



(*2*) Use the initial condition to find the constant.



(*3*) The antiderivative formula is:

$$F(x) = \boxed{\phantom{xxxxxxxxxxxxx}}$$

**S12.8.**   Evaluate the definite integral:

$$\int_{0}^{3}(e^{5x}-2x^2)dx$$

## AREA BETWEEN CURVES

**S12.9.**   Find the area between $y_1 = 3x - 2$ and $y_2 = 2 - x^2$

(*1*)   Compute the points of intersection of both functions to find the range of $x$.

(*2*)   Graph both functions to know the order of the functions in the integration.

(*3*)   Evaluate the definite integral.

## INTEGRATION BY SUBSTITUTION AND BY PARTS

**S12.10.** Use integration by substitution to determine the following indefinite integral:

$$\int \frac{30x^4}{\sqrt{6x^5 - 36}}\,dx$$

(*1*) Select a suitable $u$ function.

$$u(x) = \boxed{\phantom{xxxxxxxxxxxxxxxxxx}}$$

(*2*) Compute $du$ and find $dx$ in terms of $du$.

$$du = \boxed{\phantom{xxxxxxxx}}\ dx$$

$$dx = \boxed{\phantom{xxxxxxxx}}\ du$$

(*3*) Substitute the relationship in the integration.

(*4*) Integrate.

(*5*) Substitute back to have the integration in terms of $x$.

**S12.11.** Use integration by parts to determine the following definite integral:

$$\int_0^7 30x\sqrt{9 + x}\,dx$$

(*1*) Pick suitable $f(x)$ and $g(x)$ functions.

$$f(x) = \boxed{\phantom{xxxxxxxxxxxx}} \qquad g'(x) = \boxed{\phantom{xxxxxxxxxxxx}}$$

(2) Compute $f'(x)$ and find $g(x)$ in terms of $dx$.

Differentiate. $\qquad\qquad$ $f'(x) =$ [                    ]

Integrate. $\qquad\qquad$ $g(x) =$ [                    ]

(3) Substitute the relationship in the integration and create an indefinite integral.

[                                                                    ]

(4) Integrate the indefinite integral.

[                                                                    ]

(5) Then, include the ranges of $x$ ($0 \le x \le 7$) to convert the integrand into a definite integral.

[                                                                    ]

(6) Compute the definite integral.

[                                                                    ]

## APPLICATION IN BUSINESS AND ECONOMICS

**S12.12.** Given the demand and supply functions:

$$\text{Demand: } P = -4Q + 30 \qquad\qquad \text{Supply: } P = 2Q + 6$$

Use integration to find the consumer surplus (CS) and producer surplus (PS).

(1) Find the price $P^*$ and quantity $Q^*$ at equilibrium.

[                                                                    ]

*(2)*   Graph the supply and demand, indicating the market equilibrium and the areas for CS and PS.

*(3)*   Use definite integrals to find CS and PS.

**S12.13.** Estimate the firm's additional total revenue received from increasing sales from 5 to 8 units, given the marginal revenue function $MR = 32 - 4x$.

*(1)*   Find the total revenue $TR$ function.

$$TR = \boxed{\phantom{xxxxxxxxxxxxxxxxxxxx}}$$

*(2)*   Calculate the increase in revenue within the range of sales.

**S12.14.** Find the profit function $\pi$ given the marginal revenue function $MR = 27 - x$, marginal cost function $MC = 2x$, and a fixed cost of 200.

*(1)*   Compute the total revenue $TR$ and $TC$ functions.

$$TR = \boxed{\phantom{xxxxxxxxxxxxxx}} \qquad TC = \boxed{\phantom{xxxxxxxxxxxxxxxx}}$$

*(2)*   Find the profit function $\pi$.

$$\pi = \boxed{\phantom{xxxxxxxxxxxxxxxx}}$$

# Chapter 13

## THEORETICAL CONCEPTS

**S13.1.**  Given a differentiable function $z = f(x, y)$, consider $f_{xx} \cdot f_{yy} > (f_{xy})^2$, $f_{xx} > 0$ and $f_{yy} > 0$ at the critical value $(x^*, y^*)$. This point can be classified as the_____ point.

    *a.*  inflection                     *c.*  relative minimum

    *b.*  relative maximum            *d.*  saddle

**S13.2.**  Young's theorem states that for a differentiable function $z = f(x, y)$, it must hold $f_{xy} = f_{yx}$

                                                  _____ True  _____ False

**S13.3.**  In constrained optimization, the Lagrange function combines the objective function with the restriction.

                                                    _____ True  _____ False

**S13.4.**  For a profit function $\pi = \pi(x, y)$ to be maximized at $(x^*, y^*)$, it requires the following conditions.

$$\pi_{xx} \;\rule{2em}{0.4pt}\; 0 \qquad \pi_{yy} \;\rule{2em}{0.4pt}\; 0 \qquad \pi_{xx}\,\pi_{yy} \;\rule{2em}{0.4pt}\; (\pi_{xy})^2$$

**S13.5.**  Given a differentiable function $z = f(x, y)$, when taking the partial derivative with respect to $x$, $y$ is allowed to vary.

                                                  _____ True  _____ False

## PARTIAL DERIVATIVES

Find the first-order partial derivative with respect to $x$.

**S13.6.**  Use the basic rules to differentiate $f(x, y) = 7x^4 y + 3x^2 y^{1/2} - 4y^2 - 1000$.

**S13.7.**  Use the product rule to differentiate $f(x) = (9x^{1/3} y + 2)(10x^{-1/5} + 2y)$.

**S13.8.** Use the quotient rule to differentiate $f(x) = \dfrac{4x^2}{5x - y}$

**S13.9.** Use the generalized power rule to differentiate $f(x) = (4x^2 - 5y + 8xy)^{-1/2}$.

**S13.10.** Use the natural exponential rule to differentiate $f(x) = e^{2x}y - 3x^2 e^y$.

**S13.11.** Use the natural logarithmic rule to differentiate $f(x) = \ln(x^{1/2}y^{1/2} + 3xy^{-3})$.

## SECOND-ORDER AND CROSS-PARTIAL DERIVATIVES

**S13.12.** Compute the second-order and cross-partial derivative with respect to $x$ and $y$, and verify the Young's Theorem for the function $z = 8x^{0.5}y^{0.7} + 2x^2y + 3xy^3$

　(1)　Find the first-order derivative.

*(2)*    Find the second-order derivative.

*(3)*    Find the cross-partial derivatives and verify Young's Theorem.

## OPTIMIZATION IN BUSINESS AND ECONOMICS

**S13.13.**   A monopolist produces two goods, $x$ and $y$. Find the critical value at which the firm's profit $\pi$ is maximized for which the demand functions are

$$P_x = 55 - 4x \text{ and } P_y = 45 - 2y$$

and total cost function is

$$C = x^2 + y^2 + 5xy + 60$$

*(1)*    Establish the profit function $\pi = TR - TC$ where $TR = P_x \cdot x + P_y \cdot y$

$$\pi = \boxed{\phantom{xxxxxxxxxxxxxxxxxxxx}}$$

*(2)*    Take the first-order condition for the profit function with respect to $x$ and $y$ and find $(x^*, y^*)$.

$$\pi_x = \boxed{\phantom{xxxxxxxxxxxxxxxxxx}}$$

$$\pi_y = \boxed{\phantom{xxxxxxxxxxxxxxxxxx}}$$

$$(x^*, y^*) = \boxed{\phantom{xxxxxxxxxxxx}}$$

*(3)*    Test the second-order condition to be sure of a maximum.

(4)  Find the optimal prices $P_x^*$, $P_y^*$ and profit $\pi^*$

<br>

## CONSTRAINED OPTIMIZATION

**S13.14.** Given the function

$$z = x^2 + 3xy + y^2 - x + 3y$$

The constraint is

$$x + y = 42$$

Find the optimal combination $(x^*, y^*)$ that optimizes the Lagrange function.

(1)  Establish the Lagrange function.

$$\mathcal{L} = \underline{\hspace{5cm}}$$

(2)  Take the first-order partial derivatives for optimization to find $(x^*, y^*)$:

$$\mathcal{L}_x = \underline{\hspace{5cm}}$$

$$\mathcal{L}_y = \underline{\hspace{5cm}}$$

$$\mathcal{L}_\lambda = \underline{\hspace{5cm}}$$

$$(x^*, y^*) = \underline{\hspace{3cm}}$$

<br>

## CONSTRAINED OPTIMIZATION IN BUSINESS AND ECONOMICS

**S13.15.** A manufacturer produces $q$ amount of a good using labor $L$ and capital $K$. The *Cobb-Douglas* production function of the good is

$$q = L^{2/3} K^{1/3}$$

The budget parameters are

$$q = L^{2/3}K^{1/3}$$

Find the optimal labor ($L^*$) and capital ($K^*$) needed to minimize its production cost given the budget.

*(1)*   Elaborate the budget constraint:

$$\boxed{\phantom{xx}}\,L + \boxed{\phantom{xx}}\,K = \boxed{\phantom{xx}}$$

*(2)*   Establish the Lagrange function.

$$\mathcal{L} = \boxed{\phantom{xxxxxxxxxxxxxxxxxxxxxxxxxx}}$$

*(3)*   Take the first-order partial derivatives for optimization to find ($L^*$, $K^*$):

$$\mathcal{L}_K = \boxed{\phantom{xxxxxxxxxxxxxxxxxxxxxxxxxxxxx}}$$

$$\mathcal{L}_L = \boxed{\phantom{xxxxxxxxxxxxxxxxxxxxxxxxxxxxx}}$$

$$\mathcal{L}_\lambda = \boxed{\phantom{xxxxxxxxxxxxxxxxxxxxxxxxxxxxx}}$$

$$(L^*, K^*) = \boxed{\phantom{xxxxxxxxxxxx}}$$

# Chapter 14

## THEORETICAL CONCEPTS

**S14.1.**   An infinite geometric series _____ (converges/diverges) if the absolute value of the common ratio is less than 1.

**S14.2.**   The standard deviation is the squared value of the variance.                    _____ True    _____ False

**S14.3.**   Which of the following projects must be accepted according to the NPV rule?

    *a.*   −2500                              *c.*   −1250

    *b.*   7500                               *d.*   0

**S14.4.**   An infinite sequence diverges if the sequence approaches a specific value as *n* increases.

                                                    _____ True    _____ False

**S14.5.**   The net present value is defined as the sum of all future cash flows at nominal value.

                                                    _____ True    _____ False

## SEQUENCES AND SUMMATION

**S14.6.** Given the sequence $\{x_n\} = \{1, 9, -4, 3, 8, -1\}$, compute

$$\sum_{i=2}^{3}(x_{i-1} + x_{5-i})$$

Describe each term with its respective index.

$$\sum_{i=2}^{3}(x_{i-1} + x_{5-i}) = (\boxed{\phantom{0}} + x_{5-2}) + (\boxed{\phantom{0}} + \boxed{\phantom{0}})$$
$$\underbrace{\phantom{( + x_{5-2})}}_{i=2} \underbrace{\phantom{( + )}}_{i=3}$$

$$\sum_{i=2}^{3}(x_{i-1} + x_{5-i}) = (\boxed{\phantom{0}} + x_3) + (\boxed{\phantom{0}} + \boxed{\phantom{0}})$$

Replace the terms with their corresponding values.

$$\sum_{i=2}^{3}(x_{i-1} + x_{5-i}) = (\boxed{\phantom{0}} + \boxed{\phantom{0}}) + (\boxed{\phantom{0}} + \boxed{\phantom{0}})$$

$$\sum_{i=2}^{3}(x_{i-1} + x_{5-i}) = \boxed{\phantom{0}}$$

**S14.7.** Given the 6×7 table below,

| 10 | 7 | 1 | 12 | 96 | 23 | -4 |
|----|----|----|----|----|----|----|
| -3 | 0 | 4 | 14 | 36 | 11 | 12 |
| 8 | -2 | 53 | 11 | 23 | 16 | 16 |
| 11 | -8 | 64 | -2 | 1 | 5 | -7 |
| 17 | 11 | 75 | 88 | -1 | -43 | -21 |
| 20 | 99 | 87 | 73 | 71 | 12 | 72 |

Compute the series given by the following rule:

$$\sum_{i=3}^{5}(x_{i4} + x_{2i})$$

Describe each term with its respective index.

$$\sum_{i=3}^{5}(x_{i4} + x_{2i}) = (\boxed{\phantom{0}} + \boxed{\phantom{0}}) + (\boxed{\phantom{0}} + \boxed{\phantom{0}}) + (\boxed{\phantom{0}} + \boxed{\phantom{0}})$$
$$\underbrace{\phantom{( + )}}_{i=3} \underbrace{\phantom{( + )}}_{i=4} \underbrace{\phantom{( + )}}_{i=5}$$

Replace the terms with their corresponding values.

$$\sum_{i=3}^{5}(x_{i4} + x_{2i}) = (\boxed{\phantom{x}} + \boxed{\phantom{x}}) + (\boxed{\phantom{x}} + \boxed{\phantom{x}}) + (\boxed{\phantom{x}} + \boxed{\phantom{x}})$$

$$\sum_{i=3}^{5}(x_{i4} + x_{2i}) = \boxed{\phantom{x}}$$

## PROPERTIES AND SPECIAL FORMULAS OF SUMMATIONS

**S14.8.**  Given the table below.

| $k$ | $x_k$ |
|-----|-------|
| 1 | 8 |
| 2 | 17 |
| 3 | -5 |
| 4 | 9 |
| 5 | -3 |

Find the value of

$$\sum_{k=1}^{5}(x_k + k)^2$$

Expand the polynomial in the summation.

$$\sum_{k=1}^{5}(x_k + k)^2 = \boxed{\phantom{xxx}} + \boxed{\phantom{xxx}} + \boxed{\phantom{xxx}}$$

Expand the terms.

$$\sum_{k=1}^{5}(x_k + k)^2 = \boxed{\phantom{xxxxxxx}} + \boxed{\phantom{xxxxxxx}} + \boxed{\phantom{xxxxxxx}}$$

Use summation properties and formulas.

$$\sum_{k=1}^{5}(x_k + k)^2 = \boxed{\phantom{xxxxxxx}} + \boxed{\phantom{xxxxxxx}} + \boxed{\phantom{xxxxxxx}}$$

$$\sum_{k=1}^{5}(x_k + k)^2 = \boxed{\phantom{xxxxxxx}}$$

## MEAN AND STANDARD DEVIATION

**S14.9.** Find the mean and standard deviation of the table used in S14.8.

(*1*) Compute the mean.

$$\bar{X} = \frac{1}{\Box} \sum_{i=1}^{5} x_i = \frac{\Box + \Box + \Box + \Box + \Box}{\Box} = \Box$$

(*2*) Find the variance.

$$S_x^2 = \frac{1}{\Box} \sum_{i=1}^{5} (x_i - \bar{X})^2 = \frac{\boxed{\phantom{xx}} + \boxed{\phantom{xx}} + \boxed{\phantom{xx}} + \boxed{\phantom{xx}} + \boxed{\phantom{xx}}}{\Box}$$

$$S_x^2 = \boxed{\phantom{xxx}}$$

(*3*) Compute the standard deviation.

$$S_x = \sqrt{S_x^2} = \boxed{\phantom{xxx}}$$

## INFINITE SERIES

**S14.10.** Determine if the geometric series converges or diverges. If it converges, find the value of

$$s_n = \sum_{i=3}^{\infty} 72\left(\frac{1}{6}\right)^i$$

Here, $r =$ _____ $\boxed{\phantom{xx}}$ 1. Therefore, the geometric series _____ (converges/diverges).

However, note that the index starts at $i = 3$.

$$\sum_{i=3}^{\infty} \left(\frac{1}{6}\right)^i = \boxed{\phantom{xx}} + \boxed{\phantom{xx}} + \boxed{\phantom{xx}} + \ldots$$

In order to use the geometric formula, complete the series by adding and subtracting the first two terms needed to have the index start at $i = 0$.

$$= \left[\boxed{\phantom{x}} + \boxed{\phantom{x}} + \boxed{\phantom{x}} + \ldots\right] + \left(\boxed{\phantom{x}} + \boxed{\phantom{x}} + \boxed{\phantom{x}}\right) - \left(\boxed{\phantom{x}} + \boxed{\phantom{x}} + \boxed{\phantom{x}}\right)$$

Reorganize terms.

$$\sum_{i=3}^{\infty} 72\left(\frac{1}{6}\right)^i = \boxed{\phantom{x}}\left[\boxed{\phantom{xxxxxx}}\right] - \boxed{\phantom{x}}\left(\boxed{\phantom{xxx}}\right)$$

Use the formula

$$\sum_{i=3}^{\infty} 72\left(\frac{1}{6}\right)^i = \boxed{\phantom{x}} \left[\frac{1}{1-\boxed{\phantom{x}}}\right] - \boxed{\phantom{x}}$$

$$\sum_{i=3}^{\infty} 72\left(\frac{1}{6}\right)^i = \boxed{\phantom{x}}$$

## NET PRESENT VALUE

**S14.11.** Compute the net present value of the project given the cash flows below and determine if the project must be approved or rejected. The discount rate is 25%.

|  | Year | | | | | | |
| --- | --- | --- | --- | --- | --- | --- | --- |
| Future value | 0 | 1 | 2 | 3 | 4 | 5 | 6 |
| A | −25000 | −2500 | 0 | 5000 | 17000 | 25000 | 48000 |

Substitute in the NPV formula.

$$NPV = \boxed{\phantom{x}} + \frac{-2500}{(1+0.25)} + \boxed{\phantom{x}} + \boxed{\phantom{x}} + \boxed{\phantom{x}} + \boxed{\phantom{x}} + \boxed{\phantom{x}}$$

Calculate.

$$NPV = \boxed{\phantom{x}} \underline{\hspace{2cm}} \text{(Accept/Reject) the project as NPV} \boxed{\phantom{x}} 0$$

# Solutions to Additional Practice Problems

## Chapter 1

### THEORETICAL CONCEPTS

**S1.1.** True. The expression can be expressed as an exponential term using the property $\sqrt[n]{x^m} = x^{\frac{m}{n}}$.

**S1.2.** When a polynomial has only two terms, it is called binomial.

**S1.3.** To determine the degree, add the exponents of each term; these are 1.5, 2, and 1, respectively. The degree of the polynomial is the highest exponent among the monomials; this is 2.

**S1.4.** True. Using the property $x^{m/n} = \sqrt[n]{x^m}$, the term $x^{-1/2}$ can also be expressed as $\sqrt[2]{x^{-1}} = \sqrt{1/x}$.

**S1.5.** True.

### EXPONENTS AND FRACTIONS

Use the rules of exponents to simplify these fractions:

**S1.6.**

$$\frac{x^3 \cdot x^8}{x^7} = x^{3+8-7} = x^4$$

**S1.7.**

$$\frac{x^{1/2} \cdot x^{-1/3}}{x^{1/6}} = x^{1/2-1/3-1/6} = x^0 = 1$$

### POLYNOMIALS

Multiply these polynomials:

**S1.8.**

$$(4x - 2y)(4x^2 + 3xy + 1)$$

$$= 16x^3 + 12x^2y + 4x - 8x^2y - 6xy^2 - 2y$$
$$= 16x^3 + 4x^2y + 4x - 12x^2y - 8x^2y - 4x^2y - 6xy^2 - 2y$$

**S1.9.**

$$(3x - 7y)(3 + 6/y)$$

$$= 9x + \frac{18x}{y} - 21y - 42$$

## FACTORING

**S1.10.** Factor the following polynomial and express it as two factors:

$$2x^2 - 5x - 12$$

$$2x^2 - 5x - 12 = (a+b)(c+d)$$

1) $a \cdot b = 2$. Integer factors $1 \cdot 2$, $2 \cdot 1$.

2) $c \cdot d = -12$. Integer factors $1 \cdot -12$, $-12 \cdot 1$, $-6 \cdot 2$, $-2 \cdot 6$, $-4 \cdot 3$, $-3 \cdot 4$.

3) $ad + bc = -5$, with $a = 1$, $b = 2$, $d + 2b$ must be equal to $-5$.

$$2x^2 - 5x - 12 = (x - 4)(2x + 3)$$

## ORDER OF MATHEMATICAL OPERATIONS

**S1.11.** Simplify the following operations and express them in exponential terms:

$$\sqrt{\left[\frac{20x^5y^{-3}}{10x^2y^{-1}}\right] \cdot 2z^{\left[3\left(\frac{4}{3} + \frac{3}{4}\right)\right]}}$$

$$\sqrt{[2x^3y^{-2}] \cdot 2z^{\left[3\left(\frac{25}{12}\right)\right]}} = \sqrt{[2x^3y^{-2}] \cdot 2z^{\frac{25}{4}}} = \sqrt{4x^3y^{-2}z^{\frac{25}{4}}}$$
$$= 2x^{3/2}y^{-1}z^{25/8}$$

# Chapter 2

## THEORETICAL CONCEPTS

**S2.1.** True. A constant $k$ can be added on both sides of the equation without affecting the equality.

**S2.2.** In the fourth quadrant, the values of $x$ are positive, whereas $y$ is negative.

**S2.3.** The steepest slope is represented by the slope with the largest absolute number. Thus, the steepest positive slope has the largest positive value, which in this case is 7.5.

**S2.4.** In a linear cost function, the intercept represents the fixed cost.

**S2.5.** True.

## EQUATIONS

Solve the following linear equations by using properties of equality:

**S2.6.**

$$4x + 2 = 2(x - 1) + 10$$

$$2x = 6 \qquad \text{(Addition and subtraction)}$$

$$\boxed{x = 3} \qquad \text{(Division)}$$

**S2.7.**

$$(1/4)x - 2 = (1/5)x + 1$$

$$(1/4 - 1/5)x = 1 + 2 \qquad \text{(Moving terms)}$$

$$(1/20)x = 3 \qquad \text{(Addition and subtraction)}$$

$$\boxed{x = 60} \qquad \text{(Multiplication property)}$$

**S2.8.**

$$\frac{4}{x-2} = \frac{3}{2x}$$

$$8x = 3x - 6 \qquad \text{(Multiplication of fractions)}$$

$$5x = -6 \qquad \text{(Addition and subtraction)}$$

$$\boxed{x = -1.2} \qquad \text{(Division)}$$

## SLOPES AND INTERCEPTS

**S2.9.** Find the slope of

$$5x + 2y = 20$$

The slope can be found by converting the equation into slope-intercept form.

$$2y = 20 - 5x \qquad \text{(Moving terms)}$$

$$y = 10 - 2.5x \qquad \text{(Division)}$$

Slope: $m = -2.5$

**S2.10.** Find the $x$- and $y$-intercepts of the following equation:

$$4x - 2y = 10$$

---

The $x$-intercept occurs where the line crosses the $x$-axis, which is the point where $y = 0$. Setting $y = 0$ and solving for $y$, we have

$$4x - 2(0) = 10 \qquad\qquad x = 2.5$$

The $x$-intercept is (2.5, 0).

To find the $y$-intercept, we set $x = 0$ and solve for $y$:

$$4(0) - 2y = 10 \qquad\qquad y = -5$$

The $y$-intercept is (0, −5).

---

**S2.11.** Find the slope-intercept form of

$$3x + 7y = 21$$

---

The slope-intercept form $y = mx + b$ for the equation can be found by

| | |
|---|---|
| Moving terms | $7y = -3x + 21$ |
| Dividing by the coefficient of $y$ | $y = -\dfrac{3}{7}x + 3$ |

The slope-intercept form is $y = -\dfrac{3}{7}x + 3$.

---

## APPLICATIONS OF LINEAR EQUATIONS

**S2.12.** An agricultural firm operates in pure competition, receiving \$3 for each bag of potatoes sold. It has a variable cost of \$1.2 per item and a fixed cost of \$2,000. What is the number of bags the farmer must sell to break even (i.e., profit is neither negative nor positive)?

---

Profit $\pi$ is defined as revenue $R$ after taking out cost C: $\qquad \pi = R - C$

where $R = 3x$ and $C = 2000 + 1.2x$. Substituting:

$$\pi = 3x - (2000 + 1.2x)$$
$$\pi = 1.8x - 2000$$

Break-even occurs when $\pi = 0$:

$$0 = 1.8x - 2000$$
$$x = 2000/1.8 = 1111.11$$

Units of product must be a positive integer; therefore, the number of bags for break-even is 1,112.

**S2.13.** Ernesto has a budget of $50 that can be spent for lunch every week. He can buy either spaghetti or salad. The price of spaghetti is $8 per meal and the price of salad is $4 per serving.

I.  Consider $x$ as the number of spaghetti meals and $y$ as servings of salad per week. Write Ernesto's budget equation in standard form.

$$\boxed{8x + 4y = 50}$$

II.  Graph the budget line for Ernesto's lunch.

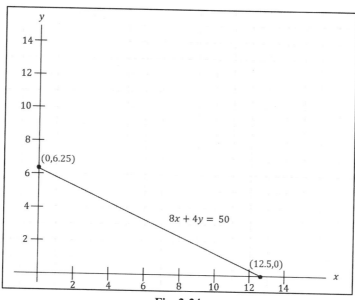

**Fig. 2-24**

# Chapter 3

## THEORETICAL CONCEPTS

**S3.1.** True. The function only needs an $x^3$ term to be called a cubic function.

**S3.2.** The coefficient $a$ of the profit function $\pi = ax^2 + bx + c$ is negative. Thus, the profit function $\pi$ has a maximum point.

**S3.3.** Here, $a = 1$, $b = -4$, and $c = 1$. The vertex $(x, y)$ is found by

$$x = -\frac{b}{2a} = \frac{-(-4)}{2(1)} \qquad\qquad y = \frac{4ac - b^2}{4a} = \frac{4(1)(1) - (-4)^2}{4(1)} = \frac{4 - 16}{4}$$

$$x = 2 \qquad\qquad\qquad y = -3$$

The coordinates of the vertex are $(2, -3)$.

**S3.4.** The vertex represents the maximum point for an inverted U-shaped parabola.

**S3.5.** True.

## THE ALGEBRA OF FUNCTIONS

**S3.6.**   Given $F(x) = x^2$ and $G(x) = x - \dfrac{1}{x}$, find $G(F(x))$ and $F(G(x))$.

$$G(F(x)) = G(x^2) = \boxed{x^2 - \dfrac{1}{x^2}} \qquad\qquad F(G(x)) = F\left(x - \dfrac{1}{x}\right) = \boxed{\left(x - \dfrac{1}{x}\right)^2}$$

**S3.7.**   Given $F(x) = 2x/(x-2)$ and $G(x) = x^2 - 4$, find $(F \cdot G)(x)$.

$$(F \cdot G)(x) = \left(\dfrac{2x}{x-2}\right)(x^2 - 4) = \left(\dfrac{2x}{\cancel{x-2}}\right)(x+2)(\cancel{x-2}) = \boxed{2x^2 + 4x}$$

**S3.8.**   Given $F(x) = 4/(x-1)$ and $G(x) = 2/(x^2 - 1)$, find $(F - G)(x)$.

$$(F - G)(x) = \dfrac{4}{x-1} - \dfrac{2}{x^2 - 1} = \dfrac{4(x+1)}{x^2 - 1} - \dfrac{2}{x^2 - 1} = \boxed{\dfrac{4x+2}{x^2 - 1}}$$

## SOLVING QUADRATIC FUNCTIONS

**S3.9.**   Graph the quadratic function indicating its vertex and intercepts:

$$F(x) = x^2 - 3x + 2$$

(a)   Here, $a = 1$, $b = -3$, and $c = 2$. With $a > 0$, the parabola opens up.

(b)   The vertex $(x, y)$ is found by:

$$x = -\dfrac{b}{2a} = \dfrac{-(-3)}{2(1)} = 1.5 \qquad\qquad y = \dfrac{4ac - b^2}{4a} = \dfrac{4(1)(2) - (-3)^2}{4(1)} = \dfrac{8 - 9}{4} = -0.25$$

   The coordinates of the vertex are $(1.5, -0.25)$.

(c)   The $x$-coordinates of the $x$-intercepts are found by setting the equation equal to zero.

---

$$x^2 - 3x + 2 = 0$$

Factoring the equation $(x - 2)(x - 1) = 0$ and setting each factor equal to zero:

$$x - 2 = 0 \qquad\qquad x - 1 = 0$$
$$x = 2 \qquad\qquad x = 1$$

The $x$-intercepts are $(2, 0)$ and $(1, 0)$.

---

(d)   The $y$-coordinate of the $y$-intercept is found by $x = 0$.

$$f(0) = 0^2 - 3(0) + 2 = 2$$

The $y$-intercept is $(0, 2)$.

(e)     The graph of the function $F(x) = x^2 - 3x + 2$ is

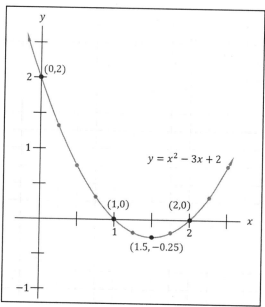

**Fig. 3-31**

## NONLINEAR FUNCTIONS IN BUSINESS, ECONOMICS, AND FINANCE

**S3.10.** Given the total revenue $TR = 90x - x^2$ and total cost $TC = 60x + 120$.

(a)   The profit function $\pi$ is the revenue left after cost is accounted for.

$$\pi = TR - TC = 90x - x^2 - (60x + 120)$$

$$\pi = -x^2 + 30x - 120$$

(b)   The maximum level of profit occurs at the vertex point. Here, $a = -1$, $b = 30$, and $c = -120$. The vertex $(x, \pi)$ is found by

$$x = -\frac{b}{2a} = \frac{-(30)}{2(-1)} = 15 \qquad \pi = \frac{4ac - b^2}{4a} = \frac{4(-1)(-120) - (30)^2}{4(-1)} = \frac{-420}{4} = 105$$

The coordinates of the vertex are (15, 105). Thus, the maximum level of profit $\pi^* = 105$ occurs when $x = 15$.

(c)   The break-even points occur at $\pi = 0$, which is equivalent to finding the $x$-intercepts of the profit function. Solving $\pi(x) = 0$ by using the quadratic formula where $a = -1$, $b = 30$, and $c = -120$:

$$x = \frac{-(30) \pm \sqrt{(30)^2 - 4(-1)(-120)}}{2(-1)} = \frac{-30 \pm \sqrt{420}}{-2} \approx \frac{-30 \pm 20.49}{-2}$$

Add and subtract 20.49 by turns in the numerator.

$$x = \frac{-30 + 20.49}{-2} = 4.75 \qquad\qquad x = \frac{-30 - 20.49}{-2} = 25.25$$

The break-even points are (4.75, 0) and (25.25, 0).

(d)  Here, $a < 0$, the parabola opens down, and the vertex of the profit function is (15, 105) and the $x$-intercepts are (4.75, 0) and (25.25, 0).

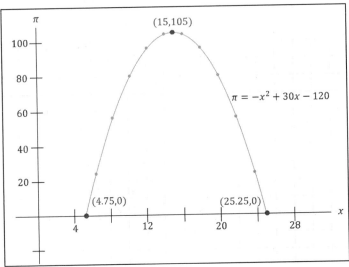

**Fig. 3-32**

# Chapter 4

## THEORETICAL CONCEPTS

**S4.1.**  True.

**S4.2.**  A $4 \times 3$ system of equations has *4* equations and *3* variables. This system is *over*constrained.

**S4.3.**  The slope of the supply equation is *negative*. This means that a higher price of the good will *increase* its quantity supplied.

**S4.4.**  False. Any quantity lower than the break-even point represents a loss for a linear profit function.

## ELIMINATION AND SUBSTITUTION METHOD

**S4.5.**
$$4x + y = 21 \qquad (4.84)$$
$$3x - 2y = 13 \qquad (4.85)$$

(a)  Elimination method
   1)  Select $y$ to be eliminated and multiply equation (*4.84*) by *2* and (*4.85*) by *1*.
$$8x + 2y = 42 \qquad (4.86)$$
$$3x - 2y = 13 \qquad (4.87)$$

   2)  Add (*4.86*) and (*4.87*) to eliminate $y$.

$$11x = 55$$

Then, the value of $x = 5$.

3) Substitute the value found in step 2) in (*4.84*).

$$4(5) + y = 21$$

The value of $y = 1$.

(*b*) Substitution method

1) Solve one of the original equations (*4.84*) for $y$.

$$y = 21 - 4x \qquad\qquad (4.88)$$

2) Substitute the value $y$ in (*4.88*) into the original equation (*4.85*).

$$3x - 2(21 - 4x) = 13$$
$$11x - 42 = 13$$
$$11x = 55$$

$x = 5$.

3) Then substituting the value of $x$ in (*4.84*).

$$4(5) + y = 21$$

We get $y = 1$.

## SUPPLY AND DEMAND

**S4.6.** The market of a product is given by these linear equations:

$$\text{Supply: } -2P + 4Q = -12 \qquad\qquad \text{Demand: } 5P + 20Q = 150$$

where $P$ is price and $Q$ is quantity of the product.

I. Find the equilibrium price ($P^*$) and quantity ($Q^*$) sold in this market.

---

Solve by elimination method:

1) Select $P$ to be eliminated and multiply the supply equation by 5 and the demand by 2:

$$-10P + 20Q = -60 \qquad\qquad (4.89)$$
$$10P + 40Q = 300 \qquad\qquad (4.90)$$

2) Add (*4.89*) and (*4.90*) to eliminate $P$.

$$60Q = 240$$

Then, the value of $Q^* = 4$.

3) Substitute the value found in step 2) in (*4.89*):

$$-10P + 20(4) = -60$$

The value of $P^* = 14$.

The market equilibrium is $P^* = 14$, $Q^* = 4$.

---

II.   Graph the supply and demand of this market, setting price in the $y$-axis and quantity in the $x$-axis.

---

The graph requires both the supply and demand to be in slope-intercept form.

1) For supply:

$$-2P = -4Q - 12$$
$$P = 2Q + 6 \tag{4.91}$$

2) For demand:

$$5P = -20Q + 150$$
$$P = -4Q + 30 \tag{4.92}$$

The market equilibrium is $P^* = 14$, $Q^* = 4$. Thus, the supply-demand graph is depicted in Fig. 4-26.

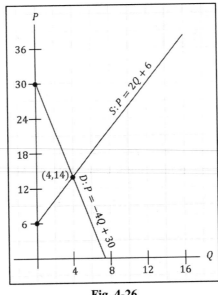

**Fig. 4-26**

---

## BREAK-EVEN ANALYSIS

**S4.7.**   Given the revenue $R(x) = 90x - x^2$ and cost $C(x) = 60x + 120$:

(a)   Find the break-even points $x^*$ by setting $R(x) = C(x)$.

$$90x - x^2 = 60x + 120$$
$$-x^2 + 30x - 120 = 0$$

Here, $a = -1$, $b = 30$, and $c = -120$. Solving by quadratic formula,

$$x = \frac{-(30) \pm \sqrt{(30)^2 - 4(-1)(-120)}}{2(-1)} = \frac{-30 \pm \sqrt{420}}{-2} \approx \frac{-30 \pm 20.49}{-2}$$

Add and subtract 20.49 by turns in the numerator:

$$x = \frac{-30 + 20.49}{-2} = 4.75 \qquad x = \frac{-30 - 20.49}{-2} = 25.25$$

> $x = 4.75$ and $x = 25.25$ are the break-even levels of output.

At $x = 4.75$, $R(x) = C(x) = 405$
At $x = 25.25$, $R(x) = C(x) = 1635$

(b) The graph of the revenue $R(x)$ and cost $C(x)$ functions are depicted in Fig. 4-27.

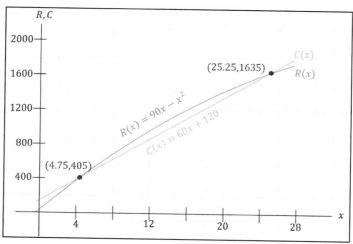

**Fig. 4-27**

## INCOME DETERMINATION MODEL

**S4.8.** Find the equilibrium level of income $Y_e$, given

$$Y = C + I + (X - Z) \qquad C = C_0 + bY_d \qquad Z = Z_0 + zY_d$$
$$I = I_0 \qquad\qquad X = X_0 \qquad\qquad Y_d = Y - T$$
$$T = T_0 + tY$$

where $C_0 = 80$, $I_0 = 100$, $X_0 = 90$, $Z_0 = 50$, $T_0 = 10$, $b = 0.8$, $z = 0.1$, $t = 0.2$.

(a) Find the reduced form of $Y_e$.

$$Y = C + I + X - Z = C_0 + b(Y - T) + I_0 + X_0 - Z_0 - z(Y - T)$$
$$Y = C_0 + b(Y - T_0 - tY) + I_0 + X_0 - Z_0 - z(Y - T_0 - tY)$$
$$Y = C_0 + bY - bT_0 - btY + I_0 + X_0 - Z_0 - zY + zT_0 + ztY$$
$$Y - bY + btY + zY - ztY = C_0 - bT_0 + I_0 + X_0 - Z_0 + zT_0$$
$$Y(1 - b + bt + z - zt) = C_0 - bT_0 + I_0 + X_0 - Z_0 + zT_0$$
$$Y_e = \frac{1}{1 - b + bt + z - zt}(C_0 - bT_0 + I_0 + X_0 - Z_0 + zT_0)$$

(b)  Determine the numerical value of $Y_e$.

$$Y_e = \frac{1}{1 - 0.8 + 0.8(0.2) + 0.1 - 0.1(0.2)}(80 - 0.8(10) + 100 + 90 - 50 + 0.1(10))$$

$$Y_e = \frac{1}{0.44}(213) = 489.09$$

(c)  Find the effect of the multiplier if an income tax T with a proportional component t is incorporated into the model.

The multiplier is changed from $1/(1 - b + z)$ to $1/(1 - b + bt + z - zt)$.

$$\frac{1}{1 - b + z} = \frac{1}{1 - 0.8 + 0.1} = \frac{1}{0.3} = 3.33$$

$$\frac{1}{1 - b + bt + z - zt} = \frac{1}{1 - 0.8 + 0.8(0.2) + 0.1 - 0.1(0.2)} = \frac{1}{0.44} = 2.72$$

This reduces the size of the multiplier.

# Chapter 5

## THEORETICAL CONCEPTS

**S5.1.**  True. If the addition is conformable, then $A + B = B + A$.

**S5.2.**  $A \times B = AB$.

$(4 \times 3) \times (3 \times 5) = (4 \times 5)$.

The matrix multiplication $AB$ is conformable: the number of columns (3) of the lead matrix is equal to the number of rows of the lag matrix.

The resultant $AB$ is a (4×5) matrix: the number of rows of the lead matrix (4) by the number of columns of the lag matrix (5).

**S5.3.**  The transpose of the matrix

$$C = \begin{bmatrix} 0 & 4 & 6 & 7 & 9 \\ 1 & 2 & 3 & 8 & 5 \\ 5 & 3 & 4 & 0 & 2 \end{bmatrix} \quad \text{is} \quad C' = \begin{bmatrix} 0 & 1 & 5 \\ 4 & 2 & 3 \\ 6 & 3 & 4 \\ 7 & 8 & 0 \\ 9 & 5 & 2 \end{bmatrix}.$$

**S5.4.**  False. The system of linear equations is converted from matrix to augmented form.

## MATRIX OPERATIONS

Refer to the following matrices in answering the questions below:

$$A = \begin{bmatrix} 12 & 8 & 3 \\ 10 & 1 & 1 \end{bmatrix} \quad B = \begin{bmatrix} 2 & -10 & 5 \\ -11 & 24 & 4 \end{bmatrix} \quad C = \begin{bmatrix} 9 & 3 \\ 7 & 6 \end{bmatrix} \quad D = \begin{bmatrix} 10 & 0 \\ -1 & 1 \end{bmatrix}$$

**S5.5.**   Compute $A + 2B$.

$$A + 2B = \begin{bmatrix} 12 & 8 & 3 \\ 10 & 1 & 1 \end{bmatrix} + 2\begin{bmatrix} 2 & -10 & 5 \\ -11 & 24 & 4 \end{bmatrix} = \begin{bmatrix} 12 & 8 & 3 \\ 10 & 1 & 1 \end{bmatrix} + \begin{bmatrix} 4 & -20 & 10 \\ -22 & 48 & 8 \end{bmatrix}$$

$$A + 2B = \begin{bmatrix} 12+4 & 8-20 & 3+10 \\ 10-22 & 1+48 & 1+8 \end{bmatrix} = \begin{bmatrix} 12 & 8 & 3 \\ 10 & 1 & 1 \end{bmatrix}$$

**S5.6.**   Compute $C - 3D$.

$$C - 3D = \begin{bmatrix} 9 & 3 \\ 7 & 6 \end{bmatrix} - 3\begin{bmatrix} 10 & 0 \\ -1 & 1 \end{bmatrix} = \begin{bmatrix} 9-3(10) & 3-3(0) \\ 7-3(-1) & 6-3(6) \end{bmatrix} = \begin{bmatrix} -21 & 3 \\ 10 & -12 \end{bmatrix}$$

**S5.7.**   Compute $AB'$

1)   The transpose of $B$, known as $B'$ is

$$B' = \begin{bmatrix} 2 & -11 \\ -10 & 24 \\ 5 & 4 \end{bmatrix}$$

2)   $A \times B' = AB$.

$(2 \times 3) \times (3 \times 2) = (2 \times 2)$.

The dimensions of $AB'$ are $(2 \times 2)$.

3)   The matrix $AB'$ is:

Here $A$ rows $(R_1, R_2)$ will be multiplied by $B'$ columns $(C_1, C_2)$ using vector multiplication, so

$$A = \begin{bmatrix} 12 & 8 & 3 \\ 10 & 1 & 1 \end{bmatrix} \quad B' = \begin{bmatrix} 2 & -11 \\ -10 & 24 \\ 5 & 4 \end{bmatrix}$$

$$AB' = \begin{bmatrix} R_1C_1 & R_1C_2 \\ R_2C_1 & R_2C_2 \end{bmatrix} = \begin{bmatrix} 12(2)+8(-10)+3(5) & 12(-11)+8(24)+3(4) \\ 10(2)+1(-10)+1(5) & 10(-11)+1(24)+1(4) \end{bmatrix}$$

$$AB' = \begin{bmatrix} 41 & 72 \\ 15 & -82 \end{bmatrix}$$

## GAUSSIAN ELIMINATION

Use Gaussian elimination methods to solve the following system of simultaneous equations:

**S5.8.** $\quad\quad\quad\quad\quad\quad 4x + y = 21$ $\hfill (4.84)$

$\hfill (4.85)$

$\quad\quad\quad\quad\quad\quad\quad 3x - 2y = 13$

1)  Set up the matrix form.

$$A \cdot X = B$$

$$\begin{bmatrix} 4 & 1 \\ 3 & -2 \end{bmatrix} \begin{bmatrix} x \\ y \end{bmatrix} = \begin{bmatrix} 21 \\ 13 \end{bmatrix}$$

2)  Express the system as an augmented matrix.

$$A \mid B = \begin{bmatrix} 4 & 1 & \bigm| & 21 \\ 3 & -2 & \bigm| & 13 \end{bmatrix}$$

3)  Apply row operations.

*3a.* Multiply row 1 by ¼ (the multiplicative inverse of the first element in row 1).

$$\begin{bmatrix} 1 & \dfrac{1}{4} & \bigm| & \dfrac{21}{4} \\ 3 & -2 & \bigm| & 13 \end{bmatrix}$$

*3b.* Subtract 3 times row 1 from row 2 (to eliminate the first element in row 2).

$$\begin{bmatrix} 1 & \dfrac{1}{4} & \bigm| & \dfrac{21}{4} \\ 0 & -\dfrac{11}{4} & \bigm| & -\dfrac{11}{4} \end{bmatrix}$$

*3c.* Multiply row 2 by −4/11 (the multiplicative inverse of the first element in row 2).

$$\begin{bmatrix} 1 & \dfrac{1}{4} & \bigm| & \dfrac{21}{4} \\ 0 & 1 & \bigm| & 1 \end{bmatrix}$$

*3d.* Subtract ¼ times row 2 from row 1 (to eliminate the second element in row 1).

$$\begin{bmatrix} 1 & 0 & \bigm| & 5 \\ 0 & 1 & \bigm| & 1 \end{bmatrix}$$

Thus, $x = 5$ and $y = 1$, since

$$\begin{bmatrix} 1 & 0 \\ 0 & 1 \end{bmatrix} \begin{bmatrix} x \\ y \end{bmatrix} = \begin{bmatrix} 5 \\ 1 \end{bmatrix}.$$

## SUPPLY AND DEMAND

**S5.9.** The market of a product is given by these linear equations:

$$\text{Supply: } -2P + 4Q = -12 \qquad\qquad \text{Demand: } 5P + 20Q = 150$$

where $P$ is price and $Q$ is quantity of the product. Find the equilibrium price ($P^*$) and quantity ($Q^*$) sold in this market by using Gaussian elimination methods.

1) Compute the matrix form.

$$A \cdot X = B$$

$$\begin{bmatrix} -2 & 4 \\ 5 & 20 \end{bmatrix} \begin{bmatrix} P \\ Q \end{bmatrix} = \begin{bmatrix} -12 \\ 150 \end{bmatrix}$$

2) Find the augmented matrix.

$$A \,|\, B = \begin{bmatrix} -2 & 4 & | & -12 \\ 5 & 20 & | & 150 \end{bmatrix}$$

3) Proceed with the row operations.

*3a.* Multiply row 1 by $-\frac{1}{2}$ (the multiplicative inverse of the first element in row 1).

$$\begin{bmatrix} 1 & -2 & | & 6 \\ 5 & 20 & | & 150 \end{bmatrix}$$

*3b.* Subtract 5 times row 1 from row 2 (to eliminate the first element in row 2).

$$\begin{bmatrix} 1 & -2 & | & 6 \\ 0 & 30 & | & 120 \end{bmatrix}$$

*3c.* Multiply row 2 by $-1/30$ (the multiplicative inverse of the first element in row 2).

$$\begin{bmatrix} 1 & -2 & | & 6 \\ 0 & 1 & | & 4 \end{bmatrix}$$

*3d.* Add 2 times row 2 to row 1 (to eliminate the second element in row 1).

$$\begin{bmatrix} 1 & 0 & | & 14 \\ 0 & 1 & | & 4 \end{bmatrix}$$

4) Since

$$\begin{bmatrix} 1 & 0 \\ 0 & 1 \end{bmatrix} \begin{bmatrix} P \\ Q \end{bmatrix} = \begin{bmatrix} 14 \\ 4 \end{bmatrix}$$

The equilibrium price $P^*$ is 14 and quantity $Q^*$ is 4.

# Chapter 6

## THEORETICAL CONCEPTS

**S6.1.** False. $|B| \neq 0$ if $B$ is a nonsingular matrix.

**S6.2.** $AA^{-1}$ results in a (4×4) identity matrix.

**S6.3.** True.

**S6.4.** True.

## MATRIX DETERMINANTS

Refer to the following matrices in answering the questions below:

$$A = \begin{bmatrix} 9 & 3 \\ 7 & 6 \end{bmatrix} \qquad B = \begin{bmatrix} 1 & 4 & 0 \\ 5 & 2 & 8 \\ 7 & 6 & 3 \end{bmatrix}$$

**S6.5.** Compute $|A|$:

$$|A| = 9(6) - 7(3) = 33$$

With $|A| \neq 0$, matrix $A$ is nonsingular.

**S6.6.** Expand the matrix by repeating the first two columns to the right of the original matrix.

$$|B| = 1 \cdot 2 \cdot 3 + 4 \cdot 8 \cdot 7 + 0 \cdot 5 \cdot 6 - [7 \cdot 2 \cdot 0 + 6 \cdot 8 \cdot 1 + 3 \cdot 5 \cdot 4]$$
$$|B| = 6 + 224 + 0 - [0 + 48 + 60] = 230 - 108 = 122$$

With $|B| \neq 0$, matrix $B$ is nonsingular.

## SOLVING SYSTEMS OF LINEAR EQUATIONS

Given the following system of simultaneous equations:

$$4x + y = 21$$
$$3x - 2y = 13$$

**S6.7.** Solve the system using inverse matrices.

(a) Find the inverse matrix $A^{-1}$.

1. Set up the augmented matrix.

$$\left[\begin{array}{cc|cc} 4 & 1 & 1 & 0 \\ 3 & -2 & 0 & 1 \end{array}\right]$$

2. Use Gaussian elimination methods to obtain the inverse.

*2a.* Multiply row 1 by ¼ to obtain 1 in the $a_{11}$ position.

$$\left[\begin{array}{cc|cc} \boxed{1} & \dfrac{1}{4} & \dfrac{1}{4} & 0 \\ 3 & -2 & 0 & 1 \end{array}\right]$$

*2b.* Subtract 3 times row 1 from row 2 to clear column 1.

$$\left[\begin{array}{cc|cc} 1 & \dfrac{1}{4} & \dfrac{1}{4} & 0 \\ \boxed{0} & -\dfrac{11}{4} & -\dfrac{3}{4} & 1 \end{array}\right]$$

*2c.* Multiply row 2 by $-4/11$ to obtain 1 in the $a_{22}$ position.

$$\left[\begin{array}{cc|cc} 1 & \dfrac{1}{4} & \dfrac{1}{4} & 0 \\ 0 & \boxed{1} & \dfrac{3}{11} & -\dfrac{4}{11} \end{array}\right]$$

*2d.* Subtract ¼ times row 2 from row 1 to clear column 2.

$$\left[\begin{array}{cc|cc} 1 & \boxed{0} & \dfrac{2}{11} & \dfrac{1}{11} \\ 0 & 1 & \dfrac{3}{11} & -\dfrac{4}{11} \end{array}\right]$$

3. The identity matrix is now on the left-hand side of the bar.

$$A^{-1} = \left[\begin{array}{cc} \dfrac{2}{11} & \dfrac{1}{11} \\ \dfrac{3}{11} & -\dfrac{4}{11} \end{array}\right]$$

(b)  The procedure to solve for the unknown variables uses the formula $X = A^{-1}B$; here $B = \begin{bmatrix} 21 \\ 13 \end{bmatrix}$.

$$X = \begin{bmatrix} \dfrac{2}{11} & \dfrac{1}{11} \\ \dfrac{3}{11} & -\dfrac{4}{11} \end{bmatrix} \begin{bmatrix} 21 \\ 13 \end{bmatrix}$$

Multiply the two matrices using ordinary row-column operations.

$$X = \begin{bmatrix} \dfrac{2}{11}(21) + \dfrac{1}{11}(13) \\ \dfrac{3}{11}(21) - \dfrac{4}{11}(13) \end{bmatrix} = \begin{bmatrix} \dfrac{55}{11} \\ \dfrac{11}{11} \end{bmatrix} = \begin{bmatrix} 5 \\ 1 \end{bmatrix}$$

The solution is $\bar{x} = 5$, $\bar{y} = 1$.

**S6.8.**  Solve the system using Cramer's rule.

1)  Express the equations in matrix form.

$$A \cdot X = B$$

$$\begin{bmatrix} 4 & 1 \\ 3 & -2 \end{bmatrix} \begin{bmatrix} x \\ y \end{bmatrix} = \begin{bmatrix} 21 \\ 13 \end{bmatrix}$$

2)  The determinant of $A$ is

$$|A| = 4(-2) - 3(1) = -11$$

3)  Replacing the first column of $A$, which contains the coefficients of $x$, with the column vector of constants $B$,

$$A_1 = \begin{bmatrix} 21 & 1 \\ 13 & -2 \end{bmatrix}$$

The determinant of $A_1$ is

$$|A_1| = 21(-2) - 13(1) = -55$$

Use Cramer's rule to find $\bar{x}$

$$\bar{x} = \frac{|A_1|}{|A|} = \frac{-55}{-11} = 5$$

4)  Replace the second column of $A$, which contains the coefficients of $y$, with the column vector of constant $B$.

$$A_2 = \begin{bmatrix} 4 & 21 \\ 3 & 13 \end{bmatrix}$$

The determinant of $A_2$ is

$$|A_2| = 4(13) - 3(21) = -11$$

Use Cramer's rule to find $\bar{y}$.

$$\bar{y} = \frac{|A_2|}{|A|} = \frac{-11}{-11} = 1$$

## SOLVING SYSTEMS OF LINEAR EQUATIONS

**S6.9.** Solve the following system of equations using Cramer's rule:

$$x_1 + 2x_2 \qquad = 8$$
$$5x_1 \qquad + 4x_3 = 14$$
$$3x_1 + 2x_2 + x_3 = 13$$

*1)* Express the equations in matrix form.

$$A \cdot X = B$$

$$\begin{bmatrix} 1 & 2 & 0 \\ 5 & 0 & 4 \\ 3 & 2 & 1 \end{bmatrix} \begin{bmatrix} x_1 \\ x_2 \\ x_3 \end{bmatrix} = \begin{bmatrix} 8 \\ 14 \\ 13 \end{bmatrix}$$

*2)* Expand the matrix to find the determinant of $A$ by repeating the first two columns to the right of the original matrix.

$$\left( \begin{matrix} 1 & 2 & 0 \\ 5 & 0 & 4 \\ 3 & 2 & 1 \end{matrix} \right. \left. \begin{matrix} 1 & 2 \\ 5 & 0 \\ 3 & 2 \end{matrix} \right)$$

$$|A| = 1 \cdot 0 \cdot 1 + 2 \cdot 4 \cdot 3 + 0 \cdot 5 \cdot 2 - [3 \cdot 0 \cdot 0 + 2 \cdot 4 \cdot 1 + 1 \cdot 5 \cdot 2]$$

$$|A| = 0 + 24 + 0 - [0 + 8 + 10] = 24 - 18 = 6$$

*3)* Solve for $x_1$.

Replace the first column of $A$ with the column vector $B$: $A_1 = \begin{bmatrix} 8 & 2 & 0 \\ 14 & 0 & 4 \\ 13 & 2 & 1 \end{bmatrix}$, then find its determinant.

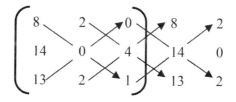

$$|A_1| = 8 \cdot 0 \cdot 1 + 2 \cdot 4 \cdot 13 + 0 \cdot 14 \cdot 2 - [13 \cdot 0 \cdot 0 + 2 \cdot 4 \cdot 8 + 1 \cdot 14 \cdot 2]$$

$$|A_1| = 0 + 104 + 0 - [0 + 64 + 28] = 104 - 92 = 12$$

and

$$\bar{x}_1 = \frac{|A_1|}{|A|} = \frac{12}{6} = 2$$

4)  Solve for $x_2$.

Replace the second column of $A$ with the column vector $B$: $A_2 = \begin{bmatrix} 1 & 8 & 0 \\ 5 & 14 & 4 \\ 3 & 13 & 1 \end{bmatrix}$, then find its determinant.

$$
\begin{pmatrix}
1 & 8 & 0 \\
5 & 14 & 4 \\
3 & 13 & 1
\end{pmatrix}
\begin{matrix}
1 & 8 \\
5 & 14 \\
3 & 13
\end{matrix}
$$

$$|A_2| = 1 \cdot 14 \cdot 1 + 8 \cdot 4 \cdot 3 + 0 \cdot 5 \cdot 13 - [3 \cdot 14 \cdot 0 + 13 \cdot 4 \cdot 1 + 1 \cdot 5 \cdot 8]$$

$$|A_2| = 14 + 96 + 0 - [0 + 52 + 40] = 110 - 92 = 18$$

and

$$\bar{x}_2 = \frac{|A_2|}{|A|} = \frac{18}{6} = 3$$

5)  Solve for $x_3$.

Replace the second column of $A$ with the column vector $B$: $A_3 = \begin{bmatrix} 1 & 2 & 8 \\ 5 & 0 & 14 \\ 3 & 2 & 13 \end{bmatrix}$, then find its determinant.

$$
\begin{pmatrix}
1 & 2 & 8 \\
5 & 0 & 14 \\
3 & 2 & 13
\end{pmatrix}
\begin{matrix}
1 & 2 \\
5 & 0 \\
3 & 2
\end{matrix}
$$

$$|A| = 1 \cdot 0 \cdot 13 + 2 \cdot 14 \cdot 3 + 8 \cdot 5 \cdot 2 - [3 \cdot 0 \cdot 8 + 2 \cdot 14 \cdot 1 + 13 \cdot 5 \cdot 2]$$

$$|A| = 0 + 84 + 80 - [0 + 28 + 130] = 164 - 158 = 6$$

and

$$\bar{x}_3 = \frac{|A_3|}{|A|} = \frac{6}{6} = 1$$

Thus, $\bar{x}_1 = 2$, $\bar{x}_2 = 3$ and $\bar{x}_3 = 1$.

**BUSINESS AND ECONOMIC APPLICATIONS: SUPPLY AND DEMAND**

**S6.10.** The market of a product is given by these linear equations:

$$\text{Supply: } -2P+4Q=-12 \qquad \text{Demand: } 5P+20Q=150$$

where $P$ is price and $Q$ is quantity of the product. Find the equilibrium price ($P^*$) and quantity ($Q^*$) sold in this market by using Cramer's rule.

*1)* Express the equations in matrix form.

$$A \cdot X = B$$

$$\begin{bmatrix} -2 & 4 \\ 5 & 20 \end{bmatrix} \begin{bmatrix} P \\ Q \end{bmatrix} = \begin{bmatrix} -12 \\ 150 \end{bmatrix}$$

*2)* Find the determinant of $A$.

$$|A| = -2(20) - 5(4) = -60$$

*3)* Replace the first column of $A$, which contains the coefficients of $P$, with the column vector of constant $B$.

$$A_1 = \begin{bmatrix} -12 & 4 \\ 150 & 20 \end{bmatrix}$$

The determinant of $A_1$ is

$$|A_1| = -12(20) - 150(4) = -840$$

Use Cramer's rule to find $P^*$.

$$P^* = \frac{|A_1|}{|A|} = \frac{-840}{-60} = 14$$

*4)* Replace the second column of $A$, which contains the coefficients of $Q$, with the column vector of constant $B$.

$$A_2 = \begin{bmatrix} -2 & -12 \\ 5 & 150 \end{bmatrix}$$

The determinant of $A_2$ is

$$|A_2| = -2(150) - 5(-12) = -240$$

Use Cramer's rule to find $Q^*$.

$$Q^* = \frac{|A_2|}{|A|} = \frac{-240}{-60} = 4$$

*5)* The equilibrium price $P^*$ is 14 and quantity $Q^*$ is 4.

# Chapter 7

## THEORETICAL CONCEPTS

**S7.1.** Assume that the constrained function can be solved by linear programming and has a unique solution. The solution will be situated in one of the corners of the feasible regions according to the extreme-point theorem.

**S7.2.** The graph of an equality constraint requires the inequality to be in slope-intercept form.

**S7.3.** A "greater than or equal to" inequality may be converted to an equation by adding a surplus variable.

**S7.4.** True.

**S7.5.** Using the basis theorem, where $n = 5$ equations, $v = 7$ variables:

$$N = \frac{v!}{n!(v-n)!} = \frac{7!}{5!(7-5)!} = \frac{7!}{5!(2)!} = \frac{7 \times 6 \times 5!}{5!(2)} = \frac{42}{2} = 21$$

## OPTIMIZATION USING GRAPHS

Use graphs to solve the following linear programming problems.

**S7.6.**   Maximize         $\pi = 12x_1 + 10x_2$
           subject to        $2x_1 + 5x_2 \le 90$      (Constraint A)
                             $14x_1 + 7x_2 \le 210$    (Constraint B)
                             $8x_1 + 10x_2 \le 200$    (Constraint C)
                             $x_1, x_2 \ge 0$

(1)   For all inequalities constraints, solve each $x_2$ in terms of $x_1$.

    1a.  For constraint A:        $x_2 = -(2/5) x_1 + 18$
    1b.  For constraint B:        $x_2 = -2 x_1 + 30$
    1c.  For constraint C:        $x_2 = -(4/5) x_1 + 20$

(2)   See Fig. 7-14(a) for the graphed constraints.

(3)   For the objective function, solve $x_2$ in terms of $x_1$.

$$x_2 = -(6/5) x_1 + \pi/10$$

The slope is $-(6/5)$.

(4)   See Fig. 7-14(b) for the feasible region and solution.

The point of tangency occurs at $x_1^* = 8.\overline{3}$, $x_2^* = 13.\overline{3}$, then $\pi^* = 12(8.\overline{3}) + 10(13.\overline{3}) = 233.\overline{3}$

    The solution is: $x_1^* = 8.\overline{3}$,   $x_2^* = 13.\overline{3}$,   $\pi^* = 233.\overline{3}$

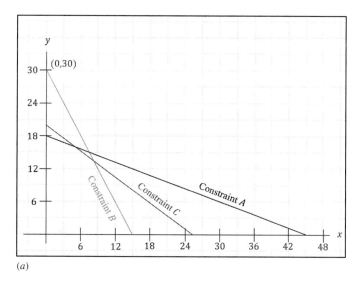

(a)

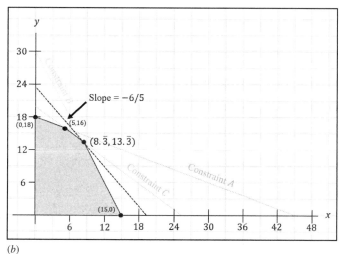

(b)

**Fig. 7-14**

**S7.7.**   Minimize       $\pi = 8x_1 + 5x_2$
subject to              $4x_1 + 3x_2 \geq 120$       (Constraint $A$)
$2x_1 + 3x_2 \geq 80$        (Constraint $B$)
$2x_1 + \phantom{3}x_2 \geq 50$        (Constraint $C$)
$x_1, x_2 \geq 0$

(1)   For all inequalities constraints, solve each $x_2$ in terms of $x_1$.

    *1a.* For constraint $A$:           $x_2 = (4/3)\, x_1 + 40$
    *1b.* For constraint $B$:           $x_2 = (2/3)\, x_1 + 80/3$
    *1c.* For constraint $C$:           $x_2 = -2x_1 \phantom{aa} + 50$

(2)   See Fig. 7-15(*a*) for the graphed constraints.

(3)   For the objective function, solve $x_2$ in terms of $x_1$.

$$x_2 = -(8/5)\, x_1 + \pi/5$$

The slope is $-(8/5)$.

(*4*)   See Fig. 7-15(*b*) for the feasible region and solution.

The point of tangency occurs at $x_1^* = 15$, $x_2^* = 20$, then $\pi^* = 8(15) + 5(20) = 220$

The solution is: $x_1^* = 15$,   $x_2^* = 20$,   $\pi^* = 220$

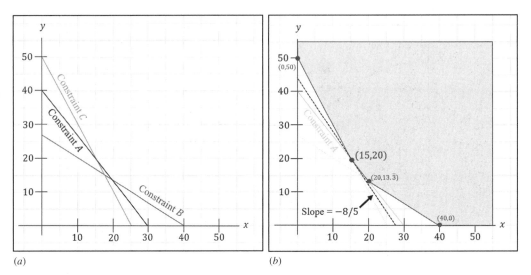

**Fig. 7-15**

## REPRESENTATION OF BUSINESS AND ECONOMICS PROBLEMS

**S7.8.**   (*1*) Represent the data as a linear programming problem and fill the missing values.

$$
\begin{array}{lll}
\text{Maximize} & \pi = 640x_1 + 360x_2 & \\
\text{subject to} & x_1 + x_2 \le 40 & \text{(Land constraint)} \\
& 900x_1 + 540x_2 \le 27000 & \text{(Capital constraint)} \\
& 12x_1 + 4x_2 \le 320 & \text{(Labor constraint)} \\
& x_1, x_2 \ge 0 &
\end{array}
$$

(2)  Figure 7-16 shows the feasible region indicating the point of tangency (solution).

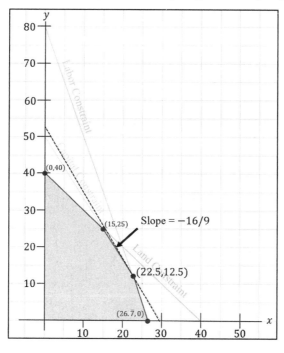

**Fig. 7-16**

The point of tangency occurs at $x_1^* = 22.5$, $x_2^* = 12.5$, then $\pi^* = 640(22.5) + 360(12.5) = 18900$

The solution is: corn = 22.5 acres, soybeans = 12.5 acres, profit = $18,900.

# Chapter 8

## THEORETICAL CONCEPTS

**S8.1.** The computational method to solve a linear programming model is called the simplex algorithm.

**S8.2.** False. The displacement ratio is found by dividing the elements of the constant column by the elements of the pivot column.

**S8.3.** The dual of a maximization problem is a minimization problem.

**S8.4.** If the second decision variable $y_2$ of the dual is zero ($y_2 = 0$), then the shadow price of the primal is zero. Therefore, the second equation of the primal (constraint $B$) is not active. Thus, the slack variable is greater than 0 ($s_2 > 0$).

**S8.5.** True.

## OPTIMIZATION USING THE SIMPLEX ALGORITHM

**S8.6.**  Maximize              $\pi = 12x_1 + 10x_2$
subject to                      $2x_1 + 5x_2 \le 90$                (Constraint $A$)
                                $14x_1 + 7x_2 \le 210$              (Constraint $B$)
                                $8x_1 + 10x_2 \le 200$              (Constraint $C$)
                                $x_1, x_2 \ge 0$

(1)  Construct the initial simplex tableau.

    *1a.* Add slack variables to the inequalities to make them equations.

$$2x_1 + 5x_2 + s_1 = 90$$
$$14x_1 + 7x_2 + s_2 = 210$$
$$8x_1 + 10x_2 + s_3 = 200$$

    *1b.* Express the constraints in matrix form.

$$\begin{bmatrix} 2 & 5 & 1 & 0 & 0 \\ 14 & 7 & 0 & 1 & 0 \\ 8 & 10 & 0 & 0 & 1 \end{bmatrix} \begin{bmatrix} x_1 \\ x_2 \\ s_1 \\ s_2 \\ s_3 \end{bmatrix} = \begin{bmatrix} 90 \\ 210 \\ 200 \end{bmatrix}$$

    *1c.* Form the initial tableau.

| $x_1$ | $x_2$ | $s_1$ | $s_2$ | $s_3$ | Constant |
|---|---|---|---|---|---|
| 2 | 5 | 1 | 0 | 0 | 90 |
| (14) | 7 | 0 | 1 | 0 | 210 |
| 8 | 10 | 0 | 0 | 1 | 200 |
| −12 | −10 | 0 | 0 | 0 | 0 |

The first basic solution is: $x_1 = x_2 = 0$, $s_1 = 90$, $s_2 = 210$, $s_3 = 200$, $\pi = 0$.

(2)  Change the basis. The negative indicator with the largest absolute value determines the pivot column, in this case −12. The displacement ratios (in order) are 45, 15, and 25. The smallest displacement ratio decides the pivot row. Thus, (14) becomes the pivot element.

(3)  Pivoting

    *3a.* Multiply the pivot row by the reciprocal of the pivot element; in other words, multiply row 2 of the initial tableau by 1/4.

| $x_1$ | $x_2$ | $s_1$ | $s_2$ | $s_3$ | Constant |
|---|---|---|---|---|---|
| 2 | 5 | 1 | 0 | 0 | 90 |
| 1 | 1/2 | 0 | 1/14 | 0 | 15 |
| 8 | 10 | 0 | 0 | 1 | 200 |
| −12 | −10 | 0 | 0 | 0 | 0 |

*3b.* Having reduced the pivot element to 1, clear the pivot column by subtracting 2 times row 2 from row 1, subtracting 8 times row 2 from row 3, and adding 12 times row 2 to row 4. This gives a second tableau.

| $x_1$ | $x_2$ | $s_1$ | $s_2$ | $s_3$ | Constant |
|---|---|---|---|---|---|
| 0 | 4 | 1 | −1/7 | 0 | 60 |
| 1 | 1/2 | 0 | 1/14 | 0 | 15 |
| 0 | ⑥ | 0 | −4/7 | 1 | 80 |
| 0 | −4 | 0 | 6/7 | 0 | 180 |

↑

The second basic feasible solution can be reached by setting equal to zero all the variable headers that are not composed of unit vectors (in this case $x_2$ and $s_2$). The second basic feasible solution is then: $x_1 = 15$, $x_2 = 0$, $s_1 = 60$, $s_2 = 0$, $s_3 = 80$, $\pi = 180$.

*3c.* Since −4 in the second column is the only negative indicator, $x_2$ is brought to the basis and column 2 becomes the pivot column. The displacement ratios (in order) are 15, 30, 13.3. Thus, ⑥ becomes the pivot element. Multiply the pivot row (in this case row 3) by 1/6.

| $x_1$ | $x_2$ | $s_1$ | $s_2$ | $s_3$ | Constant |
|---|---|---|---|---|---|
| 0 | 4 | 1 | −1/7 | 0 | 60 |
| 1 | 1/2 | 0 | 1/14 | 0 | 15 |
| 0 | 1 | 0 | −2/21 | 1/6 | 40/3 |
| 0 | −4 | 0 | 6/7 | 0 | 180 |

*3d.* Having reduced the pivot element to 1, clear the pivot column by subtracting 4 times row 3 from row 1, subtracting ½ times row 3 from row 2, and adding 4 times row 3 to row 4. This gives a third tableau.

| $x_1$ | $x_2$ | $s_1$ | $s_2$ | $s_3$ | Constant |
|---|---|---|---|---|---|
| 0 | 0 | 1 | 5/21 | 2/3 | 20/3 |
| 1 | 0 | 0 | 5/42 | 0 | 25/3 |
| 0 | 1 | 0 | −2/21 | 1/6 | 40/3 |
| 0 | 0 | 0 | 10/21 | 0 | 700/3 |

After setting all the variables heading nonunit vectors equal to zero ($s_2 = s_3 = 0$) and rearranging the unit column vectors, we see that the third feasible solution is: $x_1 = 25/3$, $x_2 = 40/3$, $s_1 = 20/3$. Since there are no negative indicators left in the last row, this is the optimal solution. Note that $s_1 = 20/3$; thus, the second constraint is not active and 20/3 units of the first input remain unused. For a graphic representation, see Problem S7.6 and its respective Fig. 7-14.

The solution is: $x_1^* = 25/3$, $x_2^* = 40/3$, $\pi^* = 700/3$.

## THE DUAL

Find the dual of the following problems:

**S8.7.** Maximize      $\pi = 12x_1 + 10x_2$
      subject to

$$2x_1 + 5x_2 \le 90 \qquad \text{(Constraint } A\text{)}$$
$$14x_1 + 7x_2 \le 210 \qquad \text{(Constraint } B\text{)}$$
$$8x_1 + 10x_2 \le 200 \qquad \text{(Constraint } C\text{)}$$
$$x_1, x_2 \ge 0$$

The dual is:

Minimize    $c = 90y_1 + 210y_2 + 200y_3$
subject to
$$2y_1 + 14y_2 + 8y_3 \geq 12 \qquad \text{(Constraint } I)$$
$$5y_1 + 7y_2 + 10y_3 \geq 10 \qquad \text{(Constraint } II)$$
$$y_1, y_2, y_3 \geq 0$$

**S8.8.**   Minimize    $c = 8x_1 + 5x_2$
subject to
$$4x_1 + 3x_2 \geq 120 \qquad \text{(Constraint } A)$$
$$2x_1 + 3x_2 \geq 80 \qquad \text{(Constraint } B)$$
$$2x_1 + x_2 \geq 50 \qquad \text{(Constraint } C)$$
$$x_1, x_2 \geq 0$$

The dual is:

Maximize    $\pi = 120y_1 + 80y_2 + 50y_3$
subject to
$$4y_1 + 2y_2 + 2y_3 \geq 8 \qquad \text{(Constraint } I)$$
$$3y_1 + 3y_2 + y_3 \geq 10 \qquad \text{(Constraint } II)$$
$$y_1, y_2, y_3 \geq 0$$

## OPTIMIZATION WITH THE DUAL

**S8.9.**   Answer the questions below based on the following linear programming problem is given

Minimize    $c = 24y_1 + 60y_2 + 90y_3$
subject to
$$y_1 + 2y_2 + 3y_3 \geq 1 \qquad \text{(Constraint } I)$$
$$y_1 + 5y_2 + 4y_3 \geq 2 \qquad \text{(Constraint } II)$$
$$y_1, y_2, y_3 \geq 0$$

(1)   Find the dual of this model and label $x$ as the decision variable and $\pi$ as the dual optimal variable.

Maximize    $\pi = x_1 + 2x_2$
subject to
$$x_1 + x_2 \leq 24 \qquad \text{(Constraint } A)$$
$$2x_1 + 5x_2 \leq 60 \qquad \text{(Constraint } B)$$
$$3x_1 + 4x_2 \leq 90 \qquad \text{(Constraint } C)$$
$$x_1, x_2 \geq 0$$

(2)   Solve the dual graphically and indicate the value of $\pi^*$.

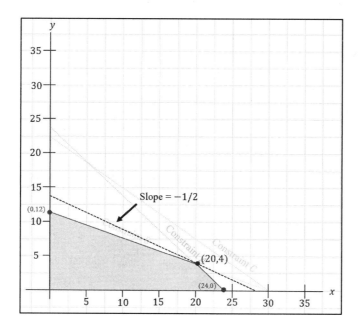

The solution is: $x_1^* = 20$, $x_2^* = 4$, $\pi^* = 28$. Considering that the objective values are the same for the dual and the primal, then $c^* = 28$.

(3) Write both the primal and the dual including slack (label them as $s$) and surplus variables (label them as $t$).

3a. Primal:

$$y_1 + 2y_2 + 3y_3 - t_1 = 1$$
$$y_1 + 5y_2 + 4y_3 - t_2 = 2$$

3b. Dual:

$$x_1 + \quad x_2 + s_1 = 24$$
$$2x_1 + 5x_2 + s_2 = 60$$
$$3x_1 + 4x_2 + s_3 = 90$$

(4) Find the value of the slack variables ($s_1$, $s_2$, and $s_3$) using the dual equation system.

Substitute the optimal values. $x_1^* = 20$, $x_2^* = 4$ in the dual:

$$20 + \quad 4 + s_1 = 24 \qquad\qquad (s_1^* = 0)$$
$$2(20) + 5(4) + s_2 = 60 \qquad\qquad (s_2^* = 0)$$
$$3(20) + 4(4) + s_3 = 90 \qquad\qquad (s_3^* = 14)$$

With this information:

Constraint $A$ is active. The value of the slack variable $= 0$.

Constraint $B$ is active. The value of the slack variable $= 0$.

Constraint $C$ is not active. The value of the slack variable $= 14$.

(5) According to the second dual theorem, which constraints are active in the dual?

The second dual theorem states that, if the decision variables of the dual ($x_1$ and $x_2$) have nonzero values, then their corresponding primal surplus/slack variables, here $t_1$ and $t_2$, must be zero. Thus, $t_1^* = t_2^* = 0$. Therefore,

> Constraint *I* is active. The value of the surplus variable is zero.

> Constraint *II* is active. The value of the surplus variable is zero.

(6)  Solve the primal system of equations.

With $s_1^* = s_2^* = 0$, according to the second dual theorem, the corresponding primal decision variables $y_1$ and $y_2$ must have nonzero values. With $s_3^* \neq 0$, then $y_3 = 0$.

Substitute both values in the primal equation

$$y_1 + 2y_2 + 3(0) - (0) = 1$$
$$y_1 + 5y_2 + 4(0) - (0) = 2$$

and solve simultaneously:

$$y_1^* = 1/3, \ y_2^* = 1/3$$

One way to check the answer is by finding the optimal value: $c = 24(1/3) + 60(1/3) + 90(0) = 28$, which is the same as the optimal value of the dual ($\pi^* = 28$), which was found graphically.

> The solution is: $y_1^* = \dfrac{1}{3}, \ y_2^* = \dfrac{1}{3}, \ y_3^* = 0, \ c^* = 28$

# Chapter 9

## THEORETICAL CONCEPTS

**S9.1.**  True.

**S9.2.**  True.

**S9.3.**  The derivative of a quadratic function is a linear function.

**S9.4.**  The technique used to take the derivative of a composite function is called the chain rule.

**S9.5.**  False. The third-order derivative can be obtained by taking thrice the derivative of the function.

## LIMITS AND DIFFERENTIATION

**S9.6.**  Find $\lim\limits_{x \to 5} f(x)$, if it exists, given

$$f(x) = \frac{2x - 10}{x^2 - 25} \qquad\qquad (x \neq 5)$$

*(1)*   Plug values of $x$ in the function, fill out the table (use three decimal points), and graph it.

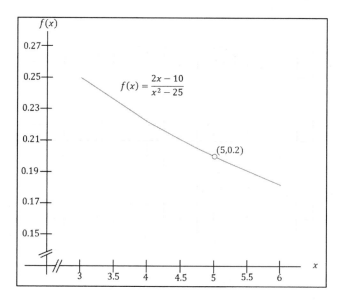

| $x$ | $f(x)$ |
|-----|--------|
| 3   | 0.250  |
| 4   | 0.222  |
| 4.5 | 0.211  |
| 4.8 | 0.204  |
| 4.9 | 0.202  |
| 5.1 | 0.198  |
| 5.2 | 0.196  |
| 5.5 | 0.190  |
| 6   | 0.182  |

**Fig. 9-8**

*(2)*   Solve the function algebraically; be careful when simplifying terms in the denominator.

$$\lim_{x \to 5} \frac{2x - 10}{x^2 - 25} = \lim_{x \to 5} \frac{2(x - 5)}{(x + 5)(x - 5)} = \lim_{x \to 5} \frac{2}{(x + 5)} = \frac{2}{7}$$

**S9.7.**   Using limits, find the derivative of $f(x) = x^2 - 2x$.

*(1)*   Employ the general formula of the derivative and replace the function.

$$f'(x) = \lim_{\Delta x \to 0} \frac{f(x + \Delta x) - f(x)}{\Delta x} = \lim_{\Delta x \to 0} \frac{(x + \Delta x)^2 - 2(x + \Delta x) - (x^2 - 2x)}{\Delta x}$$

*(2)*   Simplify the results.

$$f'(x) = \lim_{\Delta x \to 0} \frac{x^2 + 2x\Delta x + (\Delta x)^2 - 2x - 2\Delta x - x^2 + 2x}{\Delta x} = \lim_{\Delta x \to 0} \frac{(\Delta x)^2 + 2x\Delta x - 2\Delta x}{\Delta x}$$

(3)  Divide through by $\Delta x$.

$$f'(x) = \lim_{\Delta x \to 0} \frac{\Delta x(\Delta x + 2x - 2)}{\Delta x} = \lim_{\Delta x \to 0} (\Delta x + 2x - 2)$$

(4)  Take the limit of the simplified expression.

$$f'(x) = 0 + 2x - 2 = 2x - 2$$

## DERIVATIVE RULES

Find the first derivative of the following functions:

**S9.8.**  Use the basic rules to differentiate $f(x) = 12x^{1/3} - 10x^{-1/5} + 8$.

$$f'(x) = 12\left(\tfrac{1}{3}\right)x^{1/3 - 1} - 10\left(-\tfrac{1}{5}\right)x^{-1/5 - 1} + 0 = 4x^{-2/3} + 2x^{-6/5}$$

**S9.9.**  Use the product rule to differentiate $f(x) = (9x^{1/3} + 4)(10x^{-1/5} + 2x)$.

$$g(x) \quad \cdot \quad h'(x) \quad + \quad g'(x) \quad \cdot \quad h(x)$$

$$f'(x) = (9x^{1/3} + 4) \cdot (-2x^{-6/5} + 2) + (3x^{-2/3}) \cdot (10x^{-1/5} + 2x)$$

$$f'(x) = -18x^{-13/15} + 18x^{1/3} - 8x^{-6/5} + 8 + 30x^{-13/15} + 6x^{1/3}$$

Simplify algebraically.

$$f'(x) = 12x^{-13/15} + 24x^{1/3} + 8x^{-6/5} + 8$$

**S9.10.**  Use the quotient rule to differentiate $f(x) = \dfrac{4x^{5/4}}{5x - 1}$.

$$g'(x) \qquad h(x) \quad - \quad g(x) \qquad h'(x)$$

$$f'(x) = \frac{\boxed{(5x^{1/4})} \cdot \boxed{(5x - 1)} - \boxed{(4x^{5/4})} \cdot \boxed{(5)}}{\boxed{(5x - 1)^2}}$$

$$[h(x)]^2$$

Simplify algebraically.

$$f'(x) = \frac{25x^{5/4} + 5x^{1/4} - 20x^{5/4}}{(5x - 1)^2} = \frac{5x^{5/4} - 5x^{1/4}}{(5x - 1)^2}$$

$$f'(x) = = \frac{5x^{1/4}(x - 1)}{(5x - 1)^2}$$

**S9.11.** Use the generalized power rule to differentiate $f(x) = 1/\sqrt[3]{4x^2 - 5}$.

Define the function in exponential terms.

$$f(x) = (4x^2 - 5)^{-1/3}$$

Take the derivative with respect to $x$.

$$f'(x) = \boxed{(-1/3)} \cdot \boxed{(4x^2 - 5)^{-4/3}} \cdot \boxed{(8x)}$$
$$\quad n \qquad\qquad [g(x)]^{n-1} \quad g'(x)$$

Simplify algebraically.

$$f'(x) = \left(-\frac{8}{3}x\right)(4x^2 - 5)^{-4/3}$$

**S9.12.** Use the chain rule to differentiate $f(x) = (6x^{1/2} + 9x)^8$.

Let $\quad y = \boxed{u^8} \quad$ and $\quad u = \boxed{6x^{1/2} + 9x}$

Then $\quad dy/du = \boxed{u^8} \quad$ and $du/dx = \boxed{3x^{-1/2} + 9}$

Substitute in the formula.

$$\frac{dy}{dx} = \boxed{8u^7} \cdot \boxed{(3x^{-1/2} + 9)}$$

Replace $u$ in terms of $x$.

$$\frac{dy}{dx} = \boxed{8(6x^{1/2} + 9x)^7 (3x^{-1/2} + 9)}$$

## HIGHER-ORDER DERIVATIVE

**S9.13.** Given the function $f(x) = (x^2 + 8)^3$, find the second-order derivative.

(*1*) Take the first derivative of the function.

$$f'(x) = 3(x^2 + 8)^2 (2x) = 6x(x^2 + 8)^2$$

(*2*) Take the derivative of the first-order derivative to obtain the second-order derivative.

Apply the product rule to differentiate the function.

$$f''(x) = [6x]'(x^2 + 8)^2 + 6x[(x^2 + 8)^2]'$$
$$f''(x) = 6(x^2 + 8)^2 + 6x[2(2x)(x^2 + 8)]$$

Simplify expressions.

$$f''(x) = 6(x^4 + 16x^2 + 64) + 6x[4x^3 + 32x]$$
$$f''(x) = 6x^4 + 96x^2 + 384 + 24x^4 + 192x^2$$
$$f''(x) = \boxed{30x^4 + 288x^2 + 384}$$

**S9.14.** Given the function $f(x) = x^{-1/4} + x^{1/4}$, find the third-order derivative.

    (*1*)    Take the first derivative of the function.

$$f'(x) = -\frac{1}{4}x^{-5/4} + \frac{1}{4}x^{-1/4}$$

    (*2*)    Find the second derivative of the function.

$$f''(x) = \frac{5}{16}x^{-9/4} - \frac{1}{16}x^{-5/4}$$

    (*3*)    Compute the derivative of the second-order derivative to obtain the third-order derivative.

$$f'''(x) = \boxed{-\frac{45}{64}x^{-13/4} + \frac{5}{64}x^{-9/4}}$$

# Chapter 10

## THEORETICAL CONCEPTS

**S10.1.** A differentiable function $f(x)$ whose second-order derivative evaluated at $x = 5$ is negative can be classified as a concave function.

**S10.2.** False. The classification of relative extremum does not depend on the first-order derivative of the function.

**S10.3.** The inflection point is where the second derivative of the function equals zero.

**S10.4.** A function is increasing if the first derivative of the function evaluated at $x = a$ is greater than zero.

**S10.5.** True.

## INCREASING AND DECREASING FUNCTIONS

**S10.6.** Determine whether the function $f(x) = 7x^4 + 3x^2 - x - 1000$ is decreasing or decreasing at the point $x = 3$.

    (*1*)    Take the first order derivative.

$$f'(x) = 28x^3 + 6x - 1$$

    (*2*)    Substitute $x = 3$ and classify the point.

$$f'(3) = 28(3)^3 + 6(3) - 1 = 773 > 0 \qquad \text{function is increasing}$$

## CONCAVITY AND CONVEXITY

**S10.7.** Determine whether the function $y = 5x^5 + 3x^4 - x^2 - 200$ is convex or concave at the point $x = 2$.

    (1)  Find the second order derivative.

$$y' = 25x^4 + 12x^3 - 2x$$

$$y'' = 100x^3 + 36x^2 - 2$$

    (2)  Substitute $x = 2$ and classify the point.

$$y''(2) = 100(2)^3 + 36(2)^2 - 2 = 942 > 0 \qquad \text{convex}$$

## SKETCHING THE GRAPHS: RELATIVE EXTREMA AND INFLECTION POINT

**S10.8.** Graph the following function: $y = 0.25x^4 - 3x^3 + 10x^2 + 90$

    (1)  Find the $y$-intercept.

$$\text{Set } x = 0; \text{ then } y(0) = 90$$

    (2)  Find the critical values (relative extrema) and classify them as minima or maxima points.

        2a.  Take the first derivative, set it equal to zero, and solve for $x$.

$$y' = x^3 - 9x^2 + 20x = 0$$
$$\text{Factor:} \quad y' = x(x^2 - 9x + 20) = 0$$
$$= x(x-4)(x-5) = 0$$
$$x = 0 \quad x = 4 \quad x = 5 \qquad \text{critical values}$$

        2b.  Take the second derivative to evaluate the concavity of each critical value.

$$y'' = 3x^2 - 18x + 20$$

$$y''(0) = 3(0)^2 - 18(0) + 20 = 20 > 0 \qquad \text{convex, relative minimum}$$
$$y''(4) = 3(4)^2 - 18(4) + 20 = -4 < 0 \qquad \text{concave, relative maximum}$$
$$y''(5) = 3(5)^2 - 18(5) + 20 = \ \ 5 > 0 \qquad \text{convex, relative minimum}$$

    (3)  Find the inflection points and indicate if they are decreasing or increasing

        3a.  Set the second derivative equal to zero and solve for $x$.

$$y'' = 3x^2 - 18x + 20 = 0$$

Use the quadratic formula; here $a = 3$, $b = -18$, $c = 20$.

$$x = \frac{-(-18) \pm \sqrt{(-18)^2 - 4(3)(20)}}{2(3)} = \frac{18 \pm \sqrt{84}}{6}$$

Round to two decimals:   $x = 4.53$     $x = 1.47$    inflection points

*3b.* Plug the inflection points in the first derivative to evaluate if they are decreasing or increasing.

$$y' = x^3 - 9x^2 + 20x = 0$$

$$y'(4.53) = (4.53)^3 - 9(4.53)^2 + 20(4.53) = -1.13 < 0 \qquad \text{decreasing}$$

$$y'(1.47) = (1.47)^3 - 9(1.47)^2 + 20(1.47) = 13.12 > 0 \qquad \text{increasing}$$

*(4)*  Find the *y*-values for each critical and inflection point.

| $x$ | Point classification | $f(x)$ |
|---|---|---|
| 0 | Critical, relative minimum | 90.00 |
| 1.47 | Inflection, decreasing | 103.28 |
| 4 | Critical, relative maximum | 122.00 |
| 4.53 | Inflection, increasing | 121.61 |
| 5 | Critical, relative minimum | 121.25 |

*(5)*  The graph is depicted below in Fig. 10-20.

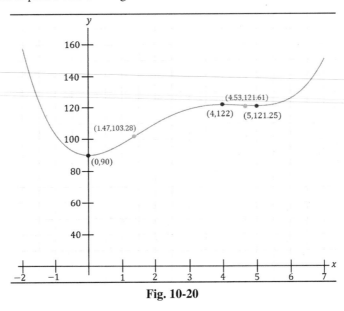

**Fig. 10-20**

## SKETCHING THE GRAPHS: MARGINAL AND AVERAGE FUNCTIONS IN ECONOMICS

**S10.9.** Given the total cost function $TC = 2Q^3 - 12Q^2 + 225Q$, create a graph with two panels. (*I*) The first one sketches the total cost curve indicating the inflection point, and (*II*) the second panel depicts the marginal and average cost curves, indicating their point of intersection and the minimum point of the MC curve.

*(1)*  Take the first and second derivative of the total cost function.

$$TC' = 6Q^2 - 24Q + 225$$
$$TC'' = 12Q - 24$$

(2) Check for (*a*) concavity and (*b*) inflection points, using the second derivative.

Find the inflection point:

$$TC''(Q) = 12Q - 24 = 0$$
$$Q = 2$$

For $\quad Q < 2, \quad TC'' < 0 \quad$ concave

For $\quad Q > 2, \quad TC'' > 0 \quad$ convex

Plug the inflection point in the first derivative.

$$TC'(2) = 6(2)^2 - 24(2) + 225 = 201 > 0 \quad \text{increasing}$$

(3) Find the average cost functions and the relative extrema.

$$AC = \frac{TC}{Q} = 2Q^2 - 12Q + 225$$

Find the relative extremum: $AC' = 4Q - 12 = 0$

$$Q = 3 \qquad \text{critical value}$$
$$AC'' = 4 > 0 \qquad \text{convex, relative minimum}$$

(4) Find the marginal cost functions and the relative extrema.

$$MC = \frac{dTC}{dQ} = 6Q^2 - 24Q + 225$$

Find the relative extremum: $MC' = 12Q - 24 = 0$

$$Q = 2 \qquad \text{critical value}$$
$$MC'' = 12 > 0 \qquad \text{convex, relative minimum}$$

(5) Verify the point of intersection between the average and the marginal cost functions (note $Q > 0$).

Set $AC = MC$: $\quad 2Q^2 - 12Q + 225 = 6Q^2 - 24Q + 225$

$$4Q^2 - 12Q = 0$$

Factor: $\quad\quad Q(4Q - 12) = 0$

Given $Q > 0$. $\quad\quad 4Q - 12 = 0$

$$Q = 3 \quad \text{point of intersection}$$

(6)    The graph is depicted here in Fig. 10-21.

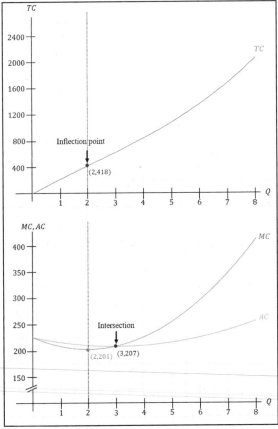

**Fig. 10-21**

## OPTIMIZATION IN BUSINESS AND ECONOMICS

**S10.10.**   Find the critical points at which profit $\pi$ is maximized given the total revenue $TR = 470Q - 3Q^2$ and total cost $TC = 32Q + 10500$.

(1)   Compute marginal revenue ($MR$) and marginal cost ($MC$) functions.

$$MR = \frac{dTR}{dQ} = 470 - 6Q \qquad MC = \frac{dTC}{dQ} = 32$$

(2)   Equate $MR = MC$ to find $Q^*$.

$$470 - 6Q = 32$$
$$Q^* = 73$$

(3)   Verify that $Q^*$ is a relative maximum point.
The objective is to maximize $\pi$; thus, plug $Q^*$ in $\pi''$ to check if it is a relative maximum.

Since $\pi = TR - TC = (470Q - 3Q^2) - (32Q + 10500)$

$$\pi = -3Q^2 + 438Q - 10500$$

Take first derivative    $\pi' = -6Q + 438$

second derivative    $\pi'' = -6 < 0$ for all values of $Q$    concave, relative maximum

(4)  Compute the maximum profit level $\pi^*$ by establishing $\pi^* = \pi(Q^*)$.

$$\pi(Q^*) = -3(73)^2 + 438(73) - 10500$$
$$\pi(73) = 5487$$

# Chapter 11

## THEORETICAL CONCEPTS

**S11.1.** True.

**S11.2.** True.

**S11.3.** False. Logarithmic functions can be negative if the term inside the logarithm is a fraction between zero and one.

**S11.4.** True.

**S11.5.** The summation of two logarithmic functions can be combined by writing only one time its base and multiplying the terms in each logarithm.

## EXPONENTIAL-LOGARITHMIC CONVERSIONS

Change each of the following functions to its corresponding inverse function.

**S11.6.** Convert $y = e^{5x+8}$ into a logarithmic function by solving for $x$.

Take natural logarithm to both sides.

$$\ln(y) = \ln(e^{5x+8})$$

Simplify    $\ln(y) = 5x + 8$

Solve for $x$    $x = (1/5)\ln(y) - 8/5$

**S11.7.** Convert $y = \ln(3x - 2)$ into an exponential function by solving for $x$.

Raise both sides to natural exponentials.

$$e^y = e^{\ln(3x-2)}$$

Simplify    $e^y = 3x - 2$

Solve for $x$    $x = \left(\dfrac{1}{3}\right)e^y + \dfrac{2}{3}$

## PROPERTIES OF LOGARITHMIC AND EXPONENTIAL FUNCTIONS

Convert the expressions to sums, differences, or products.

**S11.8.** Convert $y = \ln(5x+4) + \ln(4x-9)$ into a product.

The addition of two terms can be converted into a multiplication: $y = \ln[(5x+4)\cdot(4x-9)]$

**S11.9.** Convert $z = \ln\left(\dfrac{4e^2 x}{\sqrt[3]{y^2}}\right)$ into additions and subtractions.

Separate ratios as subtraction between terms and products as additions.

$$z = \ln\left(\frac{4e^2 x}{\sqrt[3]{y^2}}\right) = \ln(4) + \ln(e^2) + \ln(x) - \ln\left(\sqrt[3]{y^2}\right)$$

Simplify terms.

$$z = \ln(4) + 2 + \ln(x) - \frac{2}{3}\ln(y)$$

**S11.10.** Simplify $y = e^{[\ln(x^2)+x]}$.

Separating the addition of exponents as product of exponential terms

$$y = e^{[\ln(x^2)+x]} = e^{\ln(x^2)}e^x$$

Applying the property of the simplification of inverse functions $e^{\ln(w)} = w$

$$y = x^2 e^x$$

## DERIVATIVES

**S11.11.** Differentiate the logarithmic function $y = \ln\left(\dfrac{x^2}{x-2}\right)$

(1)   Use logarithmic properties to convert the expression into subtraction of terms.

$$y = \ln\left(\frac{x^2}{x-2}\right) = \ln(x^2) - \ln(x-2)$$
$$y = 2\ln(x) - \ln(x-2)$$

(2)   Take the first-order derivative.

$$y' = \frac{2}{x} - \frac{1}{(x-2)}$$

**S11.12.** Differentiate the natural exponential function $y = e^{(x^2+5)}$ by combining derivative rules.

Use the chain rule.

$$y' = (x^2+5)'e^{(x^2+5)}$$
$$y' = 2xe^{(x^2+5)}$$

**S11.13.** Differentiate the function

$$y = e^{[\ln(x-4)-\ln(x-2)]}.$$

(*1*) Simplify terms:

Use logarithmic properties to convert the expression from subtraction of terms into a ratio.

$$y = e^{[\ln(x-4)-\ln(x-2)]} = e^{\ln\left(\frac{x-4}{x-2}\right)}$$

Apply the property of the simplification of inverse functions $e^{\ln(w)} = w$.

$$y = \frac{x-4}{x-2}$$

(*2*) Take the first-order derivative

Use the quotient rule.

$$y' = \frac{(x-4)'(x-2) - (x-4)(x-2)'}{(x-2)^2} = \frac{(1)(x-2) - (x-4)(1)}{(x-2)^2} = \frac{x-2-x+4}{(x-2)^2}$$

$$y' = \frac{2}{(x-2)^2}$$

# Chapter 12

## THEORETICAL CONCEPTS

**S12.1.** True.

**S12.2.** The integral of the marginal revenue is the total revenue function.

**S12.3.** Integration by parts is the inverse rule of the product rule.

**S12.4.** True.

**S12.5.** Definite integrals may be interpreted as the area between the function and the $x$-axis.

## DEFINITE AND INDEFINITE INTEGRALS

**S12.6.** Computing the indefinite integral.

$$\int (5x^{1/2} - 2x^{-1/3})dx = 5\left(\frac{2}{3}\right)x^{3/2} - 2\left(\frac{3}{2}\right)x^{2/3} + c$$

$$= \frac{10}{3}x^{3/2} - 3x^{2/3} + c$$

**S12.7.** Find the antiderivative formula given the boundary condition $F(1) = 8$ and the function

$$f(x) = 18x^5 + 4x^{-3}$$

(*1*) Compute the indefinite integral.

$$\int (18x^5 + 4x^{-3})\,dx = 18\left(\frac{1}{6}\right)x^6 + 4\left(-\frac{1}{2}\right)x^{-2} + c$$
$$F(x) = 3x^6 - 2x^{-2} + c$$

(*2*) Use the initial condition to find the constant.

At $F(1) = 8$,         $8 = 3(1)^6 - 2(1)^{-2} + c$
$$c = 7$$

(*3*) The antiderivative formula is

$$F(x) = 3x^6 - 2x^{-2} + 7$$

**S12.8.** Evaluate the definite integral:

$$\int_0^3 (e^{5x} - 2x^2)\,dx$$

$$\int_0^3 (e^{5x} - 2x^2)\,dx = \left(\frac{1}{5}e^{5x} - \frac{2}{3}x^3\right)\Big|_0^3 = \left(\frac{1}{5}e^{5(3)} - \frac{2}{3}(3)^3\right) - \left(\frac{1}{5}e^{5(0)} - \frac{2}{3}(0)^3\right)$$

$$= \frac{1}{5}e^{15} - 18 - \left(\frac{1}{5}e^0\right) = \frac{1}{5}e^{15} - \frac{91}{5}$$

$$= \frac{1}{5}(e^{15} - 91)$$

## AREA BETWEEN CURVES

**S12.9.** Find the area between $y_1 = 26 - 3x$ and $y_2 = 30 - x^2$.

(*1*) Compute the points of intersection of both functions to find the range of $x$.

Equate $y_1 = y_2$:         $26 - 3x = 30 - x^2$
$$x^2 - 3x - 4 = 0$$
Factor:         $(x+1)(x-4) = 0$
$$x = -1 \quad x = 4$$

(2) Graphing both functions in Fig. 12-14, $y_2$ is above $y_1$ in the order for integration.

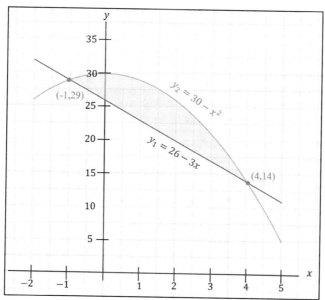

**Fig. 12-14**

(3) Evaluate the definite integral under the area in Fig. 12-14 (note $-1 \leq x \leq 4$).

$$A = \int_{-1}^{4} [(30-x^2)-(26-3x)]\,dx = \int_{-1}^{4} [-x^2+3x+4]\,dx = \left(-\frac{1}{3}x^3+\frac{3}{2}x^2+4x\right)\Big|_{-1}^{4}$$

$$A = \left(-\frac{1}{3}(4)^3+\frac{3}{2}(4)^2+4(4)\right)-\left(-\frac{1}{3}(-1)^3+\frac{3}{2}(-1)^2+4(-1)\right)$$

$$A = \frac{56}{3}+\frac{13}{6}=\frac{125}{6}$$

## INTEGRATION BY SUBSTITUTION AND BY PARTS

**S12.10.** Use integration by substitution to determine the following indefinite integral.

$$\int \frac{30x^4}{\sqrt{6x^5-36}}\,dx$$

(1) Select a suitable $u$ function.

$$u(x) = \boxed{6x^5-36}$$

(2) Compute $du$ and find $dx$ in terms of $du$.

$$du = \boxed{30x^4} \quad dx$$

$$dx = \boxed{1/(30x^4)} \quad du$$

(3)   Substitute the relationship in the integration.

$$\int \frac{30x^4}{\sqrt{6x^5-36}}dx = \int \frac{30x^4}{\sqrt{u}} \cdot \frac{du}{30x^4} = \int \frac{1}{\sqrt{u}} \cdot du = \int u^{-1/2}du$$

(4)   Integrate.

$$\int u^{-1/2}\,du = 2u^{1/2}+c$$

(5)   Substitute back to have the integration in terms of $x$.

$$\int \frac{30x^4}{\sqrt{6x^5-36}}dx = 2u^{1/2}+c = 2(6x^5-36)^{1/2}+c$$

$$\int \frac{30x^4}{\sqrt{6x^5-36}}dx = 2\sqrt{6x^5-36}+c$$

**S12.11.** Use integration by parts to determine the following definite integral.

$$\int_0^7 30x\sqrt{9+x}\,dx$$

(1)   Pick suitable $f(x)$ and $g(x)$ functions.

$$f(x)= \boxed{x} \qquad\qquad g'(x)=\boxed{\sqrt{9+x}}$$

(2)   Compute $f'(x)$ and find $g(x)$ in terms of $dx$.

Differentiate.   $f'(x)=\boxed{1}$

Integrate.        $g(x)=\boxed{\dfrac{2}{3}(9+x)^{3/2}}$

(3)   Substitute the relationship in the integration to create the indefinite integral.

First, indefinite integration is required.

$$\int [30x\sqrt{9+x}]\,dx = 30(x)\left(\frac{2}{3}\right)(9+x)^{3/2} - 30\int \left[(1)\cdot\frac{2}{3}(9+x)^{3/2}\right]dx$$

$$\int [30x\sqrt{9+x}]\,dx = 20x(9+x)^{3/2} - 20\int [(9+x)^{3/2}]\,dx$$

$$2\sqrt{6x^5-36}+c$$

(4)  Integrate the indefinite integral.

$$\int [30x\sqrt{9+x}]\,dx = 20x(9+x)^{3/2} - 20\left(\frac{2}{5}\right)(9+x)^{5/2} + c$$

$$\int [30x\sqrt{9+x}]\,dx = 20x(9+x)^{3/2} - 8(9+x)^{5/2} + c$$

(5)  Then, represent it as a definite integral in terms of $x$ (note that the constant $c$ is removed).

$$\int_0^7 [30x\sqrt{9+x}]\,dx = \left.(20x(9+x)^{3/2} - 8(9+x)^{5/2})\right|_0^7$$

(6)  Given the range of $x$ in the integrand ($0 \le x \le 7$), plug the extrema into the definite integral.

$$\int_0^7 [30x\sqrt{9+x}]\,dx = (20(7)(9+7)^{3/2} - 8(9+7)^{5/2}) - (20(0)(9+0)^{3/2} - 8(9+0)^{5/2})$$

$$= (20(7)(64) - 8(1024)) - (0 - 8(243)) = (8960 - 8192) + 1944$$

$$\int_0^7 [30x\sqrt{9+x}]\,dx = 2712$$

## APPLICATION IN BUSINESS AND ECONOMICS

**S12.12.**  Given the demand and supply functions:

$$\text{Demand: } P = -4Q+30 \qquad \text{Supply: } P = 2Q+6$$

Use integration to find the consumer surplus (CS) and producer surplus (PS).

(1)  Find the price $P^*$ and quantity $Q^*$ at equilibrium.

Equate Supply = Demand

$$-4Q+30 = 2Q+6$$

$$Q^* = 4$$

Replace $Q$ in the Supply function:

$$P^* = 2(4)+6 = 14$$

(2)   See Fig. 12-15 for the supply (S)-demand (D) diagram.

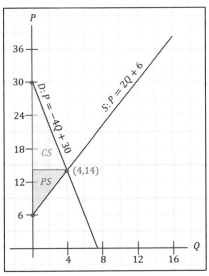

**Fig. 12-15**

(3)   Use definite integrals to find CS and PS.

$$CS = \int_{0}^{Q^*} D(Q)\,dQ - P^*Q^* = \int_{0}^{4} (-4Q+30)\,dQ - (14)(4)$$

$$= (-2Q^2 + 30Q)\Big|_{0}^{4} - 56 = (-2(4)^2 + 30(4)) - (0) - 56$$

$$CS = 88 - 56 = 32$$

$$PS = P^*Q^* - \int_{0}^{Q^*} S(Q)\,dQ = (14)(4) - \int_{0}^{4} (2Q+6)\,dQ$$

$$= 56 - (Q^2 + 6Q)\Big|_{0}^{4} = 56 - ((4)^2 + 6(4)) + (0)$$

$$PS = 56 - 40 = 16$$

**S12.13.**  Estimate the firm's additional total revenue received from increasing sales from 5 to 8 units, given the marginal revenue function $MR = 32 - 4x$.

(1)   Find the total revenue $TR$ function.

$$TR = \int (32 - 4x)\,dx = 32x - 2x^2 + c$$

Since there is no revenue when there are no sales, $TR(0) = 0$, which means that $c = 0$.

$$TR = 32x - 2x^2$$

(2)   Calculate the increase in revenue within the range of sales.

$$TR(8) - TR(5) = [32(8) - 2(8)^2] - [32(5) - 2(5)^2] = (256 - 128) - (160 - 50)$$
$$TR(8) - TR(5) = 128 - 110 = 18$$

**S12.14.** Find the profit function $\pi$ given the marginal revenue function $MR = 27 - x$, marginal cost function $MC = 2x$, and a fixed cost of 200.

(1)   Compute the total revenue $TR$ and $TC$ functions.

$$TR = \int (27 - x)dx = 27x - \frac{1}{2}x^2$$
$$TC = \int (2x)dx + 200 = x^2 + 200$$

$$TR = \boxed{27x - \frac{1}{2}x^2} \qquad TC = \boxed{x^2 + 200}$$

(2)   Find the profit function $\pi$.

$$\pi = TR - TC = 27x - \frac{1}{2}x^2 - (x^2 + 200)$$

$$\pi = \boxed{-\frac{3}{2}x^2 + 27x - 200}$$

# Chapter 13

**THEORETICAL CONCEPTS**

**S13.1.** Given a differentiable function $z = f(x, y)$, consider $f_{xx} \cdot f_{yy} > (f_{xy})^2$, $f_{xx} > 0$, and $f_{yy} > 0$ at the critical value $(x^*, y^*)$. This point can be classified as relative minimum point.

**S13.2.** True.

**S13.3.** True.

**S13.4.** For a profit function $\pi = \pi(x, y)$ to be maximized at $(x^*, y^*)$, it requires the following conditions:

$$\pi_{xx} < 0 \qquad \pi_{yy} < 0 \qquad \pi_{xx} \pi_{yy} > (\pi_{xy})^2$$

**S13.5.** False. $y$ must stay constant.

**PARTIAL DERIVATIVES**

Find the first-order partial derivative with respect to $x$.

**S13.6.** Use the basic rules to differentiate $f(x, y) = 7x^4 y + 3x^2 y^{1/2} - 4y^2 - 1000$.

> When differentiating with respect to $x$, treat all $y$ components as constants. If a monomial does not include an $x$ component, then its partial derivative with respect to $x$ is zero.
>
> $$f_x = 7[x^4]'y + 3[x^2]'y^{1/2} + 0 + 0 = 7[4x^3]y + 3[6x]y^{1/2}$$
> $$f_x = 28x^3 y + 6xy^{1/2}$$

**S13.7.** Use the product rule to differentiate $f(x) = (9x^{1/3}y + 2)(10x^{-1/5} + 2y)$.

> $$f_x = \left[\frac{d}{dx}(9x^{1/3}y + 2)\right](10x^{-1/5} + 2y) + (9x^{1/3}y + 2)\left[\frac{d}{dx}(10x^{-1/5} + 2y)\right]$$
> $$f_x = (9[x^{1/3}]'y + 0)(10x^{-1/5} + 2y) + (9x^{1/3}y + 2)(10[x^{-1/5}]' + 0)$$
> $$f_x = (3x^{-2/3}y)(10x^{-1/5} + 2y) - (9x^{1/3}y + 2)(2x^{-6/5})$$
>
> Simplify.
>
> $$f_x = 30x^{-13/5}y + 6x^{-2/3}y - 18x^{-13/5}y - 4x^{-6/5}$$
> $$f_x = 12x^{-13/5}y + 6x^{-2/3}y - 4x^{-6/5}$$

**S13.8.** Use the quotient rule to differentiate $f(x) = \dfrac{4x^2}{5x - y}$.

> $$f_x = \frac{(5x - y)(8x) - (4x^2)(5)}{(5x - y)^2} = \frac{40x^2 - 8xy - 20x^2}{(5x - y)^2}$$
> $$f_x = \frac{20x^2 - 8xy}{(5x - y)^2}$$

**S13.9.** Use the generalized power rule to differentiate $f(x) = (4x^2 - 5y + 8xy)^{-1/2}$.

> $$f_x = \left(-\frac{1}{2}\right)(4x^2 - 5y + 8xy)^{-3/2} \cdot (8x + 8y)$$
>
> Factor $(8x + 8y)$ as $8(x + y)$ and simplify.
>
> $$f_x = -4(x + y)(4x^2 - 5y + 8xy)^{-3/2}$$

**S13.10.** Use the natural exponential rule to differentiate $f(x) = e^{2x}y - 3x^2 e^y$.

Simplify.

$$f_x = [2e^{2x}]y - 3[2x]e^y$$

$$f_x = 2e^{2x}y - 6xe^y$$

**S13.11.** Use the natural logarithmic rule to differentiate $f(x) = \ln(x^{1/2}y^{1/2} + 3xy^{-3})$.

Simplify.

$$f_x = \frac{1}{(x^{1/2}y^{1/2} + 3xy^{-3})}\left[\frac{d}{dx}(x^{1/2}y^{1/2} + 3xy^{-3})\right]$$

$$f_x = \frac{1}{(x^{1/2}y^{1/2} + 3xy^{-3})}\left(\left[\frac{1}{2}x^{-1/2}\right]y^{1/2} + 3[1]y^{-3}\right)$$

$$f_x = \frac{0.5x^{-1/2}y^{1/2} + 3y^{-3}}{x^{1/2}y^{1/2} + 3xy^{-3}}$$

## SECOND-ORDER AND CROSS-PARTIAL DERIVATIVES

**S13.12.** Compute the second-order and cross partial derivative with respect to $x$ and $y$, and verify Young's Theorem for the function $z = 8x^{0.5}y^{0.7} + 2x^2y + 3xy^3$.

(1) Find the first-order derivative.

$$z_x = 4x^{-0.5}y^{0.7} + 4xy + 3y^3$$
$$z_y = 5.6x^{0.5}y^{-0.3} + 2x^2 + 9xy^2$$

(2) Find the second-order derivative.

$$z_{xx} = -2x^{-1.5}y^{0.7} + 4y$$
$$z_{yy} = -1.68x^{0.5}y^{-1.3} + 18xy$$

(3) Find the cross-partial derivatives and verify Young's Theorem.

$$z_{xy} = 2.8x^{-0.5}y^{-0.3} + 4x + 9y^2$$
$$z_{yx} = 2.8x^{-0.5}y^{-0.3} + 4x + 9y^2$$

Since $z_{xy} = z_{yx}$, Young's Theorem is verified.

## OPTIMIZATION IN BUSINESS AND ECONOMICS

**S13.13.** A monopolist produces two goods $x$ and $y$. Find the critical value at which the firm's profit $\pi$ is maximized for which the demand functions are

$$P_x = 55 - 4x \text{ and } P_y = 45 - 2y$$

and total cost function is

$$C = x^2 + y^2 + 5xy + 60$$

(1) Establish the profit function $\pi = TR - TC$, where $TR = P_x \cdot x + P_y \cdot y$.

$$\pi = (55 - 4x)x + (45 - 2y)y - (x^2 + y^2 + 5xy + 60)$$
$$\pi = 55x - 4x^2 + 45y - 2y^2 - x^2 - y^2 - 5xy - 60$$
$$\pi = \boxed{55x - 5x^2 + 45y - 3y^2 - 5xy - 60}$$

(2) Take the first-order condition for the profit function with respect to $x$ and $y$ and find $(x^*, y^*)$.

$$\pi_x = \boxed{55 - 10x - 5y = 0}$$
$$\pi_y = \boxed{45 - 6y - 5x = 0}$$
$$(x^*, y^*) = \boxed{(3, 5)}$$

(3) Test the second-order condition to be sure of a maximum.

$$\pi_{xx} = -10 < 0$$
$$\pi_{yy} = -6 < 0$$
$$\pi_{xy} = -5$$

With $\pi_{xx}$ and $\pi_{yy}$ both negative and $\pi_{xx}\pi_{yy} > (\pi_{xy})^2$, $\pi$ is maximized.

(4) Find the optimal prices $P_x^*$, $P_y^*$, and profit $\pi^*$.

$$P_x^* = 55 - 4(3) = 43$$
$$P_y^* = 45 - 2(5) = 35$$
$$\pi^* = (43)(3) + (35)(5) - (3^2 + 5^2 + 5(3)(5) + 60) = 129 + 175 - 169$$
$$\pi^* = 135$$

## CONSTRAINED OPTIMIZATION

**S13.14.** Given the function

$$z = x^2 + 3xy + y^2 - x + 3y$$

The constraint is

$$x + y = 42$$

(*1*)  Establish the Lagrange function.

$$\mathcal{L} = x^2 + 3xy + y^2 - x + 3y + \lambda(42 - x - y)$$

$$\mathcal{L} = \boxed{x^2 + 3xy + y^2 - x + 3y + 42\lambda - x\lambda - y\lambda}$$

(*2*)  Take the first-order partial derivatives for optimization to find $(x^*, y^*)$.

$$\mathcal{L}_x = \boxed{2x + 3y - 1 - \lambda = 0} \qquad \lambda = 2x + 3y - 1 \qquad\qquad (13.107)$$

$$\mathcal{L}_y = \boxed{3x + 2y - 3 - \lambda = 0} \qquad \lambda = 3x + 2y - 3 \qquad\qquad (13.108)$$

$$\mathcal{L}_\lambda = \boxed{42 - x - y = 0} \qquad\quad x + y = 42 \qquad\qquad\quad (13.109)$$

Set equation (*13.107*) equal to (*13.108*).

$$2x + 3y - 1 = 3x + 2y - 3$$

$$y = x + 2$$

Substitute in (*13.109*).

$$x + (x + 2) = 42$$

$$x = 20$$

$$y = 20 + 2 = 22$$

$$(x^*, y^*) = \boxed{(20, 22)}$$

## CONSTRAINED OPTIMIZATION IN BUSINESS AND ECONOMICS

**S13.15.** A manufacturer produces $q$ amount of a good using labor $L$ and capital $K$. The *Cobb-Douglas* production function of the good is

$$q = L^{2/3} K^{1/3}$$

The budget parameters are

$$P_K = 20 \qquad P_L = 40 \qquad B = 600$$

Find the optimal labor ($L^*$) and capital ($K^*$) needed to minimize its production cost given the budget.

(*1*)  Elaborate the budget constraint.

$$\boxed{20}\,L + \boxed{40}\,K = \boxed{600}$$

(*2*)  Establish the Lagrange function.

$$\mathcal{L} = L^{2/3} K^{1/3} + \lambda(600 - 20L - 40K)$$

$$\mathcal{L} = \boxed{L^{2/3} K^{1/3} + 600\lambda - 20L\lambda - 40K\lambda}$$

(3)   Take the first-order partial derivatives for optimization to find $(L^*, K^*)$.

$$\mathcal{L}_K = \boxed{\frac{2}{3}L^{-1/3}K^{1/3} - 20\lambda = 0} \qquad \frac{2}{3}L^{-1/3}K^{1/3} = 20\lambda \qquad (13.110)$$

$$\mathcal{L}_L = \boxed{\frac{1}{3}L^{2/3}K^{-2/3} - 40\lambda = 0} \qquad \frac{1}{3}L^{2/3}K^{-2/3} = 40\lambda \qquad (13.111)$$

$$\mathcal{L}_\lambda = \boxed{600 - 20L - 40K = 0} \qquad 20L + 40K = 600 \qquad (13.112)$$

Divide $(13.110)$ by $(13.111)$ to eliminate $\lambda$.

$$\frac{\frac{2}{3}L^{-1/3}K^{1/3}}{\frac{1}{3}L^{2/3}K^{-2/3}} = \frac{20}{40}$$

$$\frac{2K}{L} = \frac{1}{2} \quad L = 4K$$

Substitute in $(13.112)$.

$$20(4K) + 40K = 600$$
$$120K = 600$$
$$K = 5$$
$$L = 4(5) = 20$$
$$(L^*, K^*) = \boxed{(20,5)}$$

# Chapter 14

## THEORETICAL CONCEPTS

**S14.1.** An infinite geometric series converges if the absolute value of the common ratio is less than 1.

**S14.2.** False. The standard deviation is the square root of the variance.

**S14.3.** To be accepted, NPV > 0, therefore, it is 7500.

**S14.4.** False. An infinite sequence converges if the sequence approaches a specific value as $n$ increases.

**S14.5.** False. NPV is the sum of all cash flows at present value.

## SEQUENCES AND SUMMATION

**S14.6.** Given the sequence $\{x_n\} = \{1, 9, -4, 3, 8, -1\}$, compute

$$\sum_{i=2}^{3}(x_{i-1} + x_{5-i})$$

Describe each term with its respective index:

$$\sum_{i=2}^{3}(x_{i-1}+x_{5-i})=\underbrace{(x_{2-1}+x_{5-2})}_{i=2}+\underbrace{(x_{3-1}+x_{5-3})}_{i=3}$$

$$\sum_{i=2}^{3}(x_{i-1}+x_{5-i})=(x_1+x_3)+(x_2+x_2)$$

Replace the terms with their corresponding values.

$$\sum_{i=2}^{3}(x_{i-1}+x_{5-i})=(1+(-4))+(9+9)=(-3)+18$$

$$\sum_{i=2}^{3}(x_{i-1}+x_{5-i})=15$$

**S14.7.** Given the $6 \times 7$ table below,

| 10 | 7 | 1 | 12 | 96 | 23 | −4 |
|----|----|----|----|----|----|----|
| −3 | 0 | 4 | 14 | 36 | 11 | 12 |
| 8 | −2 | 53 | 11 | 23 | 16 | 16 |
| 11 | −8 | 64 | −2 | 1 | 5 | −7 |
| 17 | 11 | 75 | 88 | −1 | −43 | −21 |
| 20 | 99 | 87 | 73 | 71 | 12 | 72 |

Compute the series given by the following rule:

$$\sum_{i=3}^{5}(x_{i4}+x_{2i})$$

Describe each term with its respective index:

$$\sum_{i=3}^{5}(x_{i4}+x_{2i})=\underbrace{(x_{3,4}+x_{2,3})}_{i=3}+\underbrace{(x_{4,4}+x_{2,4})}_{i=4}+\underbrace{(x_{5,4}+x_{2,5})}_{i=5}$$

Replace the terms with their corresponding values:

$$\sum_{i=3}^{5}(x_{i4}+x_{2i})=(11+4)+(-2+14)+(88+36)$$

$$\sum_{i=3}^{5}(x_{i4}+x_{2i})=151$$

## PROPERTIES AND SPECIAL FORMULAS OF SUMMATIONS

**S14.8.** Given the table below.

| $k$ | $x_k$ |
|-----|-------|
| 1 | 8 |
| 2 | 17 |
| 3 | −5 |
| 4 | 9 |
| 5 | −3 |

Find the value of

$$\sum_{k=1}^{5}(x_k+k)^2$$

Expand the polynomial in the summation:

$$\sum_{k=1}^{5}(x_k+k)^2 = \sum_{k=1}^{5}\left(x_k^2+2x_kk+k^2\right)=\sum_{k=1}^{5}x_k^2 + 2\sum_{k=1}^{5}x_k+\sum_{k=1}^{5}k^2$$

Expand the terms:

$$\sum_{k=1}^{5}(x_k+k)^2 = \left(x_1^2+x_2^2+x_3^2+x_4^2+x_5^2\right)+2[x_1(1)+x_2(2)+x_3(3)+x_4(4)+x_5(5)]$$
$$+(1^2+2^2+3^2+4^2+5^2)$$

$$\sum_{k=1}^{5}(x_k+k)^2 = (8^2+(17)^2+(-5)^2+9^2+(-3)^2)$$
$$+2[8(1)+17(2)+(-5)(3)+9(4)+(-3)(5)]+(1^2+2^2+3^2+4^2+5^2)$$

Use summation properties and formulas:

$$\sum_{k=1}^{5}(x_k+k)^2 = 468+2(48)+\frac{5(6)(10+1)}{6}=468+96+55$$

$$\sum_{k=1}^{5}(x_k+k)^2 = 619$$

## MEAN AND STANDARD DEVIATION

**S14.9.** Find the mean and standard deviation of the table used in S14.8.

*(1)* Computing the mean

$$\overline{X}=\frac{1}{5}\sum_{i=1}^{5}x_i = \frac{8+17-5+9-3}{5}=\frac{26}{5}=5.2$$

(2)  Find the variance:

$$S_x^2 = \frac{1}{5-1}\sum_{i=1}^{5}(x_i - \overline{X})^2$$

$$S_x^2 = \frac{(8-5.2)^2+(17-5.2)^2+(-5-5.2)^2+(9-5.2)^2+(-3-5.2)^2}{4}$$

$$S_x^2 = \frac{7.84+139.24+104.04+14.44+67.24}{4}$$

$$S_x^2 = \frac{332.8}{4} = 83.2$$

(3)  Compute the standard deviation:

$$S_x = \sqrt{S_x^2} = \sqrt{83.2} = 9.1214$$

## INFINITE SERIES

**S14.10.**  Determine if the geometric series converges or diverges. If converges, find the value of

$$s_n = \sum_{i=3}^{\infty} 72\left(\frac{1}{6}\right)^i$$

Here, $r = 1/6 \boxed{<} 1$, therefore the geometric series converges.
However, note that the index starts at $i = 3$

$$\sum_{i=3}^{\infty}\left(\frac{1}{6}\right)^i = \left(\frac{1}{6}\right)^3 + \left(\frac{1}{6}\right)^4 + \left(\frac{1}{6}\right)^5 + \cdots$$

In order to use the geometric formula, complete the series by adding and subtracting the first two terms needed to have the index start at $i = 0$.

$$= \left[\left(\frac{1}{6}\right)^3 + \left(\frac{1}{6}\right)^4 + \left(\frac{1}{6}\right)^5 + \cdots\right] + \left(1 + \frac{1}{6} + \frac{1}{36}\right) - \left(1 + \frac{1}{6} + \frac{1}{36}\right)$$

Reorganize terms:

$$\sum_{i=3}^{\infty} 72\left(\frac{1}{6}\right)^i = 72\left[1+\frac{1}{6}+\left(\frac{1}{6}\right)^2+\left(\frac{1}{6}\right)^3+\left(\frac{1}{6}\right)^4+\cdots\right] - 72\left(1+\frac{1}{6}+\frac{1}{36}\right)$$

Use the formula:

$$\sum_{i=3}^{\infty} 72\left(\frac{1}{6}\right)^i = 72\left[\frac{1}{1-1/6}\right] - 72\left(\frac{43}{36}\right) = 72\left(\frac{6}{5}\right) - 2(43)$$

$$\sum_{i=3}^{\infty} 72\left(\frac{1}{6}\right)^i = 86.4 + 86 = 172.4$$

**NET PRESENT VALUE**

**S14.11.** Compute the net present value of the project given the cash flows below and determine if the project must be approved or rejected. The discount rate is 25%.

| Future value | Year | | | | | | |
|---|---|---|---|---|---|---|---|
| | 0 | 1 | 2 | 3 | 4 | 5 | 6 |
| A | −25000 | −2500 | 0 | 5000 | 17000 | 25000 | 48000 |

Substitute in the NPV formula:

$$NPV = -25000 + \frac{-2500}{(1+0.25)} + \frac{0}{(1+0.25)^2} + \frac{5000}{(1+0.25)^3}$$
$$+ \frac{17000}{(1+0.25)^4} + \frac{25000}{(1+0.25)^5} + \frac{48000}{(1+0.25)^6}$$
$$NPV = -25000 - 2000 + 0 + 6963.2 + 8192 + 12582.91$$

Calculate:

$$NPV = \boxed{3298.11} \qquad \text{Accept the project as NPV} \boxed{>} 0$$

# Index

The letter *p* following a page number refers to a Problem.

NOTES

NOTES

NOTES